Dear President Re[...],

With best wishes &
warm personal regards,

[signature]
7/18/06

FORCE MICROSCOPY

FORCE MICROSCOPY
APPLICATIONS IN BIOLOGY AND MEDICINE

Edited by

BHANU P. JENA PhD
Department of Physiology
Wayne State University School of Medicine
Detroit, Michigan

J. K. HEINRICH HÖRBER PhD
Department of Physics
University of Bristol
Bristol, United Kingdom

WILEY-LISS

A JOHN WILEY & SONS, INC., PUBLICATION

Copyright © 2006 by John Wiley & Sons, Inc. All rights reserved.

Published by John Wiley & Sons, Inc., Hoboken, New Jersey.
Published simultaneously in Canada.

No part of this publication may be reproduced, stored in a retrieval system, or transmitted in any form or by any means, electronic, mechanical, photocopying, recording, scanning, or otherwise, except as permitted under Section 107 or 108 of the 1976 United States Copyright Act, without either the prior written permission of the Publisher, or authorization through payment of the appropriate per-copy fee to the Copyright Clearance Center, Inc., 222 Rosewood Drive, Danvers, MA 01923, (978) 750-8400, fax (978) 750-4470, or on the web at www.copyright.com. Requests to the Publisher for permission should be addressed to the Permissions Department, John Wiley & Sons, Inc., 111 River Street, Hoboken, NJ 07030, (201) 748-6011, fax (201) 748-6008, or online at http://www.wiley.com/go/permission.

Limit of Liability/Disclaimer of Warranty: While the publisher and author have used their best efforts in preparing this book, they make no representations or warranties with respect to the accuracy or completeness of the contents of this book and specifically disclaim any implied warranties of merchantability or fitness for a particular purpose. No warranty may be created or extended by sales representatives or written sales materials. The advice and strategies contained herein may not be suitable for your situation. You should consult with a professional where appropriate. Neither the publisher nor author shall be liable for any loss of profit or any other commercial damages, including but not limited to special, incidental, consequential, or other damages.

For general information on our other products and services or for technical support, please contact our Customer Care Department within the United States at (800) 762-2974, outside the United States at (317) 572-3993 or fax (317) 572-4002.

Wiley also publishes its books in a variety of electronic formats. Some content that appears in print may not be available in electronic formats. For more information about Wiley products, visit our web site at www.wiley.com.

Library of Congress Cataloging-in-Publication Data:

Force microscopy : applications in biology and medicine / [edited by] Bhanu P. Jena, J.K. Heinrich Hörber.
   p. cm.
   Includes bibliographical references and index.
   ISBN-13: 978-0-471-39628-4
   ISBN-10: 0-471-39628-1
  1. Medical electronics. 2. Scanning force microscopy. 3. Scanning probe microscopy. 4. Nanotechnology. I. Jena, Bhanu P. II. Hörber, J. K. Heinrich.

R856.A2F675 2006
610.28—dc22

2005058118

Printed in the United States of America

10 9 8 7 6 5 4 3 2 1

CONTENTS

Preface vii

Contributors ix

Chapter 1. Porosome: The Universal Secretory Machinery in Cells   1
Bhanu P. Jena

Chapter 2. Molecular Mechanism of SNARE-Induced Membrane Fusion   25
Bhanu P. Jena

Chapter 3. Molecular Mechanism of Secretory Vesicle Content Expulsion During Cell Secretion   37
Bhanu P. Jena

Chapter 4. Fusion Pores in Growth-Hormone-Secreting Cells of the Pituitary Gland: An AFM Study   49
Lloyd L. Anderson and Bhanu P. Jena

Chapter 5. Properties of Microbial Cell Surfaces Examined by Atomic Force Microscopy   69
Yves F. Dufrêne

Chapter 6. Scanning Probe Microscopy of Plant Cell Wall and Its Constituents   95
Ksenija Radotić, Miodrag Mićić, and Milorad Jeremić

Chapter 7. Cellular Interactions of Nano Drug Delivery Systems   113
Rangaramanujam M. Kannan, Omathanu Pillai Perumal, and Sujatha Kannan

Chapter 8.	Adapting AFM Techniques for Studies on Living Cells	137
	J. K. Heinrich Hörber	
Chapter 9.	Intermolecular Forces of Leukocyte Adhesion Molecules	159
	Xiaohui Zhang and Vincent T. Moy	
Chapter 10.	Mechanisms of Avidity Modulation in Leukocyte Adhesion Studied by AFM	169
	Ewa P. Wojcikiewicz and Vincent T. Moy	
Chapter 11.	Resolving the Thickness and Micromechanical Properties of Lipid Bilayers and Vesicles Using AFM	181
	Guangzhao Mao and Xuemei Liang	
Chapter 12.	Imaging Soft Surfaces by SFM	201
	Andreas Janke and Tilo Pompe	
Chapter 13.	High-Speed Atomic Force Microscopy of Biomolecules in Motion	221
	Tilman E. Schäffer	
Chapter 14.	Atomic Force Microscopy in Cytogenetics	249
	S. Thalhammer and W. M. Heckl	
Chapter 15.	Atomic Force Microscopy in the Study of Macromolecular Interactions in Hemostasis and Thrombosis: Utility for Investigation of the Antiphospholipid Syndrome	267
	William J. Montigny, Anthony S. Quinn, Xiao-Xuan Wu, Edwin G. Bovill, Jacob H. Rand, and Douglas J. Taatjes	
Index		287

PREFACE

Throughout history, the development of new imaging tools has provided new insights into our perceptions of the living world and profoundly impacted human health and disease. The invention of the light microscope almost 300 years ago was the first catalyst, propelling us into an era of modern biology and medicine. Using the light microscope, a giant step into the gates of biology and medicine was made with the discovery of the unit of life, the cell. The structure and morphology of normal and diseased cells, and of disease-causing microorganisms, were revealed for the first time using the light microscope. Then in 1938, with the birth of the electron microscope (EM), a new scientific era began. Through the 1950s, a number of subcellular organelles were discovered and their functions determined using the EM. Viruses, the new life forms, were identified and observed for the first time and were implicated in diseases ranging from the common cold to autoimmune disease (AIDS). Despite the capability of the EM to image biological samples at near-nanometer resolution, sample processing resulting in morphological alterations remained a major concern. Then in the mid-1980s, scanning probe microscopy evolved, extending our perception of the living world even further, to the near-atomic realm. One such scanning probe microscope, the atomic force microscope (AFM), has helped overcome the limitations of both light and electron microscopy, enabling three-dimensional determination of the structure and dynamics of single biomolecules and live cells, at near-angstrom resolution. This unique capability of the AFM, in combination with conventional tools and techniques, has finally provided an understanding of the cell, subcellular organelles, and biomolecular structure–function, as never before.

Secretion and membrane fusion are fundamental cellular processes regulating intracellular transport, plasma membrane recycling, cell division, sexual reproduction, acid secretion, histamine release, and the release of enzymes, hormones, and neurotransmitters, to name just a few. Therefore, defects in secretion and membrane fusion lead to diabetes, Alzheimer's disease, Parkinson's disease, and a host of other diseases. The first three chapters by Jena describe the discovery of the molecular machinery and mechanism of cell secretion and membrane fusion in cells. These discoveries were made using the combined approaches of AFM, EM, x-ray diffraction, photon-correlation spectroscopy, molecular biology, biochemistry, and electrophysiology. The secretory machinery the porosome described in the first chapter, was first discovered in live secretory pancreatic acinar cells, using the AFM. Similarly, the AFM has been instrumental in the discovery of the porosome in numerous secretory cells as described in Chapter 1, including the growth-hormone-secreting cell of the pituitary gland, described in Chapter 4 by Anderson and Jena. In Chapter 5, Dufrêne describes probing the structural and physical properties of microbial cell surfaces using AFM. In Chapter 6, Radotić et al. review the current state-of-the-art of scanning probe microscopic characterization of higher plant cell wall and its components. The spatial and temporal control of the delivery of drugs at the site of action on a cellular or subcellular target is hugely important in the treatment of diseases. In recent years, nanostructured assemblies offer great promise. In Chapter 7, Kannan et al. provide case studies and discuss the chemistry of engineered dendrimers, as multifunctional nanodrug delivery particles. An overview of AFM techniques for studies on living cells is beautifully summarized in Chapter 8. Besides imaging, force microscopy has frequently been used to understand inter- and intramolecular interactions. In Chapter 9, Zhang and Moy describe how AFM can be used to probe the intermolecular forces of leukocyte adhesion molecules. Leukocytes circulating in the blood vessels need to exit the bloodstream to enter specific tissues or areas of inflammation. This find-tuned process is mediated by the interactions between leukocyte adhesion molecules and their adhesive partners expressed on the inner surface of blood vessel. If this adhesive process is not under proper control, it could lead to severe physiological disorders such as asthma and atherosclerosis. Similarly, in Chapter 10, Wojcikiewicz and Moy examine protein–protein interactions using the AFM. Study of the structure, morphology, and stability of lipid bilayers and vesicles is important to the understanding of drug delivery, gene therapy, and biosensor design. In Chapter 11 by Mao and Liang, the micromechanical properties of lipid bilayers and vesicles are examined using the AFM. Finally, in the last four chapters by Janke and Pompe (Chapter 12), Schaffer (Chapter 13), Thalhammer and Heckl (Chapter 14), and Montigny et al. (Chapter 15), the development of new approaches in the use of force microscopy in biology and medicine is discussed.

The examples provided in this book on the utility of force microscopy in biology and medicine, which by no means are exhaustive, give the reader some perspective on the power and scope of this new scientific tool. The development of the photonic force microscope discussed in Chapter 8, along with the use of force microscopy in combination with EM, electrophysiology, biochemistry, and so on (Chapter 1–3), has had enormous impact in our study of cellular and physiological processes. Nonetheless, we are just beginning to realize the enormous benefits of force microscopy to fields of biology and medicine.

Bhanu P. Jena

J. K. Heinrich Hörber

CONTRIBUTORS

Lloyd L. Anderson, Department of Animal Science, College of Agriculture, and Department of Biomedical Sciences, College of Veterinary Medicine, Iowa State University, Ames, Iowa

Edwin G. Bovill, Department of Pathology, College of Medicine, University of Vermont, Burlington, Vermont

Yves F. Dufrêne, Unité de chimie des interfaces, Université catholique de Louvain, Croix du Sud 2/18, B-1348 Louvain-la-Neuve, Belgium

W. M. Heckl, Department für Geo-und Umweltwissenschaften, University of München, Theresienstr.41, Munich, Germany

J. K. Heinrich Hörber, Department of Physics, University of Bristol, Bristol, United Kingdom

Andreas Janke, Leibriz-Institut fur Polymerforschung Dresden eV., Hohe Str. 6, 01005 Dresden, Germany

Bhanu P. Jena, Department of Physiology, Wayne State University School of Medicine, Detroit, Michigan

Milorad Jeremić, Faculty of Physical Chemistry, University of Belgrade, Studentski Trg 16, 11000 Belgrade, Serbia and Montenegro

Rangaramanujam M. Kannan, Department of Chemical Engineering and Material Science, and Biomedical Engineering, Wayne State University, Detroit, Michigan

Sujatha Kannan, Department of Critical Care Medicine, Children's Hospital of Michigan, Detroit, Michigan

Xuemei Liang, Department of Chemical Engineering and Materials Science, Wayne State University, 5050 Anthony Wayne Drive, Detroit, Michigan

Guangzhao Mao, Department of Chemical Engineering and Materials Science, Wayne State University, 5050 Anthony Wayne Drive, Detroit, Michigan

Miodrag Mićić, MP Biomedicals, Inc., 15 Morgan, Irvine, California, 92618-2005

William J. Montigny, Microscopy Imaging Center, College of Medicine, University of Vermont, Bulington, Vermont

Vincent T. Moy, Department of Physiology and Biophysics, University of Miami Miller School of Medicine, Miami, Florida

Omathanu Pillai Perumal, Department of Pharmaceutical Sciences, South Dakota State University, College of Pharmacy, Brookings, South Dakota

Tilo Pompe, Leibriz-Institut fur Polymerforschung Dresden eV., Hohe Str. 6, 01005 Dresden, Germany

Anthony S. Quinn, Department of Pathology and Microscopy Imaging Center, College of Medicine, University of Vermont, Burlington, Vermont

Ksenija Radotić, Center for Multidisciplinary Studies, University of Belgrade, Kneza Višeslava 1, 11000 Belgrade, Serbia and Montenegro

Jacob H. Rand, Department of Pathology, Montefiore Medical Center, Albert Einstein College of Medicine, Bronx, New York

Tilman E. Schäffer, Center for Nanotechnology and Institute of Physics, Westfälische Wilhelms-Universität Münster, Gievenbecker Weg 11, 48149 Münster, Germany

Douglas J. Taatjes, Department of Pathology and Microscopy Imaging Center, College of Medicine, University of Vermont, Burlington, Vermont

S. Thalhammer, Department für Geo-und Umweltwissenschaften, University of München, Theresienstr.41, Munich Germany; GSF—National Center for Environment and Health, Institute of Radiation Protection, Ingolstädter Landstrasse 1, 85764 Neuherberg, Germany

Ewa P. Wojcikiewicz, Department of Physiology and Biophysics, University of Miami Miller School of Medicine, Miami, Florida

Xiao-Xuan Wu, Department of Pathology, Montefiore Medical Center, Albert Einstein College of Medicine, Bronx, New York

Xiaohui Zhang, Department of Physiology and Biophysics, University of Miami School of Medicine, Miami, Florida

CHAPTER 1

POROSOME: THE UNIVERSAL SECRETORY MACHINERY IN CELLS

BHANU P. JENA

Department of Physiology, Wayne State University School of Medicine, Detroit, Michigan

1.1. INTRODUCTION

Secretion and membrane fusion are fundamental cellular processes regulating endoplasmic reticulum (ER)–Golgi transport, plasma membrane recycling, cell division, sexual reproduction, acid secretion, and the release of enzymes, hormones, and neurotransmitters, to name just a few. It is therefore no surprise that defects in secretion and membrane fusion give rise to diseases such as diabetes, Alzheimer's, Parkinson's, acute gastroduodenal diseases, gastroesophageal reflux disease, intestinal infections due to inhibition of gastric acid secretion, biliary diseases resulting from malfunction of secretion from hepatocytes, polycystic ovarian disease as a result of altered gonadotropin secretion, and Gitelman disease associated with growth hormone deficiency and disturbances in vasopressin secretion. Understanding cellular secretion and membrane fusion not only helps to advance our understanding of these vital cellular and physiological processes, but also helps in the development of drugs to ameliorate secretory defects, provides insight into our understanding of cellular entry and exit of viruses and other pathogens, and helps in the development of smart drug delivery systems. Therefore, secretion and membrane fusion play an important role in health and disease. Studies (Abu-Hamdah et al., 2004; Anderson, 2004; Cho et al., 2002a–f, 2004; Hörber and Miles, 2003; Jena, 1997, 2002–2004; Jena et al., 1997, 2003; Jeremic et al., 2003, 2004a,b; Kelly et al., 2004; Schneider et al., 1997) in the last decade demonstrate that membrane-bound secretory vesicles dock and transiently fuse at the base of specialized plasma membrane structures called porosomes or fusion pores, to expel vesicular contents. These studies further demonstrate that during secretion, secretory vesicles swell, enabling the expulsion of intravesicular contents through porosomes (Abu-Hamdah et al., 2004; Cho et al., 2002f; Jena et al., 1997; Kelly et al., 2004). With these findings (Abu-Hamdah et al., 2004; Anderson, 2004; Cho et al., 2002a–f, 2004; Hörber and Miles, 2003; Jena, 1997, 2002–2004; Jena et al., 1997, 2003; Jeremic et al., 2003, 2004a,b; Kelly et al., 2004; Schneider et al., 1997) a new understanding of cell secretion has emerged and confirmed by a number of laboratories (Aravanis et al., 2003; Fix et al., 2004; Lee et al., 2004; Taraska et al., 2003; Thorn et al., 2004; Tojima et al., 2000).

Force Microscopy: Applications in Biology and Medicine, edited by Bhanu P. Jena and J.K. Heinrich Hörber.
Copyright © 2006 John Wiley & Sons, Inc.

Throughout history, the development of new imaging tools has provided new insights into our perceptions of the living world and has profoundly impacted human health. The invention of the light microscope almost 300 years ago was the first catalyst, propelling us into the era of modern biology and medicine. Using the light microscope, a giant step into the gates of modern medicine was made by the discovery of the unit of life, the cell. The structure and morphology of normal and diseased cells and of disease-causing microorganisms were revealed for the first time using the light microscope. Then in 1938, with the birth of the electron microscope (EM), dawned a new era in biology and medicine. Through the mid-1940s and 1950s, a number of subcellular organelles were discovered and their functions determined using the EM. Viruses, the new life forms, were discovered and observed for the first time and were implicated in diseases ranging from the common cold to acquired immune deficiency syndrome (AIDS). Despite the capability of the EM to image biological samples at near-nanometer resolution, sample processing (fixation, dehydration, staining) results in morphological alterations and was a major concern. Then in the mid-1980s, scanning probe microscopy evolved (Binnig et al., 1986; Hörber and Miles, 2003), further extending our perception of the living world to the near atomic realm. One such scanning probe microscope, the atomic force microscope (AFM), has helped overcome both limitations of light and electron microscopy, enabling determination of the structure and dynamics of single biomolecules and live cells in 3D, at near-angstrom resolution. This unique capability of the AFM has given rise to a new discipline of "nanobioscience," heralding a new era in biology and medicine. Using AFM in combination with conventional tools and techniques, this past decade has witnessed advances in our understanding of cell secretion (Abu-Hamdah et al., 2004; Anderson, 2004; Cho et al., 2002a–f, 2004; Hörber and Miles, 2003; Jena, 1997, 2002–2004; Jena et al., 1997, 2003; Jeremic et al., 2003, 2004a,b; Kelly et al., 2004; Schneider et al., 1997) and membrane fusion (Cho et al., 2002d; Jeremic et al., 2004a,b; Weber et al., 1998), as noted earlier in the chapter.

The resolving power of the light microscope is dependent on the wavelength of the light used; therefore, 250–300 nm in lateral resolution, and much less in depth resolution, can be achieved at best. The porosome or fusion pore in live secretory cells are cup-shaped structures, measuring 100–150 nm at its opening and 15–30 nm in relative depth in the exocrine pancreas, and just 10 nm at the presynaptic membrane of the nerve terminal. As a result, it had evaded visual detection until its discovery using the AFM (Cho et al., 2002a–c, 2003, 2004; Jeremic et al., 2003; Schneider et al., 1997). The development of the AFM (Binnig et al., 1986) has enabled the imaging of live cells in physiological buffer at nanometer to subnanometer resolution. In AFM, a probe tip microfabricated from silicon or silicon nitride and mounted on a cantilever spring is used to scan the surface of the sample at a constant force. Either the probe or the sample can be precisely moved in a raster pattern using an xyz piezo tube to scan the surface of the sample (Fig. 1.1). The deflection of the cantilever measured optically is used to generate an isoforce relief of the sample (Alexander et al., 1989). Force is thus used to image surface profiles of objects by the AFM, allowing imaging of live cells and subcellular structures submerged in physiological buffer solutions. To image live cells, the scanning probe of the AFM operates in physiological buffers and may do so under two modes: contact or tapping. In the contact mode, the probe is in direct contact with the sample surface as it scans at a constant vertical force. Although high-resolution AFM images can be obtained in this mode of AFM operation, sample height information generated may not be accurate since

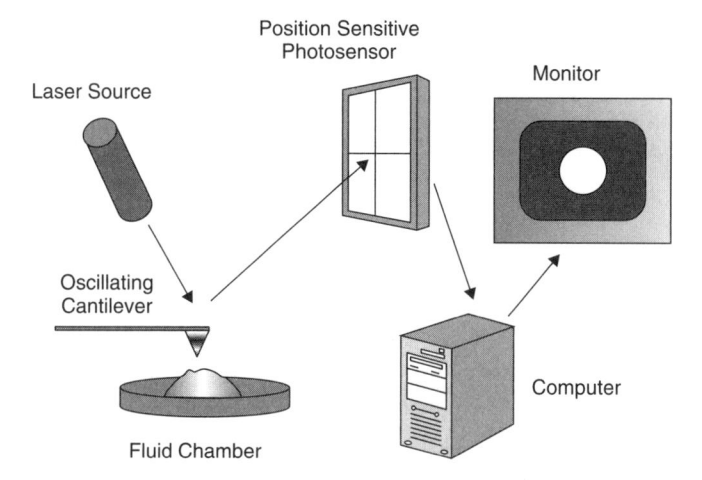

Figure 1.1. Schematic diagram depicting key components of an atomic force microscope.

the vertical scanning force may depress the soft cell. However, information on the viscoelastic properties of the cell and the spring constant of the cantilever enables measurement of the cell height. In tapping mode on the other hand, the cantilever resonates and the tip makes brief contacts with the sample. In the tapping mode in fluid, lateral forces are virtually negligible. It is therefore important that the topology of living cells be obtained using both contact and tapping modes of AFM operation in fluid. The scanning rate of the tip over the sample also plays an important role on the quality of the image. Since cells are soft samples, a high scanning rate would influence its shape. Hence, a slow tip movement over the cell would be ideal and results in minimal distortion and better image resolution. Rapid cellular events may be further monitored by using section analysis. To examine isolated cells by the AFM, freshly cleaved mica coated with Cel-Tak have also been used with great success (Cho et al., 2002a–c; Jena et al., 2003; Jeremic et al., 2003; Schneider et al., 1997). Also, to obtain optimal resolution, the contents of the bathing medium as well as the cell surface to be scanned should be devoid of any debris.

1.2. METHODS

1.2.1. Isolation of Pancreatic Acinar Cells

Acinar cells for secretion experiments, light microscopy, atomic force microscopy (AFM), and electron microscopy (EM) were isolated using minor modification of a published procedure. For each experiment, a male Sprague–Dawley rat weighing 80–100 g was euthanized by CO_2 inhalation. The pancreas was dissected and diced into 0.5-mm^3 pieces with a razor blade, mildly agitated for 10 min at 37°C in a siliconized glass tube with 5 ml of oxygenated buffer A (98 mM NaCl, 4.8 mM KCl, 2 mM $CaCl_2$, 1.2 mM $MgCl_2$, 0.1% bovine serum albumin, 0.01% soybean trypsin inhibitor, 25 mM Hepes, pH 7.4) containing 1000 units of collagenase. The suspension of acini was filtered through a 224-μm Spectra-Mesh (Spectrum Laboratory Products, Rancho Dominguez, CA) polyethylene filter to remove large clumps of acini and undissociated tissue. The acini were washed six times, 50 ml per wash, with ice-cold buffer A. Isolated rat pancreatic acini and acinar cells were plated on Cell-Tak-coated (Collaborative Biomedical Products, Bedford, MA) glass coverslips. Two to three hours after plating, cells were imaged with the AFM before and during stimulation of secretion. Isolated acinar cells and hemi-acinar preparations were used in the study because fusion of secretory vesicles at the PM in these cells occurs at the apical region facing the acinar lumen.

1.2.2. Pancreatic Plasma Membrane Preparation

Rat pancreatic PM fractions were isolated using a modification of a published method. Male Sprague–Dawley rats weighing 70–100 g were euthanized by CO_2 inhalation. Pancreas were removed and placed in ice-cold phosphate-buffered saline (PBS), pH 7.5. Adipose tissue was removed and the pancreas were diced into 0.5-mm^3 pieces using a razor blade in a few drops of homogenization buffer A (1.25 M sucrose, 0.01% trypsin inhibitor, and 25 mM Hepes, pH 6.5). The diced tissue was homogenized in 15% (w/v) ice-cold homogenization buffer A using four strokes at maximum speed of a motor-driven pestle (Wheaton overhead stirrer). One-and-a-half milliliters of the homogenate was layered over a 125-μl cushion of 2 M sucrose and 500 μl of 0.3 M sucrose was layered onto the homogenate in Beckman centrifuge tubes. After centrifugation at 145,000 × g for 90 min in a Sorvall AH-650 rotor, the material banding between the 1.2 and 0.3 M sucrose interface was collected and the protein concentration was determined. For each experiment, fresh PM was prepared and used the same day in all AFM experiments.

1.2.3. Isolation of Synaptosomes, Synaptosomal Membrane and Synaptic Vesicles

Synaptosomes, synaptosomal membrane and synaptic vesicles were prepared from rat brains (Jeong et al., 1998; Thoidis et al., 1998). Whole rat brain from Sprague–Dawley rats (100–150 g) was isolated and placed in ice-cold buffered sucrose solution (5 mM Hepes pH 7.4, 0.32 M sucrose) supplemented with protease inhibitor cocktail (Sigma, St. Louis, MO) and homogenized using Teflon-glass homogenizer (8–10 strokes). The total homogenate was centrifuged for 3 min at 2500 × g. The supernatant fraction was further centrifuged for 15 min at 14,500 × g, and the resultant pellet was resuspended in buffered sucrose solution, which was loaded onto 3–10–23% Percoll gradients. After centrifugation at 28,000 × g for 6 min, the enriched synaptosomal fraction was collected at the 10–23% Percoll gradient interface. To isolate synaptic vesicles and synaptosomal membrane (32), isolated synaptosomes were diluted with 9 vol of ice-cold H_2O (hypotonic lysis of synaptosomes to release synaptic vesicles) and immediately homogenized with three strokes in Dounce homogenizer, followed by a 30-min incubation on ice. The homogenate was centrifuged for 20 min at 25,500 × g, and the resultant pellet (enriched synaptosomal membrane preparation) and supernatant (enriched synaptic vesicles preparation) were used in our studies.

1.2.4. Preparation of Lipid Membrane on Mica and Porosome Reconstitution

To prepare lipid membrane on mica for AFM studies, freshly cleaved mica disks were placed in a fluid chamber. Two hundred microliters of the bilayer bath solution, containing 140 mM NaCl, 10 mM HEPES, and 1 mM $CaCl_2$, was placed at the center of the cleaved mica disk. Ten microliters of the brain lipid vesicles was added to the above bath solution. The mixture was then allowed to incubate for 60 min at room temperature, before washing (×10), using 100-μl bath solution/wash. The lipid membrane on mica was imaged by the AFM before and after the addition of immunoisolated porosomes.

1.2.5. Atomic Force Microscopy

"Pits" and fusion pores at the PM in live pancreatic acinar secreting cells in PBS pH

7.5 were imaged with the AFM (Bioscope III, Digital Instruments) using both contact and tapping modes. All images presented in this manuscript were obtained in the "tapping" mode in fluid, using silicon nitride tips with a spring constant of 0.06 N·m^{-1} and an imaging force of <200 pN. Images were obtained at line frequencies of 1 Hz, with 512 lines per image and constant image gains. Topographical dimensions of "pits" and fusion pores at the cell PM were analyzed using the software nanoscope IIIa4.43r8 supplied by Digital Instruments.

1.2.6. ImmunoAFM on Live Cells

Immunogold localization in live pancreatic acinar cells was assessed after a 5-min stimulation of secretion with 10 μM of the secretagogue, mastoparan. After stimulation of secretion, the live pancreatic acinar cells in buffer were exposed to at 1:200 dilution of α-amylase-specific antibody (Biomeda Corp., Foster City, CA) and 30 nm of gold conjugated secondary antibody for 1 min and were washed in PBS before AFM imaging in PBS at room temperature. "Pits" and fusion pores within, at the apical end of live pancreatic acinar cells in PBS pH 7.5, were imaged by the AFM (Bioscope III, Digital Instruments) using both contact and tapping mode. All images presented were obtained in the "tapping" mode in fluid, using silicon nitride tips as described previously.

1.2.7. ImmunoAFM on Fixed Cells

After stimulation of secretion with 10 μM mastoparan, the live pancreatic acinar cells were fixed for 30 min using ice-cold 2.5% paraformaldehyde in PBS. Cells were then washed in PBS, followed by labeling with 1:200 dilution of α-amylase-specific antibody (Biomeda Corp.) and 10 nm of gold conjugated secondary antibody for 15 min, fixed, washed in PBS, and imaged in PBS at room temperature using the AFM.

1.2.8. Isolation of Zymogen Granules

ZGs were isolated by using a modification of the method of our published procedure. Male Sprague–Dawley rats weighing 80–100 g were euthanized by CO_2 inhalation for each ZG preparation. The pancreas was dissected and diced into 0.5-mm^3 pieces. The diced pancreas was suspended in 15% (w/v) ice-cold homogenization buffer (0.3 M sucrose, 25 mM Hepes, pH 6.5, 1 mM benzamidine, 0.01% soybean trypsin inhibitor) and homogenized with a Teflon glass homogenizer. The resultant homogenate was centrifuged for 5 min at $300 \times g$ at 4°C to obtain a supernatant fraction. One volume of the supernatant fraction was mixed with 2 vol of a Percoll–Sucrose–Hepes buffer (0.3 M sucrose, 25 mM Hepes, pH 6.5, 86% Percoll, 0.01% soybean trypsin inhibitor) and centrifuged for 30 min at $16,400 \times g$ at 4°C. Pure ZGs were obtained as a loose white pellet at the bottom of the centrifuge tube.

1.2.9. Transmission Electron Microscopy

Isolated rat pancreatic acini and ZGs were fixed in 2.5% buffered paraformaldehyde (PFA) for 30 min, and the pellets were embedded in Unicryl resin and were sectioned at 40–70 nm. Thin sections were transferred to coated specimen TEM grids, dried in the presence of uranyl acetate and methyl cellulose, and examined in a transmission electron microscope.

1.2.10. Immunoprecipitation and Western Blot Analysis

Immunoblot analysis was performed on pancreatic PM and total homogenate fractions. Protein in the fractions was estimated by the Bradford method (Bradford, 1976). Pancreatic fractions were boiled in Laemmli reducing sample preparation buffer (Laemmli, 1970) for 5 min, cooled, and used for SDS-PAGE. PM proteins were resolved in a 12.5%

SDS-PAGE and electrotransferred to 0.2-μm nitrocellulose sheets for immunoblot analysis with a SNAP-23 specific antibody. The nitrocellulose was incubated for 1 h at room temperature in blocking buffer (5% nonfat milk in PBS containing 0.1% Triton X-100 and 0.02% NaN$_3$), and immunoblotted for 2 h at room temperature with the SNAP-23 antibody (ABR, Golden, CO). The primary antibodies were used at a dilution of 1:10,000 in blocking buffer. The immunoblotted nitrocellulose sheets were washed in PBS containing 0.1% Triton X-100 and 0.02% NaN$_3$ and were incubated for 1 h at room temperature in HRP-conjugated secondary antibody at a dilution of 1:2,000 in blocking buffer. The immunoblots were then washed in the PBS buffer, processed for enhanced chemiluminescence, and exposed to X-OMAT-AR film. To isolate the fusion complex for immunoblot analysis, SNAP-23 specific antibody conjugated to protein A-sepharose was used. One gram of total pancreatic homogenate solubilized in Triton/Lubrol solubilization buffer (0.5% Lubrol; 1 mM benzamidine; 5 mM ATP; 5 mM EDTA; 0.5% Triton X-100, in PBS) supplemented with protease inhibitor mix (Sigma, St. Louis, MO) was used. SNAP-23 antibody conjugated to the protein A-sepharose was incubated with the solubilized homogenate for 1 h at room temperature followed by washing with wash buffer (500 mM NaCl, 10 mM TRIS, 2 mM EDTA, pH = 7.5). The immunoprecipitated sample attached to the immuno-sepharose beads was incubated in Laemmli sample preparation buffer—prior to 12.5% SDS-PAGE, electrotransfer to nitrocellulose, and immunoblot analysis—using specific antibodies to actin (Sigma), fodrin (Santa Cruz Biotechnology Inc., Santa Cruz, CA), vimentin (Sigma, St. Louis, MO), syntaxin 2 (Alomone Labs, Jerusalem, Israel), Ca^{2+}-β 3 (Alomone Labs), and Ca^{2+}-α 1c (Alomone Labs).

1.3. POROSOME: A NEW CELLULAR STRUCTURE

Earlier electrophysiological studies on mast cells suggested the existence of fusion pores at the cell plasma membrane (PM), which became continuous with the secretory vesicle membrane following stimulation of secretion (Monck et al., 1995). AFM has confirmed the existence of the fusion pore or porosome as permanent structures at the cell plasma membrane and has revealed its morphology and dynamics in the exocrine pancreas (Cho et al., 2002a; Jena et al., 2003; Jeremic et al., 2003; Schneider et al., 1997), neuroendocrine cells (Cho et al., 2002b,c), and neurons (Cho et al., 2004), at near-nanometer resolution and in real time.

Isolated live pancreatic acinar cells in physiological buffer, when imaged with the AFM (Cho et al., 2002a; Jena et al., 2003; Jeremic et al., 2003; Schneider et al., 1997), reveal at the apical PM a group of circular "pits" measuring 0.4–1.2 μm in diameter which contain smaller 'depressions' (Fig. 1.2). Each depression averages between 100 and 150 nm in diameter, and typically 3–4 depressions are located within a pit. The basolateral membrane of acinar cells is devoid of either pits or depressions. High-resolution AFM images of depressions in live cells further reveal a cone-shaped morphology. The depth of each depression cone measures 15–30 nm. Similarly, growth hormone (GH)-secreting cells of the pituitary gland and chromaffin cells, β cells of the exocrine pancreas, mast cells, and neurons possess depressions at their PM, suggesting their universal presence in secretory cells. Exposure of pancreatic acinar cells to a secretagogue (mastoparan) results in a time-dependent increase (20–35%) in depression diameter, followed by a return to resting size on completion of secretion (Cho et al., 2002a; Jena et al., 2003; Jeremic et al., 2003; Schneider et al., 1997) (Fig. 1.3). No demonstrable change in pit size is detected following stimulation of secretion (Schneider et al.,

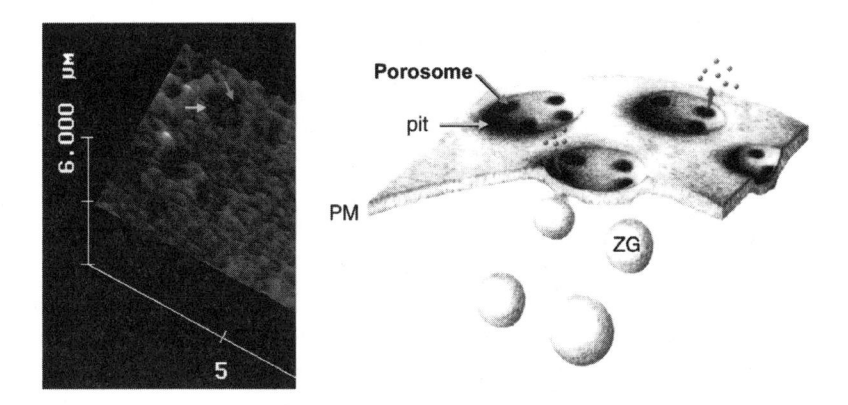

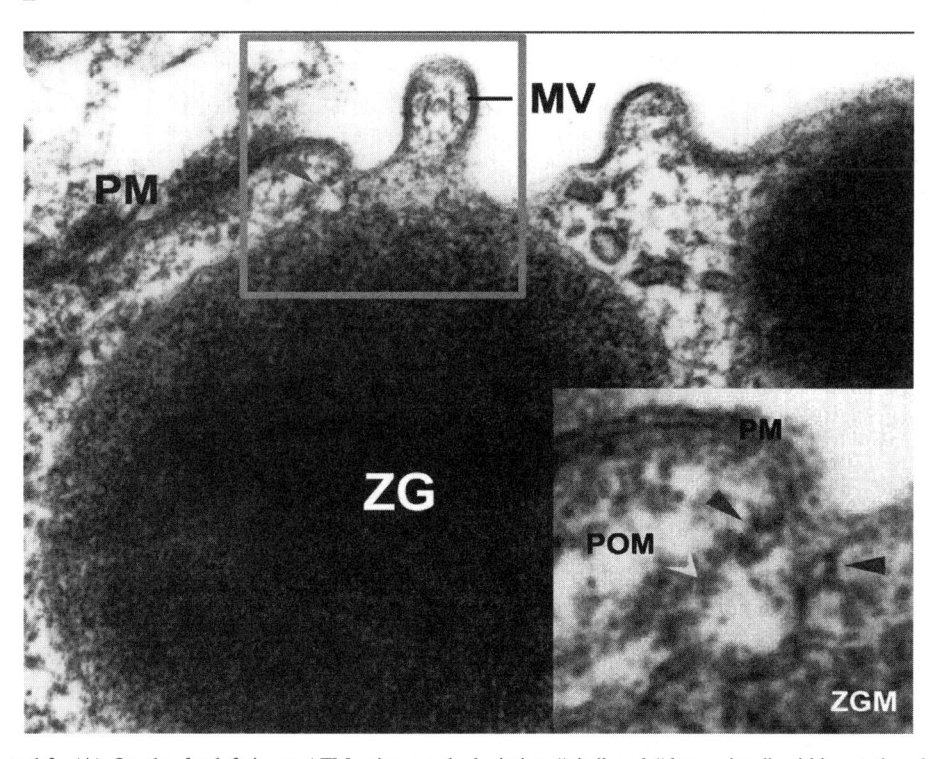

Figure 1.2. (**A**) On the far left is an AFM micrograph depicting "pits" and "depressions" within, at the plasma membrane in live pancreatic acinar cells. On the right is a schematic drawing depicting depressions, at the cell plasma membrane, where membrane-bound secretory vesicles dock and fuse to release vesicular contents (Schneider et al., 1997). (**B**) Electron micrograph depicting a porosome close to a microvilli (MV) at the apical plasma membrane (PM) of a pancreatic acinar cell. Note association of the porosome membrane and the zymogen granule membrane (ZGM) of a docked zymogen granule (ZG), the membrane-bound secretory vesicle of exocrine pancreas. Also a cross section of the ring at the mouth of the porosome is seen.

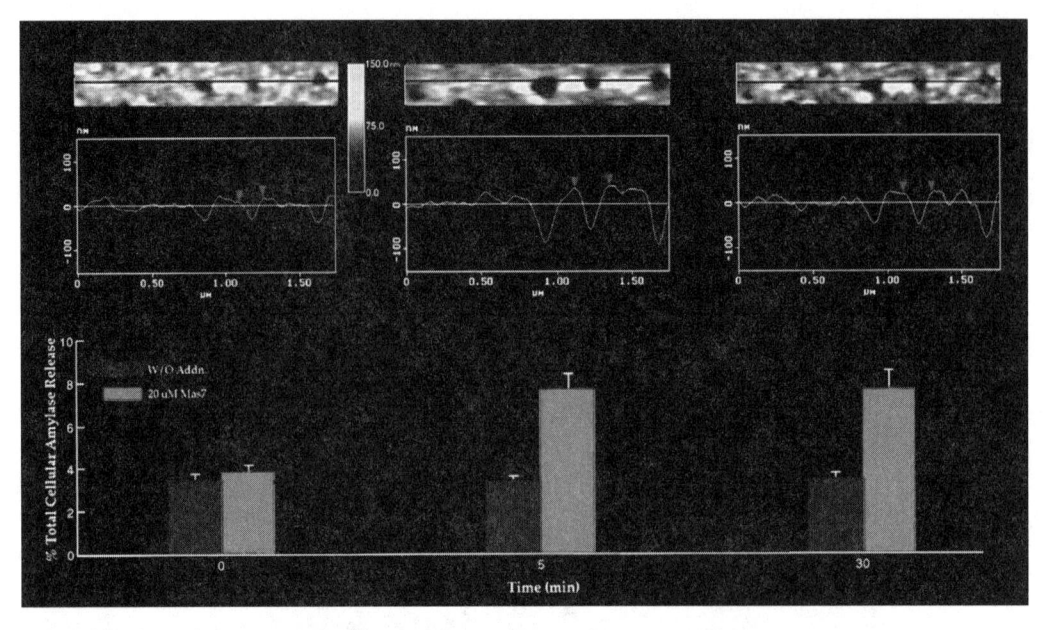

Figure 1.3. Dynamics of depressions following stimulation of secretion. The top panel shows a number of depressions within a pit in a live pancreatic acinar cell. The scan line across three depressions in the top panel is represented graphically in the middle panel and defines the diameter and relative depth of the depressions; the middle depressions are represented by arrowheads. The bottom panel represents percent of total cellular amylase release in the presence and absence of the secretagogue Mas 7. Notice an increase in the diameter and depth of depressions, correlating with an increase in total cellular amylase release at 5 min after stimulation of secretion. At 30 min after stimulation of secretion, there is a decrease in diameter and depth of depressions, with no further increase in amylase release over the 5-min time point. No significant increase in amylase secretion or depressions diameter were observed in resting acini or those exposed to the nonstimulatory mastoparan analog Mas 17 (Jena, 2002; Schneider et al., 1997).

1997). Enlargement of depression diameter and an increase in its relative depth after exposure to secretagogues correlated with increased secretion. Conversely, exposure of pancreatic acinar cells to cytochalasin B, a fungal toxin that inhibits actin polymerization, results in a 15–20% decrease in depression size and a consequent 50–60% loss in secretion (Schneider et al., 1997). Results from these studies suggested depressions to be the fusion pores in pancreatic acinar cells. Furthermore, these studies demonstrate the involvement of actin in regulation of both the structure and function of depressions. Analogous to pancreatic acinar cells, examination of resting GH-secreting cells of the pituitary (Cho et al., 2002b) and chromaffin cells of the adrenal medulla (Cho et al., 2002c) also reveal the presence of pits and depressions at the cell PM (Fig. 1.4). The presence of porosomes in neurons, in β cells of the endocrine pancreas, and in mast cells have also been demonstrated (Figs. 1.4–1.8) (Cho et al., 2004; Jena, 2004). Depressions in resting GH cells measure 154 ± 4.5 nm (mean $\pm$ SE) in diameter. Exposure of GH cells to a secretagogue results in a 40% increase in depression diameter (215 ± 4.6 nm; $p < 0.01$) but no appreciable change in pit size. The enlargement of depression diameter during secretion and the known effect that actin depolymerizing agents decrease depression size and inhibit secretion (Schneider et al., 1997) suggested depressions to be the fusion pores. However, a more direct demonstration that porosomes are functional supramolecular complexes came

POROSOME: A NEW CELLULAR STRUCTURE 9

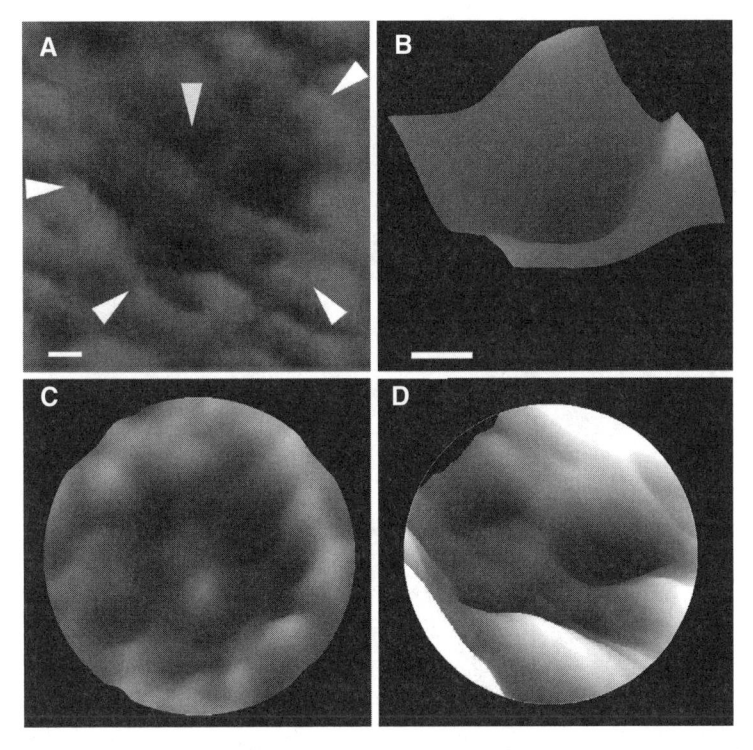

Figure 1.4. AFM micrograph of depressions or porosomes or fusion pores in the live secretory cell of the exocrine pancreas (**A, B**), the growth hormone (GH)-secreting cell of the pituitary (**C**), and in the chromaffin cell (**D**). Note the "pit" (white arrowheads at the margin) with four depressions (arrowhead at 12 o' clock). A high-resolution AFM micrograph of one porosome is shown in part B. Bars = 40 nm for parts A and B. Similarly, AFM micrographs of porosomes in β cell of the endocrine pancreas and mast cell have also been demonstrated.

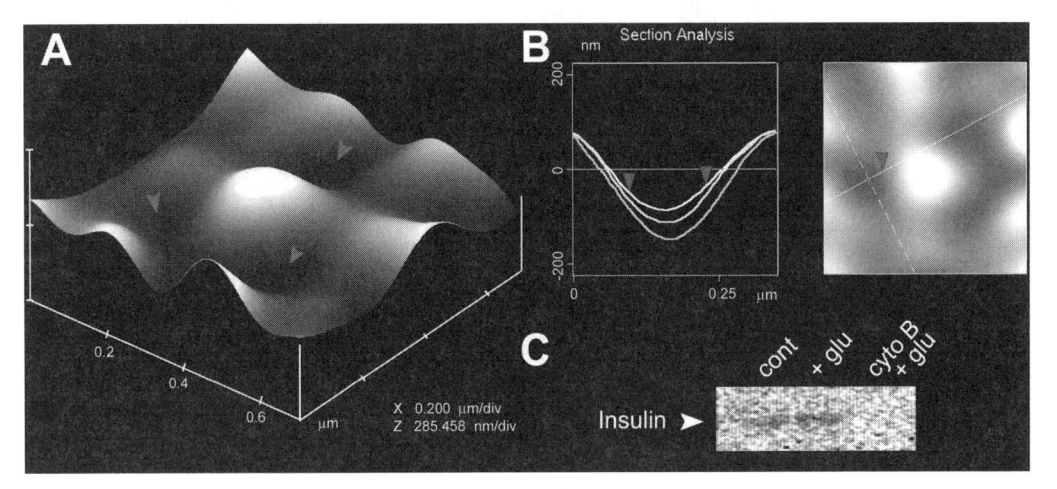

Figure 1.5. Porosomes or fusion pores in β cells of the endocrine pancreas. (**A**) AFM micrograph of a pit with three porosomes (arrowheads) at the PM in a live β cell. (**B**) Section analysis through a porosome at rest (top trace), during secretion (bottom trace), and following completion of secretion (middle trace). (**C**) Exposure of β cells to the actin depolymerizing agent cytochalasin B results in decreased "porosome" size (not shown), accompanied by a loss in insulin secretion, as detected by immunoblot analysis of the cell incubation medium.

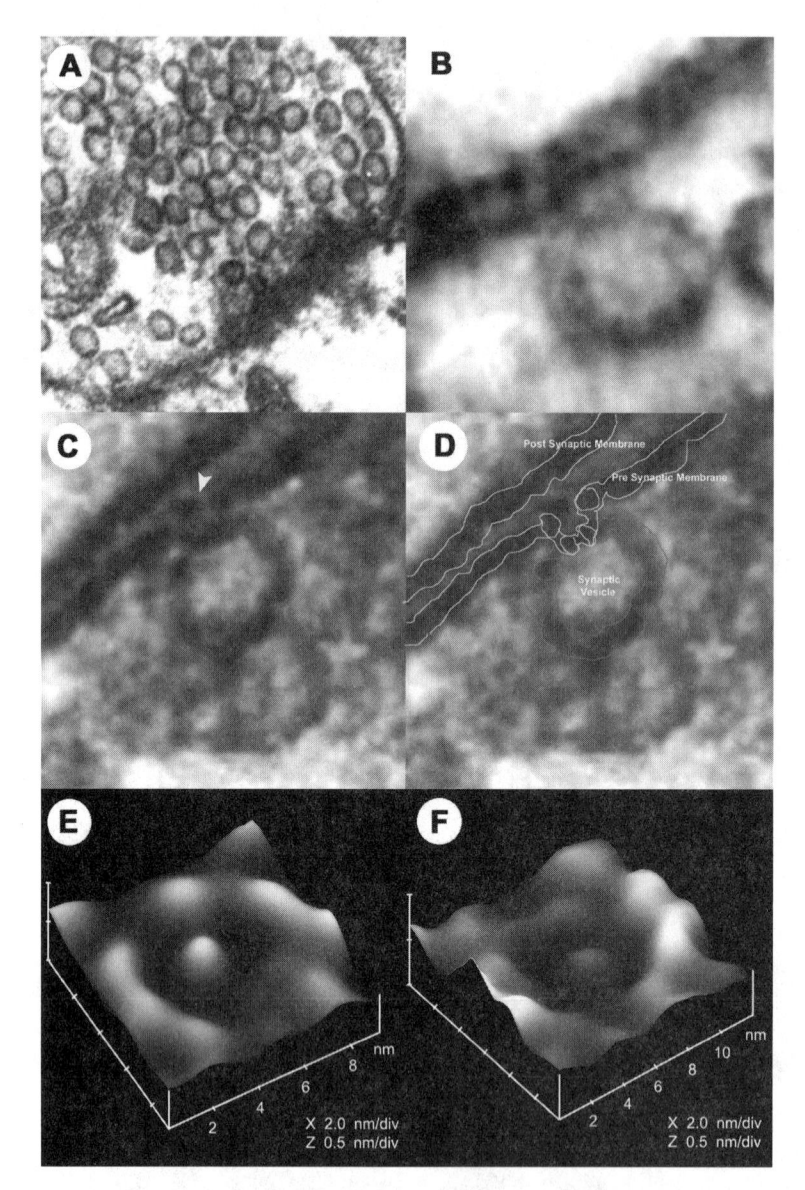

Figure 1.6. Electron micrograph of porosomes in neurons. (**A**) Electron micrograph of a synaptosome demonstrating the presence of 40- to 50-nm synaptic vesicles. (**B–D**) Electron micrograph of neuronal porosomes, which are 10- to 15-nm cup-shaped structures at the presynaptic membrane (arrowhead), where synaptic vesicles transiently dock and fuse to release vesicular contents. (**E**) Atomic force micrograph of a fusion pore or porosome at the nerve terminal in a live synaptosome. (**F**) Atomic force micrograph of isolated neuronal porosome, reconstituted into lipid membrane.

from immuno-AFM studies demonstrating the specific localization of secretory products at the porosomes, following stimulation of secretion (Figs. 1.9–1.12) (Jeremic et al., 2003). ZGs fused with the porosome-reconstituted bilayer, as demonstrated by an increase in capacitance and conductance and in a time-dependent release of the ZG enzyme amylase from cis to the trans compartment of the bilayer chamber. Amylase is detected using immunoblot analysis of the buffer in the cis and trans chambers, using

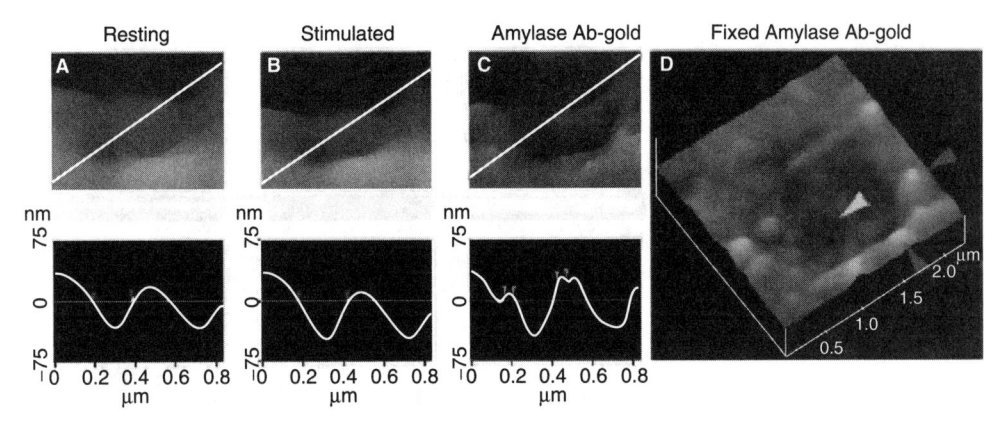

Figure 1.7. Depressions are fusion pores or porosomes. Porosomes dilate to allow expulsion of intra vesicular contents. (**A, B**) AFM micrographs and section analysis of a pit and two out of the four fusion pores or porosomes, demonstrating enlargement following stimulation of secretion. (**C**) Exposure of live cells to gold conjugated-amylase antibody (Ab) results in specific localization of immuno-gold to the porosome opening. (arrowhead) Amylase is one of the proteins within secretory vesicles of the exocrine pancreas. (**D**) AFM micrograph of a fixed pancreatic acinar cell, demonstrating a pit and porosomes within labeled with amylaze-immunogold. Arrowheads point to immunogold clusters (Cho et al., 2002a) at the pit and porosomes within.

a previously characterized amylase specific antibody (Cho et al., 2002a). As observed in immunoblot assays of isolated porosomes, chloride channel activities are also detected within the reconstituted porosome complex. Furthermore, the chloride channel inhibitor DIDS was found to inhibit current activity in the porosome-reconstituted bilayer. In summary, these studies demonstrate that the porosome in the exocrine pancreas is a 100- to 150-nm-diameter supramolecular cup-shaped lipoprotein basket at the cell PM, where membrane-bound secretory vesicles dock and fuse to release vesicular contents. Similar studies have now been performed in neurons, demonstrating both the structural (Fig. 1.6E,F) and functional reconstitution of the isolated neuronal porosome complex. The biochemical composition of the neuronal porosome has also been determined (Cho et al., 2004).

Similar to the isolation of porosomes from the exocrine pancreas, neuronal porosomes were immunoisolated from detergent-solubilized synaptosome preparations, using a SNAP-25 specific antibody. Electrophoretic resolution of the immunoisolates reveal the presence of 12 distinct protein bands, as determined by sypro protein staining of the resolved complex (Fig. 1.13A). Furthermore, electrotransfer of the resolved porosomal complex onto nitrocellulose membrane, followed by immunoblot analysis using various antibodies, revealed 9 proteins. In agreement with earlier findings, SNAP-25, the P/Q-type calcium channel, actin, syntaxin-1, synaptotagmin-1, vimentin, the *N-ethylmaleimide*-sensitive factor (NSF), the chloride channel CLC-3, and the alpha subunit of the heterotrimeric GTP-binding G_o were identified to constitute part of the complex. To test whether the complete porosome complex was immunoisolated, the immunoisolate was reconstituted into lipid membrane prepared using brain dioleoylphosphatidylcholine (DOPC) and dioleylphosphatidylserine (DOPS) in a ratio of 7:3. At low resolution, the AFM reveals the immunoisolates to arrange in L- or V-shaped structures (Fig. 1.13B, red arrowheads), which at higher resolution demonstrates the presence of porosomes in patches (Fig. 1.13C,D, green arrowhead), similar to what is observed at the presynaptic membrane in intact synaptosomes. Further imaging the reconstituted

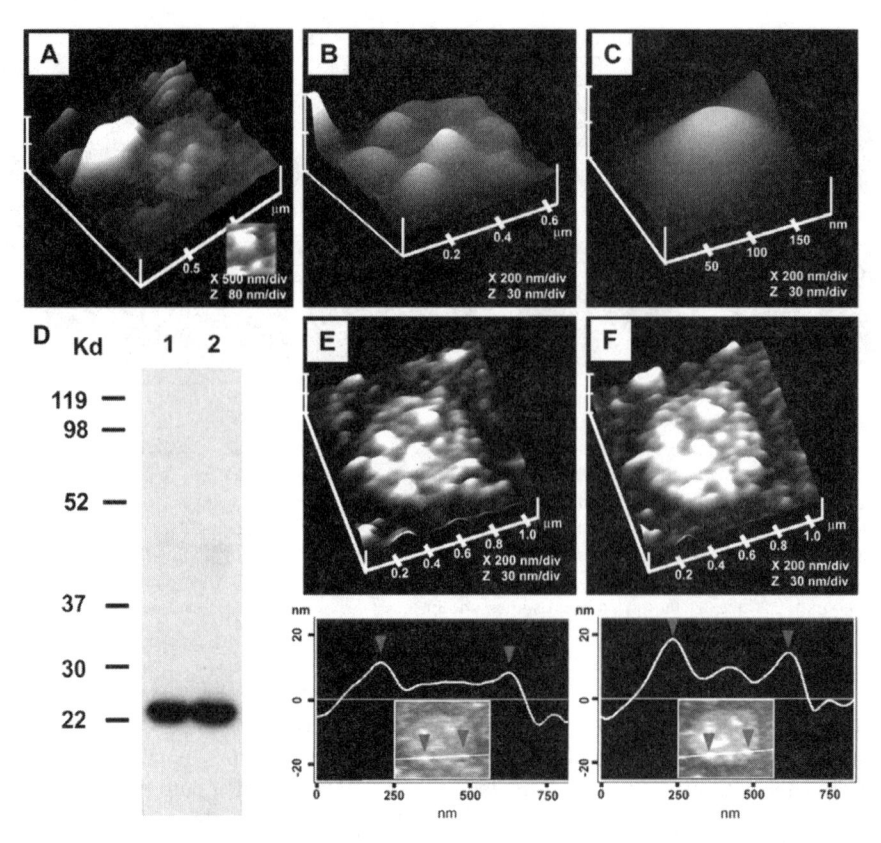

12 POROSOME: THE UNIVERSAL SECRETORY MACHINERY IN CELLS

Figure 1.8. Morphology of the cytosolic side of the porosome revealed in AFM studies on isolated pancreatic plasma membrane (PM) preparations. (**A**) AFM micrograph of isolated PM preparation reveals the cytosolic end of a pit with inverted cup-shaped structures, the porosome. Note the 600 nm in diameter ZG at the left-hand corner of the pit. (**B**) Higher magnification of the same pit showing clearly the 4 to 5 porosomes within. (**C**) The cytosolic end of a single porosome is depicted in this AFM micrograph. (**D**) Immunoblot analysis of 10 μg and 20 μg of pancreatic PM preparations, using SNAP-23 antibody, demonstrates a single 23-kDa immunoreactive band. (**E, F**) The cytosolic side of the PM demonstrating the presence of a pit with a number of porosomes within, shown prior to (**E**) and following addition of the SNAP-23 antibody (**F**) Note the increase in height of the porosome cone base revealed by section analysis (bottom panel), demonstrating localization of SNAP-23 antibody at the base of the porosome (Jeremic et al., 2003).

immunoisolate at greater resolutions using the AFM shows the presence of 8- to 10-nm porosomes (Fig. 1.13E). As observed in electron and AFM micrographs of the presynaptic membrane in synaptosomes, the immunoisolated and lipid-reconstituted porosome reveals the presence of an approximately 2 nm in diameter central plug. These studies confirm the complete isolation and structural reconstitution of the neuronal porosome in artificial lipid bilayers. To understand the structure of the porosome at the cytosolic side of the presynaptic membrane, isolated synaptosomal membrane preparations were imaged by the AFM. These studies reveal the architecture of the cytosolic part of porosomes. In synaptosomes, 40- to 50-nm synaptic vesicles are arranged in ribbons, as seen in the electron micrographs (longitudinal section, Fig. 1.14A, or in cross section Fig. 1.14B). Similarly, when the cytosolic domain of isolated synaptosomal membrane preparations were analyzed by the AFM, 40- to 50-nm synaptic vesicles

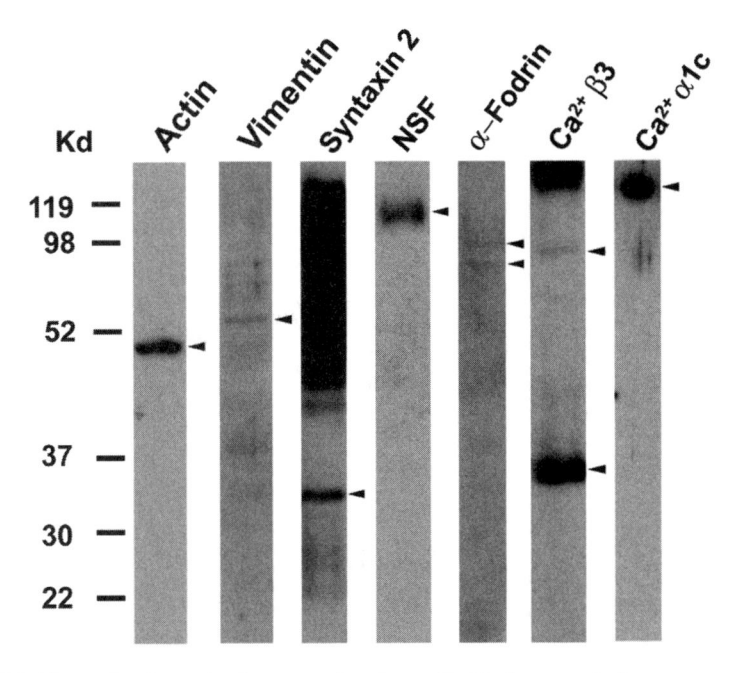

Figure 1.9. SNAP-23 associated proteins in pancreatic acinar cells. Total pancreatic homogenate was immunoprecipitated using the SNAP-23 specific antibody. The precipitated material was resolved using 12.5% SDS-PAGE, electrotransferred to nitrocellulose membrane and then probed using antibodies to a number of proteins. Association of SNAP-23 with syntaxin2, with cytoskeletal proteins actin, α-fodrin, and vimentin, and calcium channels β3 and α1c, together with the SNARE regulatory protein NSF, is demonstrated (arrowheads). Lanes showing more than one arrowhead suggest presence of isomers or possible proteolytic degradation of the specific protein (Jena et al., 2003).

were also found to be arranged in ribbons (Fig. 1.14C,D), thus confirming EM observations. On close examination (Fig. 1.14D–F) of docked synaptic vesicles (blue arrowheads), synaptic vesicles are found attached to the base of the 8- to 10-nm porosomes (red arrowheads). It is possible to see both the porosome and the attached synaptic vesicle, since while imaging with the AFM, synaptic vesicles can be gently pushed away from their docked sites to reveal the porosome lying beneath them (Fig. 1.14C–F). Additionally, bare porosomes with no synaptic vesicles attached are also found at the cytosolic side of synaptosomal membrane preparations (Fig. 1.14D).

Calcium, target membrane proteins SNAP-25 and syntaxin (t-SNARE), and secretory vesicle-associated membrane protein (v-SNARE) are the minimal machinery involved in fusion of opposing bilayers (Jeremic et al., 2004a,b). NSF is an ATPase that is suggested to disassemble the t-/v-SNARE complex in the presence of ATP (Jeong et al., 1998). To test this hypothesis and to further confirm and determine the morphology of neuronal porosomes, synaptosomal membrane preparations with docked synaptic vesicles were imaged using the AFM in the presence and absence of ATP (Fig. 1.15A). As hypothesized, addition of 50 µM ATP resulted in t-/v-SNARE disassembly and the release of docked vesicles (blue arrowhead) at the porosome patch (Fig. 1.15B,C, red arrowhead). At higher resolution, the base of the porosome is clearly revealed (Fig. 1.15D). At increased imaging forces (300–500 pN instead of <200 pN), porosome patches (Fig. 1.15E, red arrowhead) and individual porosomes

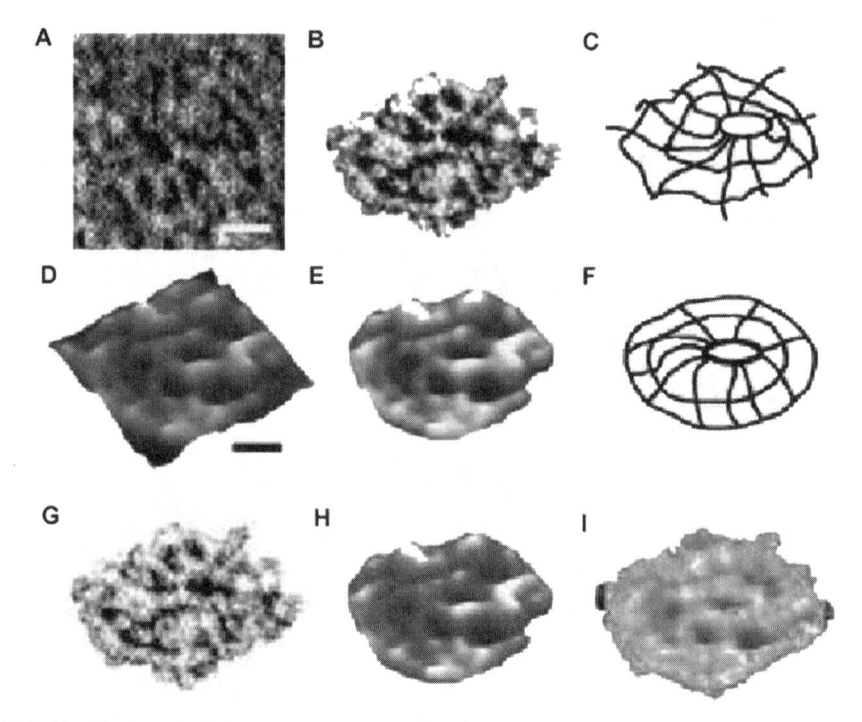

Figure 1.10. Negatively stained electron micrograph and atomic force micrograph of the immunoisolated porosome complex. (**A**) Negatively stained electron micrograph of an immunoisolated porosome complex from solubilized pancreatic plasma membrane preparations, using a SNAP-23 specific antibody. Note the three rings and the 10 spokes that originate from the inner smallest ring. This structure represents the protein backbone of the porosome complex, since the three rings and the vertical spikes are observed in electron micrographs of cells and porosome co-isolated with ZGs. Bar = 30 nm. (**B**) Electron micrograph of the fusion pore complex, cut out from (**A**). (**C**) Outline of the structure presented for clarity. (**D–F**) Atomic force micrograph of the isolated pore complex in near physiological buffer. Bar = 30 nm. Note the structural similarity of the complex, imaged both by EM (**G**) and AFM (**H**). The EM and AFM micrographs are superimposable (**I**).

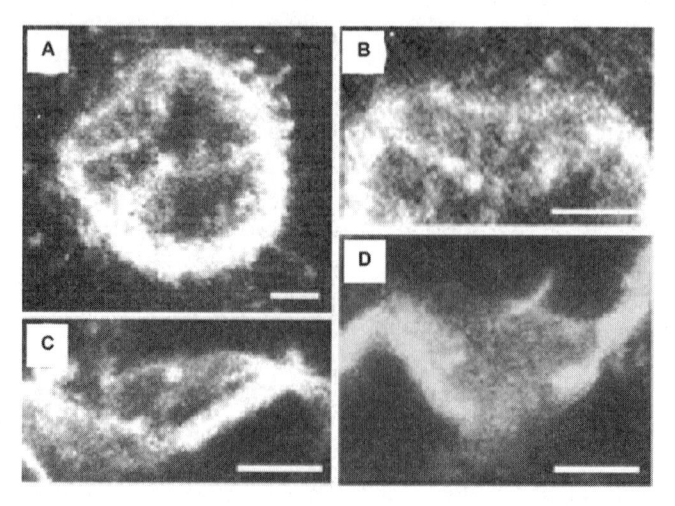

Figure 1.11. Electron micrographs of reconstituted porosome or fusion pore complex in liposomes, showing a cup-shaped basket-like morphology. (**A**) A 500-nm vesicle with an incorporated porosome is shown. Note the spokes in the complex. (**B–D**) The reconstituted complex at greater magnification is shown. Bar represents 100 nm.

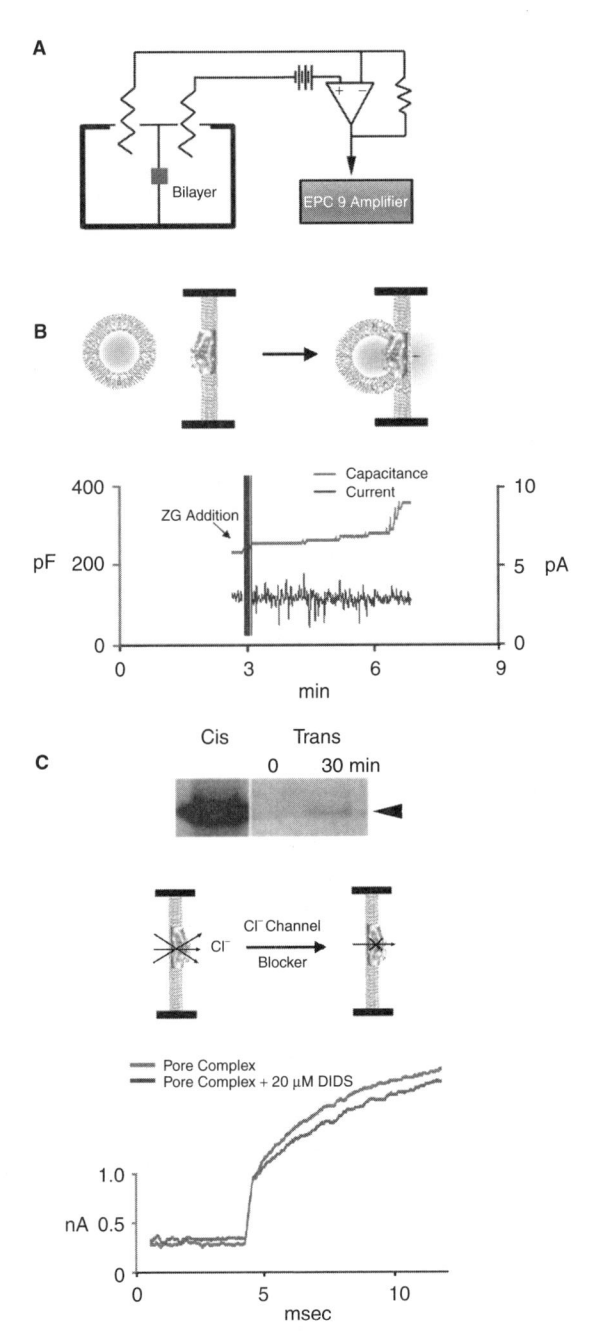

Figure 1.12. Lipid bilayer-reconstituted porosome complex is functional. (**A**) Schematic drawing of the EPC9 bilayer setup for electrophysiological measurements. (**B**) Zymogen granules (ZGs) added to the cis compartment of the bilayer chamber fuse with the reconstituted porosomes, as demonstrated by an increase in capacitance and current activities, and a concomitant time-dependent release of amylase (a major ZG content) to the trans compartment of the bilayer chamber. The movement of amylase from the cis to the trans side of the chamber was determined by immunoblot analysis of the contents in the cis and the trans chamber over time. (**C**) As demonstrated by immunoblot analysis of the immunoisolated complex, electrical measurements in the presence and absence of chloride ion channel blocker DIDS demonstrate the presence of chloride channels in association with the complex, and its role in porosome function.

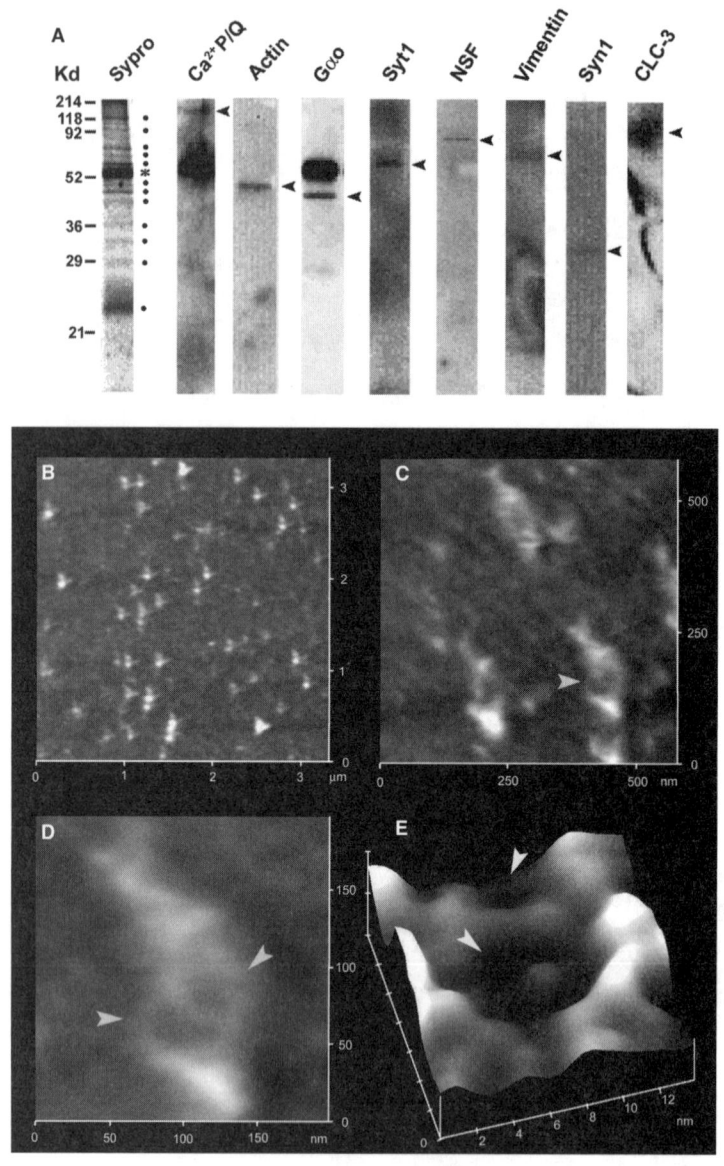

Figure 1.13. Composition of the neuronal porosome and its reconstitution in lipid membrane. **(A)** Proteins immunoisolated from detergent-solubilized synaptosomal membrane preparation, using a SNAP-25 specific antibody. Immunoisolates when resolved by SDS-PAGE, along with the resolved proteins in the gel stained using Sypro dye, reveals 12 specific bands (•), suggesting the presence of at least 12 proteins in the complex, not including the heavy chain (∗) of the SNAP-25 antibody. When the resolved proteins were electrotransferred to nitrocellulose membrane and probed with various antibodies, calcium channel P/Q, actin, Gαo, syntaxin-1 (Syn1), synaptotagmin-1 (Syt1), NSF, vimentin, and the chloride channel CLC-3 were identified (Bradford, 1976; Laemmli, 1970). **(B–E)** Atomic force micrographs at different resolution of the reconstituted immunoisolate in lipid membrane. **(B)** At low magnification, the immunoisolated complex arrange in L- or V-shaped structures (white arrowheads). **(C, D)** Within the V-shaped structures, patches (arrowheads) of porosomes are found. **(E)** Each reconstituted porosome is almost identical to the porosome observed in intact synaptosomes. The central plug is clearly seen. This AFM micrograph demonstrates the presence of two porosomes (arrowhead), although only half of the second porosome is in view.

(Fig. 1.15F,G) were further defined in greater structural detail. At 5–8 Å resolution, eight peripheral knobs at the porosomal opening (Fig. 1.15F–H, yellow arrowheads) and a central plug (green arrowhead) at the base were revealed. These studies further provide a direct demonstration of synaptic vesicles docking at these sites and confirming them to be porosomes, where synaptic vesicles dock and transiently fuse to release neurotransmitters (Aravanis et al., 2003). Hence synaptic vesicles are able to fuse transiently and successively at porosomes in the presynaptic membrane without loss of identity, as reported in earlier studies (Aravanis et al., 2003).

To be able to assess the functionality of the reconstituted porosome preparations (Figs. 1.13E and 1.16A), an electrophysiological bilayer setup was used (Fig. 1.16B). Membrane capacitance was continually monitored throughout these experiments. Following reconstitution of the bilayer membrane with porosomes, isolated synaptic vesicles were added to the cis compartment of the bilayer chamber. A large number of synaptic vesicles fused at the bilayer, as demonstrated by the significant stepwise increases in the membrane capacitance (Fig. 1.16C,D). As expected from results in Fig. 15B, addition of 50 µM ATP allowed t-/v-SNARE disassembly and the release of docked vesicles, resulting in the return of the bilayers membrane capacitance to resting levels (Fig. 1.16C,D). Addition of recombinant NSF had no further effect on membrane capacitance. Thus, the associated NSF at the t-/v-SNARE complex is adequate for complete disassembly of the SNARE complex for release of synaptic vesicles following transient fusion and the completion of a round of neurotransmitter release. To further biochemically assess the release of docked synaptic vesicles following ATP treatment, synaptosomal membrane preparations were exposed to 50 µM ATP, and the supernatant fraction was assessed for synaptic vesicles by monitoring levels of the synaptic vesicle proteins SV2 and VAMP-2 (Fig. 1.16E). Our study demonstrates that both SV2 and VAMP-2 proteins are enriched in supernatant fractions following exposure of isolated synaptosomal membrane to ATP (Fig. 1.16E). Thus, AFM, electrophysiological measurements, and immunoanalysis confirmed the dissociation of porosome-docked synaptic vesicles following ATP exposure. As previously suggested (Jena, 1997, 2002), such a mechanism may allow for the multiple transient docking-fusion and release cycles that synaptic vesicles may undergo during neurotransmission, without loss of vesicle identity (Aravanis et al., 2003). The neuronal porosome, although an order of magnitude smaller than those in the exocrine pancreas or in neuroendocrine cells, possesses many similarities both in structure and composition. Thus, nature has designed the porosome as a general secretory machinery, but has fine-tuned it to suit various secretory processes in different cells. Hence, porosome size may be a form of such fine-tuning. It is well known that smaller vesicles fuse more efficiently than larger ones, and hence the curvature of both the secretory vesicle and the porosome base would dictate the efficacy and potency of vesicle fusion at the cell plasma membrane. For example, because neurons are fast secretory cells, they possess small (40–50 nm) secretory vesicles and porosome bases (2–4 nm) for rapid and efficient fusion. In contrast, a slow secretory cell like the exocrine pancreas possesses larger secretory vesicles (1000 nm in diameter) that fuse at porosomes having bases that measure 20–30 nm in diameter.

1.4. MOLECULAR UNDERSTANDING OF CELL SECRETION

Fusion pores or porosomes as permanent structures at the cell plasma membrane are present in all secretory cells examined, ranging from exocrine, endocrine, and neuroendocrine cells to neurons, where

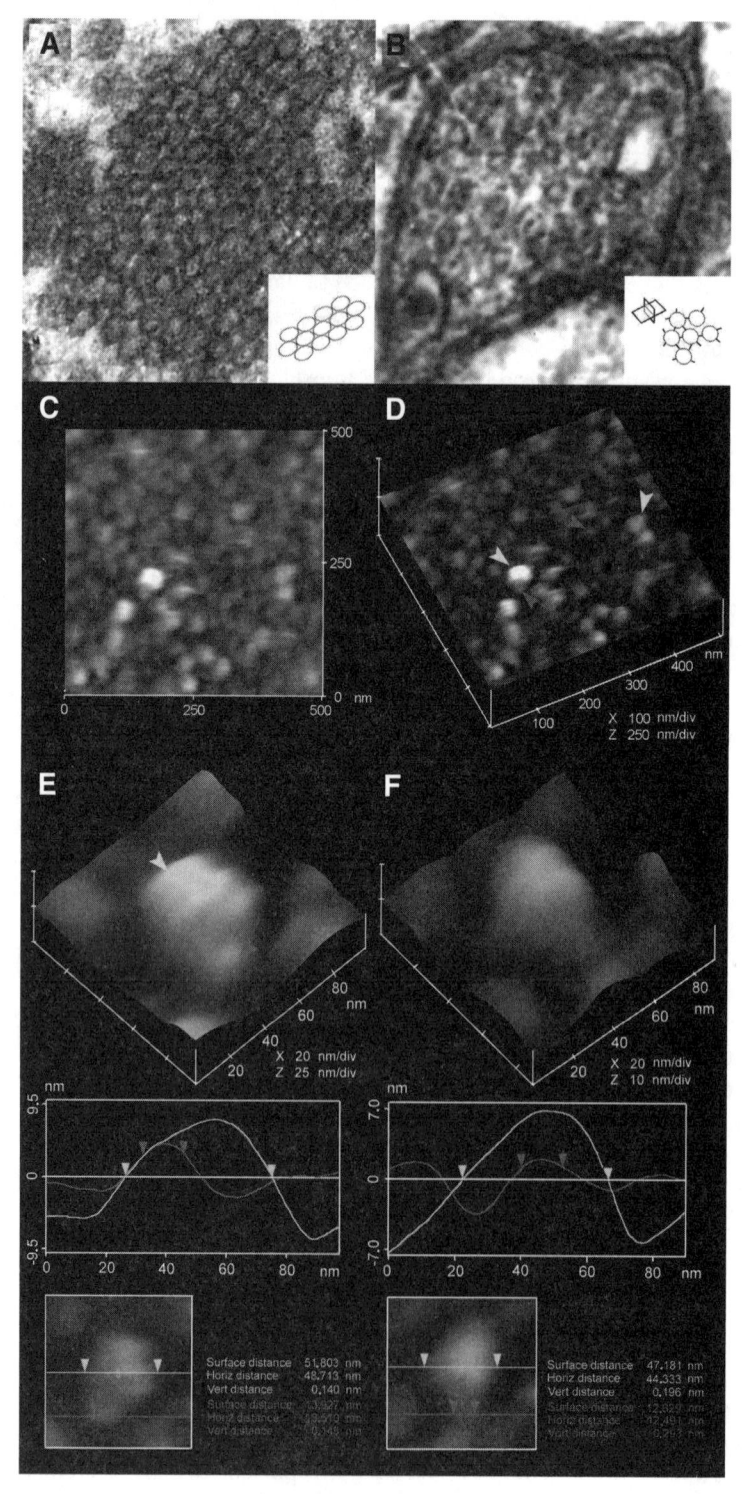

18 POROSOME: THE UNIVERSAL SECRETORY MACHINERY IN CELLS

Figure 1.14

membrane-bound secretory vesicles dock and transiently fuse to expel vesicular contents. Porosomes in pancreatic acinar or GH-secreting cells are cone-shaped structures at the plasma membrane, with a 100- to 150-nm-diameter opening. Membrane-bound secretory vesicles ranging in size from 0.2 to 1.2 μm in diameter dock and fuse at porosomes to release vesicular contents. Following fusion of secretory vesicles at porosomes, only a 20–35% increase in porosome diameter is demonstrated. It is therefore reasonable to conclude that secretory vesicles "transiently" dock and fuse at the site. In contrast to accepted belief, if secretory vesicles were to completely incorporate at porosomes, the PM structure would distend much wider than what is observed. Furthermore, if secretory vesicles were to completely fuse at the plasma membrane, there would be a loss in vesicle number following secretion. Examination of secretory vesicles within cells before and after secretion demonstrates that the total number of secretory vesicles remains unchanged following secretion (Cho et al., 2002e; Lee et al., 2004). However, the number of empty and partially empty vesicles increases significantly, supporting the occurrence of transient fusion. Earlier studies on mast cells also demonstrated an increase in the number of spent and partially spent vesicles following stimulation of secretion, without any demonstrable increase in cell size. Similarly, secretory granules are recaptured largely intact after stimulated exocytosis in cultured endocrine cells (Taraska et al., 2003). Other support in evidence of transient fusion is the presence of neurotransmitter transporters at the synaptic vesicle membrane. These vesicle-associated transporters would be of little use if vesicles were to fuse completely at the plasma membrane to be compensatorily endocytosed at a later time. In further support, a recent study reports that single synaptic vesicles fuse transiently and successively without loss of vesicle identity (Aravanis et al., 2003). Although the fusion of secretory vesicles at the cell plasma membrane occurs transiently, complete incorporation of membrane at the cell plasma membrane would occur when cells need to incorporate signaling molecules like receptors, second messengers, or ion channels. Similarly, total fusion would occur intracellularly, where during protein transport and maturation, vesicles derived from the cis Golgi would completely fuse with the trans Golgi apparatus. The discovery of the porosome, along with an understanding of the molecular mechanism of membrane fusion and the swelling of secretory vesicles required for expulsion of vesicular contents, provides an understanding of secretion and membrane fusion in cells at the molecular level. These findings have prompted many laboratories to work in the area and further confirm these findings. Thus, the porosome is a supramolecular structure universally present in secretory cells, from the exocrine pancreas to the neurons, and in the endocrine to neuroendocrine cells, where membrane-bound secretory vesicles transiently dock and fuse to expel intravesicular contents. Hence, the secretory process in

Figure 1.14. Arrangement of synaptic vesicles and porosomes at the presynaptic membrane. (A) Electron micrograph of rat brain synaptosome demonstrating the ribbon-like arrangement (inset) of 40- to 50-nm synaptic vesicles. (B) Cross section of such a ribbon (inset) reveals the interaction between the synaptic vesicles. (C, D) Examination of the presynaptic membrane from the cytosolic side (inside out) using AFM, confirmed such a ribbon arrangement of docked synaptic vesicles (white arrowheads) at porosomes (grey arrowheads). Bare porosomes (lacking docked synaptic vesicles) are also seen. The AFM micrograph in part c is a 2-D image, and the one in part d is a 3-D image. (E, F) AFM micrograph of a docked synaptic vesicle at a porosome. During imaging using the AFM, the interaction of the cantilever tip with the sample sometimes resulted in pushing away the docked synaptic vesicle, enough to expose the porosome lying beneath. AFM section analysis further reveals the size of synaptic vesicles (white section line and arrowheads) and porosomes (grey section line and arrowheads).

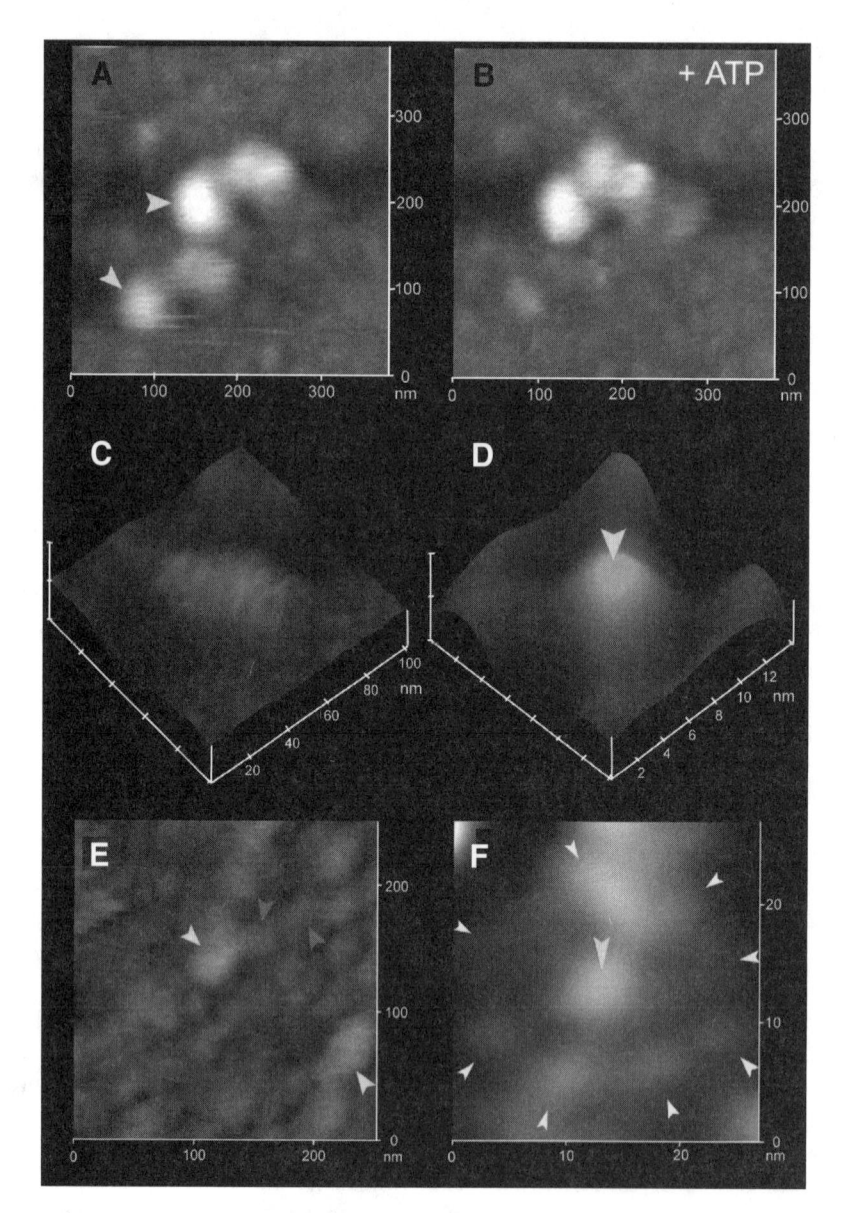

Figure 1.15. AFM micrographs revealing the dynamics of docked synaptic vesicles at porosomes and the porosome architecture at 4–5 Å resolution. (**A**) AFM micrograph of five docked synaptic vesicle at porosomes. (**B**) Addition of 50 μM ATP dislodges two synaptic vesicles at the lower left, and exposing the porosome patches. This also demonstrates that a single synaptic vesicle may dock at more than one porosome complex. (**C–G**) AFM micrographs obtained at higher imaging forces (300–500 pN rather than <200 pN) reveal porosomes architecture at greater detail. (**C**) AFM micrograph of one of the porosome patches where a synaptic vesicle was docked prior to ATP exposure. (**D**) Base of a single porosome. (**E**) High-force AFM micrograph of the cytosolic face of the presynaptic membrane, demonstrating the ribbon arrangement of porosome patches (grey arrowhead) and docked synaptic vesicles (white arrowheads). Note how the spherical synaptic vesicles are compressed and flattened at higher imaging forces. (**F, G**) At such higher imaging forces, porosomes reveal the presence of eight globular structures (white arrowhead) surrounding a central plug (grey arrowhead), as demonstrated in the (**H**) schematic diagram.

MOLECULAR UNDERSTANDING OF CELL SECRETION 21

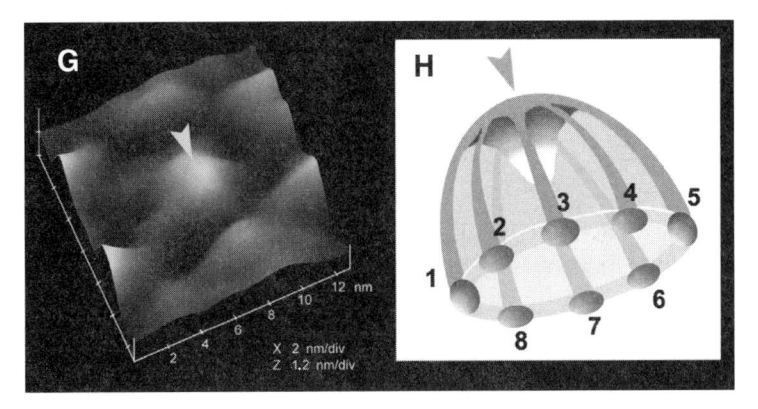

Figure 1.15. (*continued*)

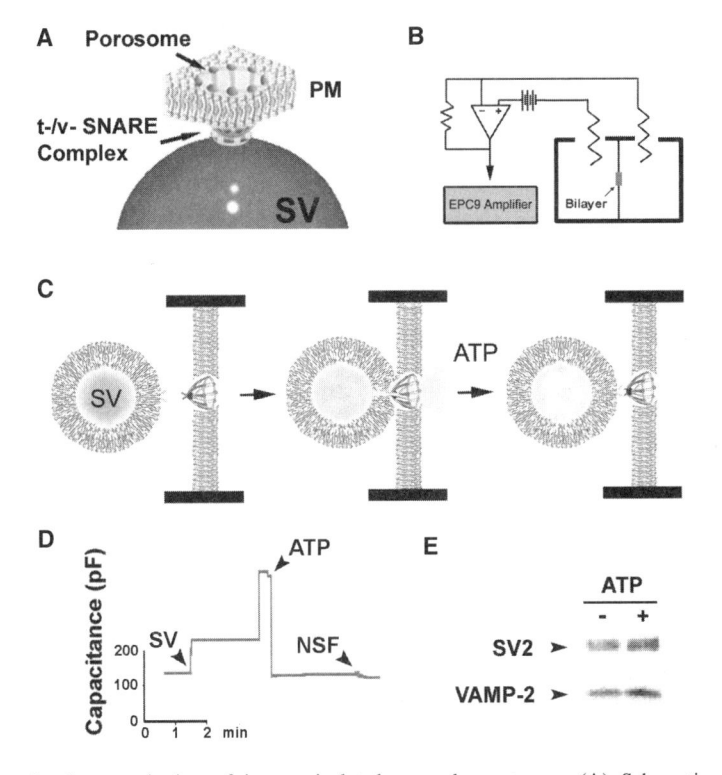

Figure 1.16. Functional reconstitution of immunoisolated nuronal porosomes. (**A**) Schematic representation of a porosome at the presynaptic membrane, with a docked synaptic vesicle (SV) at its base. (**B**) Schematic drawing of an EPC9 electrophysiological bilayers apparatus, to continually monitor changes in the capacitance of porosome-reconstituted membrane, when synaptic vesicles are introduced into the cis bilayers chamber followed by ATP and purified recombinant NSF protein. (**C**) Schematic representation of SV docking at the base of a porosome, fusing to release its contents, and disengaging in the presence of ATM. (**D**) Capacitance measurements of porosome-reconstituted bilayers support the experiment in Fig. 1.15B and the schematic diagram in part c. Exposure of the reconstituted bilayers to SVs results in a dramatic increase in membrane capacitance, which drops to baseline following exposure to 50 μM ATP. Recombinant NSF has no further effect ($n = 6$). (**E**) Similarly, in agreement, exposure of isolated synaptosomal membrane preparations to 50 μM ATP results in the release of SVs from the membrane into the incubation medium, as demonstrated by immunoblot analysis of the incubating medium using the SV-specific protein antibodies, SV2 and VAMP-2.

cells is a highly regulated event, orchestrated by a number of ions and biomolecules.

ACKNOWLEDGMENT

Supported by Grants DK-56212 and NS-39918 from the National Institutes of Health (BPJ).

REFERENCES

Abu-Hamdah, R., Cho, W. J., Cho, S. J., Jeremic, A., Kelly, M., Ilie, A. E., and Jena, B. P. (2004). Regulation of the water channel aquaporin-1: Isolation and reconstitution of the regulatory complex. *Cell Biol. Int.* **28**:7–17 (published on-line 2003).

Alexander, S., Hellemans, L., Marti, O., Schneir, J., Elings, V., and Hansma, P. K. (1989). An atomic resolution atomic force microscope implemented using an optical lever. *J. Appl. Phys.* **65**:164–167.

Anderson, L. L. (2004). Discovery of a new cellular structure—the porosome: Elucidation of the molecular mechanism of secretion. *Cell Biol. Int.* **28**:3–5.

Aravanis, A. M., Pyle, J. L., and Tsien, R. W. (2003). Single synaptic vesicles fusing transiently and successively without loss of identity. *Nature* **423**:643–647.

Bennett, V. (1990). Spectrin-based membrane skeleton: A multipotential adaptor between plasma membrane and cytoplasm. *Physiol. Rev.* **70**:1029–1065.

Binnig, G., Quate, C. F., and Gerber, C. H. (1986). Atomic force microscope. *Phys. Rev. Lett.* **56**:930–933.

Bradford, M. M. (1976). A rapid and sensitive method for the quantitation of microgram quantities of protein utilizing the principle of protein-dye binding. *Anal. Biochem.* **72**:248–254.

Cho, S. J., Quinn, A. S., Stromer, M. H., Dash, S., Cho, J., Taatjes, D. J., and Jena, B. P. (2002a). Structure and dynamics of the fusion pore in live cells. *Cell Biol. Int.* **26**:35–42.

Cho, S. J., Jeftinija, K., Glavaski, A., Jeftinija, S., Jena, B. P., and Anderson, L. L. (2002b). Structure and dynamics of the fusion pores in live GH-secreting cells revealed using atomic force microscopy. *Endocrinology* **143**: 1144–1148.

Cho, S. J., Wakade, A., Pappas, G. D., and Jena, B. P. (2002c). New structure involved in transient membrane fusion and exocytosis. *Ann. New York Acad. Sci.* **971**:254–256.

Cho, S. J., Kelly, M., Rognlien, K. T., Cho, J., Hoerber, J. K. H., and Jena, B. P. (2002d). SNAREs in opposing bilayers interact in a circular array to form conducting pores. *Biophys. J.* **83**:2522–2527.

Cho, S. J., Cho, J., and Jena, B. P. (2002e). The number of secretory vesicles remains unchanged following exocytosis. *Cell Biol. Int.* **26**:29–33.

Cho, S. J., Sattar, A. K., Jeong, E-H., Satchi, M., Cho, J. A., Dash, S., Mayes, M. S., Stromer, M. H., and Jena, B. P. (2002f). Aquaporin 1 regulates GTP-induced rapid gating of water in secretory vesicles. *Proc. Natl. Acad. Sci. USA* **99**:4720–4724.

Cho, W. J., Jeremic, A., Rognlien, K. T., Zhvania, M. G., Lazrishvili, I., Tamar, B., and Jena, B. P. (2004). Structure, isolation, composition and reconstitution of the neuronal fusion pore. *Cell Biol. Int.* **28**:699–708 (published on-line August 25, 2004).

Faigle, W., Colucci-Guyon, E., Louvard, D., Amigorena, S., and Galli, T. (2000). Vimentin filaments in fibroblasts are a reservoir for SNAP-23, a component of the membrane fusion machinery. *Mol. Biol. Cell.* **11**:3485–3494.

Fix, M., Melia, T. J., Jaiswal, J. K., Rappoport, J. Z., You, D., Söllner, T. H., Rothman, J. E., and Simon, S. M. (2004). Imaging single membrane fusion events mediated by SNARE proteins. *Proc. Natl. Acad. Sci. USA* **101**: 7311–7316.

Gaisano, H. Y., Sheu, L., Wong, P. P., Klip, A., and Trimble, W. S. (1997). SNAP-23 is located in the basolateral plasma membrane of rat pancreatic acinar cells. *FEBS Lett.* **414**:298–302.

Goodson, H. V., Valetti, C., and Kreis, T. E. (1997). Motors and membrane traffic. *Curr. Opin. Cell Biol.* **9**:18–28.

Hörber, J. K. H., and Miles, M. J. (2003). Scanning probe evolution in biology. *Science* **302**:1002–1005.

Jena, B. P. (1997). Exocytotic Fusion: Total or transient. *Cell Biol. Int.* **21**:257–259.

Jena, B. P. (2002). Fusion pore in live cells. *NIPS* **17**:219–222.

Jena, B. P. (2003). Fusion Pore: Structure and dynamics. *J. Endo.* **176**:169–174.

Jena, B. P. (2004). Discovery of the Porosome: Revealing the molecular mechanism of secretion and membrane fusion in cells. *J. Cell Mol. Med.* **8**:1–21.

Jena, B. P., Schneider, S. W., Geibel, J. P., Webster, P., Oberleithner, H., and Sritharan, K. C. (1997). G_i regulation of secretory vesicle swelling examined by atomic force microscopy. *Proc. Natl. Acad. Sci. USA* **94**:13317–13322.

Jena, B. P., Cho, S. J., Jeremic, A., Stromer, M. H., and Abu-Hamdah, R. (2003). Structure and composition of the fusion pore. *Biophys. J.* **84**:1337–1343.

Jeong, E.-H., Webster, P., Khuong, C. Q., Sattar, A. K. M. A., Satchi, M., and Jena, B. P. (1998). The native membrane fusion machinery in cells. *Cell Biol. Int.* **22**:657–670.

Jeremic, A., Kelly, M., Cho, S. J., Stromer, M. H., and Jena, B. P. (2003). Reconstituted fusion pore. *Biophys. J.* **85**:2035–2043.

Jeremic, A., Kelly, M., Cho, W. J., Cho, S. J., Horber, J. K. H., and Jena, B. P. (2004a). Calcium drives fusion of SNARE-apposed bilayers. *Cell Biol. Int.* **28**:19–31 (published on-line 2003).

Jeremic, A., Cho, W. J., and Jena, B. P. (2004b). Membrane fusion: What may transpire at the atomic level. *J. Biol. Phys. Chem.* **4**:139–142.

Kelly, M., Cho, W. J., Jeremic, A., Abu-Hamdah, R., and Jena, B. P. (2004). Vesicle swelling regulates content expulsion during secretion. *Cell Biol. Int.* **28**:709–716 (published on-line August 25, 2004).

Laemmli, U. K. (1970). Cleavage of structural proteins during the assembly of the head of bacteriophage T4. *Nature.* **227**:680–685.

Lee, J. S., Mayes, M. S., Stromer, M. H., Scanes, C. G., Jeftinija, S., and Anderson, L. L. (2004). Number of secretory vesicles in growth hormone cells of the pituitary remains unchanged after secretion. *Exp. Biol. Med.* **229**:291–302.

Monck, J. R., Oberhauser, A. F., and Fernandez, J. M. (1995). The exocytotic fusion pore interface: A model of the site of neurotransmitter release. *Mol. Membr. Biol.* **12**:151–156.

Nakano, M., Nogami, S., Sato, S., Terano, A., and Shirataki, H. (2001). Interaction of syntaxin with α-fodrin, a major component of the submembranous cytoskeleton. *Biochem. Biophys. Res. Commun.* **288**:468–475.

Ohyama, A., Komiya, Y., and Igarashi, M. (2001). Globular tail of myosin-V is bound to vamp/synaptobrevin. *Biochem. Biophys. Res. Commun.* **280**:988–991.

Schneider, S. W., Sritharan, K. C., Geibel, J. P., Oberleithner, H., and Jena, B. P. (1997). Surface dynamics in living acinar cells imaged by atomic force microscopy: Identification of plasma membrane structures involved in exocytosis. *Proc. Natl. Acad. Sci. USA* **94**:316–321.

Taraska, J. W., Perrais, D., Ohara-Imaizumi, M., Nagamatsu, S., and Almers, W. (2003). Secretory granules are recaptured largely intact after stimulated exocytosis in cultured endocrine cells. *Proc. Natl. Acad. Sci. USA* **100**:2070–2075.

Thoidis, G., Chen, P., Pushkin, A. V., Vallega, G., Leeman, S. E., Fine, R. E., and Kandror, K. V. (1998). Two distinct populations of synaptic-like vesicles from rat brain. *Proc. Natl. Acad. Sci. USA.* **95**:183–188.

Thorn, P., Fogarty, K. E., and Parker, I. (2004). Zymogen granule exocytosis is characterized by long fusion pore openings and preservation of vesicle lipid identity. *Proc. Natl. Acad. Sci. USA* **101**:6774–6779.

Tojima,T., Yamane, Y., Takagi, H., Takeshita, T., Sugiyama, T., Haga, H., Kawabata, K., Ushiki, T., Abe, K., Yoshioka, T., and Ito, E. (2000). Three-dimensional characterization of interior structures of exocytotic apertures of nerve cells using atomic force microscopy. *Neuroscience* **101**:471–481.

Weber, T., Zemelman, B. V., McNew, J. A., Westerman, B., Gmachi, M., Parlati, F., Söllner, T. H., and Rothman, J. E. (1998). SNAREpins: Minimal machinery for membrane fusion. *Cell* **92**:759–772.

CHAPTER 2

MOLECULAR MECHANISM OF SNARE-INDUCED MEMBRANE FUSION

BHANU P. JENA

Department of Physiology, Wayne State University School of Medicine, Detroit, Michigan

2.1. INTRODUCTION

Cell secretion involves the sequential interaction of proteins in opposing bilayers. The classical concept of fusion is a three-step process of cell excitation, docking, and fusion, in which docking may occur before cell excitation. Membrane fusion in cells has been known to be mediated via soluble N-ethylmaleimide-sensitive factor (NSF)-attachment protein receptors (SNAREs) (Weber et al., 1998). SNAREs are classified as v-SNARE and t-SNAREs, depending on their primary location either in secretory vesicle (v-) membrane or in target (t-) membranes (Rothman, 1994). Studies demonstrate that t- and v-SNARE-reconstituted lipid vesicles can fuse with one another, suggesting SNAREs to be the minimal membrane fusion machinery (Weber et al., 1988). The structure of the SNARE complex formed by interacting native (Jeong et al., 1998), or recombinant (Hanson et al., 1997; Sutton et al., 1998) t- and v-SNAREs, has been examined using electron microscopy (Jeong et al., 1998; Hanson et al., 1997) and x-ray crystallography (Sutton et al., 1998). However, the molecular mechanism of SNARE-induced membrane fusion was unknown. Studies in the past 4–5 years, using recombinant SNARE proteins reconstituted in lipid membrane, have for the first time unraveled the molecular mechanism of membrane fusion in cells (Cho et al., 2002; Jeremic et al., 2004a,b). These discoveries were made utilizing new approaches of atomic force microscopy (AFM), in combination with conventional techniques like electrophysiology, electron microscopy (EM), x-ray diffraction, light scattering, and photon correlation spectroscopy.

2.2. t-/v-SNAREs IN APPOSING BILAYERS INTERACT IN A CIRCULAR ARRAY TO FORM CONDUCTING CHANNELS

Although the partial structure (so-called core domain) of the SNARE complex formed by interacting soluble t- and v-SNAREs had been examined using x-ray crystallography (Sutton et al., 1998), the structure and arrangement of the t-/v-SNARE complex formed when full-length t-SNARE and v-SNARE present in opposing bilayers meet, was unknown. Examination of

---

Force Microscopy: Applications in Biology and Medicine, edited by Bhanu P. Jena and J.K. Heinrich Hörber.
Copyright © 2006 John Wiley & Sons, Inc.

the structure and arrangement of purified recombinant t- and v-SNAREs in artificial lipid bilayers, using AFM, has solved this riddle. To further evaluate the establishment of SNARE-induced continuity between opposing bilayers, conductance and capacitance of membranes in the presence and absence of SNARE proteins was examined (Cho et al., 2002). If pore structures were to form by direct addition of SNAREs to a single membrane, an increase in conductance would be observed due to the movement of ions through the pore. To determine the interaction between t- and v-SNAREs present in opposing bilayers, v-SNARE-reconstituted artificial lipid vesicles were exposed to t-SNARE-reconstituted lipid membranes. The structure and arrangement of the SNARE complex formed as a result, and any changes in capacitance or conductance were recorded using AFM and an EPC9 electrophysiological bilayer apparatus (Cho et al., 2002). Results from the study demonstrate that t-SNAREs and v-SNARE, when present in opposing bilayers, interact in a circular array, and in the presence of Ca^{2+}, form conducting channels (Cho et al., 2002). The interaction of t-/v-SNARE proteins to form a conducting pore or channel was found to be strictly dependent on the presence of t-SNAREs and v-SNARE in opposing bilayers. Addition of purified recombinant v-SNARE to a t-SNARE-reconstituted lipid membrane increased only

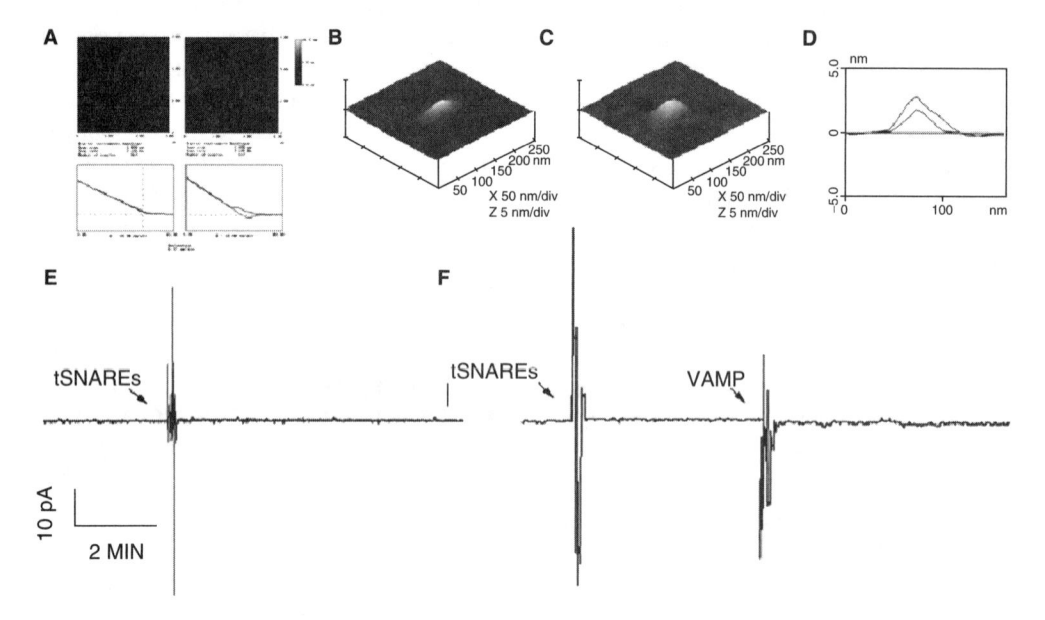

Figure 2.1. AFM micrographs and force plots of mica and lipid surface and of SNAREs on lipid membrane. (**A**) AFM performed on freshly cleaved mica (*left*) and on lipid membrane formed on the same mica surface (*right*), demonstrating differences in the force–distance curves. Note the curvilinear shape exhibited in the force–distance curves of the lipid surface in contrast to mica. Three-dimensional AFM micrographs of neuronal t-SNAREs deposited on the lipid membrane (**B**), and after the addition of v-SNARE (**C**). (**D**) Section analysis of the SNARE complex in parts B and C is depicted. Note that the smaller curve belonging to the t-SNARE complex in part B is markedly enlarged after addition of v-SNARE. Artificial bilayer lipid membranes are nonconducting either in the presence or absence of SNAREs (**E, F**). Current–time traces of bilayer membranes containing proteins involved in docking and fusion of synaptic vesicles while the membranes are held at −60 mV (current/reference voltage). (**E**) When t-SNAREs are added to the planar lipid bilayer containing the synaptic vesicle protein, VAMP-2, no occurrence of current spike for fusion event at the bilayer membrane is observed ($n = 7$). (**F**) Similarly, no current spike is observed when t-SNAREs (syntaxin 1A-1 and SNAP25) are added to the cis side of a bilayer chamber, following with VAMP-2. Increasing the concentration of t-SNAREs and VAMP-2 protein.

the size of the globular t-SNARE oligomer without influencing the electrical properties of the membrane (Cho et al., 2002). However, when v-SNARE vesicles are added to t-SNARE reconstituted membrane, SNAREs assembles in a ring pattern (Figs. 2.1 and 2.2) and a stepwise increase in capacitance and conductance is observed (Fig. 2.3). Thus, t- and v-SNAREs are required to reside in opposing bilayers to allow appropriate t-/v-SNARE interactions leading to the establishment of continuity between the opposing bilayers, only in the presence of calcium (Cho et al., 2002).

2.3. Ca^{2+} AND SNAREs ARE THE MINIMAL FUSION MACHINERY IN CELLS

Studies using SNARE-reconstituted liposomes and bilayers (Jeremic et al., 2004a) demonstrate a low fusion rate ($\tau = 16$ min) between t- and v-SNARE-reconstituted liposomes in the absence of Ca^{2+}; they also show that exposure of t-/v-SNARE liposomes to Ca^{2+} drives vesicle fusion on a near-physiological relevant timescale ($\tau \sim 10$ s), demonstrating an essential role of Ca^{2+} in membrane fusion. The Ca^{2+} effect on membrane fusion in SNARE-reconstituted liposomes is downstream of SNAREs, suggesting a regulatory role for Ca^{2+}-binding proteins in membrane fusion in the physiological state (Jeremic et al., 2004a). It is further demonstrated from these studies that in the physiological state in cells, both SNAREs and Ca^{2+} operate as the minimal fusion machinery (Figs. 2.4 and 2.5) (Jeremic et al., 2004a). Native and synthetic vesicles exhibit a significant negative surface charge primarily due to the polar phosphate head groups. These polar head groups produce a repulsive force, preventing aggregation and fusion of apposing vesicles. SNAREs bring opposing bilayers closer to within a distance of 2–3 Å (Fig. 2.6), allowing Ca^{2+} to bridge them (Jeremic et al., 2004a,b). The bound Ca^{2+} then leads to the expulsion of water between the bilayers at the bridging site, allowing lipid mixing and membrane fusion. Hence SNAREs, besides bringing opposing bilayers closer, dictate the site and size of the fusion area during cell secretion. The size of the t-/v-SNARE complex forming the channel is dictated by the curvature of the opposing membranes, hence it depends on the size of t-/v-SNARE-reconstituted vesicles. The smaller the vesicles, the smaller the channel formed (unpublished observation).

2.4. AT THE ATOMIC LEVEL, HOW DOES Ca^{2+} PARTICIPATE IN MEMBRANE FUSION?

This important question has been resolved by membrane fusion studies utilizing t-/v-SNARE reconstituted vesicles and membrane bilayers (Jeremic et al., 2004b). Calcium ion [Ca^{2+}] is essential to life's processes, and is present in every living cell. Ca^{2+} participates in diverse cellular processes, such as metabolism, secretion, proliferation, muscle contraction, cell adhesion, learning, and memory. Although calcium is present in abundance within cells, it is well sequestered and is available only on demand. Upon certain cellular stimulus for instance, Ca^{2+} concentration at specific locations (i.e., nano-environment) within the cell is elevated by several orders of magnitude within a brief period (some in < 1 ms). This prompt mobilization of Ca^{2+} is essential for many physiological processes, such as the release of neurotransmitters or cell signaling. A unique set of chemical and physical properties of the Ca^{2+} ion makes it ideal for performing these biochemical reactions. Ca^{2+} ion exists in its hydrated state within cells. The properties of hydrated calcium have been extensively studied using x-ray diffraction, neutron scattering, in combination with molecular dynamics simulations (Bako et al., 2002; Chialvo and Simonson, 2003; Licheri et al., 1976; Schwenk et al., 2001). The molecular dynamic simulations include

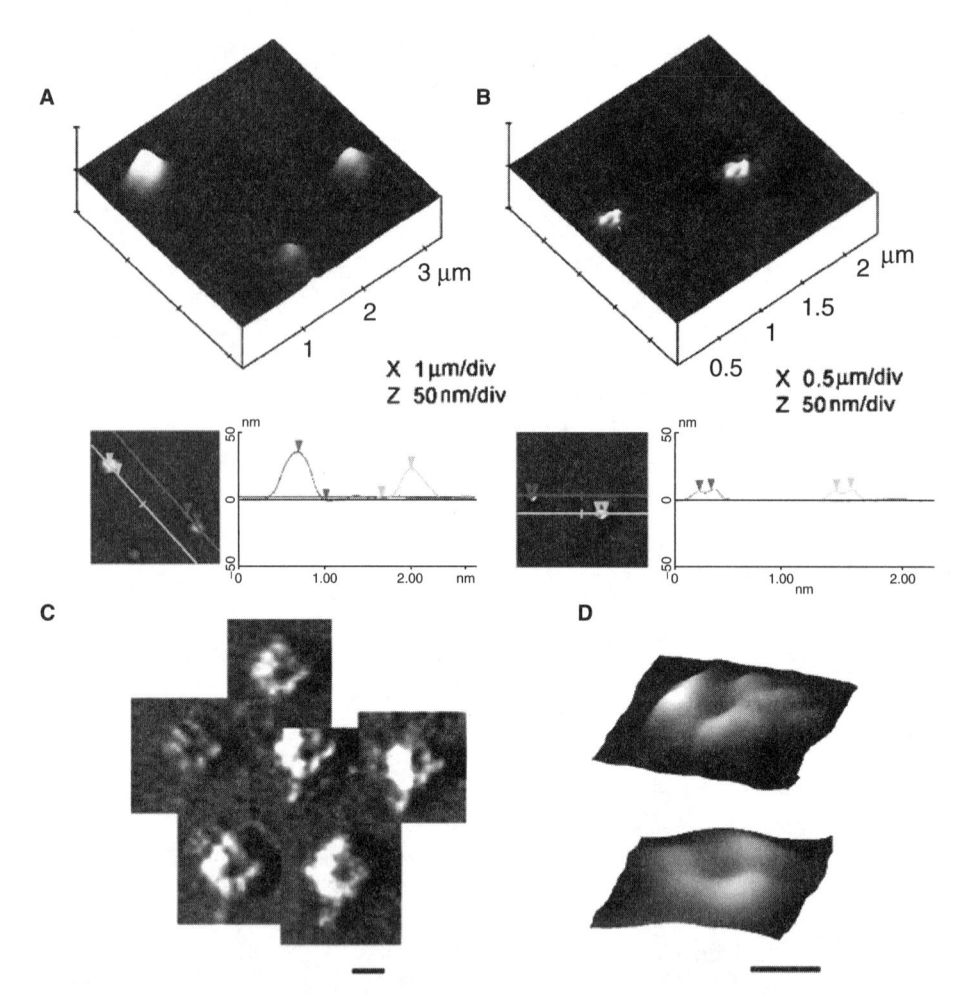

Figure 2.2. Pore-like structures are formed when t-SNAREs and v-SNARE in opposing bilayers interact. (**A**) Unfused v-SNARE vesicles on t-SNARE reconstituted lipid membrane. (**B**) Dislodgement or fusion of v-SNARE-reconstituted vesicles with a t-SNARE-reconstituted lipid membrane exhibits formation of pore-like structures due to the interaction of v- and t-SNAREs in a circular array. The size of the pores range between 50 and 150 nm (**B–D**). Several 3D AFM amplitude images of SNAREs arranged in a circular array (**C**) and some at higher resolution (**D**), illustrating a pore-like structure at the center is depicted. Scale bar is 100 nm. Recombinant t-SNAREs and v-SNARE in opposing bilayers drive membrane fusion. (**E**) When t-SNARE vesicles were exposed to v-SNARE-reconstituted bilayers, vesicles fused. Vesicles containing nystatin/ergosterol and t-SNAREs were added to the cis side of the bilayer chamber. Fusion of t-SNARE containing vesicles with the membrane was observed as current spikes that collapse as the nystatin spreads into the bilayer membrane. To determine membrane stability, the transmembrane gradient of KCl was increased, allowing gradient-driven fusion of nystatin-associated vesicles.

three-body corrections compared with ab initio quantum mechanics/molecular mechanics molecular dynamics simulations. First principles molecular dynamics has also been used to investigate the structural, vibrational, and energetic properties of $[Ca(H_2O)_n]^{2+}$ clusters and the hydration shell of calcium ion. These studies demonstrate that hydrated calcium $[Ca(H_2O)_n]^{2+}$ has more than one shell around the Ca^{2+}, with the first hydration shell around the Ca^{2+} having six water molecules in an octahedral arrangement (Chialvo and Simonson, 2003). In studies using light scattering and x-ray diffraction

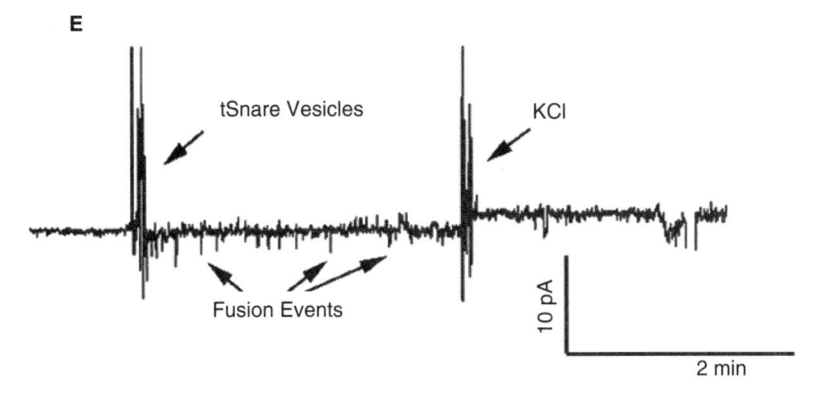

AT THE ATOMIC LEVEL, HOW DOES Ca^{2+} PARTICIPATE IN MEMBRANE FUSION?   29

Figure 2.2. (*continued*)

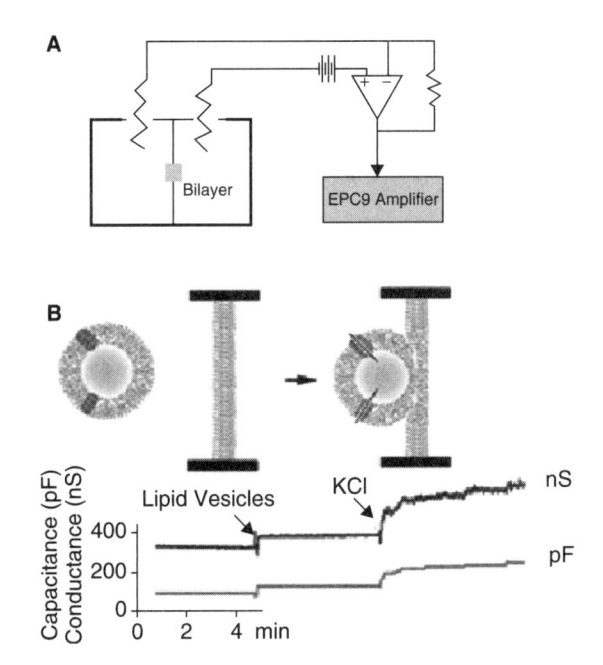

Figure 2.3. Opposing bilayers containing t- and v-SNAREs, respectively, interact in a circular array to form conducting pores. (**A**) Schematic diagram of the bilayer–electrophysiology setup. (**B**) Lipid-vesicle-containing nystatin channels and both vesicles and membrane bilayer without SNAREs demonstrate no significant changes in capacitance and conductance. Initial increase in conductance and capacitance may be due to vesicle–membrane attachment. To demonstrate membrane stability (both bilayer membrane and vesicles), the transmembrane gradient of KCl was increased to allow gradient-driven fusion and a concomitant increase of conductance and capacitance. (**C**) When t-SNARE vesicles were added to a v-SNARE membrane support, the SNAREs in opposing bilayers formed a ring pattern, thereby forming pores (as seen in the AFM micrograph on the extreme right); stepwise increases in capacitance and conductance (−60-mV holding potential) were seen. Docking and fusion of the vesicle at the bilayer membrane opens vesicle-associated nystatin channels and SNARE-induced pore formation, allowing conductance of ions from cis to the trans side of the bilayer membrane. Then further addition of KCl to induce gradient-driven fusion resulted in little or no further increase in conductance and capacitance, demonstrating that docked vesicles have already fused.

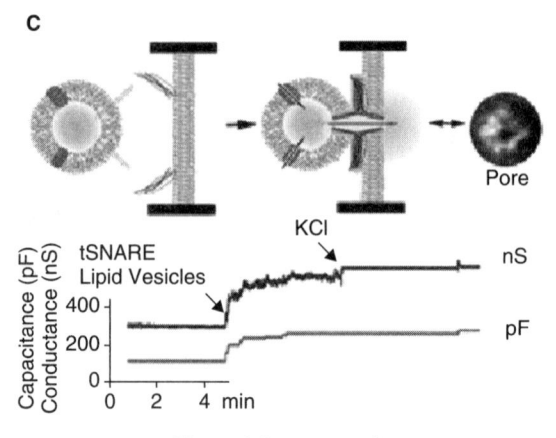

Figure 2.3. (*continued*)

of SNARE-reconstituted liposomes, it was demonstrated that fusion proceeds only when Ca^{2+} ions are available between the t- and v-SNARE-apposed bilayers (Fig. 2.7) (Jeremic et al., 2004b).

To monitor interaction(s) between Ca^{2+} ions and phosphate on the lipid membrane head groups, the x-ray diffraction method has been utilized (Jeremic et al., 2004a,b). This experimental approach for monitoring interbilayers contacts essentially requires the presence of (i) highly concentrated lipid suspensions (>10 mM) favoring a multitude of intervesicular contacts and (ii) fully hydrated liposomes, where vesicles have total freedom to interact with each other in solution, hence establishing confined hydrated areas between adjacent bilayers. This small fluid space could arise from interbilayer hydrogen bond formation through water molecules (McIntosh, 2000) in addition to bridging forces contributed by trans-SNARE complex formation (Cho et al., 2002; Jeremic et al., 2004a). When these two conditions are met, liposomes diffract as shown (Fig. 2.6). Mixing of t- and v-SNARE liposomes in the absence of Ca^{2+} leads to a diffuse and asymmetric diffractogram (depicted by the gray trace in Fig. 2.6), a typical characteristic of short-range ordering in a liquid system. In contrast, mixing the t-SNARE and v-SNARE liposomes in the presence of Ca^{2+} leads to a more structured diffractogram (depicted by the black trace in Fig 2.6) with an approximately 12% increase in x-ray scattering intensity, pointing to an increase in the number of contacts between apposing bilayers established presumably by calcium–PO bridges, as previously suggested (Portis et al., 1979). The ordering effect of Ca^{2+} on interbilayer contacts observed in x-ray studies (Jeremic et al., 2004b) is in good agreement with recent light, AFM, and spectroscopic studies, suggesting close apposition of PO lipid head groups in the presence of Ca^{2+} ions followed by formation of Ca^{2+}–PO bridges between adjacent bilayers (Jeremic et al., 2004a; Laroche et al., 1991). X-ray study shows that effect of Ca^{2+} on bilayer orientation and interbilayer contacts is most prominent in the area of 3 Å, with the appearance of an additional peak (shoulder) at 2.8 Å (depicted by the arrow in Fig 2.6), both of which are within the ionic radius of Ca^{2+} (Jeremic et al., 2004b). These studies suggest that the ionic radius of Ca^{2+} may play an important role in membrane fusion. But there remained a major spatial problem, which was recently resolved (Jeremic et al., 2004b). As discussed earlier, calcium ions [Ca^{2+}] exist in their hydrated state within cells. Hydrated calcium [Ca(H$_2$O)$_n$]$^{2+}$

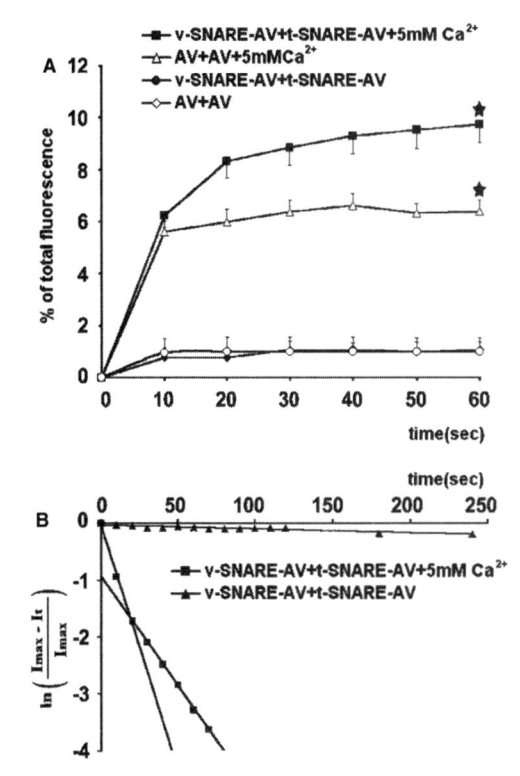

Figure 2.4. Fluorimetric fusion assays demonstrate the ability of Ca^{2+} to induce rapid lipid mixing of plain (AV) and SNARE-associated vesicles. Addition of 5 mM Ca^{2+} to liposomal solution significantly increases the fusion of plain and SNARE-associated vesicles (+P<0.05, Student t-test between AV and AV + Ca^{2+} or t-/v-SNARE-AV + Ca^{2+}, $n = 5$). Note the inability of SNAREs in the absence of Ca^{2+} to significantly induce vesicle fusion (P > 0.1, Student t-test between AV and t-/v-SNARE-AV, $n = 5$). Incorporation of t-/v-SNAREs at the vesicles membrane increases the overall yield but does not alter the rate of Ca^{2+}-induced membrane fusion (**A**). The graph depicts the first-order kinetics of SNAREs vesicle fusion in the presence and absence of Ca^{2+} (**B**).

has more than one shell around the Ca^{2+}, with the first hydration shell having six water molecules in an octahedral arrangement (Chialvo and Simonson, 2003), and measuring >6 Å (Fig. 2.7). Studies reveal that for hydrated Ca^{2+} ion, depending on its coordination number, the nearest average neighbor Ca^{2+}–O and Ca^{2+}–H distances are at $r \sim 2.54$ Å and $r \sim 3.2$ Å in the first hydration shell, respectively. How then, would a hydrated calcium ion measuring >6 Å fit between the 2.8- to 3-Å space established by t-/v-SNAREs, between the apposing vesicle bilayers? One possibility would be that calcium has to be present in the buffer solution when t-SNARE vesicles and v-SNARE vesicles meet. If t- and v-SNARE vesicles are allowed to mix in a calcium-free buffer, prior to the addition of calcium, no fusion should occur, since the 2.8- to 3-Å space would not be able to accommodate the hydrated calcium. This hypothesis was tested (Jeremic et al., 2004b) and was found to be indeed correct. Light scattering experiments (Fig. 2.7) were performed on t-SNARE- and v-SNARE-reconstituted phospholipids vesicles, in the presence and absence of calcium and in the presence of NSF + ATP. NSF (N-ethylmalemide-sensitive factor) is an ATPase that is known to disassemble the t-/v-SNARE complex. Using light scattering measurements, aggregation and membrane fusion of lipid vesicles can be monitored on the second timescale (Jeremic et al., 2004a; Wilschut et al., 1980). The initial rapid increase in intensity of light scattering was found to be initiated on exposure of t-SNARE vesicles to v-SNARE vesicles in solution, followed by a slow decay of light scattering (Fig. 2.7) due to vesicle–vesicle interaction and settling. These studies show that if t-SNARE vesicles and v-SNARE vesicles are allowed to interact prior to calcium addition (depicted by arrow, Fig. 2.7), no significant change in light scattering is observed (there is no significant decrease in scattering, attributed to little fusion between the vesicle suspension). On the contrary, when calcium is present in the buffer solution prior to addition of the t-SNARE and v-SNARE vesicles, there is a marked drop in light scattering, as a result of vesicle aggregation and fusion (Fig. 2.7). However, in the presence of NSF-ATP in the assay buffer containing calcium, a significant inhibition

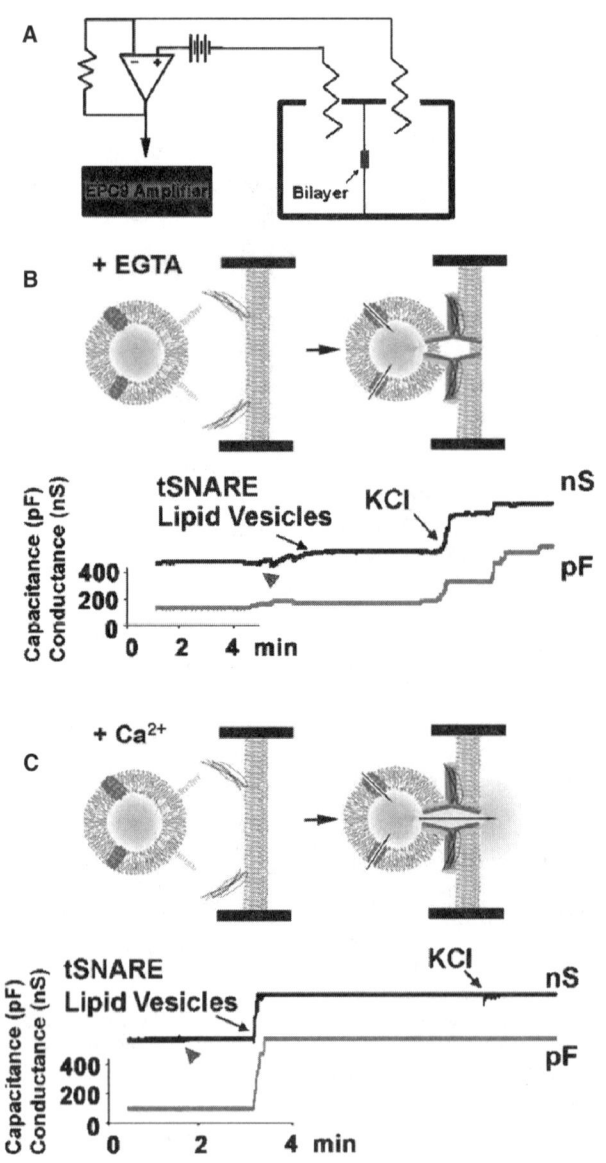

Figure 2.5. Conductance and capacitance measurements of SNARE-reconstituted lipid bilayers. The EPC9-electrophysiological setup is shown (**A**). In the presence of 5 mM EGTA, t-SNARE-associated vesicles containing nystatin channels at their membrane interact with v-SNARE-reconstituted lipid bilayer without fusing (**B**). Note no change in conductance or capacitance following exposure of SNARE-associated lipid vesicles to the bilayer. The vesicles fuse, however, when 3 mM KCl is applied, demonstrating fusion of docked vesicles and presence of an intact bilayer (**B**). The arrowhead indicates when the stirring is switched on to mix the addition. In the presence of 1 mM CaCl$_2$, the t-SNARE-associated vesicles fuse with the v-SNARE-reconstituted bilayer as depicted in a consequent increase in conductance and capacitance. Since a large majority of docked vesicles have fused, addition of 3 mM KCl has no further effect (**C**). Traces B and C are representative profiles from one of five separate experiments.

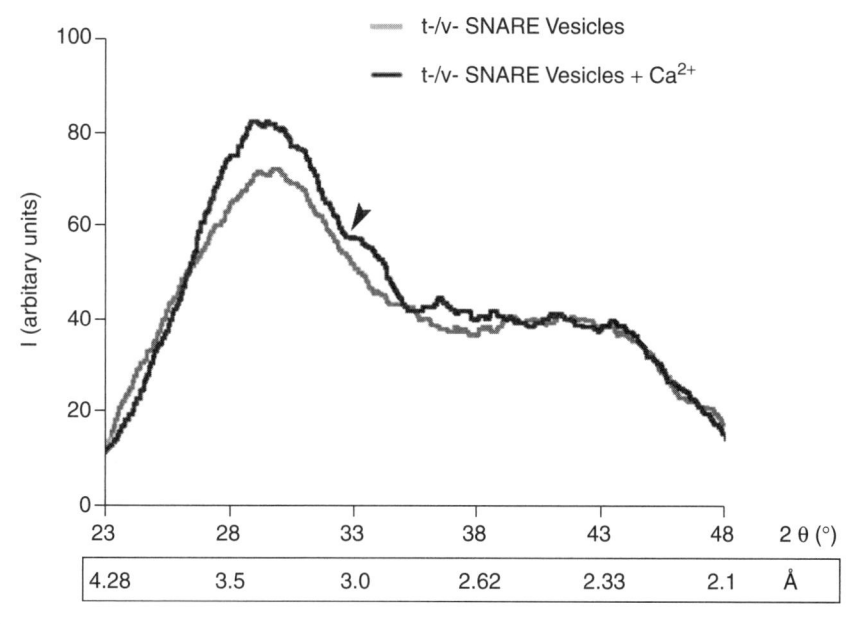

Figure 2.6. Wide-angle x-ray diffraction patterns of interacting SNARE-vesicles. Representative diffraction profiles from one of four separate experiments using t- and v-SNARE-reconstituted lipid vesicles, in the presence or absence of 5 mM Ca^{2+}, is shown. Arrows mark appearance of new peaks in the x-ray diffractogram, following addition of calcium.

in aggregation and fusion of proteoliposomes is observed (Fig. 2.7). NSF, in the absence of ATP, has no effect on the light scattering properties of the vesicle mixture. These results demonstrate that NSF-ATP disassembles the SNARE complex, thereby reducing the number of interacting vesicles in solution. In addition, disassembly of trans-SNARE complex will then leave apposed bilayers widely separated, out of reach for the formation of Ca^{2+}–PO bridges, preventing membrane fusion (Fig. 2.7). Similarly, if the restricted area between adjacent bilayers delineated by the circular arrangement of the t-/v-SNARE complex (Cho et al., 2002) is preformed, then hydrated Ca^{2+} ions are too large (Fig. 2.7) to be accommodated between bilayers and, hence, subsequent addition of Ca^{2+} would have no effect (Fig 2.7). However, when t-SNARE vesicles interact with v-SNARE vesicles in the presence of Ca^{2+}, the t-/v-SNARE complex formed allow formation of calcium–phosphate bridges between opposing bilayers, leading to the expulsion of water around the Ca^{2+} ion to enable lipid mixing and membrane fusion (Fig. 2.7). Thus, x-ray and light scattering studies (Jeremic et al., 2004b) demonstrate that calcium bridging of the apposing bilayers is required to enable membrane fusion. This calcium bridging of apposing bilayers allows for the release of water from the hydrated Ca^{2+} ion, leading to bilayers destabilization and membrane fusion. In addition, the binding of calcium to the phosphate head groups of the apposing bilayers may also displace the loosely coordinated water at the PO groups, further contributing to the destabilization of the lipid bilayer, leading to membrane fusion.

ACKNOWLEDGMENT

Supported by Grants DK-56212 and NS-39918 from the National Institutes of Health (BPJ).

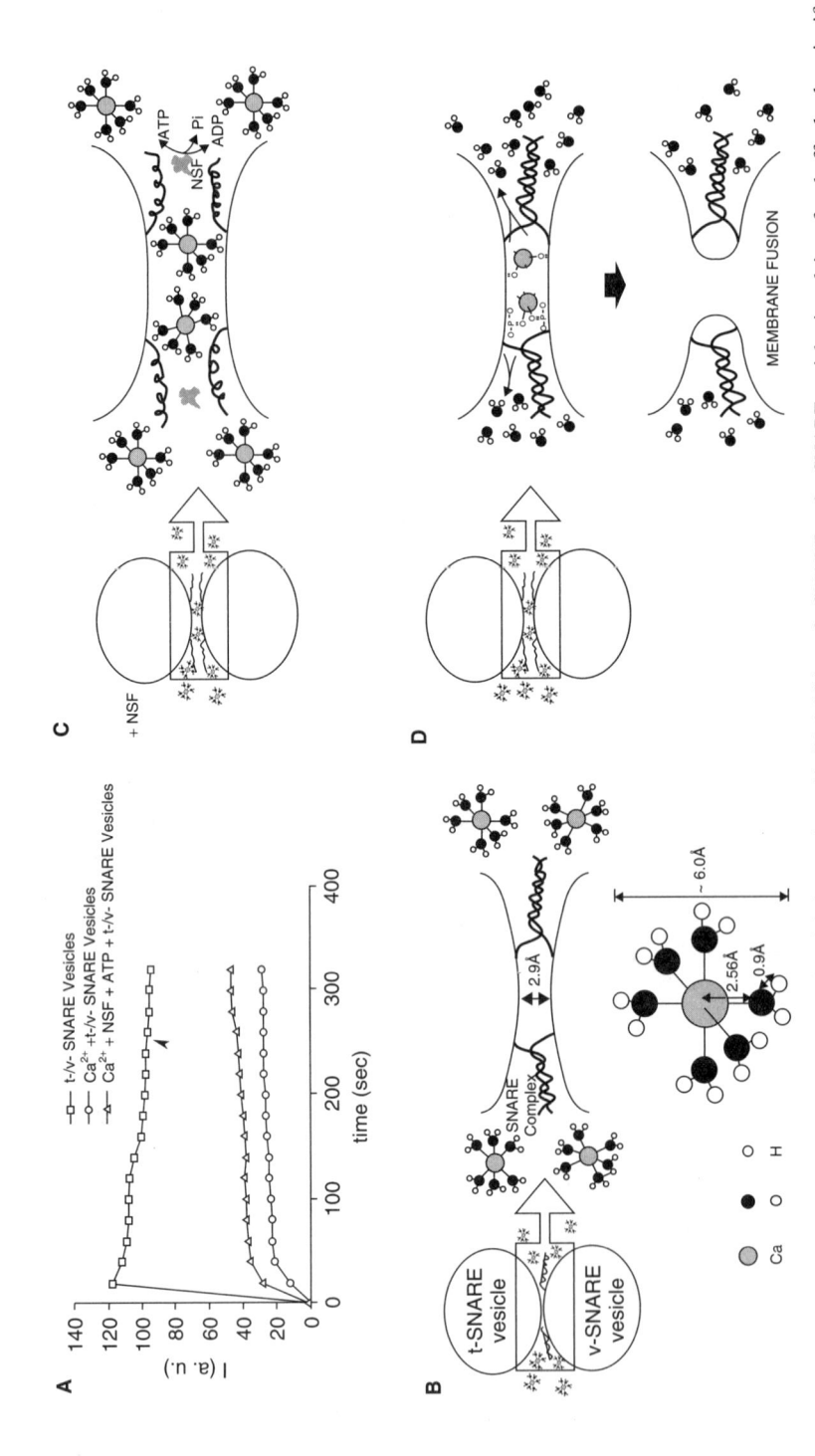

Figure 2.7. Light scattering profiles of SNARE-associated vesicle interactions. (**A, B**) Addition of t-SNARE and v-SNARE vesicles in calcium-free buffer lead to significant increase in light scattering. Subsequent addition of 5 mM Ca^{2+} (marked by arrowhead) does not have any significant effect on light scattering (□). (**A, C**) In the presence of *NSF*-ATP (1 μg/ml) in assay buffer containing 5 mM Ca^{2+}, significantly inhibited vesicle aggregation and fusion (△). (A, D) When the assay buffer was supplemented with 5 mM Ca^{2+}, prior to addition of t- and v-SNARE vesicles, it led to a four-fold decrease in light scattering intensity due to Ca^{2+}-induced aggregation and fusion of t-/v-SNARE apposed vesicles (○). Light scattering profiles shown are representatives of four separate experiments.

REFERENCES

Bako, I., Hutter, J., and Palinkas, G. (2002). Car–Parrinello molecular dynamics simulation of the hydrated calcium ion. *J. Chem. Phys.* **117**:9838–9843.

Chialvo, A. A., and Simonson, J. M. (2003). The structure of CaCl$_2$ aqueous solutions over a wide range of concentration. Interpretation of diffraction experiments via molecular simulation. *J. Chem. Phys.* **119**:8052–8061.

Cho, S. J., Kelly, M., Rognlien, K. T., Cho, J., Hoerber, J. K. H., and Jena, B. P. (2002). SNAREs in opposing bilayers interact in a circular array to form conducting pores. *Biophys. J.* **83**:2522–2527.

Hanson, P. I., Roth, R., Morisaki, H., Jahn, R., and Heuser, J. E. (1997). Structure and conformational changes in NSF and its membrane receptor complexes visualized by quick-freeze/deep-etch electron microscopy. *Cell* **90**:523–535.

Jeong, E.-H., Webster, P., Khuong, C. Q., Sattar, A. K. M. A., Satchi, M., and Jena, B. P. (1998). The native membrane fusion machinery in cells. *Cell Biol. Int.* **22**:657–670.

Jeremic, A., Kelly, M., Cho, W. J., Cho, S. J., Horber, J. K. H., and Jena, B. P. (2004a). Calcium drives fusion of SNARE-apposed bilayers. *Cell. Biol. Int.* **28**:19–31.

Jeremic, A., Cho, W. J., and Jena, B. P. (2004b). Membrane fusion: What may transpire at the atomic level. *J. Biol. Phys. Chem.* **4**:139–142.

Laroche, G., Dufourc, E. J., Dufoureq, J., and Pezolet, M. (1991). Structure and dynamics of dimyristoylphosphatidic acid/calcium complex by 2H NMR, infrared, spectroscopies and small-angle x-ray diffraction. *Biochemistry* **30**:3105–3114.

Licheri, G., Piccaluga, G., and Pinna, G. (1976). X-ray diffraction study of the average solute species in CaCl$_2$ aqueous solutions. *J. Chem. Phys.* **64**:2437–2446.

McIntosh, T. J. (2000). Short-range interactions between lipid bilayers measured by X-ray diffraction. *Curr. Opin. Struct. Biol.* **10**: 481–485.

Portis, A., Newton, C., Pangborn, W., and Papahadjopoulos, D. (1979). Studies on the mechanism of membrane fusion: Evidence for an intermembrane Ca^{2+}–phospholipid complex, synergism with Mg^{2+}, and inhibition by spectrin. *Biochemistry* **18**:780–790.

Rothman, J. E. (1994). Mechanism of intracellular protein transport. *Nature* **372**: 55–63.

Schwenk, C. F., Loeffler, H. H., and Rode, B. M. (2001). Molecular dynamics simulations of Ca^{2+} in water: Comparison of a classical simulation including three-body corrections and Born–Oppenheimer ab initio and density functional theory quantum mechanical/molecular mechanics simulations. *J. Chem. Phys.* **115**:10808–10813.

Sutton, R. B., Fasshauer, D., Jahn, R., and Brunger, A. T. (1998). Crystal structure of a SNARE complex involved in synaptic exocytosis at 2.4 Å resolution. *Nature* **395**:347–353.

Weber, T., Zemelman, B. V., McNew, J. A., Westerman, B., Gmachl, M., Parlati, F., Sollner, T. H. and Rothman, J. E. (1998). SNARE-pins: Minimal machinery for membrane fusion. *Cell* **92**:759–772.

Wilschut, J., Duzgunes, N., Fraley, R., and Papahadjopoulos, D. (1980). Studies on the mechanism of membrane fusion: Kinetics of calcium ion induced fusion of phosphatidylserine vesicles followed by a new assay for mixing of aqueous vesicle content. *Biochemistry* **19**:6011–6021.

CHAPTER 3

MOLECULAR MECHANISM OF SECRETORY VESICLE CONTENT EXPULSION DURING CELL SECRETION

BHANU P. JENA
Department of Physiology, Wayne State University School of Medicine, Detroit, Michigan

3.1. INTRODUCTION

In the last decade, the molecular mechanism of vesicle swelling (Abu-Hamdah et al., 2004; Cho et al., 2002c; Jena et al., 1997) and its involvement in the regulated expulsion of vesicular contents (Kelly et al., 2004) has been determined. Secretory vesicle swelling is critical for secretion (Alvarez de Toledo et al., 1993; Curran and Brodwick, 1991; Monck et al., 1991; Sattar et al., 2002), however, the underlying mechanism of vesicle swelling was largely unknown until recently (Abu-Hamdah et al., 2004; Cho et al., 2002; Jena et al., 1997). In mast cells, an increase in secretory vesicle volume after stimulation of secretion has previously been suggested from electrophysiological measurements (Fernandez et al., 1991). However, direct evidence of secretory vesicle swelling in live cells was first demonstrated in pancreatic acinar cells using the AFM (Fig. 3.1) (Cho et al., 2002a). Isolated zymogen granules (ZGs), the membrane-bound secretory vesicles in exocrine pancreas and parotid glands, possess Cl^-- and ATP-sensitive, K^+-selective ion channels at the vesicle membrane, whose activities have been implicated in vesicle swelling (Jena et al., 1997). Additionally, secretion of ZG contents in permeabilized pancreatic acinar cells requires the presence of both K^+ and Cl^- ions. In vitro ZG-pancreatic plasma membrane fusion assays further demonstrate potentiation of fusion in the presence of GTP (Sattar et al., 2002). $G_{\alpha i}$ protein has been implicated in the regulation of both K^+ and Cl^- ion channels in a number of tissues. Analogous to the regulation of K^+ and Cl^- ion channels at the cell plasma membrane, their regulation at the ZG membrane by a $G_{\alpha i3}$ protein was demonstrated (Jena et al., 1997). Isolated ZGs from exocrine pancreas swell rapidly in response to GTP (Fig. 3.2) (Jena et al., 1997). These studies suggested the involvement of rapid water entry into ZGs following exposure to GTP. Therefore, when the possible involvement of water channels or aquaporins in ZG swelling was explored (Cho et al., 2002c), results from the study demonstrate the presence of aquaporin-1 (AQP1) at the ZG membranes and its participation in GTP-mediated water entry and swelling (Fig. 3.3) (Cho et al., 2002c). To further understand the molecular mechanism

---

Force Microscopy: Applications in Biology and Medicine, edited by Bhanu P. Jena and J.K. Heinrich Hörber.
Copyright © 2006 John Wiley & Sons, Inc.

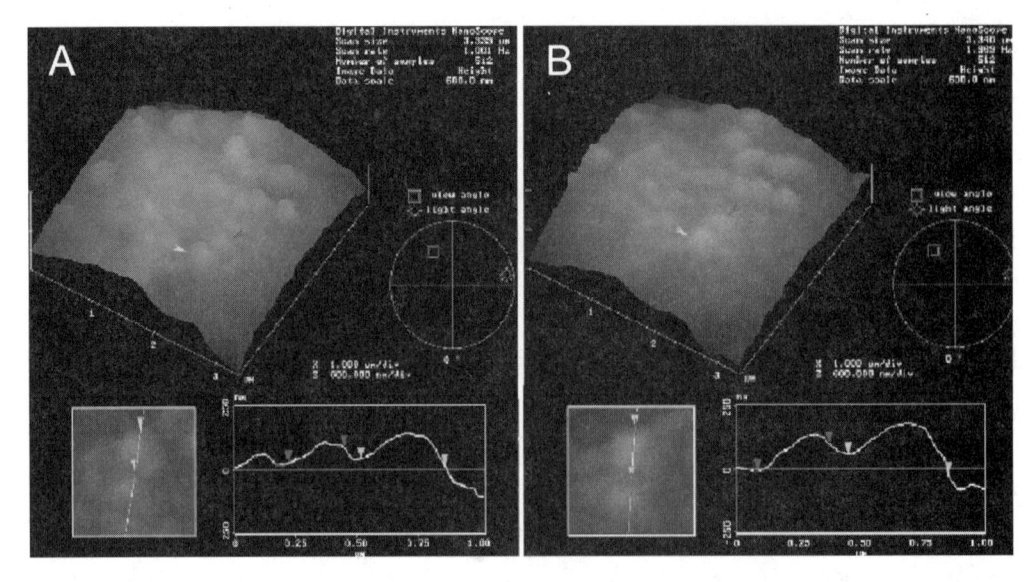

Figure 3.1. AFM micrograph or live pancreatic acinar cell demonstrating the size of ZGs within the cell, in resting (**A**) and following stimulation of secretion (**B**). Note the increase in size of the same granules immediately following a secretory stimuli (section analysis of two such ZGs are shown). There is no loss of secretory vesicles following completion of secretion.

of secretory vesicle swelling, the regulation of AQP1 in the ZG was determined (Abu-Hamdah et al., 2004). Detergent-solubilized ZGs immunoprecipitated with monoclonal AQP-1 antibody co-isolates AQP-1, PLA2, $G_{\alpha i3}$, potassium channel IRK-8, and the chloride channel ClC-2 (Abu-Hamdah et al., 2004). Exposure of ZGs to either the potassium channel blocker glyburide or the PLA2 inhibitor ONO-RS-082 blocked GTP-induced ZG swelling. RBC, known to possess AQP-1 at the plasma membrane, also swell on exposure to the $G_{\alpha i}$-agonist mastoparan and respond similarly to ONO-RS-082 and glyburide, as do ZGs. Additionally, liposomes reconstituted with the AQP-1 immunoisolated complex from solubilized ZGs, also swell in response to GTP. Glyburide or ONO-RS-082 abolished the GTP effect in reconstituted liposomes. Furthermore, immunoisolate-reconstituted planar lipid membrane demonstrate conductance, which is sensitive to glyburide and to an AQP-1 specific antibody. These results demonstrate a $G_{\alpha i3}$-PLA2-mediated pathway and potassium channel involvement in AQP-1 regulation (Fig. 3.4) (Abu-Hamdah et al., 2004), contributing to our understanding of the molecular mechanism of ZG swelling. Studies in both slow and fast secretory cells (neurons) further reveal that secretory vesicle swelling is a requirement for the regulated expulsion of intravesicular contents during cell secretion (Kelly et al., 2004).

3.2. VESICLE SWELLING IN A SLOW SECRETORY CELL (PANCREATIC ACINAR) IS REQUIRED IN INTRAVESICULAR CONTENT EXPULSION DURING SECRETION

Pancreatic acinar cells (Fig. 3.5A) have been used in studies that demonstrate that vesicle swelling is a requirement for the expulsion of intravesicular contents during cell secretion (Kelly et al., 2004). Isolated live pancreatic acinar cells in near-physiological buffer imaged using the atomic force microscope (AFM) at higher force (200–300 pN) is able

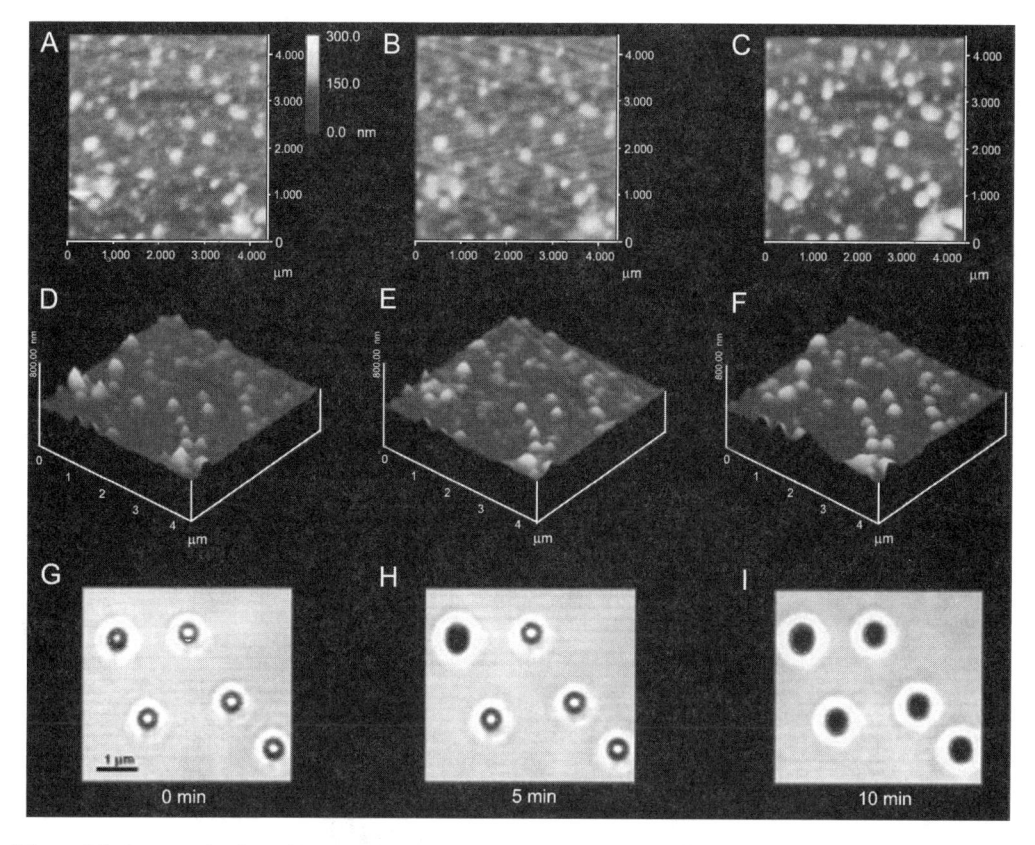

Figure 3.2. Increase in size of isolated ZGs in the presence of GTP. **(A–C)** Two-dimensional AFM images of zymogen granules (ZG), the membrane-bound secretory vesicles of the exocrine pancreas, after exposure to 20 μM GTP at time 0 min **(A)**, 5 min **(B)**, and 10 min **(C)**. **(D–F)** The same granules are shown in three dimensions at 0, 5, and 10 min, respectively, following exposure to GTP. **(G–I)** The GTP-induced increase in size of another group of ZGs observed by confocal microscopy is shown at time 0, 5, and 10 min following GTP addition. (Bar represents 1 μm.) Values represent one of three representative experiments.

to reveal the structure and dynamics of ZGs, the membrane-bound secretory vesicles of the exocrine pancreas, lying immediately below the apical plasma membrane (PM) of the cell (Fig. 3.5B). Within 2.5 min of exposure to a physiological secretory stimulus (1 μM carbamylcholine), the majority of ZGs within cells were found to swell (Fig. 3.5C), followed by a decrease in ZG size (Fig. 3.5D) by which time most of the release of secretory products from within ZGs had occurred (Fig. 3.5E). These studies (Kelly et al., 2004) reveal, for the first time in live cells, intracellular swelling of secretory vesicles following stimulation of secretion and their deflation following partial discharge of vesicular contents. Measurements of intracellular ZG size further reveal that different vesicles swell differently, following a secretory stimulus. For example, the ZG marked by the red arrowhead swelled to show a 23–25% increase in diameter, in contrast to the green arrowhead-marked ZG which increased by only 10–11% (Fig. 3.5B,C). This differential swelling among ZGs within the same cell may explain why following stimulation of secretion, some intracellular ZGs demonstrate the presence of less vesicular content than others, and hence have

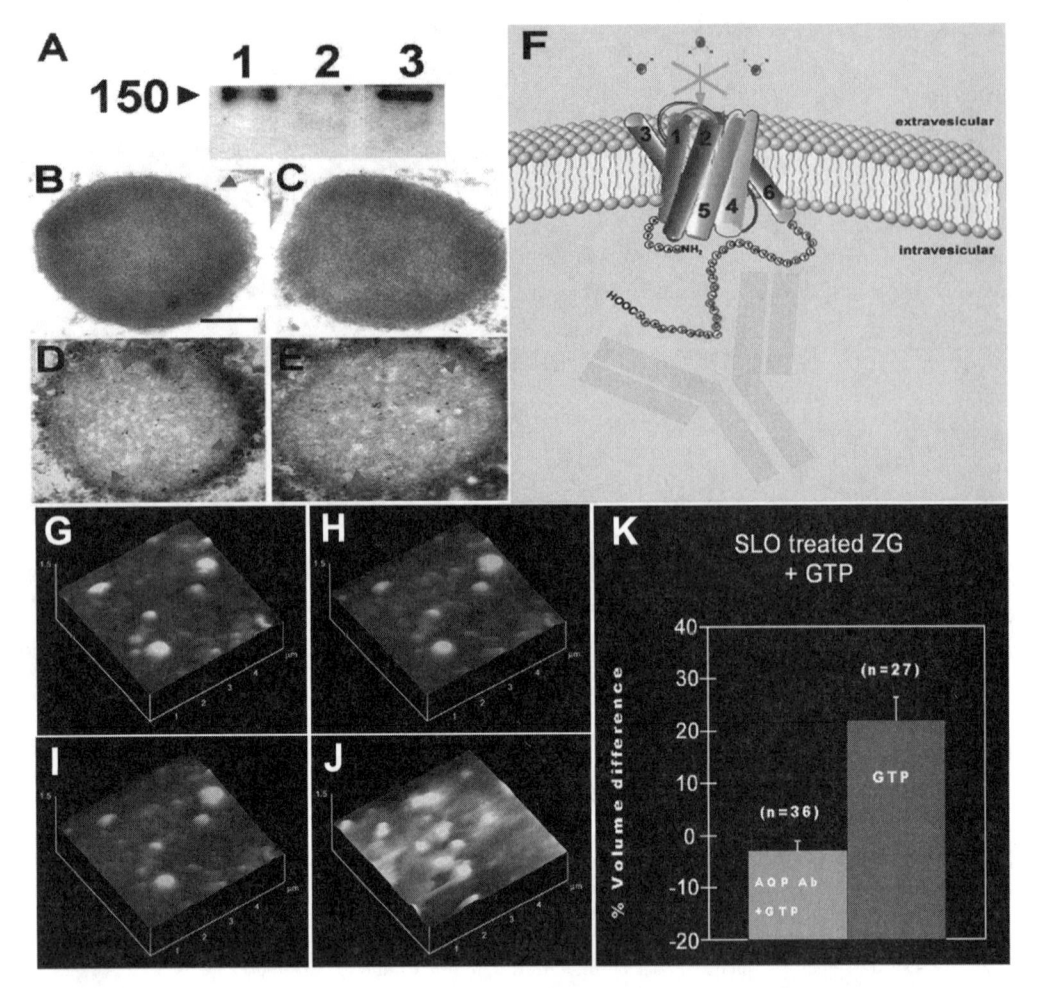

Figure 3.3. AQP1-specific antibody binds to the ZG membrane and blocks water entry. (**A**) Immunoblot assay demonstrating the presence of AQP1 antibody in SLO-permeabilized ZG. Lanes: 1, AQP1 antibody alone; 2, nonpermeable ZG exposed to antibody; 3, permeable ZG exposed to AQP1 antibody. Immunoelectron micrographs of intact ZGs exposed to AQP1 antibody demonstrate little labeling (**B, C**). (Bar = 200 nm.) Contrarily, SLO-treated ZG demonstrate intense gold labeling at the luminal side of the ZG membrane (**D, E**). AQP1 regulates GTP-induced water entry in ZG. (**F**) Schematic diagram of ZG membrane depicting AQP1-specific antibody binding to the carboxyl domain of AQP1 at the intragranular side to block water gating. (**G,H, K**) AQP1 antibody introduced into ZG blocks GTP-induced water entry and swelling (from **G** to **H**, after GTP exposure). (**I–K**) However, only vehicle introduced into ZG retains the stimulatory effect of GTP (from **I** to **J**, after GTP exposure).

discharged more of their contents (Cho et al., 2002a; Kelly et al., 2004).

To determine precisely the role of swelling in vesicle–plasma membrane fusion and in the expulsion of intravesicular contents, an electrophysiological ZG-reconstituted lipid bilayer fusion assay (Cho et al., 2002b; Jeremic et al., 2003) has been employed.

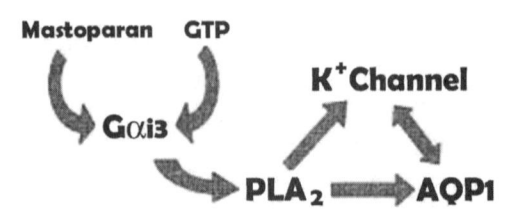

Figure 3.4. $G_{\alpha i}$-PLA_2-mediated pathway and potassium involvement in AQP-1 regulation.

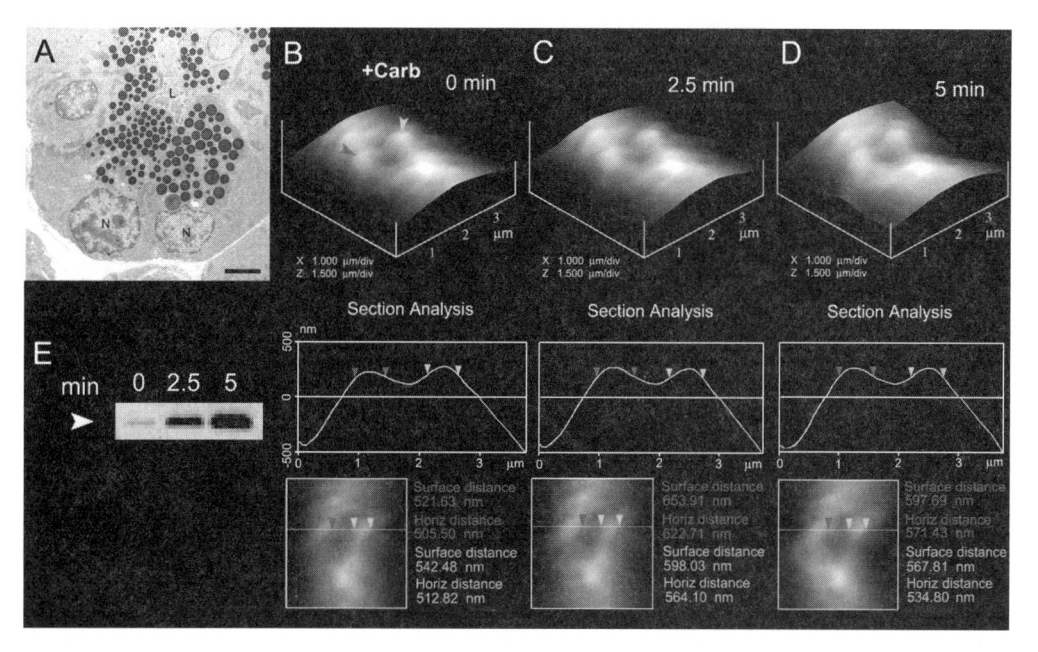

Figure 3.5. The swelling dynamics of ZGs in live pancreatic acinar cells. **(A)** Electron micrograph of pancreatic acinar cells showing the basolaterally located nucleus (N) and the apically located ZGs. The apical end of the cell faces the acinar lumen (L). Bar = 2.5 µm. **(B–D)** The apical ends of live pancreatic acinar cells were imaged by AFM, showing ZGs (grey and white arrowheads) lying just below the apical plasma membrane. Exposure of the cell to a secretory stimulus using 1 µM carbamylcholine resulted in ZG swelling within 2.5 min, followed by a decrease in ZG size after 5 min. The decrease in size of ZGs after 5 min is due to the release of secretory products such as α-amylase, as demonstrated by the immunoblot assay **(E)**.

The ZGs used in the bilayer fusion assays were characterized for their purity and their ability to respond to a swelling stimulus. ZGs were isolated (Jena et al., 1997) and their purity assessed using electron microscopy (Fig. 3.6A). As previously reported (Abu-Hamdah et al., 2004; Cho et al., 2002c; Jena et al., 1997), exposure of isolated ZGs (Fig. 3.6B) to GTP resulted in ZG swelling (Fig. 3.6C). Once again, similar to what is observed in live acinar cells (Fig. 3.5), each isolated ZG responded differently to the same swelling stimulus (Kelly et al., 2004). For example, the red arrowhead points to a ZG whose diameter increased by 29% as opposed to the green arrowhead pointing ZG that increased only by a modest 8%. The differential response of isolated ZGs to GTP was further assessed by measuring changes in the volume of isolated ZGs of different sizes (Fig. 3.6D). ZGs in the exocrine pancreas range in size from 0.2 to 1.3 µm in diameter (Jena et al., 1997). Not all ZGs were found to swell following a GTP challenge (Jena et al., 1997; Kelly et al., 2004). Most ZG volume increases were between 5% and 20%; however, larger increases up to 45% were observed only in ZGs ranging from 250 to 750 nm in diameter (Fig. 3.6D) (Kelly et al., 2004). These studies (Kelly et al., 2004) demonstrate that following stimulation of secretion, ZGs within pancreatic acinar cells swell, followed by (i) a release of intravesicular contents through porosomes (12) at the cell plasma membrane and (ii) a return to resting size on completion of secretion. On the contrary, isolated ZGs stay swollen following exposure to GTP, since there is no outlet for the release of the intravesicular contents. In acinar cells, little or no secretion is detected 2.5 min following stimulation of secretion, although the

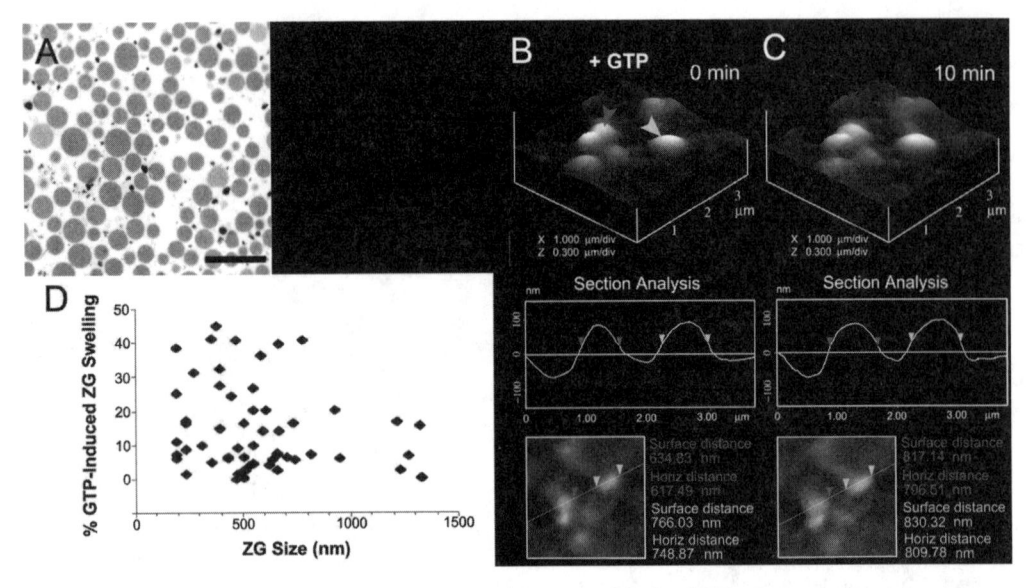

Figure 3.6. Swelling of isolated ZGs. (**A**) Electron micrograph of isolated ZGs demonstrating a homogeneous preparation. Bar = 2.5 μm. (**B, C**) Isolated ZGs, on exposure to 20 μM GTP, swell rapidly. Note the enlargement of ZGs as determined by AFM section analysis of two vesicles (grey and white arrowheads). (**D**) Percent ZG volume increase in response to 20 μM GTP. Note how different ZGs respond to the GTP-induced swelling differently.

ZGs within them were completely swollen (Fig. 3.5C). However, at 5 min following stimulation, ZGs deflate and the intravesicular α-amylase released from the acinar cell could be detected, suggesting the involvement of ZG swelling in cell secretion (Kelly et al., 2004).

In electrophysiological bilayer fusion assays, immunoisolated fusion pores or porosomes from the exocrine pancreas, functionally reconstituted into the lipid membrane of the bilayer apparatus, have been used in our studies (Jeremic et al., 2003). In these experiments, membrane conductance and capacitance is continually monitored (Fig. 3.7A). Reconstitution of the porosome into the lipid membrane resulted in a small increase in capacitance (Fig. 3.7B), possibly due to increase in membrane surface area contributed by the incorporation of porosomes, ranging in size from 100 to 150 nm in diameter (Jeremic et al., 2003). Isolated ZGs when added to the cis compartment of such a porosome-reconstituted bilayer chamber, fuse at the base of porosomes in the lipid membrane (Fig. 3.7A) and is detected as a step increase in membrane capacitance (Fig. 3.7B). However, even after 15 min of ZG addition to the cis compartment of the bilayer chamber, little or no release of the intravesicular enzyme α-amylase was detected in the trans compartment of the chamber (Fig. 3.7C, D). On the contrary, exposure of ZGs to 20 μM GTP induced swelling (Abu-Hamdah et al., 2004; Cho et al., 2002c; Jena et al., 1997), and it results in both (a) the potentiation of fusion and (b) a robust expulsion of α-amylase into the trans compartment of the bilayer chamber (Fig. 3.7C, D). These studies demonstrated that during secretion, secretory vesicle swelling is required for the efficient expulsion of intravesicular contents (Kelly et al., 2004).

Within minutes or even seconds following stimulation of secretion, empty and partially empty secretory vesicles accumulate within cells (Cho et al., 2002b; Lawson et al., 1975; Plattner et al., 1997). There may be two pos

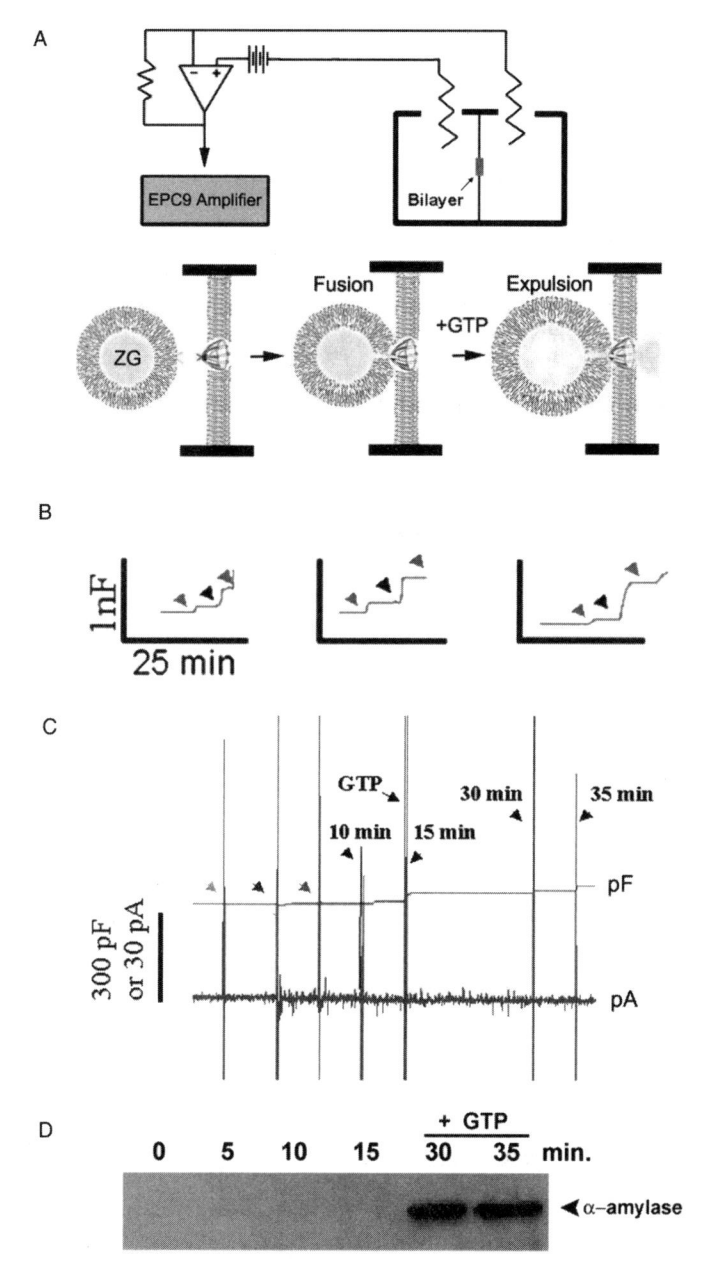

VESICLE SWELLING IN A SLOW SECRETORY CELL (PANCREATIC ACINAR) 43

Figure 3.7. Fusion of isolated ZGs at porosome-reconstituted bilayer and GTP-induced expulsion of α-amylase. (**A**) Schematic diagram of the EPC9 bilayer apparatus showing the cis and trans chambers. Isolated ZGs, when added to the cis chamber, fuse at the bilayers-reconstituted porosome. Addition of GTP to the cis chamber induces ZG swelling and expulsion of its contents such as α-amylase to the trans bilayers chamber. (**B**) Capacitance traces of the lipid bilayer from three separate experiments following reconstitution of porosomes, following addition of ZGs to the cis chamber, and at the 5-min time point after ZG addition. Note the small increase in membrane capacitance following porosome reconstitution, along with a greater increase when ZGs fuse at the bilayers. (**C**) In a separate experiment, 15 min after addition of ZGs to the cis chamber, 20 μM GTP was introduced. Note the increase in capacitance, demonstrating potentiation of ZG fusion. Flickers in current trace represent current activity. (**D**) Immunoblot analysis of α-amylase in the trans chamber fluid at different times following exposure to ZGs and GTP. Note the undetectable levels of α-amylase even up to 15 min following ZG fusion at the bilayer. However, following exposure to GTP, significant amounts of α-amylase from within ZGs were expelled into the trans bilayers chamber.

sible explanations for such accumulation of partially empty vesicles. Following fusion at the porosome, secretory vesicles may either (a) remain fused for a brief period and therefore time would be the limiting factor for partial release of intravesicular contents, or alternately, inadequate vesicle swelling would be unable to generate the required intravesicular pressure for complete discharge of contents. Results in Fig. 3.5 suggests that it would be highly unlikely that generation of partially empty vesicles result from brief periods of vesicle fusion at porosomes. After addition of ZGs to the cis chamber of the bilayer apparatus, membrane capacitance continues to increase (demonstrating the establishment of continuity between the opposing bilayers), however, little or no detectable secretion occurred even after 15 min (Fig. 3.5), suggesting that either variable degrees of vesicle swelling or repetitive cycles of fusion and swelling of the same vesicle, or both, may operate during cell secretion. Under these circumstances, empty and partially empty vesicles could be generated within cells following secretion. To test this hypothesis, two key parameters have been examined (Kelly et al., 2004): (i) whether the extent of swelling is same for all ZGs exposed to a certain concentration of GTP and (ii) whether ZG is capable of swelling to different degrees. And, if so, whether there is a correlation between extent of swelling and the quantity of intravesicular contents expelled. The answer to the first question is clear: Different ZGs respond to the same stimulus differently (Fig. 3.5) (Kelly et al., 2004). As previously discussed, studies (Kelly et al., 2004) reveal that different ZGs within cells or in isolation undergo different degrees of swelling, even though they are exposed to the same stimuli (carbamylcholine for live pancreatic acinar cells) or GTP for isolated ZGs (Figs. 3.5B–D, 3.6B–D). The requirement of ZG swelling for expulsion of vesicular contents has been further confirmed, since GTP dose-dependently increases in ZG swelling (Fig. 3.8A–C) translated into increased secretion of the intravesicular α-amylase enzyme (Fig. 3.8D). Although higher GTP concentrations elicit an increased ZG swelling, the extent of swelling between ZGs once again is varies (Kelly et al., 2004).

3.3. SYNAPTIC VESICLE SWELLING IS ALSO REQUIRED FOR INTRAVESICULAR CONTENT EXPULSION DURING SECRETION IN NEURONS

To determine if a similar or an alternate mechanism is responsible for the release of secretory products in a fast secretory cell, synaptosomes and synaptic vesicles from rat brain was used in our studies (Kelly et al., 2004). Since synaptic vesicle membrane is known to possess both Gi and Go proteins, it was hypothesized that GTP and the G_i-agonist (mastoparan) may mediate vesicle swelling (Kelly et al., 2004). To test this hypothesis, isolated synaptosomes (Fig. 3.9A) were lysed to obtain synaptic vesicles and synaptosomal membrane (Kelly et al., 2004). Isolated synaptosomal membrane, when placed on mica and imaged by the AFM in near-physiological buffer, reveal on the cytosolic compartment 40–50 nm in diameter synaptic vesicles still docked to the presynaptic membrane. Similar to ZGs, exposure of synaptic vesicles (Fig. 3.9B) to 20 μM GTP (Fig. 3.9C) resulted in an increase in synaptic vesicle swelling. However, exposure to Ca^{2+} resulted in the transient fusion of synaptic vesicles at the presynaptic membrane, expulsion of intravesicular contents, and the consequent decrease in size of the synaptic vesicle (Fig. 3.9D,E). In Fig. 3.9B–D, the blue arrowhead points to a synaptic vesicle undergoing this process. Additionally, as observed in ZGs of the exocrine pancreas, not all synaptic vesicles were found to swell; if they did, they swelled to different extents even though they had been exposed to the same stimulus (Kelly et al., 2004).

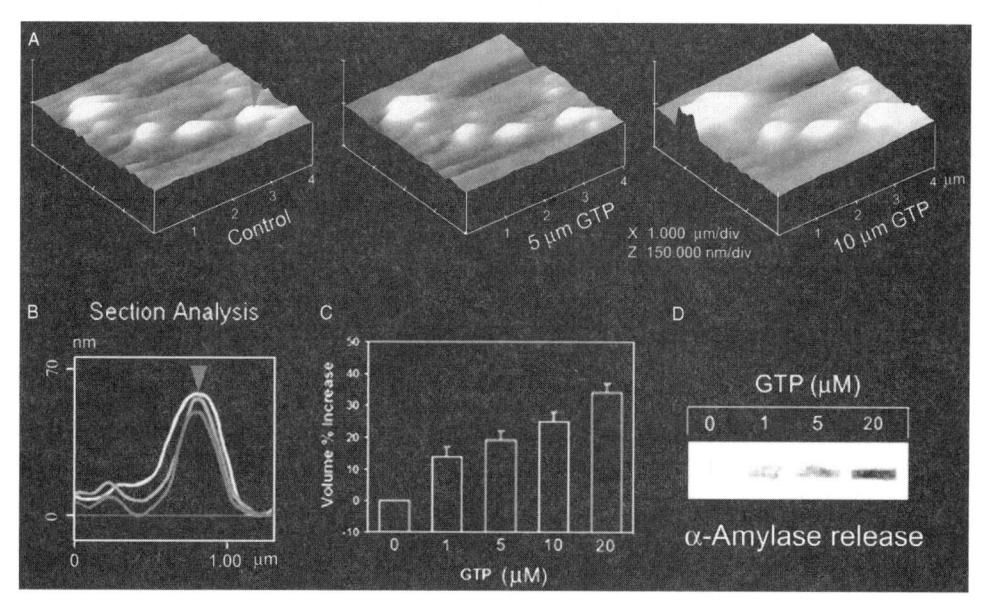

Figure 3.8. The extent of ZG swelling is directly proportional to the amount of intravesicular contents released. (**A**) AFM micrographs showing the GTP dose-dependent increase in swelling of isolated ZGs. (**B**) Note the AFM section analysis of a single ZG (grey arrowhead), showing the height and relative width at resting (control, grey outline), following exposure to 5 μM GTP (middle white outline) and 10 μM GTP (outer white outline). (**C**) Graph demonstrating the GTP dose-dependent percent increase in ZG volume. Data are expressed as mean ± SEM. (**D**) Immunoblot analysis of α-amylase in the trans chamber fluid of the bilayers chamber following exposure to different doses of GTP. Note the GTP dose-dependent increase in α-amylase release from within ZGs fused at the cis side of the reconstituted bilayer.

This differential response of synaptic vesicles within the same nerve ending may dictate and regulate the potency and efficacy of neurotransmitter release at the nerve terminal. To further confirm synaptic vesicle swelling and determine swelling rate, light scattering experiments have been performed. Light scattering studies demonstrate a mastoparan-dose-dependent increase in synaptic vesicle swelling (Fig. 7.9F). Mastoparan (20 μM) induces a time-dependent (in seconds) increase in synaptic vesicle swelling (Fig. 7.9G), as opposed to the control peptide (Mast-17) (Kelly et al., 2004).

These studies demonstrate that following stimulation of secretion, ZGs (the membrane-bound secretory vesicles in exocrine pancreas) swell. Different ZGs swell differently, and the extent of their swelling dictates the amount of intravesicular contents to be expelled. ZG swelling is therefore a requirement for the expulsion of vesicular contents in the exocrine pancreas. Similar to ZGs, synaptic vesicles also swell, enabling the expulsion of neurotransmitters at the nerve terminal. This mechanism of vesicular expulsion during cell secretion may explain why partially empty vesicles are generated in secretory cells (Cho et al., 2002b; Lawson et al., 1975; Plattner et al., 1997) following secretion. The presence of empty secretory vesicles could result from multiple rounds of fusion–swelling–expulsion, which a vesicle may undergo during the secretory process. These results reflect the precise and regulated nature of cell secretion, from the exocrine pancreas to neurons.

ACKNOWLEDGMENT

This work was supported by NIH grants DK56212 and NS39918 (B.P.J).

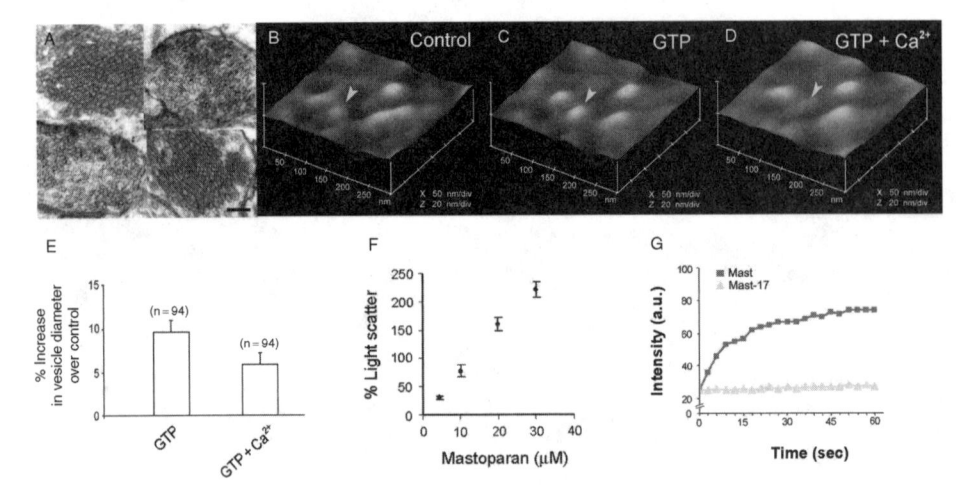

Figure 3.9. Synaptic vesicles swell in response to GTP and mastoparan, and vesicle swelling is required for neurotransmitter release. (**A**) Electron micrographs of brain synaptosomes, demonstrating the presence of 40- to 50-nm synaptic vesicles within. Bar = 200 nm. (**B**) AFM micrographs of synaptosomal membrane, demonstrating the presence of 40- to 50-nm synaptic vesicles docked to the cytosolic face of the presynaptic membrane. (**C**) Exposure of the synaptic vesicles to 20 μM GTP results in vesicle swelling (arrowhead). (**D, E**) Further addition of calcium results in the transient fusion of the synaptic vesicles at porosomes in the presynaptic membrane of the nerve terminal, along with expulsion of intravesicular contents. Note the decrease in size of the synaptic vesicle following content expulsion. (**F**) Light scattering assays on isolated synaptic vesicles demonstrate the mastoparan dose-dependent increase in vesicle swelling ($n = 5$), and they further confirm the AFM results. (**G**) Exposure of isolated synaptic vesicles to 20 μM mastoparan demonstrates a time-dependent (in seconds) increase in their swelling. Note that the control peptide mast-17 has little or no effect on synaptic vesicles.

REFERENCES

Abu-Hamdah, R., Cho, W.-J., Cho, S.-J., Jeremic, A., Kelly, M., Ilie, A. E., and Jena, B. P. (2004). Regulation of the water channel aquaporin-1: Isolation and reconstitution of the regulatory complex. *Cell Biol. Int.* **28**:7–17 (published on-line 2003).

Alvarez de Toledo, G., Fernandez-Chacon, R., and Fernandez, J. M. (1993). Release of secretory products during transient vesicle fusion. *Nature (London)* **363**:554–558.

Cho, S.-J., Cho, J., and Jena, B. P. (2002a). The number of secretory vesicles remains unchanged following exocytosis. *Cell Biol. Int.* **26**:29–33.

Cho, S. J., Kelly, M., Rognlien, K. T., Cho, J., Hoerber, J. K. H., and Jena, B. P. (2002b). SNAREs in opposing bilayers interact in a circular array to form conducting pores. *Biophys. J.* **83**:2522–2527.

Cho, S. J., Sattar, A. K., Jeong, E. H., Satchi, M., Cho, J. A., Dash, S., Mayes, M. S., Stromer, M. H., and Jena, B. P. (2002c). Aquaporin 1 regulates GTP-induced rapid gating of water in secretory vesicles. *Proc. Natl. Acad. Sci. USA* **99**:4720–4724.

Curran, M. J., and Brodwick, M. S. (1991). Ionic control of the size of the vesicle matrix of beige mouse mast cells. *J. Gen. Physiol.* **98**:771–790.

Fernandez, J. M., Villalon, M., and Verdugo, P. (1991). Reversible condensation of the mast cell secretory products *in vitro*. *Biophys. J.* **59**:1022–1027.

Jena, B. P., Schneider, S. W., Geibel, J. P., Webster, P., Oberleithner, H., and Sritharan, K. C. (1997). Gi regulation of secretory vesicle swelling examined by atomic force microscopy. *Proc. Natl. Acad. Sci. USA* **94**:13317–13322.

Jeremic, A., Kelly, M., Cho, S. J., Stromer, M. H., and Jena, B. P. Reconstituted fusion pore. *Biophys. J.* **85**:2035–2043.

Kelly, M., Cho, W. J., Jeremic, A., Abu-Hamdah, R., and Jena, B. P. (2004). Vesicle swelling regulates content expulsion during secretion. *Cell Biol. Int.* **28**:709–716.

Lawson, D., Fewtrell, C., Gomperts, B., and Raff, M. C. (1975). Anti-immunoglobulin-induced histamine secretion by rat peritoneal mast cells studied by immunoferritin electron microscopy. *J. Exp. Med.* **142**:391–401.

Monck, J. R., Oberhauser, A. F., Alvarez de Toledo, G., and Fernandez, J. M. (1991). Is swelling of the secretory granule matrix the force that dilates the exocytotic fusion pore? *Biophys. J.* **59**:39–47.

Plattner, H., Artalejo, A. R., and Neher, E. (1997). Ultrastructural organization of bovine chromaffin cell cortex—analysis by cryofixation and morphometry of aspects pertinent to exocytosis. *J. Cell Biol.* **139**:1709–1717.

Sattar, A. K. M., Boinpally, R., Stromer, M. H., and Jena, B. P. (2002) Gαi3 in pancreatic zymogen granule participates in vesicular fusion. *J. Biochem.* **131**:815–820.

CHAPTER 4

FUSION PORES IN GROWTH-HORMONE-SECRETING CELLS OF THE PITUITARY GLAND: AN AFM STUDY

LLOYD L. ANDERSON
Department of Animal Science, College of Agriculture, and Department of Biomedical Sciences, College of Veterinary Medicine, Iowa State University, Ames, Iowa

BHANU P. JENA
Department of Physiology, Wayne State University School of Medicine, Detroit, Michigan

4.1. INTRODUCTION

This will briefly review the physiology of growth hormone (GH) secretion *in vivo* focused on an animal model, the pig. There is increasing knowledge of the subcellular events leading to GH release. This includes signal transduction mechanisms and the physical nature of GH secretory granules fusing transiently with the cell membrane at the nanoscale. Hypothalamic hormones that influence GH release and synthesis with particular attention to other peptides as well as GH-releasing hormone (GHRH), somatostatin (SRIF), and the natural GH-secretagogue, ghrelin, will be considered. The immunohistochemical distribution and differential immunoreactivity of somatotrophs in porcine anterior pituitary across ages may reflect changes in the number of GH vesicles per cell or heterogeneity of somatotrophs. Finally, a major focus on the cell biology for controlled GH secretion involving movement of secretory granules and fusion pores and secretory pits of somatotrophs utilizing atomic force microscopy (AFM) and transmission electron microscopy (TEM) will be examined.

4.2. NEUROENDOCRINE REGULATION OF GH SECRETION *IN VIVO*

A series of stimulatory and inhibitory releasing hormones of hypothalamic and peripheral origins controls the release of growth hormone (GH) from somatotrophs. Until recently, the consensus was of two hypothalamic releasing hormones for GH, respectively GH-releasing hormone (GHRH; stimulatory GH) and somatostatin (SRIF; inhibitory) with GH release and synthesis reflecting a balance between these.

In vivo approaches such as stalk sectioning and hypothalamic deafferentation provide

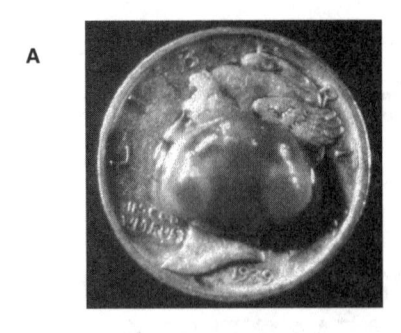

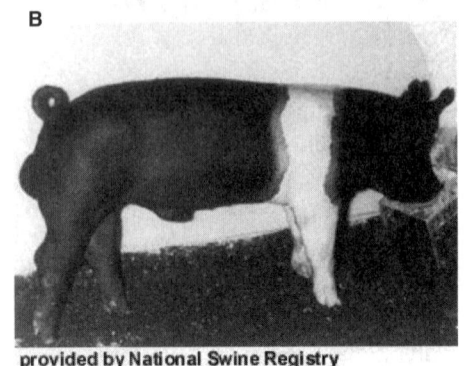

Figure 4.1. Adult pig and its pituitary gland. (A) Porcine pituitary gland (∼250 mg), resting on a dime, is located at the base of the brain and connected to the hypothalamus by an hypophyseal stalk. (B) Adult Hampshire male pig (∼120-kg body weight).

strong evidence for the physiological control of GH release. The pituitary gland in the adult pig weighs about 250 mg (Fig. 4.1). The anterior lobe of the gland contains somatotrophs that release GH into peripheral blood in a pulsatile pattern (Fig. 4.2). After hypophyseal stalk transection (HST), the pulsatile GH secretion pattern is obliterated (Klindt et al., 1983). Control of GH secretion was investigated in pigs by comparing the effects of anterior (AHD), complete (CHD), and posterior (PHD) hypothalamic deafferentation with sham-operated controls (SOC) (Molina et al., 1986). A camera lucida drawing of the sagittal view of the porcine thalamus and hypothalamus with a depiction of the hypothalamic areas isolated by the knife is shown in Fig. 4.3. Mean serum concentrations of GH after AHD, CHD, and PHD were reduced ($P < 0.01$) when compared with SOC gilts (Fig. 4.4). Furthermore, episodic GH release evident in SOC animals was obliterated after hypothalamic deafferentation. Thus, maintenance of episodic GH secretion depends upon its neural connections traversing the anterior and posterior aspects of the hypothalamus in the pig.

There is at least one other hypothalamic releasing hormone for GH, with the identification by Kojima and colleagues of ghrelin as the natural ligand for the GH secretagogue (GHS) receptor (GHS-R; Kojima et al., 1999). GH secretion is influenced by peptides/proteins produced in the hypothalamus and in the periphery, including gonadotropin-releasing hormone (GnRH), insulin-like growth factor 1 (IGF-1), leptin, pituitary adenylate cyclase activating polypeptide (PACAP), and thyrotropin-releasing hormone (TRH). There is increasing knowledge of the subcellular events leading to GH release. This includes signal transduction mechanisms and the physical nature of GH secretory granules fusing transiently with the cell membrane.

4.3. HYPOTHALAMIC HORMONES CONTROLLING GH RELEASE AND SYNTHESIS

4.3.1. Growth-Hormone-Releasing Factor (GHRH)

The major function of GHRH, released from neurosecretory terminals in the median eminence, is to stimulate the release and synthesis of GH. GHRH is expressed in the arcuate nucleus of the hypothalamus together with other tissues—for example, intestine, gonads, immune tissues, and the placenta. GHRH-immunoreactive (-IR) neurons have been identified in coronal and sagittal frozen sections of bovine and porcine hypothalami (Leshin et al., 1994). Rounded bipolar GHRH-IR perikarya are localized in ventrolateral regions of the arcuate nucleus (ARC) in both species. GHRH-immunoreactive fibers

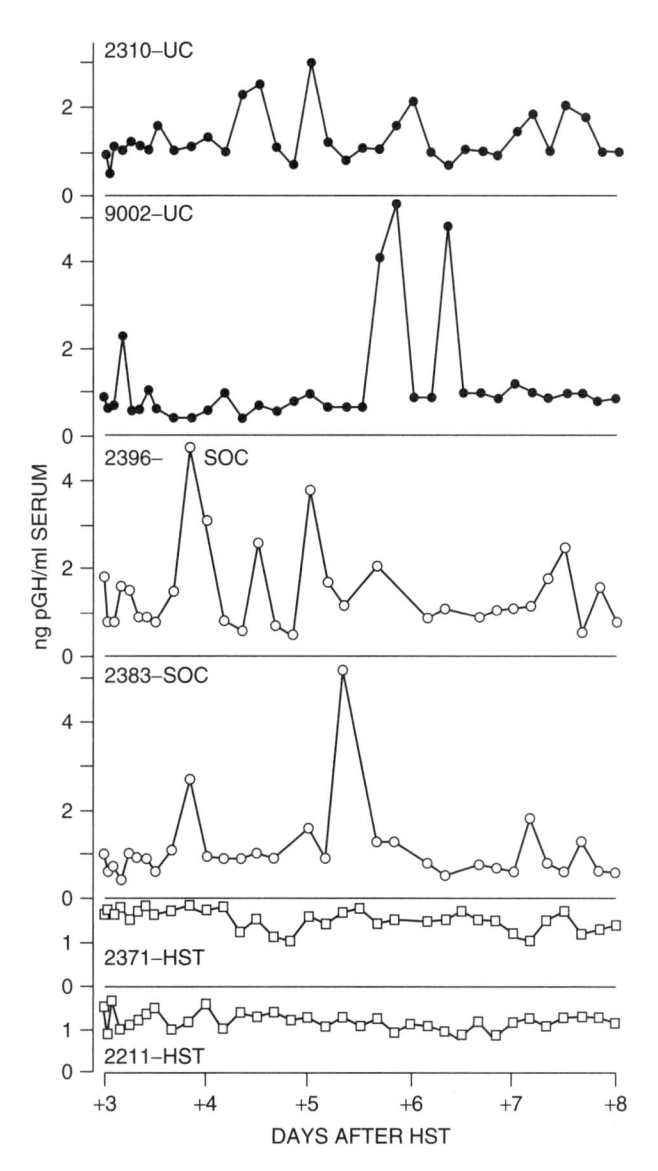

Figure 4.2. Sequential profiles of peripheral serum concentrations of GH in two gilts from each treatment group (UC, unoperated control: SOC, sham operated control: HST, hypophyseal stalk transection) during a 120-h period from Day +3 to Day +8. HST or SOC was performed on Day 0. Four-digit numbers designate individual gilts. Symbols indicate unoperated control (●), sham-operated control (○), and hypophyseal (□) animals. (Reprinted from *Proceedings of the Society for Experimental Biology and Medicine*, Vol. 172, Klindt, J., Ford, J. J., Beradinelli, G., and Anderson, L. L. Growth hormone secretion after hypophyseal stalk transection in pigs, pp. 503–513. Copyright 1983, with permission from the Society for Experimental Biology and Medicine, Maywood, NJ.)

projected ventrally into the median eminence (ME) in both species.

Signal transduction involves Gs-protein, adenylate cyclase (isoform II and/or IV), cyclic 3′, 5′-adenosine monophosphate (cAMP), and protein kinase A. There is a cAMP response element (CRE) upstream from the coding region of the GH gene. GHRH increases GH expression (e.g., cattle, Silverman et al., 1988; chickens, Vasilatos-Younken

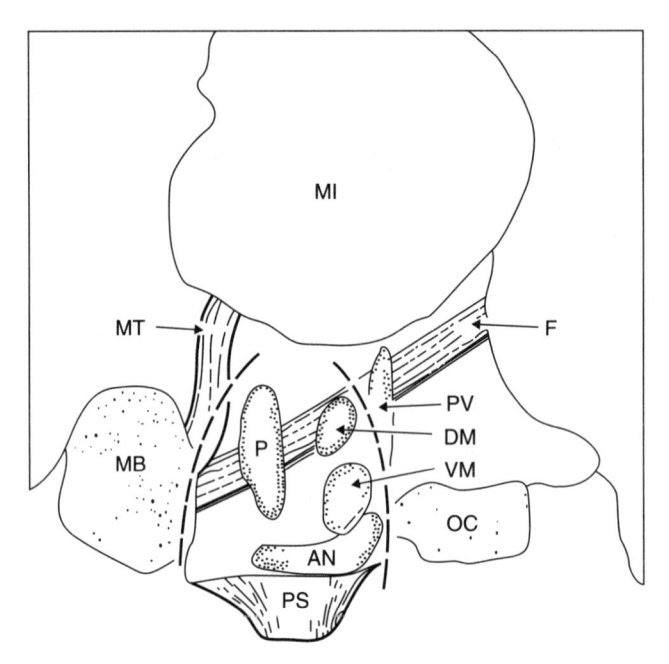

Figure 4.3. A camera lucida drawing of the sagittal view of the porcine thalamus and hypothalamus with a depiction of the areas isolated by the knife. Interrupted lines define the arc for position of anterior and posterior knife cuts. The mammillary bodies (MB), mammillothalamic tract (MT), fornix (F), dorsomedial nucleus (DM), ventromedial nucleus (VM), arcuate nucleus (AN), posterior nucleus (P), optic chiasm (OC), messa intermedia (MI), and pituitary stalk (PS) are indicated. ×2.7. (Reprinted from *Proceedings of the Society for Experimental Biology and Medicine*, Vol. 183, Molina, J. R., Klindt, J., Ford, J. J., and Anderson, L. L. Growth hormone and prolactin secretion after hypothalamic deafferentation in pigs, pp. 163–168. Copyright 1986, with permission from the Society for Experimental Biology and Medicine, Maywood, NJ.)

et al., 1992; Radecki et al., 1994). Intracellular Ca^{2+} concentrations are increased by GHRH. This involves both influx of Ca^{2+} (via L- and T-type voltage-sensitive Ca^{2+} channels) and phospholipase C hydrolysis of phosphatidyl inositol, leading to mobilizing intracellular Ca^{2+} stores.

4.3.2. Somatostatin (SRIF)

Somatostatin (SRIF) is a cyclic 14- or 28-amino-acid residue containing peptide. A major function of SRIF is suppressing the release, but not synthesis, of GH by the somatotroph (reviewed in Tannenbaum and Epelbaum, 1999). SRIF-IR neurons have been identified in bovine and porcine hypothalamic tissue (Leshin et al., 1994). Bipolar SRIF-IR perikarya are located about the third ventricle in the periventricular nucleus. SRIF-IR fibers project ventrally into the ME. SRIF-IR fibers densely innervate the ventromedial and arcuate (ARC) nuclei in pigs, but are not as distinguishable as in cattle (Leshin et al., 1994). Double immunostaining reveals a close apposition of SRIF-IR fibers with GHRH-IR perikarya in the ARC and ventromedial nuclei and apposition of SRIF- and GHRH-IR varicosities in the ME (Leshin et al., 1994).

The SRIF receptor (sstr) is encoded by five different genes (sstr1-5; reviewed in Tannenbaum and Epelbaum, 1999). The dominant sstr influencing GH release from the somatotroph appears to be sstr-2 (Reed et al., 1999). Signal transduction involves G-protein-coupled reduction in L- and T-type voltage-sensitive Ca^{2+} influx/channels and increased K^+ channels (reviewed in Tannenbaum and Epelbaum, 1999).

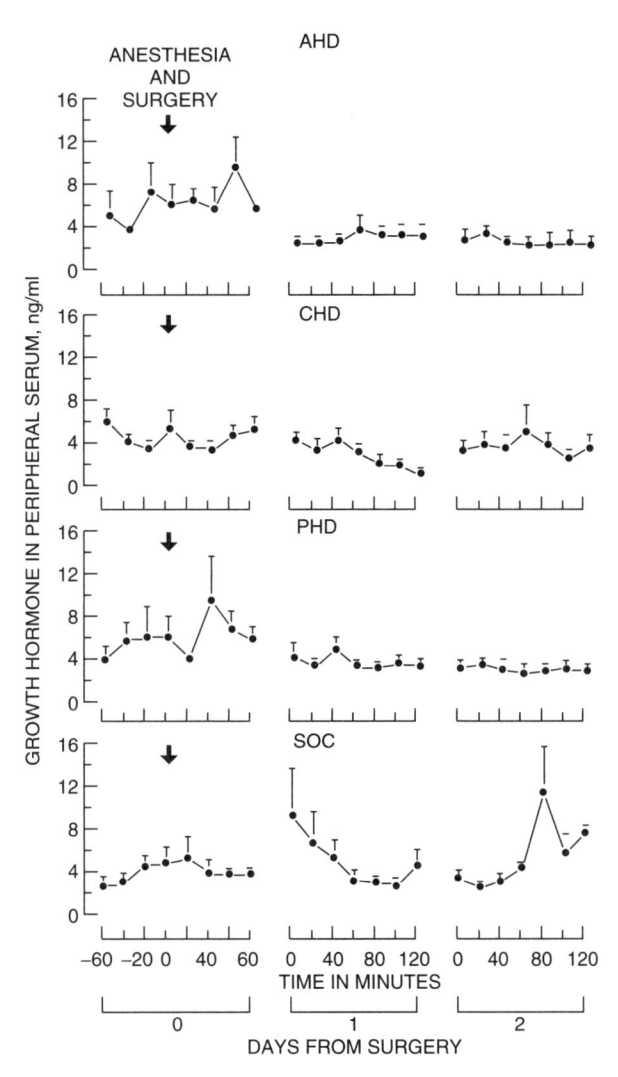

Figure 4.4. GH concentrations in peripheral serum in ovariectomized prepuberal gilts during anesthesia and early recovery (Day 0) and 24 and 48 h after hypothalamic deafferentation (Day 1 and 2, respectively). Groups consisted of 4 AHD, 5 CHD, 4 PHD, and 4 SOC gilts. Values are expressed as means ± SEM. (Reprinted from *Proceedings of the Society for Experimental Biology and Medicine*, Vol. 183, Molina, J. R., Klindt, J., Ford, J. J., and Anderson, L. L. Growth hormone and prolactin secretion after hypothalamic deafferentation in pigs, pp. 163–168. Copyright 1986, with permission from the Society for Experimental Biology and Medicine, Maywood, NJ.)

4.3.3. Ghrelin

Ghrelin is a natural ligand for the GH-secretagogue receptor (GHS-R; Lee et al., 2002). It is produced by the stomach, small intestine, and central nervous system—for example, the hypothalamus, specifically including the ARC (Kojima et al., 1999; Lee et al., 2002; Masuda et al., 2000).

Ghrelin acts by binding to the GHS receptor. The GHS-R is a seven-transmembrane domain G-protein-coupled receptor (Howard et al., 1996). Signal transduction involves activation of phospholipase C (via G protein), generation of inositol phosphate and diacyl glycerol, and increased intracellular Ca^{2+} (reviewed in Bowers, 1999). Ghrelin stimulates GH release both directly (acting at the level of the anterior pituitary gland) and by enhancing GHRH release. The GHS-R is expressed in various hypothalamic and thalamic nuclei, the dentate gyrus, substantial nigra, ventral tegmentum, and facial nuclei of the brainstem, thus implicating a possible central role for ghrelin (Bennet et al., 1997; Guan et al., 1997; Yokote et al., 1998; Mitchell et al., 2001). For instance, ghrelin plays a critically important role in the control of appetite, stimulating food intake (Lawrence et al., 2002).

Ghrelin, when injected intracerebroventricularly into rodents, stimulates feeding behavior and an increase in body weight (Nakazato et al., 2001; Tschöp et al., 2000). Synthetic GHSs and the endogenous hormone, ghrelin, induce immediate gene *c-fos* only in the ARC, the site that also contains neuropeptide Y (NPY; Dickson et al., 1993; Dickson and Luckman, 1997; Hewson and Dickson, 2000). Additionally, the ARC and NPY neurons possess GHS-R mRNA, and these neurons also project to orexinand melanin-concentrating hormone (MCH)-containing neurons in the lateral hypothalamus, neurons that affect feed intake (Broberger et al., 1998; Horvath et al., 1999). The effects of ghrelin on feed intake appear to be mediated via NPY. Preadministration of the Y1 NPY receptor antagonist (BIB03304) blocks both (a) the stimulatory effects of intracerebroventricular (icv) injection of ghrelin or GHRP-6 on food intake in rats and (b) the decreased body-core temperature (Shintani et al., 2001; Brown et al., 1973). Thus, NPY neurons of the ARC are likely a primary site for ghrelin increasing food intake.

It might be speculated that motilin may have analogous effects to ghrelin. Motilin is a highly conserved, 22-amino-acid polypeptide (Brown et al., 1973) secreted by enterochromaffin cells of the small intestine. This peptide stimulates gastrointestinal motor activity and seems to play a physiological role in the regulation of fasting motility patterns (Yanaihara et al., 1990). Both IR motilin and mRNA expression are found in the gastrointestinal tract, brain, nerves, and other endocrine glands in several species, including monkey, man, pig, sheep, and rabbit (Huang et al., 1998). The motilin receptor, designated motilin-R1A (MTLR1A), is a heterotrimeric, guanosine triphosphate-binding, protein (G protein)-coupled receptor. It was isolated from human stomach, and its amino acid sequence was found to be 52% identical to the human GHSR (Feighner et al., 1999). The high amino-acid-sequence identity between MTL-R1A and the GHSR suggests a role for motilin in the control of GH secretion.

4.3.4. Other Hypothalamic Peptides Influencing GH Release Directly

1. Thyrotropin-Releasing Hormone (TRH). The modified tripeptide, TRH, can stimulate the release of GH in many species under some, but not under all, physiological circumstances.
2. Pituitary Adenylate Cyclase-Activating Peptide (PACAP). Pituitary adenylate cyclase-activating peptide (PACAP) is a potent GH secretagogue, and has even been proposed as the ancestral releasing factor for GH (reviewed in Sherwood et al., 2000).
3. Gonadotropin-Releasing Hormone. In higher vertebrates, gonadotropin-releasing hormone (GnRH) does not stimulate GH release. However, in some species, at specific physiological phases, GnRH can evoke GH secretion.
4. Leptin. The effects of leptin on GH release appear to be predominantly stimulatory. The presence of leptin receptors in/on somatotropes is well established (e.g., Shimon et al, 1998; Iqbal et al., 2000; Lin et al., 2000, 2003).
5. Others (FMRF Amide, Corticotropin-Releasing Hormone, Galanin, and NPY). Studies, particularly in lower

vertebrate species, suggest that somatotropes are promiscuous in terms of the peptides that will stimulate GH release. In some species of lower vertebrates, corticotropin-releasing hormone (CRH) has been reported to stimulate GH secretion [e.g., reptiles (Denver and Licht, 1990) and fish (Rousseau et al., 1999)]. Neuropeptide Y has a direct effect of GH release in some species of fish (Peng et al., 1993a,b). In contrast, galanin can inhibit GH release (Arvat et al., 1995; Giustina et al., 1996, 1997). Moreover, peptides related to the neuropeptide Phe-Met-Arg-Phe-NH$_2$ (FM-RF amide), originally described in mollusks, stimulate GH release in amphibians. Similar peptides have been identified in brains of vertebrates (i.e., rat, chicken, frog, carp) with a similar RF amide motif at their C-termini (Dockray et al., 1983; Chartrel et al., 2002; Fujimoto et al., 1998; Ukena et al., 2002). A novel hypothalamic RF amide peptide localized in the hypothalamus of the bullfrog has GH-releasing activity and was designated frog GH-releasing peptide [fGRP (Koda et al., 2002)].

4.4. SIGNAL TRANSDUCTION MECHANISM SYSTEMS

4.4.1. Adenylate Cyclase/cAMP/ Protein Kinase A and Phospholipase C-IP3-Protein Kinase C

Somatotrophs, excitable cells that exhibit spontaneous action potential, when treated with GHRH, depolarize the cell membrane potential and stimulate Ca^{2+} influx via transmembrane Ca^{2+} channels, resulting in increases in intracellular free Ca^{2+} ([Ca^{2+}]; Chen and Clarke, 1995; Chen, 2002). This involves both an influx of Ca^{2+} (via L- and T-type voltage-sensitive Ca^{2+} channels) and phospholipase C hydrolysis of phosphatidyl inositol, leading to mobilizing intracellular Ca^{2+} stores (reviewed in Frohman and Kineman, 1999). It is well established that the signal transduction for GHRH, and probably PACAP, involves the following:

1. Binding to specific receptors in the plasma membrane of the somatotropes.
2. Activation of adenylate cyclase (isoform II and/or IV) via the coupling protein, Gs-protein cyclic 3′,5′-adenosine monophosphate (cAMP).
3. Activation of protein kinase A. In addition, there is activation of phospholipase C-IP3-protein kinase C and voltage-sensitive calcium channels.

Signal transduction for ghrelin and synthetic analogues or GHS (e.g., GHRP-6 and nonpeptidergic GHS such as L-692, 585, MK-677) involves activation of phospholipase C (via G protein), generation of inositol phosphate and diacyl glycerol, and increased intracellular Ca^{2+} (reviewed in Bowers, 1999). Modification of the somatotroph's electrophysiological properties leads to changes in sensitivity to ghrelin or its analogues.

We have recently demonstrated, using pharmacological blockers, that both ghrelin and a synthetic GHS can increase intracellular calcium concentrations in porcine somatotrope (Glavaski-Joksimovic et al., 2002, 2003). This two-phase phenomenon involves, first, mobilization of intracellular calcium, followed by calcium entry via L channels with somatotrope depolarization (see Fig. 4.5; Glavaski-Joksimovic et al., 2002). The effects of GHS are blocked by inhibitors of adenylate cyclase and phospholipase C, supporting the involvement of both these, adenylate cyclase/cAMP/protein kinase A and phospholipase C; in the generation of inositol phosphate and diacyl glycerol and increased intracellular Ca^{2+}.

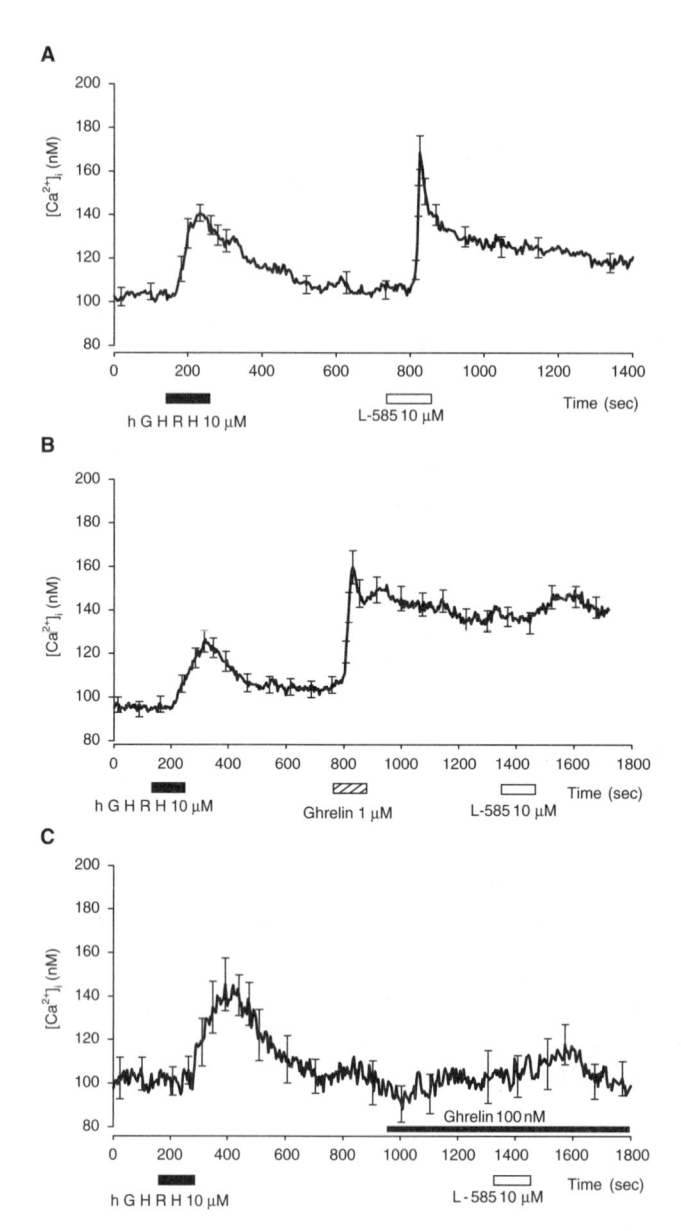

Figure 4.5. Inhibitory effects of ghrelin on calcium transients evoked by L-585 in isolated porcine somatotrophs. **(A)** Control studies on calcium transients in isolated porcine somatotropes evoked by a 2-min application of human growth hormone-releasing hormone (hGHRH) (10 μM) and L-585 (10 μM; $n = 29$). **(B)** Application of L-585 (10 μM) 10 min after administration of ghrelin (1 μM) did not have an additive effect on $[Ca^{2+}]_i$ in isolated porcine somatotropes ($n = 58$). **(C)** Ghrelin (100 μM) alone did not affect $[Ca^{2+}]_i$, but the changes in $[Ca^{2+}]_i$ evoked by 10 μM L-585 were almost completely blocked ($n = 7$). (Reprinted from *Neuroendocrinology*, Vol. 77, Glavaski-Joksimovic, A., Jeftinija, K., Scanes, C. G., Anderson L. L., and Jeftinija, S. Stimulatory effect of ghrelin on isolated porcine somatotropes, pp. 366–378. Copyright 2003, with permission from Karger, Basel.)

4.5. IMMUNOCYTOCHEMICAL DISTRIBUTION OF SOMATOTROPHS IN PORCINE ANTERIOR PITUITARY

Growth hormone (GH) is a primary regulator that plays an important role in determining body composition to maintain a beneficial ratio between skeletal muscle and fat (Anderson et al., 2004). Animals deficient in GH lack normal skeletal and muscular development (Ford and Anderson, 1967). At 40 days of gestation (term = 114 days), porcine fetuses have measurable concentrations of GH which increase 40-fold to peak values at about 90 days of gestation in an episodic secretion pattern (Klindt et al., 1983; Klindt and Stone, 1984). Hypophyseal stalk transection in pigs and cattle obliterates pulsatile GH secretion; however, basal GH concentrations are sufficient for continued growth (Klindt et al., 1983; Plouzek et al., 1988; Anderson et al., 1999). Immediately after birth, serum GH concentrations decline abruptly, followed by elevated hormone secretion 3 to 5 weeks after birth (Sun et al., 2002). These postnatal rises in serum GH secretion occur before weaning; thus dietary changes are not the cause of this increase (Owens et al., 1991). The objective of this immunohistochemical study was to identify the spatial distribution patterns of growth hormone secreting cells (somatotrophs) in the newborn and prepubertal porcine pituitary.

(a) Immunohistochemistry. The immunoperoxidase/avidin-biotin system was used to identify immunoreactive GH cells in the anterior pituitary. Tissue sections were washed in 50 mM potassium phosphate buffered saline (KPBS) and then treated with 0.3% hydrogen peroxide to neutralize endogenous peroxidase activity. The sections were incubated for 1 h at room temperature with normal goat serum to block nonspecific binding. The slides were incubated overnight with monkey antiporcine GH (pGH) antibody (1:500,000 for day 1 and day 42 pigs, 1:100,000 for day 100 pigs). The sections were thoroughly washed and treated for 1 h with biotinylated goat antihuman IgG (1:500) and then treated with avidin–peroxidase compound. Peroxidase activity was detected by using 0.04% 3,3′-diaminobenzidine tetrahydrochloride, nickel sulfate (2.5%), and 0.1% hydrogen peroxide in 0.1 M sodium acetate for 6 min to visualize immunoreactive GH cells as a black-colored reaction product. Tissue sections then were dehydrated. Sections were mounted in acrytol (Surgipath Medical Industries, Graylake, IL) mounting medium. Negative control immunohistochemical tests were done by substituting normal goat serum or KPBS for the primary antisera on randomly selected sections of the serial sets. No staining was observed in any negative control section.

(b) Quantitative Analysis. Ten sections of each serial set were preliminarily examined at the lowest magnification ($\times 10$ objective) to confirm that the change of cell distribution pattern was consistent from proximal to distal levels. Three sections per pituitary gland were selected in order from proximal (nearest to the brain), middle (largest part of gland), and distal (farthest from the brain) levels. Anterior lobes containing somatotrophs were divided into five radial regions (regions 1 and 5 in the lateral wings of anterior lobe, regions 2 and 4 in the shoulder areas, and region 3 in the center) with a $\times 10$ objective (final magnification $\times 125$). Three different positions (position a—proximal to the intermediate lobe; position b—middle; position c—nearest the outer surface of the pituitary) were photographed (unit area 30,495 μm^2) along each of the five regions with a $\times 40$ objective (final magnification $\times 500$) for quantification. Identifying regions and positions within regions provided a method to compare the distribution patterns of immunoreactive somatotrophs. Somatotrophs were counted at 15 positions (5 regions $\times$ 3 positions per region) in three sections (proximal, middle, and distal levels) from each pituitary. The number of somatotrophs in each position of the same age group was averaged.

No differences were found among the total somatotrophs across the age groups (1, 42, and 100 days old). A distinctive pattern was found in somatotroph distribution throughout the gland in the three age groups. Characteristics of the pattern included a high population of somatotrophs (44 ± 1.2; mean ± standard error of the mean per 30,495 μm^2) in regions 1 and 5 and a low population (22 ± 1.4) in regions 2 and 4 at each level ($P < 0.05$) (Fig. 4.6). Somatotrophs increased 55% in region 3 from proximal to distal levels at all ages. Somatotrophs in region 3 at the proximal level decreased 33% as age increased.

The present study with immunocytochemical staining of somatotrophs from different age groups required different optimal concentrations of the primary antibody, porcine GH antibody raised in monkey. Different immunoreactive density of somatotrophs with increasing age of the pig may reflect changes in the number of GH vesicles or heterogeneity of the somatotrophs. Our data suggest that there may be regional specificity of cellular differentiation and transformation to facilitate GH secretion in the need for endocrine regulation during rapid growing period in young pig. An imaging technique for three-dimensional reconstruction of distribution pattern with serial sections of large size models would improve visualizing spatial characteristics. Because the pituitary is an endocrine gland with a complex and heterogeneous distribution of cells which functionally communicate with neighboring endocrine cells, a temporal correlation of changes in GH production with changes in other hypothalamic/pituitary hormones must be taken into account to increase our understanding of pituitary function. The characteristics of spatial distribution patterns of specific endocrine cells in pituitary gland may be applied to examine therapeutic strategies for the management of pituitary disorders in both physiological and pathological condition.

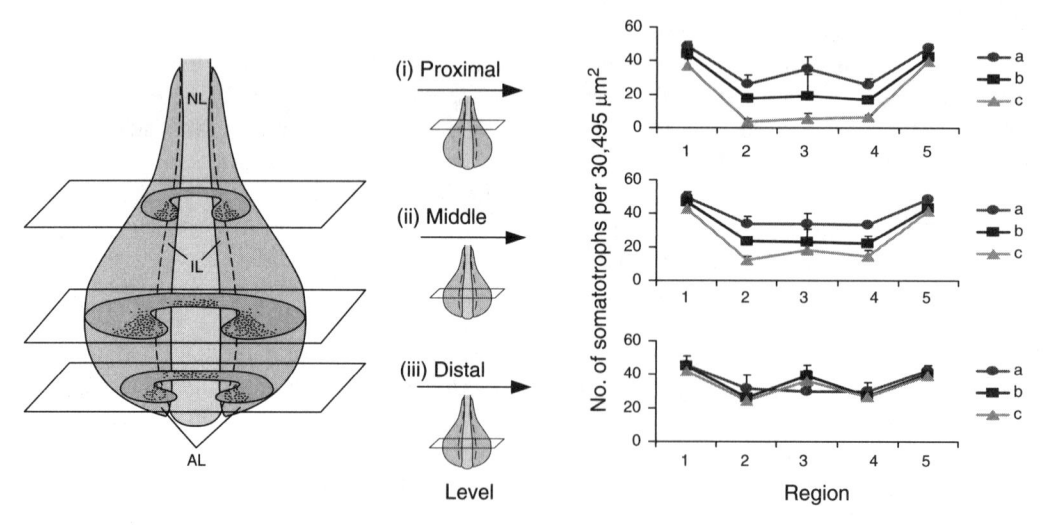

Figure 4.6. Schematic diagram depicts a general spatial distribution pattern of porcine somatotrophs in all age groups as indicated by gray dots (*left*). NL, neural lobe (posterior pituitary); IL, intermediate lobe; AL, anterior lobe (anterior pituitary). Proximal (nearest to the brain), middle (largest part of gland), and distal (farthest from the brain) levels. Graphs (*left*) indicated mean numbers of somatotrophs per 30,495 μm^2 ± standard error of the mean.

4.6. STRUCTURE AND DYNAMIC OF THE FUSION PORES IN LIVE GH-SECRETING CELLS REVEALED USING ATOMIC FORCE MICROSCOPY

Until recently, our understanding of secretory vesicle fusion at the cell plasma membrane was obtained from morphological, electrophysiological, and biochemical studies all suggesting the presence of "fusion pores" at the cell plasma membrane (Anderson, 2004). Using atomic force microscopy (AFM), the structure and dynamics of the "fusion pore" were first revealed and examined in live pancreatic acinar cells. In live resting pancreatic acinar cells, "pits" measuring 0.5–2 μm and containing 3–20 "depressions" of 100- to 180-nm diameter were identified only at the apical region of these cells where membrane-bound secretory vesicles are known to dock and fuse. Following stimulation of secretion, only "depression" enlarged and returned to resting size following completion of secretion. Exposure of acinar cells to cytochalasin B, a fungal toxin that inhibits actin polymerization, resulted in a 50–60% loss of stimulated amylase secretion. A significant decrease in depression diameter was also observed in acinar cells exposed to the fungal toxin. To determine if similar structures are present in neuroendocrine cells, the GH-secreting cell of the pig pituitary was studied using AFM and transmission electron microscopy (TEM). Results from this study demonstrate the presence of pits and depressions in GH-secreting cells of the pituitary and their involvement in hormone release.

Reverse hemolytic plaque assay on isolated GH-secreting cells of the pig pituitary demonstrated GH release following exposure to a nonpeptidyl GH secretagogue (L-692,585) (Fig. 4.7). Examination of resting GH cells revealed the presence of "pits" and "depressions" at the plasma membrane.

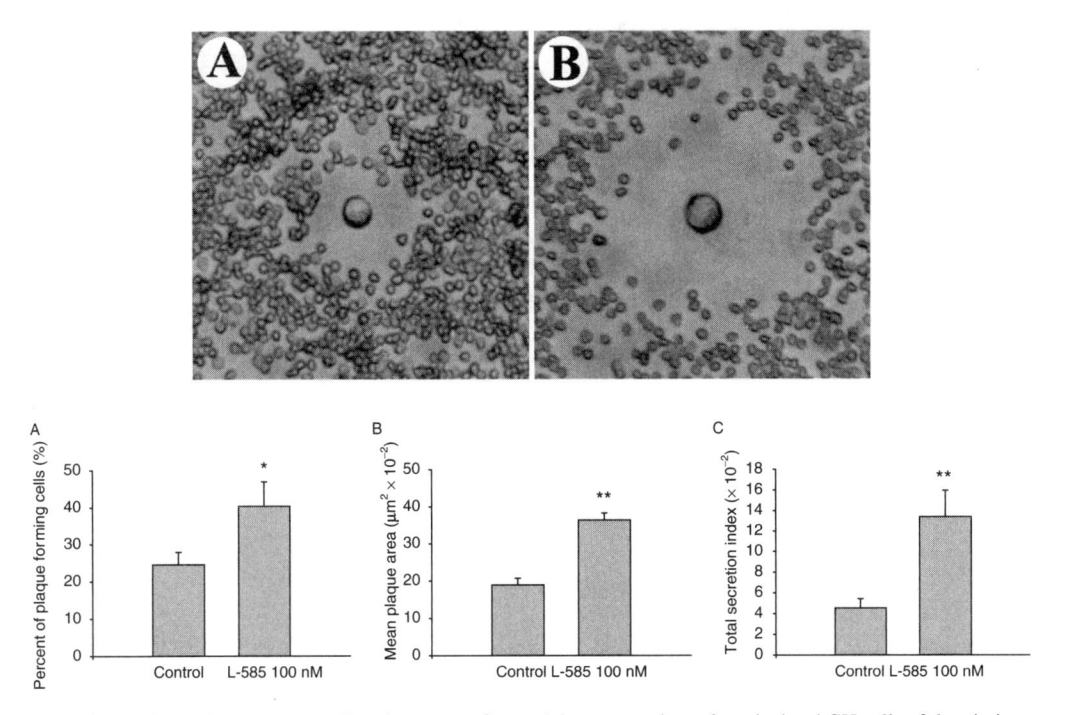

Figure 4.7. Light microscopy revealing the extent of growth hormone release from isolated GH cells of the pituitary gland, in a typical reverse hemolytic plaque assay. The larger the plaque (clear area), the greater the release. Note the presence of a small plaque in a resting GH cell (**A**), compared to an L-692,585 stimulated cell (**B**).

spectroscopy is used to image objects using the AFM, less elastic samples are better resolved. From studies using that pancreatic acinar cells, it was determined that 1 h treatment of ice-cold 2.5% paraformaldehyde in phosphate buffered saline, pH 7.5, was ideal in retaining the structural integrity of "pits" and "depressions." No detectable changes were identified in live cells following fixation. To obtain high-resolution images of the cells and of the "pits" and "depressions" and to determine the distribution of 30-nm GH-immunogold labeling, stimulated GH cells were fixed following immunogold labeling. In conformation with our observation in live GH cells, AFM images of the immunolabeled fixed cells demonstrate specific localization of immunogold at "depressions" (Fig. 4.8), implicating them to be secretory sites at the cell plasma membrane. In agreement with earlier studies in pancreatic acinar cells, the GH-secreting neuroendocrine cells of the pituitary demonstrate the presence of "pits" and "depressions" at the plasma membrane, where secretory vesicles dock and fuse to release vesicular contents. The presence of "pits" and "depressions" in both exocrine and neuroendocrine cells suggests that these structures may be universal to secretory cells, where exocytosis occurs (Fig. 4.9).

4.7. NUMBER OF SECRETORY VESICLES IN GROWTH HORMONE CELLS OF THE PITUITARY REMAINS UNCHANGED AFTER SECRETION

Physiological processes such as neurotransmission and the secretion of enzymes or hormones require fusion of membrane-bounded secretory vesicles at the cell plasma membrane and the consequent release of vesicular contents. It has been commonly accepted that exocytosis requires the total incorporation of secretory vesicle membrane into the cell plasma membrane for release of vesicular contents; however, studies in the last decade (Cho et al., 2002c–e; Jena et al., 2003;

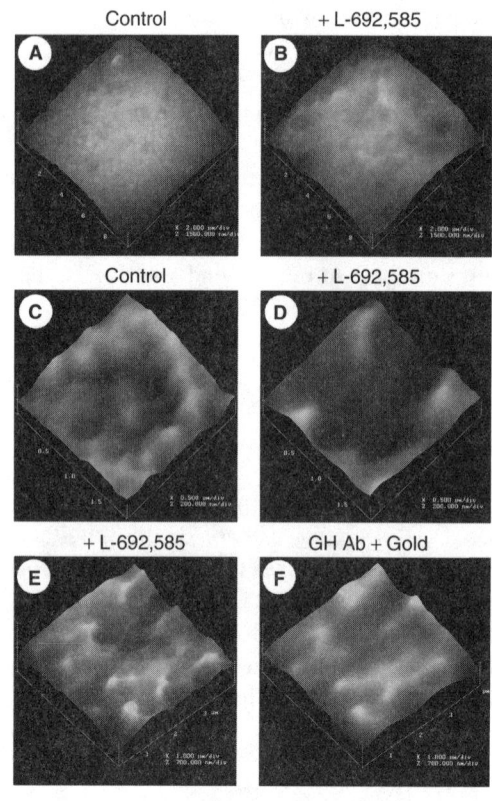

Figure 4.8. High-resolution atomic force microscopy (AFM) performed on resting and stimulated GH-secreting cells, after fixation. Note a large area of a cell surface in resting (**A**) and in a stimulated (**B**) cell. AFM micrographs of a "pit" with "depressions," in resting cell (**C**), again clearly demonstrating the enlargement of porosomes or fusion pores following stimulation of secretion (**D**). Exposure of "pits" in a stimulated cell (**E**) to 30-nm gold-tagged GH-antibody results in binding of released growth hormone at the site to 30-nm gold-tagged GH-antibody (**F**). Note the large amounts of gold-tagged antibody binding at these sites obscuring the porosome opening.

Depressions in resting cells measure 154 ± 4.5 nm. However, following exposure of GH cells to the secretagogue L-692,585, a marked increase in the size of "depressions" is demonstrated (Fig. 4.8). When stimulated live cells were exposed to 30-nm gold-tagged GH-antibody, gold particles were found to decorate "pit" and "depression" structures (Fig. 4.8). Aldehydes fixation of biological samples is known to result in decreased elasticity and increased hardness. Since force

TEM on stimulated and resting bovine chromaffin cells of the adrenal cortex showed no significant change in the number of peripheral dense-core vesicles after stimulation of secretion (Plattner et al., 1997). Similarly, combined studies using atomic force microscopy (AFM) and TEM clearly demonstrate no change in the total number of secretory vesicles following secretion in pancreatic acinar cells (Cho et al., 2002b). Fusion pores or depressions in pancreatic acinar- or growth hormone (GH)-secreting cells are cone-shaped structures at the plasma membrane, with a 100- to 150-nm-diameter opening (Cho et al., 2002c; Schneider et al., 1997). Membrane-bounded secretory vesicles range in diameter from 0.2 to 1.2 μm dock and fuse at depressions to release vesicular contents. After fusion of secretory vesicles at depressions, a 20–40% increase in depression diameter has been demonstrated. It has therefore been concluded that secretory vesicles "transiently" dock and fuse at depressions (Cho et al., 2002a; Jena, 2002; Jena et al., 1997). In contrast to accepted belief, if secretory vesicles were to completely incorporate at depressions, the fusion pore would distend much wider than observed. Additionally, if secretory vesicles were to completely fuse at the plasma membrane, there would be a decrease in total number of vesicles after secretion.

Although the fusion of secretory vesicles at the cell plasma membrane occurs transiently, complete incorporation of membrane at the cell plasma membrane takes place when cells need to incorporate signaling molecules like receptors, second messengers, and ion channels. Therefore, in GH-secreting cells, transient fusion is suggested and no change in total number of vesicles is hypothesized. The present study has been undertaken to test this hypothesis (Lee et al., 2004).

Control (resting) pituitary cells exposed to PBS contained more than twice as many filled vesicles than did the stimulated

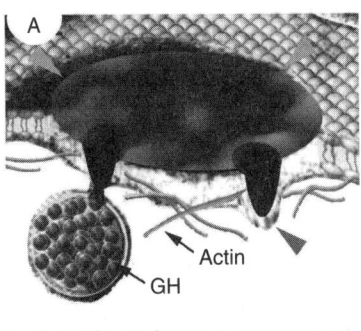

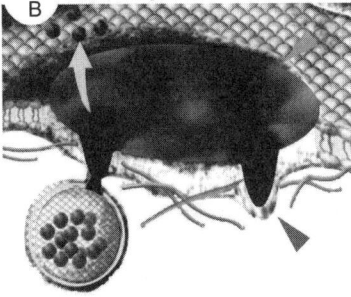

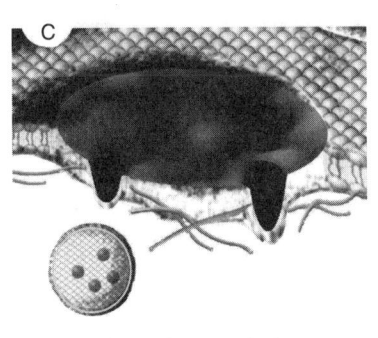

Figure 4.9. Schematic diagrams depict cross sectional views at the cell plasma membrane of a "pit" and with porosomes, and a vesicle containing growth hormone attached to one. Immediately following GH-secretagogue stimulation, the secretory vesicle containing hormone docks and fuses with the porosome (A) and then releases hormone (B) through the porosome. After hormone release from the vesicle (C) the porosome opening becomes smaller and the vesicle disassociates from it.

Jeremic et al., 2003; Lawson et al., 1975; Platter et al., 1997; Schneider et al., 1997) clearly demonstrate otherwise. Transmission electron microscopy (TEM) studies on mast cells demonstrate that, after stimulation of secretion, several intact as well as empty and partly empty secretory vesicles are present (Cho et al., 2002f). Quantitative cells exposed to the GH secretagogue, L-692,585 (4.9 ± 0.21 in control, 2.3 ± 0.23 in stimulated; $P < 0.001$). Stimulated cells contained nearly twice as many empty vesicles (0.6 ± 0.13 in control, 1.2 ± 0.16 in stimulated; $P < 0.05$) and 2.5 times more partly empty vesicles than did control cells (1.1 ± 0.08 in control, 2.6 ± 0.12 in stimulated; $P < 0.001$). However, there was no significant difference in total number of vesicles between control and stimulated pituitary cells. The remarkable increase in number of empty and partly empty vesicles in stimulated cells compared with control cells is evident. The intracellular distribution of empty or partially empty vesicles seemed random in stimulated cells. Immunogold labeling with GH antibody occurs only in electron dense GH vesicles in both control and stimulated cells. There was a complete absence of immunogold labeling with GH antibody in empty vesicles or other areas of the cytoplasm. Brief exposure of cultured porcine pituitary cells to the nonpeptidyl L-692,585, ghrelin, or human GH-releasing hormone (hGHRH) evoked a marked increase in intracellular calcium $[Ca^{2+}]_i$ concentration (Glavaski-Joksimovic et al., 2002, 2003). The result of stimulating live GH cells for 90 s with L-692,585 is that filled secretory vesicles empty very rapidly, most likely by docking at the plasma membrane. After releasing vesicular contents, the empty or partly empty vesicles return to the cytoplasmic compartments of the cell.

From our study on GH cells, 90-s exposure to 20 mM L-692,585 (GH secretagogue) for stimulation of GH secretion was ideal in inducing rapid and effective exocytosis. L-692,585 intravenously administered to pigs also causes an immediate peak release of GH > 80 ng/ml peripheral plasma within 10 min compared with pulsatile GH peaks of 6–10 ng/ml in placebo treated controls (Fig. 4.10). Intracellular signal transduction of GH-secretagogue undergoes a phosphoinositol-protein kinase C pathway that induces intracellular Ca^{2+} accumulation and depolarization, leading to exocytosis of GH-containing vesicles. The step increase in plasma membrane capacitance of GH cells, therefore, may result from a rapid transient fusion which is consistent with the AFM observation that shows the presence of "pits" which contain "depressions" or fusion pores (porosomes) at the GH cell plasma membrane that increase markedly in diameter after L-692,585 exposure.

The data reported here clearly show that the total number of secretory vesicles

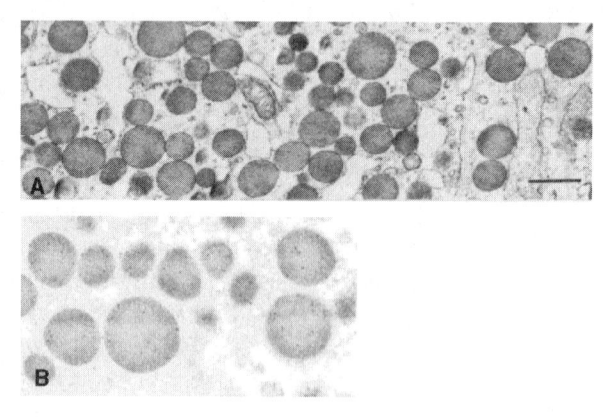

Figure 4.10. Electron micrograph of secretory vesicles in porcine pituitary gland. **(A)** Typical structure of tissue embedded in Epon-Araldite. **(B)** Secretory granules embedded in unicryl, sectioned and specifically labeled with anti-GH and a gold-conjugated second antibody. Note the immunogold labeled vesicles, demonstrating the presence of growth hormone within.

in porcine pituitary cells is not decreased after exocytosis. Such a decrease would be required if the secretory vesicle membrane were to fuse with the plasma membrane (as in the generally accepted model for secretion). The present data are consistent with a mechanism where vesicles transiently dock and fuse at the fusion pore (porosome) to release vesicular contents after stimulation of secretion. Based on these and other supporting findings, transient fusion of secretory vesicles at the porosome on the plasma membrane may be universal in the process of exocytosis.

ACKNOWLEDGMENT

This work was supported by research grant USDA NRI 2003-35206-12817 (L.L.A.) and the Iowa Agriculture and Home Economics Experiment Station, Ames, Iowa; by Hatch Act and State of Iowa funds; and by NIH grants DK56212 and NS39918 (B.P.J.).

REFERENCES

Anderson, L. L. (2004). Discovery of a new cellular structure: The porosome, elucidation of the molecular mechanism of secretion. *Cell Biol. Int.* **28**:3–5.

Anderson, L. L., Hard, D. L., Trenkle, A. H., and Cho, S.-J. (1999). Long term growth after hypophyseal stalk transection and hypophysectomy of beef calves. *Endocrinology* **140**:2405–2414.

Anderson, L. L., Jeftinija, S., and Scanes, C. G. (2004). Growth hormone secretion: Molecular and cellular mechanisms and *in vivo* approaches. *Exp. Biol. Med.* **229**:291–302.

Arvat, E., Gianotti, L., Ramunni, J, Grottoli, S., Brossa, P. C., Bertagna, A., Camanni, F., and Ghigo, E. (1995). Effect of galanin on basal and stimulated secretion of prolactin gonadotropins, thyrotropin, adrenocorticotropin and cortisol in humans. *Eur. J. Endocrinol.* **133**:300–304.

Bennet, P. A., Thomas, G. B., Howard, A. D., Feighner, S. D., Van der Ploeg, L. H. T.,
Smith, R. G., and Robinson, I. C. (1997). Hypothalamic growth hormone secretagogue-receptor (GHS-R) expression is regulated by growth hormone in the rat. *Endocrinology* **138**:4522–4557.

Bowers, C. Y. (1999). Growth hormone-releasing peptides. In: Handbook of Physiology, Section 7: *The Endocrine System*, Vol. V: *Hormonal Control of Growth* (J. L. Kostyo and H. M. Goodman, eds.), pp. 187–219, American Physiological Society, Oxford University Press, New York.

Broberger, C., DeLecea, L., Sutcliffe, J. C., and Horfelt, T. (1998). Hypocretin/orexin- and melanin-concentrating hormone-expressing cells from distinct populations in the rodent lateral hypothalamus relationship to the neuropeptide Y and agouti gene-related protein systems. *J. Comp. Neurol.* **402**:406–474.

Brown, J. C., Cook, M. A., and Dryburgh, T. (1973). Motilin, a gastric motor activity stimulating polypeptide: The complete amino acid sequence. *Can. J. Biochem.* **51**:533–537.

Chartrel, N., Dujardin, C., Leprince, J., Desrues, L., Tonon, M. C., Cellier, E., Cosette, P., Jouenne, T., Simonnet, G., and Vaudry, H. (2002). Isolation, characterization, and distribution of a novel neuropeptide, *Rana* RFamide (R-RFa), in the brain of the European green frog *Rana esculenta*. *J. Comp. Neurol.* **448**:111–127.

Chen, C. (2002). The effect of two-day treatment of primary cultured ovine somatotropes with GHRP-2 on membrane voltage-grated K^+ currents. *Endocrinology* **143**:2659–2663.

Chen, C., and Clarke, I. J. (1995). Modulation of Ca^{2+} influx in the ovine somatotroph by growth hormone-releasing factor. *Am. J. Physiol.* **268**:E204–E212.

Cho, S.-J., Abdus Sattar, A. K. M., Jeong, E. H., Satchi, M., Cho, J., Dash, S., Mayes, M. S., Stromer, M. H., and Jena B. P. (2002a). Aquaporin 1 regulates GTP induced rapid gating of water in secretory vesicles. *Proc. Natl. Acad. Sci. USA* **99**:4720–4724.

Cho, S.-J., Cho, J., and Jena, B. P. (2002b). The number of secretory vesicles remains unchanged following exocytosis. *Cell Biol. Int.* **26**:29–33.

Cho, S.-J., Jeftinija, K., Glavaski, A., Jeftinija, S., Jena, B. P., and Anderson, L. L. (2002c). Structure and dynamics of the fusion pores in live

GH-secreting cells revealed using atomic force microscopy. *Endocrinology* **143**:1144–1148.

Cho, S.-J., Kelly, M., Rognlien, K. T., Cho, J., Horber, J. K., and Jena, B. P. (2002d). SNAREs in opposing bilayers interact in a circular array to form conducting pores. *Biophys. J.* **83**:2522–2527.

Cho, S.-J., Quinn, A. S., Stromer, M. H., Dash, S., Cho, J., Taatjes, D. J., and Jena, B. P. (2002e). Structure and dynamics of the fusion pore in live cells. *Cell Biol. Int.* **26**:35–42.

Cho, S.-J., Wakade, A., Pappas, G. D., and Jena, B. P. (2002f). New structure involved in transient membrane fusion and exocytosis. *Ann. New York Acad. Sci.* **971**:254–256.

Denver, R. J., and Licht, P. (1990). Modulation of neuropeptide-stimulated pituitary hormone secretion in hatching turtles. *Gen. Comp. Endocrinol.* **77**:107–115.

Dickson, S. L., and Luckman, S. M. (1997). Induction of c-fos messenger ribonucleic acid in neuropeptide Y and growth hormone (GH)-releasing factor neurons in the rat arcuate nucleus following systemic injection of the GH secretagogue, GH-releasing-peptide-6. *Endocrinology* **138**:771–777.

Dickson, S. L., Ling, G., and Robinson, I. C. (1993). Systemic administration of growth hormone-releasing peptide activates hypothalamic arcuate neurons. *Neuroscience* **53**:303–306.

Dockray, G. J., Reeve, J. R., Jr, Shively, J., Gayton, R. J., and Barnard, C. S. (1983). A novel active pentapeptide from chicken brain identified by antibodies to FMRF-amide. *Nature* **305**:328–330.

Feighner, S. D., Tan, C. P., McKee, K. K., Palyha, O C., Hreniuk, D. L., Pong, S. S., Austin, C. P., Figueroa, D., MacNeil, D., Cascieri, M. A., Nargund, R., Bakshi, R., Abramovitz, M., Stocco, R., Kargman, S., O'Neill, G., Van Der Ploeg, L. H. D., Evans, J., Patchett, A. A., Smith, R. G., and Howard, A. D. (1999). Receptor for motilin identified in the human gastrointestinal system. *Science* **284**:2184–2188.

Ford, J. J., and Anderson, L. L. (1967). Growth in immature hypophysectomized pigs. *J. Endocrinol.* **37**:347–348.

Frohman, L. A., and Kineman, R. D. (1999). Growth hormone-releasing hormone: Discovery, regulation, and actions. In: Handbook of Physiology, Section 7: *The Endocrine System*, Vol. V: *Hormonal Control of Growth* (J. L. Kostyo and H. M. Goodman, eds.), pp. 187–219, American Physiological Society, Oxford University Press, New York.

Fujimoto, M., Takeshita, K., Wang, X., Takabatake, I., Fujisawa, Y., Teranishi, H., Ohtani, M., Muneoka, Y., and Ohta, S. (1998). Isolation and characterization of a novel bioactive peptide, *Carassius* Rfamide (C-Rfa), from the brain of the Japanese crucian carp. *Biochem. Biophys. Res. Commun.* **242**:436–440.

Giustina, A., Monfanti, C., Licini, M., Stefana, B., Ragni, G., and Turano, A. (1996). Effect of galanin on growth hormone (GH) response to thyrotropin releasing hormone of rat pituitary GH-secreting adenomatous cells (GH1) in culture. *Life Sci.* **58**:83–90.

Giustina, A., Ragni, G., Bollati, A., Cozzi, R., Licini, M., Poiesi, C., Turazzi, S., and Monfanti, C. (1997). Inhibitory effects of galanin on growth hormone (GH) release in cultured GH-secreting adenoma cells: comparative study with octreotide, GH-releasing hormone, and thyrotropin-releasing hormone. *Metabolism* **46**:425–430.

Glavaski-Joksimovic, A., Jeftinija, K., Jeremic, A., Anderson, L. L., and Jeftinija, S. (2002). Mechanism of action of the growth hormone secretagogue, L-692,585, on isolated porcine somatotropes. *J. Endocrinol.* **175**:625–636.

Glavaski-Joksimovic, A., Jeftinija, K., Scanes, C. G., Anderson, L. L., and Jeftinija, S. (2003). Stimulatory effect of ghrelin on isolated porcine somatotropes. *Neuroendocrinology* **77**:366–378.

Guan, X. M., Yum, H., Palyha, O. C., McKee, K. K., Feighner, S. D., Sirinath-Singhji, D. J., Smith, R. G., Van der Ploeg, L. H. T., and Howard, A. D. (1997). Distribution of mRNA encoding the growth hormone secretagogue receptor in brain and peripheral tissues. *Brain Res. Mol. Brain Res.* **48**:23–29.

Hewson, A. K., and Dickson, S. L. (2000). Systemic administration of ghrelin induces *Fos* and Egr-1 proteins in the hypothalamic arcuate nucleus of fasted and fed rats. *J. Neuroendocrinol.* **12**:1047–1049.

Horvath, T. L., Diano, S., and van Den Pol, A. N. (1999). Synaptic interaction between hypocretin (orexin) and neuropeptide Y cells in the rodent and primate hypothalamus: A novel circuit implicated in metabolic and endocrine regulations. *J. Neurosci.* **19**:1072–1087.

Howard, A. D., Feighner, S. D., Cully, D. F., Arena, J. P., Liverator, P. A., Rosenblum, C. I., Hamelin, M., Hreniuk, D. L., Palyha, O. C., Anderson, J., Paress, P. S., Diax, C, Dhou, M, Liu, K. K., McKee, K. K., Pong, S. S., Chaung, L. Y., Elverecht, A., Dashkevicz, M., Heavens, R., Rigby, M., Sirinath-Singhji, D. J., Dean, D. C., Melillo, D. G., Patchett, A. A., Nargund, R., Griffion, P. R., DeMaritino, J. A., Gupta, S. K., Schaeffer, J. M., Smith, R. G., and Van der Ploeg, L. H. T. (1996). A receptor in pituitary and hypothalamus that functions in growth hormone release. *Science* **273**:974–977.

Huang, Z., De Clercq, P., Depoortere, I., and Peeters, T. L. (1998). Isolation and sequence of cDNA encoding the motilin precursor from monkey intestine. Demonstration of the motilin precursor in the monkey brain. *FEBS Lett.* **435**:149–152.

Iqbal, J., Pompolo, S., Considine, R. V., and Clarke, I. J. (2000). Localization of leptin receptor-like immunoreactivity in the corticotropes, somatotropes, and gonadotropes in the ovine anterior pituitary. *Endocrinology* **141**:1515–1520.

Jena, B. P. (2002). Fusion pore in live cells. *News Physiol. Sci.* **17**:219–222.

Jena, B. P., Schneider, S. W., Geibel, J. P., Webster, P., Oberleithner, H., and Sritharan, K. C. (1997). Gi regulation of secretory vesicle swelling examined by atomic force microscopy. *Proc. Natl. Acad. Sci. USA* **94**:13317–13322.

Jena, B. P., Cho, S.-J., Jeremic, A., Stromer, M. H., and Abu-Hamdah, R. (2003). Structure and composition of the fusion pore. *Biophys. J.* **84**:1337–1343.

Jeremic, A., Kelly, M., Cho, S.-J., Stromer, M. H., and Jena, B. P. (2003). Reconstituted fusion pore. *Biophys. J.* **85**:2035–2043.

Klindt, J., and Stone, R. T. (1984). Porcine growth hormone and prolactin: Concentrations in the fetus and secretory patterns in the growing pig. *Growth* **48**:1–15.

Klindt, J., Ford, J. J., Beradinelli, G., and Anderson, L. L. (1983). Growth hormone secretion after hypophyseal stalk transection in pigs. *Proc. Soc. Exp. Biol. Med.* **172**:503–513.

Koda, A., Ukena, K., Teranishi, H., Ohta, S., Yamamoto, K., Kikuyama, S., and Tsutsui, K. (2002). A novel amphibian hypothalamic neuropeptide: Isolation, localization, and biological activity. *Endocrinology* **143**:411–419.

Kojima, M., Hosoda, H., Date, Y., Nakazato, M., Matsuo, H., and Kangawa, K. (1999). Ghrelin is a growth-hormone-releasing peptide from stomach. *Nature* **402**:656–660.

Lawrence, C. B., Snape, A. C., Baudoin, F. M. H., and Luckman, S. M. (2002). Acute central ghrelin and GH secretagogues induce feeding and activate brain and appetite centers. *Endocrinology* **143**:155–162.

Lawson, D., Fewtrell, C., Gomperts, B., and Raff, M. C. (1975). Anti-immunoglobulin-induced histamine secretion by rat peritoneal mast cells studied by immunoferritin electron microscopy. *J. Exp. Med.* **142**:391–401.

Lee, H. M., Wahng, G., Englander, E. W., Kojima, M., and Greeley, G. H., Jr. (2002). Ghrelin, a new gastrointestinal endocrine peptide that stimulates insulin secretion: Enteric distribution, ontogeny, influence of endocrine, and dietary manipulations. *Endocrinology* **143**: 185–190.

Lee, J.-S., Mayes, M. S., Stromer, M. H., Scanes, C. G., Jeftinija, S., and Anderson, L. L. (2004). Number of secretory vesicles in growth hormone cells of the pituitary remains unchanged after secretion. *Exp. Biol. Med.* **229**:632–639.

Leshin, L. S., Barb, C. R., Kiser, T. E., Rampacek, G. B., and Kraeling, R. R. (1994). Growth hormone-releasing hormone and somatostatin neurons within the porcine and bovine hypothalamus. *Neuroendocrinology* **59**:251–264.

Lin, J. Barb, C. R., Matteri, R. L., Kraeling, R. R., Chen, S., Meinersmann, R. J., and Rampacek, G. B. (2000). Long from leptin receptor mRNA expression in the brain, pituitary, and other tissues in the pig. *Domest. Anim. Endocrinol.* **19**:53–61.

Lin, J. Barb, C. R., Matteri, R. L., Kraeling, R. R., and Rampacek, G. B. (2003). Growth hormone releasing factor decreased long form leptin receptor expression in porcine anterior pituitary cells. *Domest. Anim. Endocrinol.* **24**:95–101.

Masuda, Y., Tanaka, T., Inomata, N., Ohnuma, N., Tanaka, S., Itoh, Z., Hosoda, H., Kojima, M., and Kangawa, K. (2000). Ghrelin stimulates gastric acid secretion and motility in rats. *Biochem. Biophys. Res. Commun.* **276**:905–908.

Mitchell, V., Bouret, S., Meauvillian, J. C., Schilling, A., Perret, M., Kordan, C., and Epelbaum, J. (2001). Comparative distribution of mRNA encoding the growth hormone secretagogue-receptor (GHS-R) in *Microcebus murinus* (primate, lemurian) and rat forebrain and pituitary. *J. Comp. Neurol.* **429**:469–489.

Molina, J. R., Klindt, J., Ford, J. J., and Anderson, L. L. (1986). Growth hormone and prolactin secretion after hypothalamic deafferentation in pigs. *Proc. Soc. Exp. Biol. Med.* **183**:163–168.

Nakazato, M., Murakami, N., Date, Y., Kojima, M., Matsuo, H., Kangawa, K., and Matsukura, S. (2001). A role for ghrelin in the central regulation of feeding. *Nature* **409**:194–198.

Owens, P. C., Campbell, R. G., Johnson, R. G., King, R., and Ballard, F. J. (1991). Developmental change in growth hormone, insulin-like growth factors (IGF-I and IGF-II) and IGF-binding proteins in plasma of young growing pigs. *J. Endocrinol.* **128**:439–447.

Peng, C., Chang, J. P., Yu, K. L., Wong, A. O., Van Goor, R., Peter, R. E., and Rivier, J. E. (1993a). Neuropeptide-Y stimulates growth hormone and gonadotropin-II secretion in the goldfish pituitary: Involvement of both presynaptic and pituitary cell actions. *Endocrinology* **132**:1820–1829.

Peng, C., Humphries, S., Peter, R. E., Rivier, J. E., Blomqivist, A. G., and Larhammar, D. (1993b). Actions of goldfish neuropeptide Y on the secretion of growth hormone and gonadotropin-II in female goldfish. *Gen. Comp. Endocrinol.* **90**:306–317.

Plattner, H., Artalejo, A. R., and Neher, E. (1997). Ultrastructural organization of bovine chromaffin cell cortex—analysis by cryofixation and morphometry of aspects pertinent to exocytosis. *J. Cell Biol.* **139**:1709–1717.

Plouzek, C. A., Molina, J. R., Hard, D. L., Vale, W., Rivier, J., Trenkle, A., and Anderson, L. L. (1988). Effect of growth hormone-releasing factor and somatostatin on growth hormone secretion in hypophysial stalk transected beef calves. *Proc. Soc. Exp. Biol. Med.* **189**:158–167.

Radecki, S. V., Deaver, D. R., and Scanes, C. G. (1994). Triiodothyronine reduces growth hormone secretion and pituitary growth hormone mRNA in the chicken, *in vivo* and *in vitro*. *Proc. Soc. Exp. Biol. Med.* **205**:340–346.

Reed, D. K, Korytko, A. I., Hipkin, R. W., Wehernberg, W. B, Schonbrunn, A., and Cuttler, L. (1999). Pituitary somatostatin receptor (sst)1-5 expression during rat development: age-dependent expression of sst2. *Endocrinology* **140**:4739–4744.

Rousseau, K., Le Belle, N, Marchelidon, J., and Dufour, S. (1999). Evidence that corticotropin-releasing hormone acts as growth hormone-releasing factor in a primitive teleost, the European eel (*Anguilla anuilla*). *J. Neuroendocrinol.* **11**:385–392.

Sherwood, N. M, Krueckl, S. L., and McRory, J. E. (2000). The origin and function of the pituitary adenylate cyclase-activation polypeptide (PACAP)/glucagons superfamily. *Endocr. Rev.* **21**:619–670.

Shimon, I., Yan, X., Magoffin, D. A, Friedman, T. C., and Melmed, S. (1998). Intact leptin receptor is selectively expressed in human fetal pituitary and pituitary adenomas and signals human fetal pituitary growth hormone secretion. *J. Clin. Endocrinol. Metab.* **83**:4059–4064.

Schneider, S. W., Sritharan, K. C., Geibel, J. P., Oberleithner, H., and Jena, B. P. (1997). Surface dynamics in living acinar cells imaged by atomic force microscopy: Identification of plasma membrane structures involved in exocytosis. *Proc. Natl. Acad. Sci. USA* **94**:316–321.

Shintani, M., Ogawa, Y., Ebihara, K., Aiszawa-Abe, M., Miyanaga, F., Takaya, K, Hayashi, T., Inove, G., Hosoda, K, Kojima, M., Kangawa, K., and Nakao, K. (2001). Ghrelin, an endogenous growth hormone secretagogue, is a novel orexigenic peptide that antagonizes leptin action through activation of hypothalamic neuropeptide Y/Y1 receptor pathway. *Diabetes* **50**: 227–232.

Silverman, B. L., Kaplan, S. L., Grumbach, M. M., and Miller, W. L. (1988). Hormonal regulation of growth hormone secretion and messenger ribonucleic acid accumulation in cultured bovine pituitary cells. *Endocrinology* **122**:1236–1241.

Sun, H. S., Anderson, L. L., Yu, T.-P., Kim, K.-S., Klindt, J., and Tuggle, C. K. (2002). Neonatal Meishan pigs show POU1F1 genotype effects on plasma GH and PRL concentration. *Anim. Reprod. Sci.* **69**:223–237.

Tannenbaum, G. S., and Epelbaum, J. (1999). Somatostatin. In: *Handbook of Physiology, Section 7: The Endocrine System*, Vol. V: *Hormonal Control of Growth* (J. L. Kostyo and H. M. Goodman, eds.), pp. 221–265. American Physiological Society, Oxford University Press, New York.

Tschöp, M., Smiley, D. L., and Heiman, M. L. (2000). Ghrelin induces adiposity in rodents. *Nature* **407**:903–913.

Ukena, K., Iwakoshi, E., Minakata, H., and Tsutsui, K. (2002). A novel rat hypothalamic Rfamide-related peptide identified by immunoaffinity chromatography and mass spectrometry. *FEBS Lett.* **512**:255–258.

Vasilatos-Younken, R., Tsao, P. H., Foster, D. N., Smiley, D. L., Bryant, H., and Heiman, M. L. (1992). Restoration of juvenile baseline growth hormone secretion with preservation of the ultradian growth hormone rhythm by continuous delivery of growth hormone-releasing factor. *J. Endocrinol.* **135**:371–382.

Yanaihara, N., Yanaihara, C., Mochizuki, T., Iguchi, K, and Hoshino, M. (1990). Chemical synthesis, radioimmunoassay, and distribution of immuno-reactivity of motilin. In: *Motilin* (Z. Itoh, ed.), pp. 31–46. Academic Press, Tokyo.

Yokote, R., Sato, M., Matsubara, S., Ohye, H., Niimi, M., Murao, K., and Takahara, J. (1998). Molecular cloning and gene expression of growth hormone-releasing peptide receptor in rat tissue. *Peptides* **19**:15–20.

CHAPTER 5

PROPERTIES OF MICROBIAL CELL SURFACES EXAMINED BY ATOMIC FORCE MICROSCOPY

YVES F. DUFRÊNE

Unité de chimie des interfaces, Université catholique de Louvain, Croix du Sud 2/18, B-1348 Louvain-la-Neuve, Belgium

5.1. INTRODUCTION

During the past 40 years, the importance of the microbial cell surface in biology, medicine, industry, and ecology has been increasingly recognized. Because they constitute the frontier between the cells and their environment, microbial cell walls play several key functions: supporting the internal turgor pressure of the cell, protecting the cytoplasm from the outer environment, imparting shape to the organism, and acting as a molecular sieve and controlling interfacial interactions (i.e., molecular recognition, cell adhesion, and aggregation). These biointerfacial processes have major consequences, which can be either beneficial, such as in biotechnology (wastewater treatment, bioremediation, immobilized cells in reactors), or detrimental, such as in industrial systems (biofouling, contamination) and in medicine (interactions of pathogens with animal host tissues; accumulation on implants and prosthetic devices) (Savage and Fletcher, 1985). Hence, there is considerable interest in improving our current understanding of the functions of microbial cell surfaces.

The functions of the cell surface are directly related to structure. As opposed to animal cells, microbial cells are surrounded by thick, mechanically strong cell walls (Hancock, 1991). The wall mechanical strength in eubacteria is provided by a layer of peptidoglycan consisting of glycan chains of alternating N-acetylglucosamine and N-acetylmuramic acid residues, which are cross-linked by short peptide chains. Archaebacteria possess stress-bearing wall components which may have different forms: peptidoglycan-like polymers, proteinaceous sheats, crystalline glycoprotein arrays (S-layers). In gram-positive bacteria, the peptidoglycan is bound to anionic polymers (e.g., teichoic acid) while in gram-negative bacteria it is overlayed by an outer membrane—that is, an asymmetrical bilayer of phospholipids and lipopolysaccharides containing membrane proteins (e.g., porins). For many bacterial strains, these cell wall constituents are covered by additional surface layers in the form of polysaccharide capsules, surface appendages (fimbriae, pili, fibrils, flagella), or regular crystalline arrays of (glyco)proteins, referred to as S-layers. Strong cell walls are formed in yeasts and filamentous fungi by the aggregation of polysaccharide polymers. In yeasts, these are made of a microfibrillar array of $\beta 1-3$ glucan, overlaid by $\beta 1-6$ glucan and mannoproteins. The walls

---

Force Microscopy: Applications in Biology and Medicine, edited by Bhanu P. Jena and J.K. Heinrich Hörber.
Copyright © 2006 John Wiley & Sons, Inc.

of fungal hyphae consist of microfibrillar polysaccharides, chitin or cellulose, covered by layers of proteins and glucans. For fungal spores, the wall is often covered by an outer layer of regularly arranged proteinaceous rodlets having a periodicity of about 10 nm.

Understanding the functions of microbial cell surfaces requires determination of their structural and physical properties. Various characterization methods have been developed toward this goal (Beveridge and Graham, 1991; Mozes et al., 1991): separation and chemical analysis of wall constituents, serological procedures, binding studies (dyes, antibodies and lectins), selective degradation by enzymes, cell wall mutants, modifications by antibiotics, electron microscopy techniques combined with freeze-fracture and surface replica or negative staining, x-ray photoelectron spectroscopy, infrared spectroscopy, and contact-angle and electrophoretic mobility measurements. These methods often require cell manipulation prior to examination (extraction, drying, staining) and in many cases provide averaged information obtained on a large ensemble of cells. This indicates that nondestructive, high-resolution tools are needed to probe the microbial cell surface.

With atomic force microscopy (AFM) (Binnig et al., 1986), unprecedented possibilities are now offered to microscopists to explore microbial surfaces: imaging the surface ultrastructure of cell surface layers under physiological conditions and with subnanometer resolution, monitoring conformational changes of individual membrane proteins, examining bacterial biofilms, providing high-resolution images of the surface structure of living cells, and following dynamic cellular processes. But AFM is actually much more than a microscope in that it also enables forces to be probed quantitatively. With an AFM, one can measure surface forces and produce spatially resolved maps of the physical properties of the cell surface, including surface hydrophobicity, surface charges, adhesion, and elasticity. It is also possible to measure the elasticity of single surface molecules in relation with function. These measurements provide new insight into the molecular and cellular structure–function relationships of microbial surfaces (molecular recognition, protein folding, conformational changes, cell adhesion). The aim of this contribution is to illustrate these unique capabilities. The chapter is primarily intended for biologists from various horizons (microbiology, biophysics, biomedicine) and does not require advanced understanding of physical science principles.

5.2. IMAGING SURFACE STRUCTURE

This section focuses on the use of AFM for imaging the structure of microbial cell surfaces. A brief description of appropriate sample preparation procedures and operating conditions is provided, followed by a discussion of imaging applications dealing with reconstituted cell surface layers and cells.

5.2.1. Methods

5.2.1.1. Substrates. An important requirement for AFM is that the sample must be attached to a solid substrate. For microbial layers, mica, glass, and silicon oxide are the most widely used substrates because they are flat and are easy to handle, prepare, and clean (Müller et al., 1997b; Colton et al., 1998). Mica is easily cleaved—for instance, using an adhesive tape—to produce clean surfaces that are flat over large areas. Glass coverslips, however, are always coated with organic contaminants and particles that should be removed before use. This can be achieved by washing in concentrated acidic solution followed by ultrasonication in several water solutions. In some cases, best results may be obtained using hydrophobic substrates, such as highly oriented pyrolytic graphite (HOPG). For some applications, substrates with well-defined surface chemistries can be designed;

a widely used approach is the functionalization of gold-coated surfaces with self-assembled monolayers (SAMs) of organic alkanethiols. Finally, as we shall see below, isoporous polymer membranes are valuable substrates for immobilizing cells.

5.2.1.2. Immobilization. Three immobilization strategies have been developed for AFM of reconstituted surface layers (Pum et al., 1993; Müller et al., 1997a; Colton et al., 1998): (i) physical adsorption from aqueous solution in the presence of appropriate electrolytes which has been employed successfully with purple membranes, OmpF porin and aquaporin crystals; (ii) recrystallization on a lipid monolayer, a procedure initially developed for *Bacillus* S-layers; (iii) covalent immobilization, which has been used for the hexagonally packed intermediate (HPI) layer.

Immobilizing native, living cells is often a challenge. Microbial cells have a well-defined shape and have no tendency to spread over substrates. As a result, the contact area between the cell and substrate is very small, often leading to cell detachment by the scanning probe. A second problem is the vertical motion of the sample (or probe), typically limited to a few micrometers, which makes it difficult to image the surface of large objects such as microbial cells. To circumvent these problems, several approaches have been proposed, including pretreatments such as air-drying and chemical fixation to promote cell attachment. These treatments may cause significant rearrangement/denaturation of the surface molecules, thereby compromising the biological significance of the results. An attractive alternative is to perform cell immobilization by mechanical trapping in porous membranes (Kasas and Ikai, 1995; Dufrêne et al., 1999). This approach offers two advantages; that is, it is fairly simple and straightforward and it does not involve drying, coating, or chemical fixation. However, an important limitation is that it essentially works for spherical cells (some bacteria, fungal spores, yeasts) but not for rod-like cells. Figure 5.1 illustrates how this approach can be used to immobilize and image individual cells under aqueous conditions. The two height images clearly show major differences depending on the cell type: While the yeast *Saccharomyces cerevisiae* (Fig. 5.1A) showed a very smooth surface, the spore from the fungus *Rhizopus oryzae* displayed a rough surface morphology with large fiber-like protrusions (Fig. 5.1B). This peculiar surface morphology may play a role in determining the biological function of fungal spores. Note that the cells are surrounded by artifactual structures resulting from the contact between the AFM probe and the edges of the cell and of the pore.

5.2.1.3. Imaging Conditions. In contact mode, the most widely used imaging mode, sample topography can be measured in two ways: (1) the constant-height mode, in which the cantilever deflection is recorded while the sample is scanned horizontally, and (2) the constant-deflection mode, in which the sample height is adjusted to keep the deflection of the cantilever constant using a feedback loop. The feedback output is used to display a true "height image" (Fig. 5.1), while the error signal can also be used to generate a so-called "deflection image." Both height and deflection images are often useful when dealing with microbial cells. Height images provide quantitative height measurements, thus allowing an accurate measure of cell dimensions and of the surface roughness. Deflection images do not reflect true height variations, but because they are more sensitive to fine surface details they are useful for revealing the surface ultrastructure of curved, rough samples such as microbial cells.

Selection of an appropriate imaging environment is critical for high-resolution imaging because it controls the force acting between probe and sample. When imaging in air, the layer of water and contaminant molecules on both probe and sample

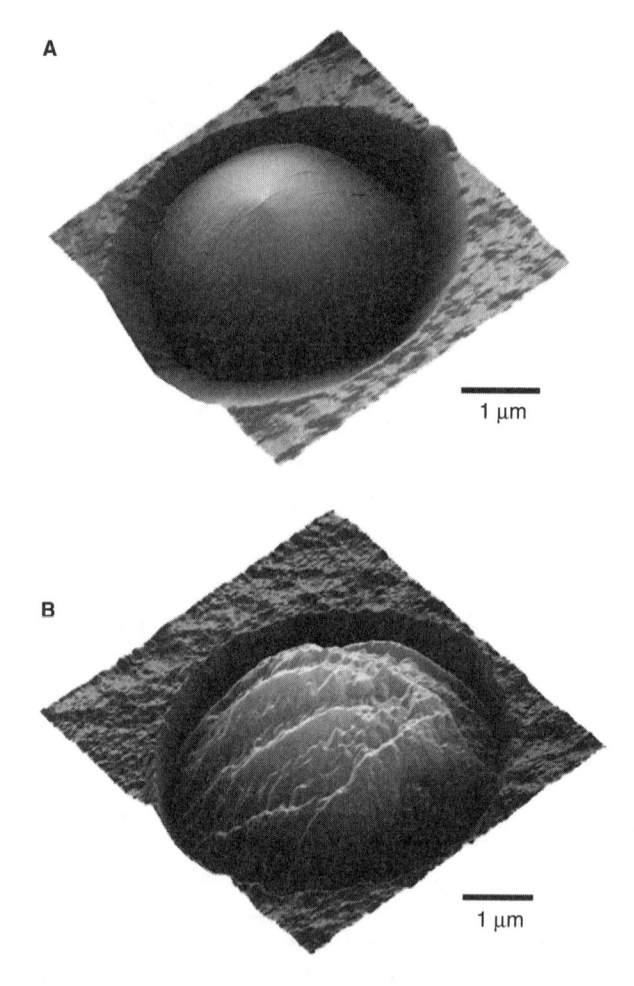

Figure 5.1. Immobilization of single microbial cells for AFM imaging under physiological conditions. 3-D AFM height images, under aqueous solution, showing a yeast cell (*Saccharomyces cerevisiae*; **A**) and a fungal spore (*Rhizopus oryzae*; **B**) trapped in a porous polymer membrane. The two cells showed very different surface morphologies: the yeast surface was very smooth, while the spore displayed a rough surface morphology with large fiber-like protrusions.

forms a meniscus pulling the two together. The resulting strong attractive force makes high-resolution imaging difficult and sometime causes sample damage. This can be eliminated by performing the imaging in aqueous solution. By selecting appropriate buffer conditions, it is generally possible to maintain an applied force in the range of 0.1–0.5 nN. For purple membrane and OmpF porin surfaces (Müller et al., 1999b), it was possible to improve the spatial resolution by changing the pH and electrolyte concentration, and the best results were obtained with 150 mM KCl or NaCl (in 10 mM Tris-HCl, pH 7.6).

For imaging applications, excellent results are obtained using cantilevers made of silicon (Si) or silicon nitride (Si_3N_4). Although the probe radii are in the range of 10–50 nm, topographs of flat cell surface layers have been acquired routinely with a resolution of 1 nm. Nanoscale protrusions extending from the probe are thought to be responsible for this high resolution.

5.2.2. Imaging Cell Surface Layers

Microbial cell surface layers made of two-dimensional (2-D) protein crystals, including *Bacillus* S-layers, hexagonally packed intermediate (HPI) layers, purple membranes, and porin OmpF and aquaporin crystals, have proved to be excellent systems for high-resolution AFM imaging. What do we learn from these images? Structural information can be directly obtained to a resolution of 0.5–1 nm, under physiological conditions, which makes AFM a complementary tool to x-ray and electron crystallography. In addition, the technique can be used for modifying single-membrane proteins or monitoring conformational changes that are related to functions.

5.2.2.1. Bacillus S-Layers.
Reconstituted S-layers of *Bacillus coagulans* E38-66 and *Bacillus sphaericus* CCM2177 imaged by AFM in buffer solution showed oblique and square lattices, respectively, with lattice parameters in good agreement with transmission electron microscopy data (Ohnesorge et al., 1992). AFM was used to investigate the recrystallization of S-layer proteins of a *Bacillus stearothermophilus* strain on untreated and modified silicon surfaces (Pum and Sleytr, 1995). The oblique lattice of the S-layer was visualized to a lateral resolution of about 1.5 nm. Recrystallization was shown to occur only at hydrophobic surfaces, the S-layer being always oriented with its more hydrophobic outer face against the interface. Both S-layers of *B. coagulans* E38-66 and *B. sphaericus* CCM2177 were recrystallized on silanized silicon substrates and imaged under liquid with nanometer lateral resolution (Pum and Sleytr, 1996).

5.2.2.2. HPI Layers.
Among the various cell surface layers, the hexagonally packed intermediate (HPI) layer of *Deinococcus radiodurans* has been the subject of considerable research efforts. AFM was used to image the surface of the HPI layer in buffer solution with a lateral resolution of 1 nm (Karrasch et al., 1994). The images agreed within a few Angströms with data obtained by electron microscopy. More recent studies have revealed that the inner surface of the HPI layers has a protruding core with a central pore exhibiting two conformations, one with and the other without a central plug. Individual pores were observed to switch from one state to the other, indicating that conformational changes can be detected at subnanometer resolution (Müller et al., 1996, 1997b). Remarkably, the AFM could also be used as a nanotool to induce structural alterations in the HPI layer (Müller et al., 1999a). Figure 5.2A shows a high resolution image of the inner surface of the HPI layer, consisting of hexameric cores with central channels that are connected by slender arms. As shown in Figure 5.2B, retraction of the AFM stylus resulted in the zipping out of an entire hexameric complex. These data illustrate the power of AFM for imaging cell surface layers at high resolution and for modifying their individual constituents, providing new opportunities for studying the mechanical stability of supramolecular structures in relation with function.

5.2.2.3. Purple Membranes.
Purple membranes from the archeon *Halobacterium* represent another class of well-studied microbial layers. These membranes contain bacteriorhodopsin, a light-driven proton pump, in the form of highly ordered 2-D lattices. In 1990, the hexagonal symmetry of purple membranes adsorbed on mica was already imaged in aqueous conditions, the lattice constant being in agreement with electron diffraction data (Butt et al., 1990). Later, conditions were established to reproducibly acquire topographs of purple membranes at subnanometer resolution (Müller et al., 1995b), and force-induced conformational changes of bacteriorhodopsin were directly visualized by AFM (Müller et al., 1995a, 1997b). Upon lowering the imaging forces from 300 pN to 100 pN, donut-shaped trimers reversibly transformed into

structures exhibiting three protrusions. The standard deviation of the sample height measured for many independent protein units was shown to provide quantitative information on the flexibility of the protein domains (Müller et al., 1998). Tapping mode AFM (TMAFM) is an interesting alternative to contact mode because it has been shown to faithfully record high-resolution images of purple membranes and to have sufficient sensitivity to contour individual peptide loops without detectable deformations (Möller et al., 1999). Structural changes of purple membrane during photobleaching in the presence of hydroxylamine were monitored using AFM (Möller et al., 2000), providing novel insights into factors triggering purple membrane formation and structure. Crystals of halorhodopsin from the overexpressing strain *Halobacterium salinarum* D2 showed an orthogonal structure and the orientation of the molecules showed p42(1)2 symmetry (Persike et al., 2001). The crystal surface was found to display different structures depending on the imaging force used, indicating that some parts of the molecule were more rigid but others more compressible. Helix-connecting loops in single molecules of halorhodopsin were assigned. The images indicated that the large extracellular loop covered the whole molecule and was very flexible.

5.2.2.4. Porin Crystals. Gram-negative bacteria are protected by an outer membrane in which trimeric channels, the porins, facilitate the passage of small solutes. As shown in Fig. 5.3 (left panels), topographs of porin OmpF crystals of *Escherichia coli* reconstituted in the presence of lipids were recorded in aqueous solution to a lateral resolution of 1 nm and a vertical resolution of 0.1 nm (Schabert et al., 1995). To assess the accuracy of the AFM topographs, averaged surface contours were directly compared with the protein structure from x-ray analysis (Fig. 5.3, right panels), with excellent agreement being obtained between the subtle features of the topographs and the models. More recently, voltage and pH-dependent conformational changes of the extracellular loops were demonstrated (Müller and Engel, 1999). To this end, porin membranes adsorbed to HOPG were imaged while applying a voltage with a closely apposed platinum wire. The observed

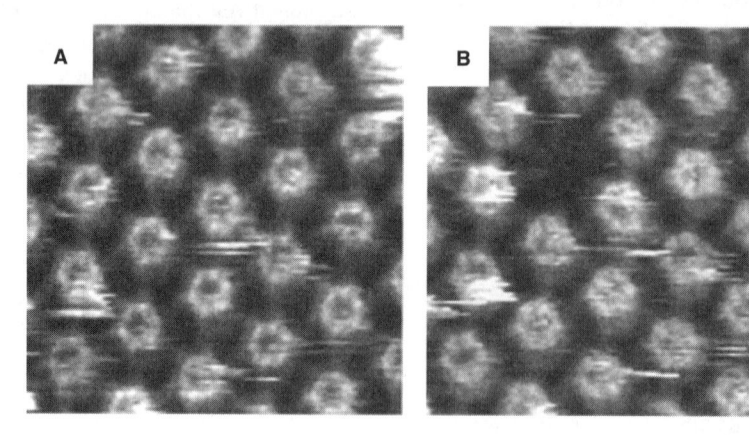

Figure 5.2. Imaging and manipulating the HPI bacterial surface layer at the molecular level (Müller et al., 1999a). (**A**) AFM topograph recorded in buffer solution for the inner surface of the HPI layer of *Deinococcus radiodurans*; the distance between the core centers is 18 nm. (**B**) The same inner surface area imaged after recording a force curve, showing that a molecular defect the size of a hexameric HPI protein complex had been created. (Used with permission from Prof. A. Engel, Universität Basel, Switzerland; copyright 1999, National Academy of Sciences, USA.)

conformational changes were correlated with the closure of the channel entrance, suggesting that this is a mechanism to protect the cells from drastic changes of the environment.

Aquaporins are ubiquitous membrane channels in bacteria, fungi, plants and animals. They are highly specific for water or small uncharged hydrophilic solutes and are involved in osmoregulation. For *E. coli* aquaporin Z, AFM images revealed 2-D crystals with p42(1)2 and p4 symmetry (Scheuring et al., 1999). Imaging both crystal types before and after cleavage of the N-termini allowed the cytoplasmic surface to be identified. Flexibility mapping and volume calculations identified the longest loop at the extracellular surface, this loop exhibiting a reversible force-induced conformational change.

5.2.3. Imaging Cells

While the potential of AFM for probing living animal cells was recognized quite early (Häberle et al., 1991), attempts to perform such measurements on microbial cells have been more limited, due in part to the difficulties associated with the sample preparation. However, recent studies have demonstrated that AFM can provide a wealth of information on the microbial surface.

5.2.3.1. Microbial Biofilms. The adhesion of microorganisms to solid surfaces is a widespread phenomenon that plays a crucial role in the natural environment (self-purification of natural water, symbiotic interactions), in industrial processes (biocorrosion, biofouling of industrial equipments, bioremediation, cells in bioreactors),

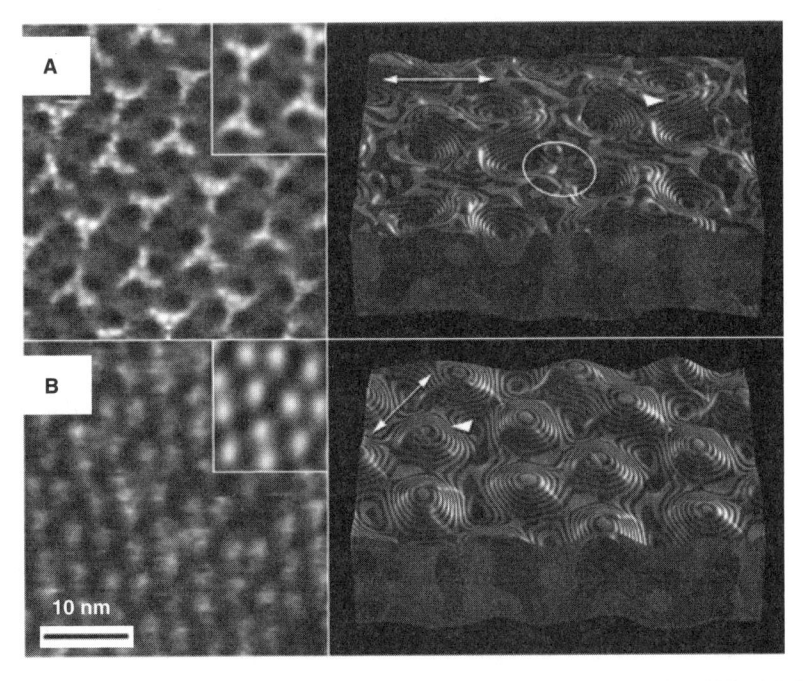

Figure 5.3. Visualizing reconstituted porin surfaces at high resolution (Schabert et al., 1995). AFM topographs of the periplasmic (**A**, left panel) and extracellular (**B**, left panel) surface of 2-D crystals of the channel forming protein porin OmpF from *Escherichia coli* reconstituted in the presence of lipids. **Insets**: averages calculated from 25 translationally aligned subframes. Comparison with the protein structure from x-ray analysis (right panels); zebra-like contours mark zones of identical altitude, with a height difference between contours of 0.1 nm. (Used with permission from Prof. A. Engel, Universität Basel, Switzerland; copyright 1995, American Association for the Advancement of Science.)

and in medicine (accumulation on biomaterials, infections). Despite their great significance, the mechanisms of cell adhesion and biofilm formation remain poorly understood at the molecular level. In recent years, AFM has helped to shed new light on these mechanisms.

The instrument has proven useful in biocorrosion studies. While imaging the topography of hydrated bacterial biofilms on a copper surface (Bremer et al., 1992), bacterial cells were shown to be associated with pits on the copper surface, supporting previous studies related to the pitting corrosion of copper. In order to elucidate the process of stainless steel corrosion, AFM was used to observe bacteria and bacterial exopolymers in their hydrated forms (Steele et al., 1994). AFM was shown to allow estimation of the width and height of bacterial cells, estimation of the thickness and width of exopolymeric capsule and bacterial flagella, and characterization of substrate roughness, including measurements of depth and diameter of individual corrosion pits (Beech et al., 1996). Corrosion, biofilm formation, and the adhesion of different, corrosion-enhancing microorganisms to different surfaces (iron, copper, pyrite) were studied in aqueous environment by AFM (Telegdi et al., 1998).

TMAFM in liquids was used to investigate the effect of iron coatings on the interactions of *Shewanella putrefaciens* with silica glass surfaces (Grantham and Dove, 1996). The results suggested that adhesion, motility, and iron surface chemistry are interrelated in environments where Fe-reducing microorganisms are present. Structures of sizes comparable to large proteins were resolved on *Pseudomonas putida* surfaces as well as flagella trapped within biofilms (Gunning et al., 1996). Biofilms formed by *Pseudomonas* sp. on hematite and goethite minerals were examined in the dried state using TMAFM (Forsythe et al., 1998).

Unsaturated biofilms of *Pseudomonas putida* (i.e., biofilms grown in humid air) were analyzed by AFM to determine surface morphology, roughness, and adhesion forces in the outer and basal cell layers of fresh and desiccated biofilms (Auerbach et al., 2000). In sharp contrast to the effects of drying on biofilms grown in fluid, drying caused little change in morphology, roughness, or adhesion forces in these unsaturated biofilms. The extracellular polymeric substances (EPS) formed "mesostructures" that were much larger than the discrete polymers of glycolipids and proteins that have been previously characterized on the outer surface of these gram-negative bacteria.

With the aim to gain insight into the supramolecular organization of bacterial EPS, AFM was used to probe, under aqueous conditions, the nanoscale morphology and molecular interactions of polystyrene substrates after adhesion of the gram-negative bacterium *Azospirillum brasilense* (van der Aa and Dufrêne, 2002). After cell adhesion under favorable conditions, topographic images revealed that the substrate surface was covered by a continuous layer of adsorbed substances, $\sim$2 nm thick, from which supramolecular aggregates were protruding. Correlations were found between the AFM data and the cell adhesion behavior, providing evidence for the involvement of EPS in the adhesion process.

5.2.3.2. Cell Surface Nanostructures.

Immobilization of single microbial cells—for example, using porous membranes (Fig. 5.1)—makes it possible to record high-resolution images of the surface. As an example, Fig. 5.4C shows a high-resolution image under aqueous solution of the surface of a spore of the fungus *Aspergillus oryzae*. The surface was covered with a crystalline-like array of rodlets, 10 ± 1 nm in diameter, consistent with electron microscopy data. Chemical analysis and enzymatic digestion have shown rodlet layers to be essentially made of proteins. Regular arrays of rodlets are known to cover the surface of spores of many fungi, including ascomycetes, basidiomycetes, and zygomycetes, and are thought

IMAGING SURFACE STRUCTURE 77

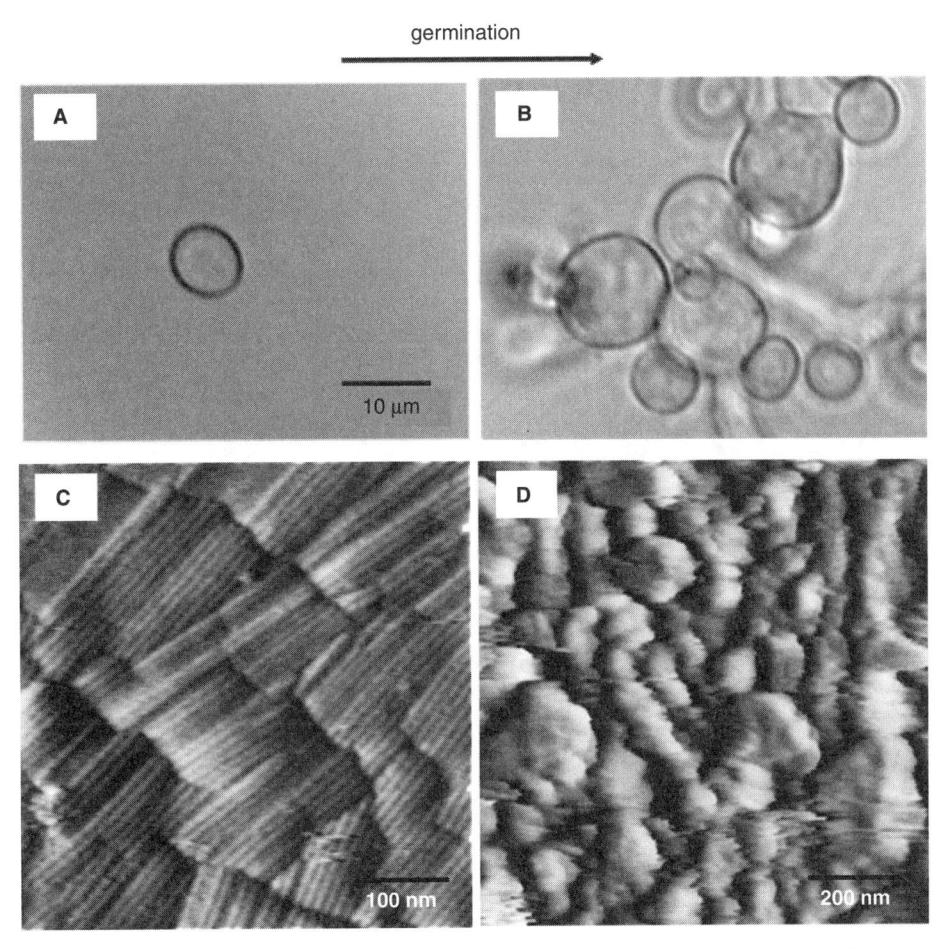

Figure 5.4. Imaging cell surface nanostructure: changes upon germination. Optical micrographs of fungal spores from *Aspergillus oryzae* in the dormant state (**A**) and after 7 h of incubation in agitated liquid medium (**B**). AFM deflection images recorded in aqueous solution for the surface of dormant (**C**) and germinating (**D**) spores revealing that, upon germination, the crystalline rodlet layer changed into a layer of soft material.

to play different biological functions such as spore dissemination and protection.

By contrast, *S. cerevisiae* cells showed a remarkably smooth (root-mean-square roughness ~1 Å over 200-nm × 200-nm areas) and uniform nanoscale surface morphology, consistent with the presence of glucan and mannoproteins (Fig. 5.5A). For bacteria, it is worth emphasizing that high-resolution imaging in contact mode is often challenging due to sample deformation. By way of example, high-resolution AFM images of *Lactococcus lactis* bacteria showed a sponge-like network of small holes together with grooves attributed to strong probe–cell interactions (Boonaert et al., 2002). In future research, the use of dynamic imaging modes may help to circumvent this kind of problems.

Appendages (fimbriae, fibrils, flagella) belong to a class of surface structures often found on bacterial cells which may play various important functions including cell adhesion, motility, and conjugation. Electron microscopy is a well-established technique to visualize appendages at high resolution. AFM may bring complementary information by allowing direct observation in the hydrated state. For instance, fibrillar

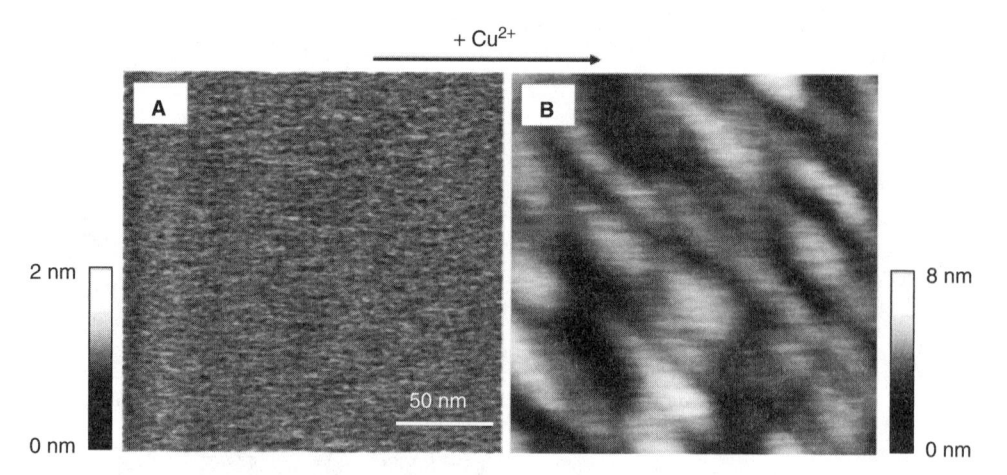

Figure 5.5. Imaging cell surface nanostructure: Influence of external agents. High-resolution AFM height images recorded in 1 mM KNO$_3$ solution for the surface of the yeast *Saccharomyces cerevisiae* prior to (**A**) and after (**B**) exposure to copper (10 mM Cu^{2+}, 10 min). Clearly, addition of copper caused a significant increase of surface roughness, from 1 Å to 7 Å. Note that the AFM probe was functionalized with COOH groups.

structures were detected on the cell surface of *Streptococcus salivarius* HB, while *S. salivarius* HBC12 appeared to have a bald cell surface, in agreement with electron microscopy data (van der Mei et al., 2000). However, due to the mobility and softness of these structures, the image quality was poor, emphasizing the need to use alternative preparation procedures (isolation, drying) and imaging modes (dynamic modes).

5.2.3.3. Physiological Changes.

One application of particular interest is the ability to study, under native conditions, the changes of cell surface structure associated with physiological processes. An example of such process is the germination of fungal spores. As illustrated in Fig. 5.4A and 5.4B, incubating *A. oryzae* spores in liquid medium for a few hours lead to their germination and, in turn, to their aggregation, a process that is of both natural and biotechnological significance. High-resolution images (Fig. 5.4C and 5.4D) showed that dramatic changes of the spore surface ultrastructure occurred upon germination, with the rodlet layer changing into a layer of soft material (van der Aa et al., 2001). On close examination, this material showed streaks oriented in the scanning direction, suggesting that soft, loosely bound material was interacting strongly with the scanning probe. These direct observations are in good agreement with previous structural and chemical studies which showed that germination of *Aspergillus* spores results in the disruption of the proteinaceous rodlet layer and reveals inner spore walls which are essentially composed of polysaccharides.

Another example is the growth of yeasts cells. Using a soft, deformable immobilization matrix made of agar gel, the growth processes of *S. cerevisiae* cells were investigated *in situ* and micrometer-size bud scars resulting from the detachment of budding daughter cells were visualized (Gad and Ikai, 1995).

5.2.3.4. Effect of External Agents.

The possibility to directly visualize the effect of external agents such as solvents, ions, chemicals, enzymes, and antibiotics on the structure of microbial cells opens the door to new applications in biotechnology and biomedicine—for example, for the rapid screening of microorganisms and for the rationale design of new antimicrobial drugs. The action of penicillin on the morpholgy of *Bacillus subtilis* cells was investigated by

AFM in air (Kasas et al., 1994). Bacteria grown in the presence of the antibiotic showed dramatic changes: Modification in cell length was first noted, followed by release of material into the environment. Changes in the morphology of *E. coli* induced by cefodizime were demonstrated using AFM in air (Braga and Ricci, 1998). Holes and rough patches were observed at the cell surface after treatment with suprainhibitory concentrations of cefodizime, while filamentation was induced by subinhibitory concentrations. In the same way, morphological alterations in *Helicobacter pylori* were visualized after exposure to rokitamycin (Braga and Ricci, 2000). Various changes were noted (i.e., cell enlargement, cell clustering, increased roughness), depending on the antibiotic concentration. The effect of vancomycin, a glycopeptide antibiotic that kills gram-positive bacteria by interfering with the synthesis of peptidoglycan, was also investigated (Boyle-Vavra et al., 2000). The fine structural detail provided by AFM extended previous topographic studies of vancomycin-susceptible and vancomycin-resistant *Staphylococcus aureus* by showing novel morphological changes occurring in the cell surface.

The effect of the denaturant LiCl on the surface of *Lactobacillus helveticus* cells was investigated by AFM (Sokolov et al., 1996). While images with a lateral resolution of 2 nm were recorded on native cells, exposure to LiCl changed the surface morphology into a pattern of holes. The influence of various chemicals (Tween 20, heparin, disodium tetraborate, sodium pyrophosphate, lysozyme, and EDTA) on the surface topography of bacterial cells was investigated using TMAFM (Camesano et al., 2000); most treatments were found to increase the surface roughness, and a change of cell shape was often noted.

Bacteria, fungi and yeasts are capable of removing heavy metals from aqueous solution in substantial amounts. Consequently, microbial biomass represents a new, attractive technology for the treatment of wastewaters (Kapoor and Viraraghavan, 1995). Despite the vast amount of literature available on biosorption processes, little is known at the molecular level due to the lack of appropriate tools. Figure 5.5 shows that the nanoscale morphology of *S. cerevisiae* cells changed dramatically upon addition of copper (10 mM for 10 min), the surface roughness increasing from 1 Å to 7 Å (root-mean-square roughness over 200-nm × 200-nm areas). Copper binding was further evidenced by force measurements, with significant adhesion being measured between treated cells and negatively charged probes. These data are consistent with biosorption studies showing that *S. cerevisiae* cells remove copper by a two-step process consisting of an initial rapid surface binding of copper ions, followed by a slower intracellular uptake (Kapoor and Viraraghavan, 1995).

Accordingly, these experiments demonstrate that AFM has a great potential in biotechnology and biomedicine studies to probe the interactions between cell surfaces and external agents at the molecular level. A unique advantage over classical microscopies is that subtle changes can be directly visualized at the cell surface without any pretreatment (drying, fixation).

5.3. MEASURING PHYSICAL PROPERTIES

A key advantage of AFM over classical microscopies is that it can measure forces and produce quantitative, spatially resolved maps of physical properties. Furthermore, the very high sensitivity of force measurements makes it possible to stretch single molecules to learn about their elastic behavior and conformational transitions. This section introduces the principles and interpretation of force measurements by AFM and their applications to microbial cell surfaces.

5.3.1. Principles of Force Measurements

5.3.1.1. Force–Distance Curves.
How does AFM measure forces? By monitoring, at a given x, y location, the cantilever deflection, expressed as a voltage, as a function of the vertical displacement of the piezoelectric scanner (z), a raw force–distance curve is constructed (Butt et al., 1995). Using the slope of the retraction force curve in the region where probe and sample are in contact, the photodiode voltage can be converted into a cantilever deflection (d). The cantilever deflection is then converted into a force (F) using Hooke's law: $F = -k \times d$, where k is the cantilever spring constant. The curve can be corrected by plotting F as a function of ($z - d$). The zero separation distance is then determined as the position of the vertical linear parts of the curve in the contact region. In addition to single force curves, spatially resolved mapping can be performed by recording arrays of force curves in the x, y plane (Heinz and Hoh, 1999).

The different parts of a force–distance curve can provide a wealth of information (Butt et al., 1995; Dufrêne et al., 2001). At large probe–sample separation distances, the force experienced by the probe is null. As the probe approaches the surface, the cantilever may bend upwards due to repulsive forces until it jumps into contact when the gradient of attractive forces exceeds the spring constant plus the gradient of repulsive forces. The approach portion of the force–distance curve can be used to measure surface forces, including van der Waals and electrostatic, solvation, hydration and steric/bridging forces. When the force is increased in the contact region, the shape of the approach curve may provide direct information on the elasticity of the sample. Upon retracting the probe from the surface, the curve often shows a hysteresis referred to as the adhesion "pull-off" force, which can be used to estimate the surface energy of solids or the binding forces between complementary biomolecules. In the presence of long, flexible molecules, an attractive force, referred to as an elongation force, may develop nonlinearly due to macromolecular stretching. In this case, it is common to present the force curve as the positive pulling force *vs* extension.

There are several limitations to keep in mind when dealing with AFM force–distance curves. First, as opposed to the surface-forces apparatus, AFM does not provide an independent measure of the probe–sample separation distance. As a result, in the presence of long-range surface forces and high sample elasticity, defining the point of contact (i.e., the zero separation) is difficult. Second, for soft samples such as many microbial surfaces, separating the relative contributions of repulsive surfaces forces and sample deformation is a delicate task, and quantitative interpretation of the curves—using, for example, the DLVO theory or the Hertz model—may be meaningless. Third, artefacts are frequently encountered (Heinz and Hoh, 1999): (a) periodic oscillations in the non-contact region (optical interference) and (b) offsets between the lines in the contact (nonlinearities of the piezo, friction) and non-contact (viscous drag) regions.

5.3.1.2. Functionalized AFM Probes.
Another issue is the control of the probe surface chemistry, which is a prerequisite for reliable, quantitative force measurements. Most commercial AFM probes have a poorly defined surface chemistry: The surface of silicon oxynitride probes is composed of ionizable silanol and silylamine groups that vary in density with pH, which is further complicated by the fact that microfabrication results in contamination by gold and other materials. To overcome this problem, different approaches have been developed to create probes of well-defined chemistry. Among these, functionalizing sharp AFM probes with self-assembled monolayers (SAM) of organic alkanethiols combines the advantage of strong chemical sensitivity with

high spatial resolution (Frisbie et al., 1994; Dufrêne, 2000). The procedure involves coating microfabricated cantilevers with a thin adhesive layer (Cr or Ti), followed by a 15- to 100-nm-thick Au layer and immersing the coated cantilevers in dilute (0.1–1 mM) ethanol solutions of the selected alkanethiol. As we shall see, thiols with different terminal functionalities (e.g., OH, CH$_3$, COOH, NH$_2$) can be used to map the physicochemical properties of microbial surfaces.

Different approaches are also available to attach biomolecules on the AFM probe: nonspecific adsorption (Lee et al., 1994b), covalent linkage on aminosilane-functionalized silica surfaces (Lee et al., 1994a), binding of proteins onto carboxyl functions using 1-ethyl-3-(3-dimethylaminopropyl)carbodiimide (EDC) and N-hydroxysuccinimide (NHS) (Benoit et al., 2000), and grafting of antibodies using polyethylene glycol derivatives with an amine-reactive end and a thiol-reactive end (Hinterdorfer et al., 1996). Because they allow measuring intermolecular forces between individual ligand-receptors, these approaches have a tremendous potential for molecular mapping of microbial surfaces.

5.3.2. Physicochemical Properties and Molecular Interactions

Physicochemical properties (surface energy, surface charges) and molecular interactions (solvation forces, van der Waals and electrostatic forces, and steric/bridging forces) are highly relevant to many cellular processes (microbial adhesion and aggregation, molecular recognition). Clearly, there is considerable challenge in using AFM to probe directly these properties at high spatial resolution.

5.3.2.1. Adhesion Mapping.
Force–volume imaging consists in recording arrays of force curves in the x, y plane (Heinz and Hoh, 1999). Such spatially resolved force measurements are very useful to map physical heterogeneities at cell surfaces. An example demonstrating the power of this approach in microbiology is given in Fig. 5.6. It can be seen that after germination, the spores from the fungus *Phanerochaete chrysosporium* showed a heterogeneous surface morphology made of smooth zones partially covered with rough granular structures (Fig. 5.6A). This was directly correlated with differences in the

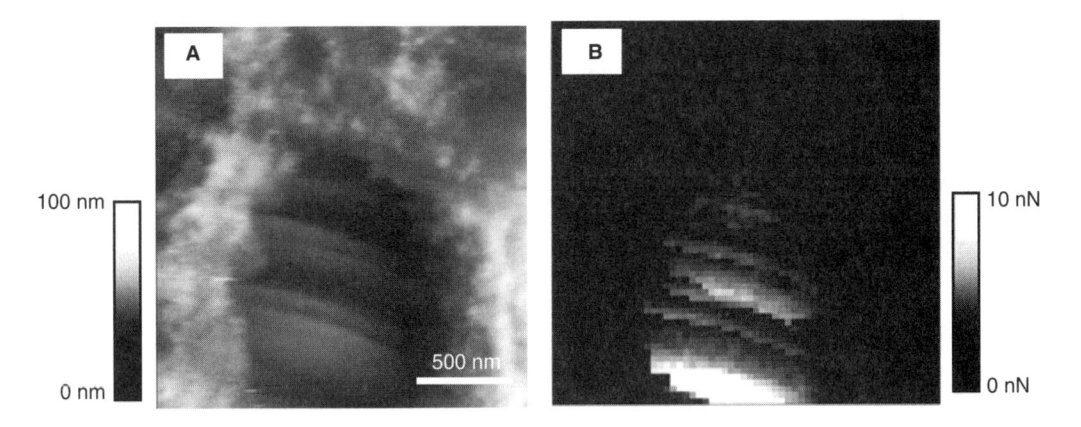

Figure 5.6. Spatially resolved force measurements. AFM height image (**A**) and adhesion map (**B**) acquired with a silicon nitride probe on the same area of a germinating spore of *Phanerochaete chrysosporium*. The adhesion map was obtained by recording 64 × 64 force–distance curves, calculating the adhesion force for each force curve and displaying adhesion force values as gray levels. The smoother area in the center of the image showed strong adhesion forces, thought to be responsible for spore aggregation.

adhesion map (Fig. 5.6B): While no adhesion forces were sensed between the silicon nitride probe and the granular material, strong adhesion forces of 9 ± 2 nN magnitude were measured on the smooth zone (Dufrêne, 2001). The measured adhesion forces may be of great biological significance in that they may be responsible for cell aggregation observed during germination. However, the origin of the forces is unclear due to the poorly defined probe chemistry. As discussed below, probes functionalized with well-defined chemical groups can provide more detailed information about cell surface properties and molecular interactions.

5.3.2.2. Surface Energy and Solvation Interactions.
During the past decades, microbial cell surface hydrophobicity (surface energy) has attracted considerable attention in view of its role in microbial adhesion processes (Doyle and Rosenberg, 1990). A variety of experimental methods (van der Mei et al., 1991) have been developed to assess microbial hydrophobicity, including water contact-angle measurements on cell lawns, adhesion to hydrocarbons, partitioning in aqueous two-phase systems, and hydrophobic interaction chromatography.

AFM with chemically functionalized probes can yield high-resolution information on surface hydrophobicity (surface energy) directly under physiological conditions, which makes it a complementary tool to these approaches (Dufrêne, 2000). Using OH- and CH_3-terminated probes, patterns of rodlets, ~10 nm in diameter, were visualized, under physiological conditions, at the surface of dormant spores of *P. chrysosporium*, demonstrating that high resolution can be obtained with the modified probes (Fig. 5.7A). Multiple (1024) force–distance curves recorded over 500-nm × 500-nm areas at the spore surface, either in deionized water or in 0.1 M NaCl solutions, always showed no adhesion for both OH- and CH_3-terminated probes (Fig. 5.7B).

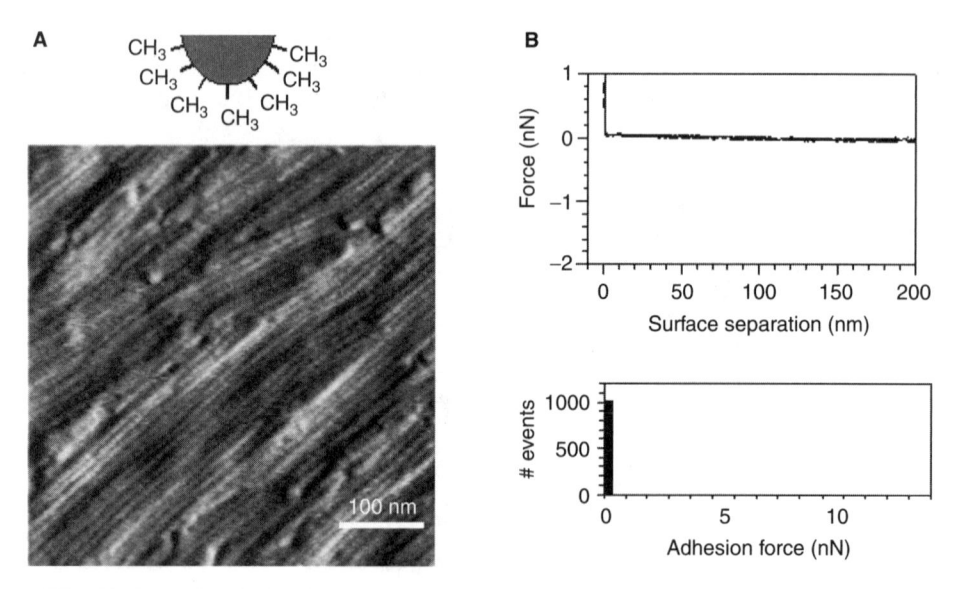

Figure 5.7. Mapping cell surface hydrophobicity (surface energy) using chemically functionalized probes. AFM deflection image (**A**) and force–distance curve and adhesion force histogram (**B**) obtained, under aqueous solution, for the surface of *P. chrysosporium* spores with a hydrophobic, CH_3-terminated probe. The adhesion force histogram was generated by recording 1024 force–distance curves over 500-nm × 500-nm areas; the narrow distribution shows that the surface was chemically homogeneous. Similar results were obtained with both CH_3- and OH-terminated probes, indicating that the spore surface was hydrophilic.

Comparison with adhesion data obtained on functionalized model substrates prior to and after the cell measurements led to the conclusion that the spore surface is uniformly hydrophilic. The highly hydrophilic, nonadhesive character of the fungal spore surface may play an important role in determining the spore biological functions, namely, protection and dispersion.

5.3.2.3. Surface Charges and Electrostatic Interactions. Force–distance curves can also be used to probe surface charges and electrostatic forces at microbial cell surfaces. A number of studies have concentrated on microbial layers. In an early work, electrostatic forces were measured at different pH values between standard silicon nitride probes and the surface of purple membranes (Butt, 1992). By comparing the approach force curves measured on both purple membranes and bare alumina, the surface charge density of the former was estimated to be -0.05 C/m^2. The interaction forces between a silica sphere and two-dimensional bacteriorhodopsin crystals were also measured at different pH values and at two electrolyte concentrations (Hartley et al., 1998). The surface potential of the crystal surface could be determined, and the isoelectric point was found to be about 4. Force–distance curves were recorded, at different electrolyte concentrations, between silicon nitride probes and the surface of purple membranes, OmpF porins, and HPI layers (Müller et al., 1999b). In the presence of 20 mM KCl, long-range repulsion forces were measured for the surface of purple membranes and OmpF porins, reflecting the electrostatic double-layer repulsion. These forces could be adjusted by changing the pH or electrolyte concentration to optimize the image resolution. No repulsion force was found for the HPI layers, indicating that the surface charge of the latter was either too small to be detected or positive.

For living microbial cells, the presence of ionizable surface groups such as amino, carboxyl, and phosphate is classically established by microelectrophoresis, titration, ion-exchange chromatography, and surface conductivity (James, 1991). With its high spatial resolution, AFM with chemically functionalized probes appears as a powerful, complementary approach to these global methods. Figure 5.8 illustrates how the presence of ionizable groups and the local isoelectric point of microbial cell surfaces can be determined using this approach. The force curves recorded between an AFM probe functionalized with COOH groups and the surface of *S. cerevisiae* cells showed tremendous changes according to pH (Fig. 5.8A): While no adhesion was observed at pH 10, multiple adhesion forces were observed at pH 4. Force mapping revealed similar results at different locations of the cell surface, indicating that the latter was fairly homogeneous as far as electrical properties are concerned. High-resolution imaging was used to confirm that the surface nanostructure was not affected by pH variations, with smooth surfaces being observed at both pH values (Fig. 5.5A). Adhesion force measurements as a function of pH, referred to as "nanotitration" curves, exhibited a sharp increase in the adhesion force at pH values around 4, with the transition occurring within 2 pH units (Fig. 5.8B).

The adhesion observed at low pH can be attributed to hydrogen bonding between the uncharged COOH probe and neutral cell surface macromolecules, while the lack of adhesion at pH values >6 would essentially reflect electrostatic repulsion between the negatively charged surfaces. The position of the maximum of the nanotitration curve at pH ~ 4 was very close to the isoelectric point of the cell surface measured by microelectrophoresis measurements. Indeed, the mobility–pH curve obtained previously for this strain was consistent with the presence of an amino-carboxyl surface (Dengis et al., 1995): At pH values smaller than 4, the isoelectric point, the surface was positively charged due to the presence of

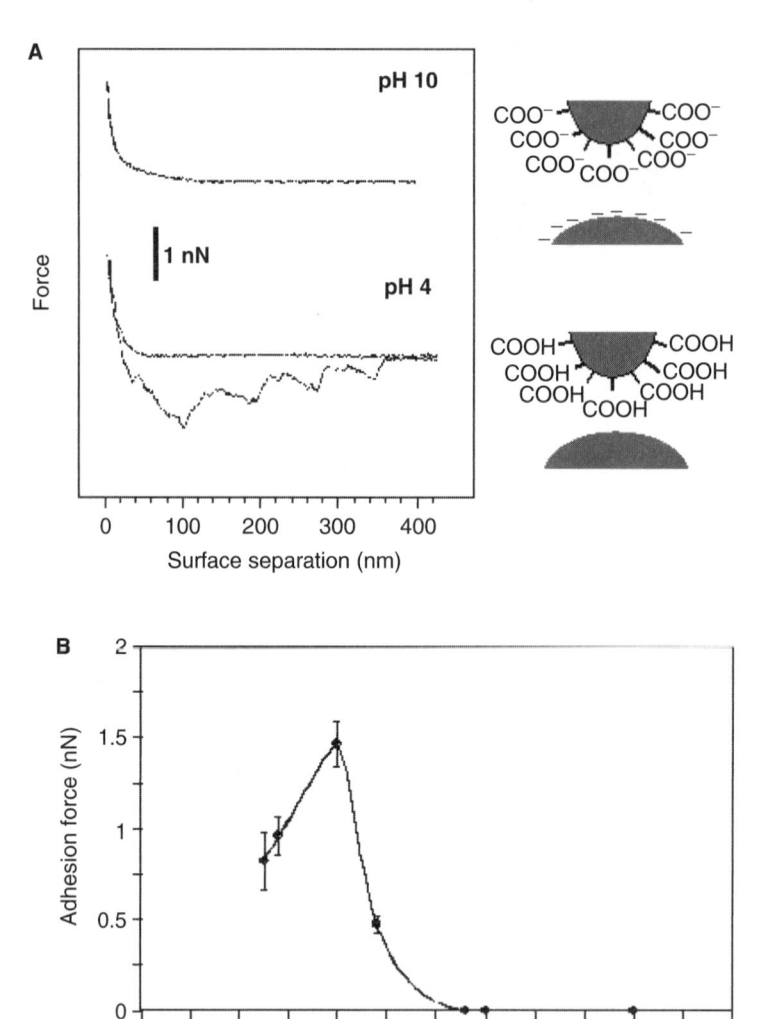

Figure 5.8. Nanotitration of cell surface ionizable functional groups using chemically functionalized probes. Force–distance curves (**A**) and adhesion force–pH curve (**B**) obtained for the surface of *S. cerevisiae* cells with COOH-terminated probes (1 mM KNO$_3$ solutions; pH adjusted with KOH and HNO$_3$). Multiple force measurements performed at different locations of the cell surface showed only small variations, indicating that the surface properties were fairly homogeneous. The adhesion maximum at pH 4 was correlated with the isoelectric point of the cell surface. Note that following treatment of the cell surface with Cu^{2+} ions, significant adhesion forces were measured at neutral pH values, supporting the notion that surface charges were actually probed.

NH$_3^+$ groups; above pH 4, an increase in the negative value of the mobility was noted due to dissociation of the weak COOH groups, until a plateau value was reached at about pH 6. The nanotitration curve was fully consistent with these microscopic data, indicating that functionalized probes represent a powerful approach to map local ionizable functional groups and isoelectric points.

The question why a significant repulsion was observed upon approach even at the cell surface isoelectric point may be raised. Presumably, this reflected the contribution of nonelectrostatic interactions—that is, steric repulsion associated with hydrated cell surface polymers (glucan and mannoproteins for yeast cells). The relative contributions of electrostatic and steric interactions associated with negatively charged bacterial strains were recently investigated (Camesano and Logan, 2000). The forces measured with silicon nitride probes as a function of pH and ionic strength were represented well by an electrosteric repulsion model but were much larger in magnitude and extended over longer distances than predicted by DLVO theory. Partially removing polysaccharides from the bacterial surfaces resulted in lower repulsive forces that decayed much more rapidly.

5.3.2.4. The "Cell Probe" Approach.

An alternative approach to probe molecular interactions associated with microbial cells is to immobilize them on the AFM probe and measure the forces between the modified probes and solid substrates. In this way, the forces between *E. coli*-coated probes and solids of different surface hydrophobicity were measured (Ong et al., 1999). Both attractive forces and cell adhesion behavior were promoted by substrate surface hydrophobicity, pointing to the role of hydrophobic interactions. Cell-coated probes were used to measure the forces between *E. coli* strains and polymer biomaterials (Razatos et al., 1998) and substrates coated with block copolymers (Razatos et al., 2001). Polymeric brush layers appeared to not only block the long-range attractive forces of interaction between bacteria and substrates but also introduce repulsive steric effects. The adhesion of metabolically active *Saccharomyces cerevisiae* cells was quantified at a hydrophilic mica surface, a mica surface with a hydrophobic coating, and a protein-coated mica surface in an aqueous environment (Bowen et al., 2001). Greatest cell adhesion was measured at the hydrophobic surface. Prior adsorption of a bovine serum albumin protein layer at the hydrophilic surface did not significantly affect cell adhesion. Changes in yeast surface hydrophobicity and zeta potential with yeast cell age were correlated with differences in adhesion.

5.3.3. Cell Surface Elasticity

How stiff are cell wall components and surface appendages? This highly relevant question can, to a certain extent, be addressed on a local scale with AFM. Note, however, that in view of the complex nature of microbial cell surfaces and of the limitations of force curves, distinguishing true sample deformation from repulsive surface forces can be a delicate task, especially when working in aqueous solution.

The mechanical properties of gram-negative magnetic bacteria of the species *Magnetospirillum gryphiswaldense* were investigated by AFM in buffer solution (Arnoldi et al., 1998). Multiple force–distance curves obtained for the substrate and for the bacteria made it possible to determine the effective compressibility of the cell wall, which was about 42 mN/m. For the same species, a theoretical expression was derived for the force exerted by the wall on the cantilever as a function of the depths of indentation generated by the AFM tip (Arnoldi et al., 2000). Evidence was provided that this reaction force is a measure for the turgor pressure of the bacterium. Making use of experimental and of theoretical results, the turgor pressure was determined to be in the range of 85–150 kPa.

The elastic properties of isolated bacterial cell wall components can also be probed using a so-called "depression technique" (Xu et al., 1996; Yao et al., 1999). A solution containing the cell wall components (e.g., archeal sheaths, peptidoglycan sacculi) is placed on a hard substrate and allowed to dry. The substrate contains grooves that are narrow compared to the width or the length of the material to be investigated so that single wall components bridging one or more grooves can be obtained. The force measurement consists in placing the probe at the midpoint of the unsuspended wall component region and increasing/decreasing the force as

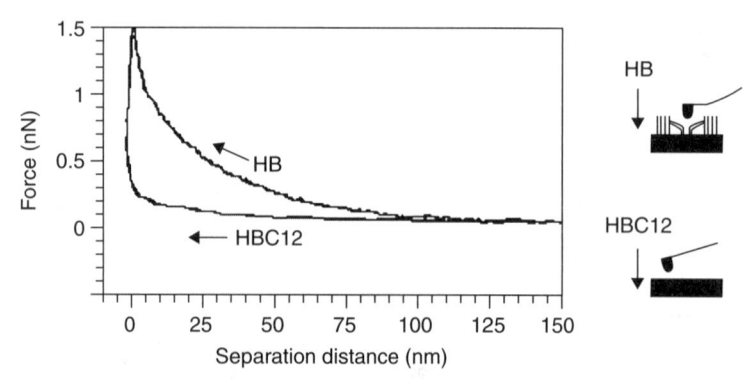

Figure 5.9. Probing the surface softness of bacterial cells. Approach force–distance curves recorded under water between a silicon nitride probe and *Streptococcus salivarius* HB or *S. salivarius* HBC12. The long-range repulsion on the fibrillated strain was attributed to the compression of a soft layer of appendages (fibrils) present at the cell surface. In view of the complexity of the surfaces and difficulty to define the point of contact, no attempt was made to fit the curves with models from contact mechanics.

a linear function of time. Comparison of the cantilever deflections between specimen and hard substrate allows an accurate determination of the specimen elasticity. Applying this approach to the proteinaceous sheath of the archeon *Methanospirillum hungatei* GP1 yielded an elastic modulus of $2-4 \times 10^{10}$ N/m², indicating the sheath could withstand an internal pressure of 400 atm, well beyond that needed for an eubacterial envelope to withstand turgor pressure. For murein sacculi of gram-negative bacteria in the hydrated state, elastic moduli of 2.5×10^7 N/m² were measured, in excellent agreement with theoretical calculation of the elasticity of the peptidoglycan network (Boulbitch et al., 2000).

Surface appendages may cause a significant increase of cell surface softness, as illustrated in Fig. 5.9, which presents the approach curves recorded on the fibrillated *S. salivarius* HB bacterial strain and on the bald *S. salivarius* HBC12 strain (van der Mei et al., 2000). Upon approach of *S. salivarius* HB, a long-range (~100 nm) repulsion force decaying exponentially with distance was detected. By contrast, the repulsion force found for *S. salivarius* HBC12 was of much shorter range (~20 nm). In agreement with the topographic images, the occurrence of a long-range repulsion on the fibrillated strain was attributed to the compression of the soft layer of fibrils present at the cell surface. In future experiments, it would be interesting to establish the role that cell surface softness may play in controlling cellular events such as cell growth and cell adhesion.

5.3.4. Elasticity of Single Macromolecules

When pulling an AFM probe away from a macromolecular system, the retraction curve often shows attractive elongation forces developing nonlinearly, which reflect the stretching of the macromolecules. The high force resolution and positional precision of AFM allows measurements at the single-molecule level. In recent years, these "single-molecule force spectroscopy" experiments have tremendously improved our current understanding of the nanomechanical properties of single biomolecules, including DNA, proteins and polysaccharides (for reviews see Clausen-Schaumann et al., 2000; Janshoff et al., 2000).

5.3.4.1. Theoretical Models. When stretching long, flexible polymers, two restoring forces may occur: At small displacements, reduction in the number of conformations gives rise to entropic elasticity forces, while at large extensions, tension in the molecular backbone may lead to enthalpic elasticity effects (bond deformation, rupture of intramolecular hydrogen bonds, and even conformational changes of the entire molecule).

Two models from statistical mechanics are often used to analyze quantitatively elongation forces: the worm-like chain (WLC) and the freely jointed chain (FJC) models (Janshoff et al., 2000). The WLC model describes the polymer as an irregular curved filament, which is linear on the scale of the persistence length, a parameter that represents the stiffness of the molecule. Hence, molecules with low persistence length have a tendency to form coils. Entropic and enthalpic contributions are combined, but extension is limited by the contour length of the molecule—that is, the length of the linearly extended molecule without stretching the molecular backbone (for relevant equations see Janshoff et al., 2000). The WLC model has been successfully applied to describe the stretching of single DNA and protein molecules (Oesterhelt et al., 2000).

In the FJC model, the polymer has a purely entropic elastic response and is considered as consisting of a series of rigid, orientationally independent statistical (Kuhn) segments, connected through flexible joints. The segment length, referred to as the Kuhn length, is a direct measure of the chain stiffness. An extended FJC model has been developed in which Kuhn segments can stretch and align under force. The polymer consists of a series of elastic springs characterized by a given segment elasticity. The extended FJC model has been successfully applied to describe the elastic behavior of polysaccharides, amylose and dextran, as probed by single-molecule force spectroscopy (Rief et al., 1997; Marszalek et al., 1998).

5.3.4.2. Application to Cell Surface Layers. Using combined AFM imaging and force spectroscopy, bacterial proteins from the HPI layer of *D. radiodurans* were unzipped (Müller et al., 1999a). Force–extension curves recorded for the inner surface of the HPI layer showed saw-tooth patterns with six force peaks of about 300 pN (Fig. 5.10). This behavior, well-fitted with a WLC model, was attributed to the sequential pulling out of the protomers of the hexameric HPI protein complex. After recording the force curve, a molecular defect the size of a hexameric complex was clearly visualized by means of high-resolution imaging (Fig. 5.2). Hence, AFM allows to correlate force measurements with resulting structural changes.

In another study combining AFM imaging and force spectroscopy, the unfolding pathways of individual bacteriorhodopsins were unraveled (Oesterhelt et al., 2000). Molecules were individually extracted from the membrane, with anchoring forces of 100–200 pN being found for the different helices. Upon retraction, the helices were found to unfold and the force spectra revealed the individuality of the unfolding pathways.

5.3.4.3. Application to Cells. Stretching single molecules directly on living cells is very challenging due to the complex and dynamic nature of the surface macromolecules. However, recent data indicate that progresses have been made in this direction.

Mannan polymers were stretched at the surface of *S. cerevisiae* cells using probes that were functionalized with the lectin concanavalin A (Gad et al., 1997). Rupture lengths up to several hundred nanometers were recorded, and the distribution of the polymers was mapped using force volume imaging. No attempt was made to fit the curves with a statistical mechanics model. The forces between living *Shewanella oneidensis* bacteria and goethite, both commonly found in Earth near-surface environments, were measured quantitatively (Lower et al., 2001). Energy values derived from these measurements showed that the affinity

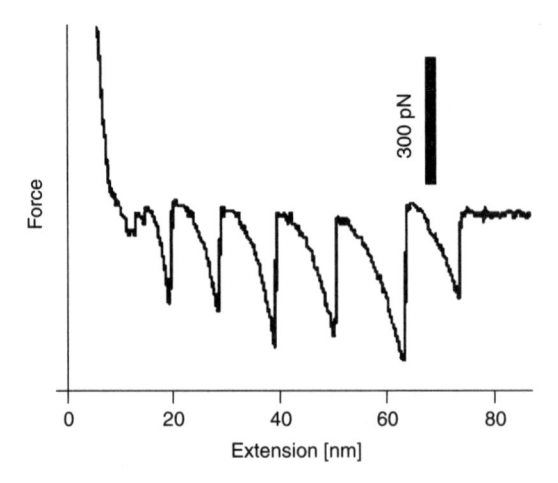

Figure 5.10. Using force spectroscopy to unzip single bacterial surface proteins (Müller et al., 1999a). Force–extension curve recorded for the inner surface region of the HPI layer showing a saw-tooth pattern with six force peaks of about 300 pN. The curves were well-fitted with a WLC model and compared favorably to those of titin molecules. As can be seen in Fig. 5.2B, the inner surface area imaged after recording the force curve showed a molecular defect the size of a hexameric HPI protein complex. (Used with permission of Prof. A. Engel, Universität Basel, Switzerland; copyright 1999, National Academy of Sciences, USA.)

between *S. oneidensis* and goethite rapidly increases by two to five times under anaerobic conditions in which electron transfer from bacterium to mineral is expected. Specific signatures in the force–extension curves were well-fitted with a WLC model and suggested to reflect the unfolding of a 150-kDa putative iron reductase.

AFM imaging and force spectroscopy were applied to *A oryzae* spores with the aim to gain insight into cellular structure–function relationships (van der Aa et al., 2001). Germination of *A. oryzae* spores in liquid medium leads to spore aggregation (Fig. 5.4A and 5.4B), a process which is still poorly understood at the molecular level. Topographic imaging revealed that, upon germination, the spore surface changed into a layer of soft granular material attributed to cell surface polysaccharides (Fig. 5.4C and 5.4D). As can be seen in Fig. 5.11, force–extension curves recorded on this surface showed attractive forces of 400 ± 100 pN magnitude, along with characteristic elongation forces and rupture lengths ranging from 20 to 500 nm. Elongation forces were well-fitted with an extended FJC model, using fitting parameters that were consistent with the stretching of individual dextran molecules. This supports the idea that polysaccharides were actually stretched. Whether single molecules were addressed remains unclear. However, given the size of the AFM probe and the wide range of observed contour lengths, it seems unlikely that a large number of polysaccharide molecules would adsorb at the same time on the probe with the same contour length so that they would give a single, discrete elongation event upon extension. The macromolecular adhesion and elasticity measured for the germinating spores may play a key role in the aggregation process; the sticky and flexible nature of long macromolecules, presumably polysaccharides, may indeed promote bridging interactions between spores. In future cell studies, developments will clearly be needed to establish unambiguously the exact biochemical nature of the macromolecules that are being stretched (specific binding using functionalized AFM probes, mutants lacking specific cell surface macromolecules).

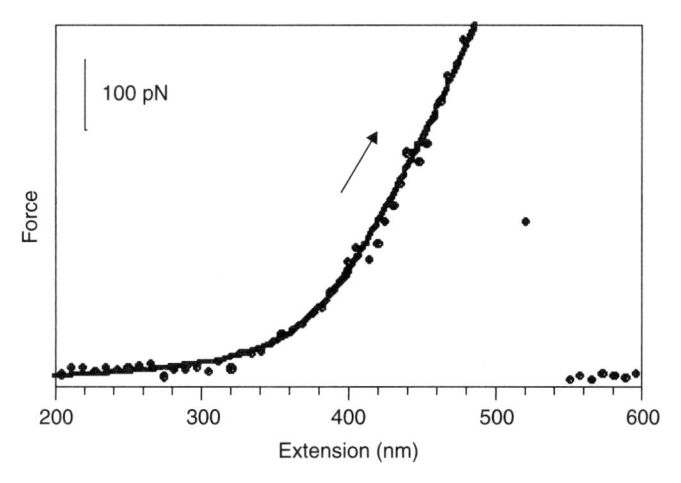

Figure 5.11. Stretching cell surface macromolecules. Typical force–extension curve recorded on a germinating spore of *A. oryzae*. The elongation force was well-fitted with an extended FJC model (continuous line), with parameters similar to values reported for the elastic deformation of single dextran and amylose polysaccharides.

5.4. CONCLUSIONS AND FUTURE PROSPECTS

The data reviewed in this chapter demonstrate that AFM is becoming a key technique in microbiology, biophysics, and biomedicine to probe the structural and physical properties of microbial surfaces. The structure of microbial cell surface layers made of 2-D protein crystals, including *Bacillus* S-layers, HPI layers, purple membranes, and porin OmpF and aquaporin crystals, can be routinely imaged at a lateral resolution of 0.5–1 nm. Remarkably, conformational changes induced by a physiological signal can be monitored in single-membrane proteins providing novel insight into structure–function relationships. Although high-resolution imaging of living cells remains challenging, a wealth of novel structural information can be obtained using AFM: observing hydrated microbial biofilms in relation with natural or industrial issues, visualizing surface ultrastructure (rodlets, appendages), following physiological changes (germination, growth), and monitoring the effect of external agents (antibiotics, metals). These studies open the door to new practical applications in biotechnology and biomedicine, such as the rapid detection of microorganisms and the rationale design of drugs.

Force–distance curves can be combined with high-resolution imaging to map the physical properties of microbial surfaces under native conditions, thereby helping to elucidate molecular and cellular functions. Provided that the experiments are carefully conducted (appropriate sample preparation procedure, probes of well-defined surface chemistry, control experiments, validation by independent techniques), force–distance curves allow one to map molecular interactions (solvation forces, electrostatic interactions) and physicochemical properties (surface energy, surface charges), to measure the stiffness of cell wall components and surface appendages and to probe the elasticity of single macromolecules.

There are many open challenges for future research. Progress in sample preparation techniques, instrumentation, and recording conditions for living cells is still needed to obtain surface topographs of living cells with subnanometer resolution and to monitor surface molecular conformational changes. One may anticipate that the development of new instruments, such as the cryogenic AFM operating in liquid nitrogen vapor, will reveal new structural details on soft microbial specimens with unprecedented resolution. Another issue is the mapping of specific interaction forces associated with antibodies, lectins, and adhesins, which are involved in many important functions. Mechanical property measurements on microbial cell walls open the exciting prospect of measuring the turgor pressure in living cells, which will provide a better understanding of microbial growth and division processes. Single-molecule force spectroscopy on living cells will make it possible to probe conformational transitions as well as the forces required to unfold proteins and to dissociate supramolecular assemblages.

REFERENCES

Arnoldi, M., Kacher, C. M., Bäuerlein, E., Radmacher, M., and Fritz, M. (1998). Elastic properties of the cell wall of *Magnetospirillum gryphiswaldense* investigated by atomic force microscopy. *Appl. Phys. A* **66**:S613–S617.

Arnoldi, M., Fritz, M., Bauerlein, E., Radmacher, M., Sackmann, E., and Boulbitch, A. (2000). Bacterial turgor pressure can be measured by atomic force microscopy. *Phys. Rev. E* **62**:1034–1044.

Auerbach, I. D., Sorensen, C., Hansma, H. G., and Holden, P. A. (2000). Physical morphology and surface properties of unsaturated *Pseudomonas putida* biofilms. *J. Bacteriol.* **182**:3809–3815.

Beech, I. B., Cheung, C. W. S., Johnson, D. B., and Smith, J. R. (1996). Comparative studies of bacterial biofilms on steel surfaces using atomic force microscopy and environmental scanning electron microscopy. *Biofouling* **10**:65–77.

Benoit, M., Gabriel, D., Gerisch, G., and Gaub, H. E. (2000). Discrete interactions in cell adhesion measured by single-molecule force spectroscopy. *Nature Cell Biol.* **2**:313–317.

Beveridge, T. J., and Graham, L. L. (1991). Surface layers of bacteria. *Microbiol Rev* **55**:684–705.

Binnig, G., Quate, C. F., and Gerber, C. (1986). Atomic force microscope. *Phys. Rev. Lett.* **56**:930–933.

Boonaert, C. J. P., Toniazzo, V., Mustin, C., Dufrêne, Y. F., and Rouxhet, P. G. (2002). Deformation of *Lactococcus lactis* surface in atomic force microscopy study. *Colloids Surfaces B: Biointerfaces* **23**:201–211.

Boulbitch, A., Quinn, B., and Pink, D. (2000). Elasticity of the rod-shaped gram-negative eubacteria. *Phys. Rev. Lett.* **85**:5246–5249.

Bowen, W. R., Lovitt, R. W., and Wright, C. J. (2001). Atomic force microscopy study of the adhesion of *Saccharomyces cerevisiae*. *J. Colloid Interface Sci.* **237**:54–61.

Boyle-Vavra, S., Hahm, J., Sibener, S. J., and Daum, R. S. (2000). Structural and topological differences between a glycopeptide-intermediate clinical strain and glycopeptide-susceptible strains of *Staphylococcus aureus* revealed by atomic force microscopy. *Antimicrob. Agents Chemother.* **44**:3456–3460.

Braga, P. C., and Ricci, D. (1998). Atomic force microscopy: Application to investigation of *Escherichia coli* morphology before and after exposure to cefodizime. *Antimicrob. Agents Chemother.* **42**:18–22.

Braga, P. C., and Ricci, D. (2000). Detection of rokitamycin-induced morphostructural alterations in *Helicobacter pylori* by atomic force microscopy. *Chemotherapy* **46**:15–22.

Bremer, P. J., Geesey, G. G., and Drake, B. (1992). Atomic force microscopy examination of the topography of a hydrated bacterial biofilm on a copper surface. *Curr. Microbiol.* **24**:223–230.

Butt, H. J., Downing, K. H., and Hansma, P. K. (1990). Imaging the membrane protein bacteriorhodopsin with the atomic force microscope. *Biophys. J.* **58**:1473–1480.

Butt, H. J. (1992). Measuring local surface charge densities in electrolyte solutions with a scanning force microscope. *Biophys. J.* **63**:578–582.

Butt, H. J., Jaschke, M., and Ducker, W. (1995). Measuring surface forces in aqueous electrolyte solution with the atomic force microscope. *Bioelectrochem. Bioenerg.* **38**:191–201.

Camesano, T. A., and Logan, B. E. (2000). Probing bacterial electrosteric interactions using atomic force microscopy. *Environ. Sci. Technol.* **34**:3354–3362.

Camesano, T. A., Natan, M. J., and Logan, B. E. (2000). Observation of changes in bacterial cell morphology using tapping mode atomic force microscopy. *Langmuir* **16**:4563–4572.

Clausen-Schaumann, H., Seitz, M., Krautbauer, R., and Gaub, H. E. (2000). Force spectroscopy with single bio-molecules. *Curr. Opin. Chem. Biol.* **4**:524–530.

Colton, R. J., Engel, A., Frommer, J. E., Gaub, H. E., Gewirth, A. A., Guckenberger, R., Rabe, J., Heckel, W. M., and Parkinson, B. (eds.) (1998). Procedures in Scanning Probe Microscopies, John Wiley & Sons, Chichester.

Dengis, P. B., Nélissen, L. R., and Rouxhet, P. G. (1995). Mechanisms of yeast flocculation: Comparison of top- and bottom-fermenting strains. *Appl. Environ. Microbiol.* **61**:718–728.

Doyle, R. J., and Rosenberg, M. (eds.) (1990). Microbial Cell Surface Hydrophobicity, American Society for Microbiology, Washington, D.C.

Dufrêne, Y. F. (2000). Direct characterization of the physicochemical properties of fungal spores using functionalized AFM probes. *Biophys. J.* **78**:3286–3291.

Dufrêne, Y. F. (2001). Application of atomic force microscopy to microbial surfaces: From reconstituted cell surface layers to living cells. *Micron* **32**:153–165.

Dufrêne, Y. F., Boonaert, C. J. P., Gerin, P. A., Asther, M., and Rouxhet, P. G. (1999). Direct probing of the surface ultrastructure and molecular interactions of dormant and germinating spores of *Phanerochaete chrysosporium*. *J. Bacteriol.* **181**:5350–5354.

Dufrêne, Y. F., Boonaert, C. J. P., van der Mei, H. C., Busscher, H. J., and Rouxhet, P. G. (2001). Probing molecular interactions and mechanical properties of microbial cell surfaces by atomic force microscopy. *Ultramicroscopy* **86**:113–120.

Forsythe, J. H., Maurice, P. A., and Hersman, L. E. (1998). Attachment of a *Pseudomonas* sp. to Fe(III)-(hydr)oxide surfaces. *Geomicrobiology* **15**:293–308.

Frisbie, C. D., Rozsnyai, L. F., Noy, A., Wrighton, M. S., and Lieber, C. M. (1994). Functional group imaging by chemical force microscopy. *Science* **265**:2071–2074.

Gad, M., and Ikai, A. (1995). Method for immobilizing microbial cells on gel surface for dynamic AFM studies. *Biophys. J.* **69**:2226–2233.

Gad, M., Itoh, A., and Ikai, A. (1997). Mapping cell wall polysaccharides of living microbial cells using atomic force microscopy. *Cell Biol. Int.* **21**:697–706.

Grantham, M. C., and Dove, P. M. (1996). Investigation of bacterial-mineral interactions using fluid tapping mode atomic force microscopy. *Geochim. Cosmochim. Acta* **60**:2473–2480.

Gunning, P. A., Kirby, A. R., Parker, M. L., Gunning, A. P., and Morris, V. J. (1996). Comparative imaging of *Pseudomonas putida* bacterial biofilms by scanning electron microscopy and both DC contact and AC non-contact atomic force microscopy. *J. Appl. Bacteriol.* **81**:276–282.

Häberle, W., Hörber, J. K. H., and Binnig, G. (1991). Force microscopy on living cells. *J. Vac. Sci. Technol. B* **9**:1210–1213.

Hancock, I. C. (1991). Microbial cell surface architecture. In: Microbial Cell Surface Analysis: Structural and Physicochemical Methods, N. Mozes, P. S. Handley, H. J. Busscher, and P. G. Rouxhet, (eds.), pp. 21–59, VCH Publishers, New York.

Hartley, P., Matsumoto, M., and Mulvaney, P. (1998). Determination of the surface potential of two-dimensional crystals of bacteriorhodopsin by AFM. *Langmuir* **14**:5203–5209.

Heinz, W. F., and Hoh, J. H. (1999). Spatially resolved force spectroscopy of biological surfaces using the atomic force microscope. *Tibtech* **17**:143–150.

Hinterdorfer, P., Baumgartner, W., Gruber, H. J., Schilcher, K., and Schindler, H. (1996). Detection and localization of individual antibody–antigen recognition events by atomic force microscopy. *Proc. Natl. Acad. Sci USA* **93**:3477–3481.

James, A. M. (1991). Charge properties of microbial cell surfaces. In: N. Mozes, P. S. Handley, H. J. Busscher, and P. G. Rouxhet, (eds.), Microbial Cell Surface Analysis: Structural and Physicochemical Methods, pp. 221–262, VCH Publishers, New York.

Janshoff, A., Neitzert, M., Oberdörfer, Y., and Fuchs, H. (2000). Force spectroscopy of molecular systems—single molecule spectroscopy of polymers and biomolecules. *Angew. Chem. Int. Edit.* **39**:3213–3237.

Kapoor, A., and Viraraghavan, T. (1995). Fungal biosorption—an alternative treatment option for heavy metal bearing wastewaters: A review. *Bioresource Technol.* **53**:195–206.

Karrasch, S., Hegerl, R., Hoh, J., Baumeister, W., and Engel, A. (1994). Atomic force microscopy produces faithful high-resolution images of protein surfaces in an aqueous environment. *Proc. Natl. Acad. Sci. USA* **91**:836–838.

Kasas, S., Fellay, B., and Cargnello, R. (1994). Observation of the action of penicillin on *Bacillus subtilis* using atomic force microscopy: Technique for the preparation of bacteria. *Surf. Interface Anal.* **21**:400–401.

Kasas, S., and Ikai, A. (1995). A method for anchoring round shaped cells for atomic force microscope imaging. *Biophys. J.* **68**:1678–1680.

Lee, G. U., Chrisey, L. A., and Colton, R. J. (1994a). Direct measurement of the forces between complementary strands of DNA. *Science* **266**:771–773.

Lee, G. U., Kidwell, D. A., and Colton, R. J. (1994b). Sensing discrete streptavidin–biotin interactions with atomic force microscopy. *Langmuir* **10**:354–357.

Lower, S. K., Hochella, M. F., and Beveridge, T. J. (2001). Bacterial recognition of mineral surfaces: Nanoscale interactions between Shewanella and α-FeOOH. *Science* **292**:1360–1363.

Marszalek, P. E., Oberhauser, A. F., Pang, Y. P., and Fernandez, J. M. (1998). Polysaccharide elasticity governed by chair–boat transitions of the glucopyranose ring. *Nature* **396**:661–664.

Möller, C., Allen, M., Elings, V., Engel, A., and Müller, D. J. (1999). Tapping-mode atomic force microscopy produces faithful high-resolution images of protein surfaces. *Biophys J* **77**:1150–1158.

Möller, C., Büldt, G., Dencher, N. A., Engel, A., and Müller, D. J. (2000). Reversible loss of crystallinity on photobleaching purple membrane in the presence of hydroxylamine. *J. Mol. Biol.* **301**:869–879.

Mozes, N., Handley, P. S., Busscher, H. J., and Rouxhet, P. G. (eds.) (1991). *Microbial Cell Surface Analysis: Structural and Physicochemical Methods*. VCH Publishers, New York.

Müller, D. J., and Engel, A. (1999). Voltage and pH-induced channel closure of porin OmpF visualized by atomic force microscopy. *J. Mol. Biol.* **285**:1347–1351.

Müller, D. J., Büldt, G., and Engel, A. (1995a). Force-induced conformational change of bacteriorhodopsin. *J. Mol. Biol.* **249**:239–243.

Müller, D. J., Schabert, F. A., Büldt, G., and Engel, A. (1995b). Imaging purple membranes in aqueous solutions at subnanometer resolution by atomic force microscopy. *Biophys. J.* **68**:1681–1686.

Müller, D. J., Baumeister, W., and Engel, A. (1996). Conformational change of the hexagonally packed intermediate layer of *Deinococcus radiodurans* monitored by atomic force microscopy. *J. Bacteriol.* **178**:3025–3030.

Müller, D. J., Engel, A., and Amrein, M. (1997a). Preparation techniques for the observation of native biological systems with the atomic focre microscope. *Biosens. Bioelectron.* **12**:867–877.

Müller, D. J., Schoenenberger, C. A., Schabert, F., and Engel, A. (1997b). Structural changes in native membrane proteins monitored at subnanometer resolution with the atomic force microscope: A review. *J. Struct. Biol.* **119**:149–157.

Müller, D. J., Fotiadis, D., and Engel, A. (1998). Mapping flexible protein domains at subnanometer resolution with the atomic force microscope. *FEBS Lett.* **430**:105–111.

Müller, D. J., Baumeister, W., and Engel, A. (1999a). Controlled unzipping of a bacterial surface layer with atomic force microscopy. *Proc. Natl. Acad. Sci. USA* **96**:13170–13174.

Müller, D. J., Fotiadis, D., Scheuring, S., Müller, S. A., and Engel, A. (1999b). Electrostatically balanced subnanometer imaging of biological specimens by atomic force microscope. *Biophys. J.* **76**:1101–1111.

Oesterhelt, F., Oesterhelt, D., Pfeiffer, M., Engel, A., Gaub, H. E., and Müller, D. J. (2000). Unfolding pathways of individual bacteriorhodopsin. *Science* **288**:143–146.

Ohnesorge, F., Heckl, W. M., Häberle, W., Pum, D., Sára, M., Schindler, H., Schilcher, K., Kiener, A., Smith, D. P. E., Sleytr, U. B., and Binnig, G. (1992). Scanning force microscopy studies of the S-layers from *Bacillus coagulans* E38-66, *Bacillus sphaericus* CCM2177 and of an antibody binding process. *Ultramicroscopy* **42–44**:1236–1242.

Ong, Y. L., Razatos, A., Georgiou, G., and Sharma, M. M. (1999). Adhesion forces between

E. coli bacteria and biomaterial surfaces. *Langmuir* **15**:2719–2725.

Persike, N., Pfeiffer, M., Guckenberger, R., Radmacher, M., and Fritz, M. (2001). Direct observation of different surface structures on high-resolution images of native halorhodopsin. *J. Mol. Biol.* **310**:773–780.

Pum, D., Weinhandl, M., Hödl, C., and Sleytr, U. B. (1993). Large-scale recrystallization of the S-layer of Bacillus coagulans E38-66 at the air/water interface and on lipid films. *J. Bacteriol.* **175**:2762–2766.

Pum, D., and Sleytr, U. B. (1995). Monomolecular reassembly of a crystalline bacterial cell surface layer (S-layer) on untreated and modified silicon surfaces. *Supramolecular Sci.* **2**:193–197.

Pum, D., and Sleytr, U. B. (1996). Molecular nanotechnology and biomimetics with S-layers. In: Crystalline Bacterial Cell Surface Proteins (U. B. Sleytr, P. Messner, D. Pum, and M. Sára, eds.), pp. 175–209, R. G. Landes, Austin, TX.

Razatos, A., Ong, Y. L., Sharma, M. M., and Georgiou, G. (1998). Molecular determinants of bacterial adhesion monitored by atomic force microscopy. *Proc. Natl. Acad. Sci. USA* **95**:11059–11064.

Razatos, A., Ong, Y. L., Boulay, F., Elbert, D. L., Hubbell, J. A., Sharma, M. M., and Georgiou, G. (2001). Force measurements between bacteria and poly(ethylene glycol)-coated surfaces. *Langmuir* **16**:9155–9158.

Rief, M., Oesterhelt, F., Heymann, B., and Gaub, H. E. (1997). Single molecule force spectroscopy on polysaccharides by atomic force microscopy. *Science* **275**:1295–1297.

Savage, D. C., and Fletcher, M. (eds.) (1985). *Bacterial Adhesion*, Plenum Press, New York.

Schabert, F. A., Henn, C., and Engel, A. (1995). Native *Escherichia coli* OmpF porin surfaces probed by atomic force microscopy. *Science* **268**:92–94.

Scheuring, S., Ringler, P., Borgnia, M., Stahlberg, H., Müller, D. J., Agre, P., and Engel, A. (1999). High resolution AFM topographs of the *Escherichia coli* water channel aquaporin Z. *EMBO J.* **18**:4981–4987.

Sokolov, I. Y., Firtel, M., and Henderson, G. S. (1996). *In situ* high-resolution atomic force microscope imaging of biological surfaces. *J. Vac. Sci. Technol. A* **14**:674–678.

Steele, A., Goddard, D. T., and Beech, I. B. (1994). An atomic force microscopy study of the biodeterioration of stainless steel in the presence of bacterial biofilms. *Int. Biodet. Biodeg.* **34**:35–46.

Telegdi, J., Keresztes, Z., Pálinkás, G., Kálmán, E., and Sand, W. (1998). Microbially influenced corrosion visualized by atomic force microscopy. *Appl. Phys. A* **66**:S639–S642.

van der Aa, B. C., Michel, R. M., Asther, M., Zamora, M. T., Rouxhet, P. G., and Dufrêne, Y. F. (2001). Stretching cell surface macromolecules by atomic force microscopy. *Langmuir* **17**:3116–3119.

van der Aa, B. C., and Dufrêne, Y. F. (2002). *In situ* characterization of bacterial extracellular polymeric substances by AFM. *Colloids Surf. B: Biointerfaces* **23**:173–182.

van der Mei, H. C., Rosenberg, M., and Busscher, H. J. (1991). Assessment of microbial cell surface hydrophobicity. In: *Microbial Cell Surface Analysis: Structural and Physicochemical Methods* (N. Mozes, P. S. Handley, H. J. Busscher, P. G. Rouxhet, eds.), pp. 263–287, VCH Publishers, New York.

van der Mei, H. C., Busscher, H. J., Bos, R., de Vries, J., Boonaert, C. J. P., and Dufrêne, Y. F. (2000). Direct probing by atomic force microscopy of the cell surface softness of a fibrillated and non-fibrillated oral streptococcal strain. *Biophys. J.* **78**:2668–2674.

Xu, W., Mulhern, P. J., Blackford, B. L., Jericho, M. H., Firtel, M., and Beveridge, T. J. (1996). Modeling and measuring the elastic properties of an archaeal surface, the sheath of *Methanospirillum hungatei*, and the implication for methane production. *J. Bacteriol.* **178**:3106–3112.

Yao, X., Jericho, M., Pink, D., and Beveridge, T. (1999). Thickness and elasticity of gram-negative murein sacculi measured by atomic force microscopy. *J. Bacteriol.* **181**:6865–6875.

CHAPTER 6

SCANNING PROBE MICROSCOPY OF PLANT CELL WALL AND ITS CONSTITUENTS

KSENIJA RADOTIĆ
Center for Multidisciplinary Studies, University of Belgrade, Kneza Višeslava 1, 11000 Belgrade, Serbia and Montenegro

MIODRAG MIĆIĆ
MP Biomedicals, Inc., 15 Morgan, Irvine, California, 92618-2005

MILORAD JEREMIĆ
Faculty of Physical Chemistry, University of Belgrade, Studentski Trg 16, 11000 Belgrade, Serbia and Montenegro

6.1. INTRODUCTION

In the majority of plant cells, the cell wall is a compartment lying outside the plasmalemma. It is involved in the determination of both the morphology and function of the plant cell. Since the cell wall is the outermost barrier of the plant cell, its major biological function is mechanical support and stress protection. Walls or selected wall components have chemical roles too, such as ion exchangers, bacterial agglutinins, and sources of messages (the oligosaccharines) to instruct the protoplast. Cell wall is built from lignin, polysaccharides, and proteins. The general architectural plan of the cell wall involves a fibrous material embedded in an amorphous matrix. The chemical and physical structure of the wall depends on the plant group and cell type, but a general characteristic of all plant cell walls is that they are not chemically and physically homogeneous (Goodwin and Mercer, 1985; Varner and Lin, 1989).

Polysaccharides in the cell wall can be subdivided in two categories: those that exist within the wall in crystalline form and those that form rather amorphous phase. The former polysaccharides are very long, unbranched molecules aggregated together in bundles called microfibrils. The microfibrils consists of individual cellulose molecules (a β-4-linked D-glucan) associated into crystalline or near-crystalline arrays that are nearly free of water (Roelofsen, 1965). Xylans (β-4-linked xylosyl residues) are also present in form of microfibrils in the cell walls. The microfibrils are embedded in a matrix composed of the noncrystalline polysaccharides. These are highly branched macromolecules containing several different species of monosaccharide residues. They are covalently or hydrogen bonded to the other polysaccharide and

---

Force Microscopy: Applications in Biology and Medicine, edited by Bhanu P. Jena and J.K. Heinrich Hörber.
Copyright © 2006 John Wiley & Sons, Inc.

nonpolysaccharide components of the cell wall. Matrix polysaccharides are comprised of hemicelluloses (xylans, xyloglucans, mannans, galactanes) and pectins (mostly built from polyuronic acid). Lignin, a branched polymer of phenolic alcohols, has an important role in mechanical support and stress protection in the cell. It strengthens the wall by forming a ramified network throughout the matrix, thus anchoring the cellulose microfibrils more firmly. Certain proteins in the cell walls have structural role (most abundant is extensin), while some of them are enzymes. Water is an extremely important constituent of the cell walls. It permeates the matrix, but not the crystalline regions of the microfibrils. The amount of water in the wall depends on the nature of the matrix polysaccharides, with which it forms close intermolecular associations and gel-like structures, and on the degree of lignification (the more lignin, the less water is present). Lignin and cellulose are considered as molecules important in water transport through the xylem of the higher plants (Laschimke, 1989).

The processes of synthesis and secretion of wall components are mostly known, but assembly of these components into a growing wall is poorly understood. It is conceivable that the addition of noncellulosic components to an existing wall is entirely by self-assembly, with each component designed to fit appropriately with others. Cellulose microfibrils are laid down by apposition on the inner surface of the existing wall. The matrix components—hemicelluloses, pectins, and proteins—are added by intussusceptions, and they can occur at any place in the existing wall (Goodwin and Mercer, 1985).

6.2. AFM EXAMINATION OF CELL WALLS

Atomic force microscopy (AFM) may contribute in revealing the general architecture of the cell walls, as well as in the way particular components in the whole wall structure are assembled. Using AFM, it has been possible to image the architecture of isolated cell wall fragments, as in case of the Chinese water chestnut cell walls (Morris et al., 1997). The cell wall is revealed as a laminated fibrous structure. Molecular structure was clearly represented throughout the topographical image. Using different image processing methods, it was possible to obtain the wall thickness. Error signal mode imaging effectively flattens the rough surface, thereby generating a better visual impression of the structure within the surface. However, it is not possible to obtain measurements of layer thickness or fiber diameter from the error signal mode images. Molecular structure is clearly represented throughout the topographical image, but the number of the gray levels required to represent this image is too large to be perceived by the eye. Thus regions of the image appear devoid of the structure. It is necessary to isolate the high-frequency structural information from the low-frequency curvature of the cell wall surface. One approach is to use a high-pass frequency filter. This improves the image, but there is some leeway in the choice of the cutoff frequency. An alternative is to locally average or smooth the surface. This can be subtracted from the topographical image yielding a "flattened" projection of the surface. The measured fiber thickness is smaller for fibers packed into arrays than for isolated fibers due to a reduction in probe broadening. Cellulose microfibrils were distinguished within the whole wall fragments, but hemicelluloses or pectin networks could not be distinguished. Disruption of the wall fragments may produce some artifacts, but reveals laminated nature of the cell wall. Using AFM, the laminated structure of the cell walls isolated from maize roots has been observed (Radotić, 1994). The fibrous structures in cell wall fragments were ascribed to cellulose, according to their dimensions in the Radotić study.

In a paper by Round et al. (1996), the general conditions and difficulties in AFM contact and error mode imaging of the plant cell walls has been addressed, the most important being sample hydrations. When the sample is imaged in the hydrated state, it does not stick to the substrate, and the force of the AFM tip is enough to peel it off, making imaging very difficult, if not impossible. On the other hand, dehydrated sample imaging in air is very difficult and introduces artifacts, as the dry samples become extremely rough, and additionally, any moisture that is present on the surface of the sample could induce artifacts by both pulling the tip down and by playing with refraction of the laser beam. Later, studies determined that butanol is the best solvent for imaging plant cell wall samples. The observed structure of the primary cell wall associated with plasmalemma expressing polysaccharides fibril structure packed into the closed arrays is very similar to the one previously observed by TEM (Blankey et al., 1983; McCann et al., 1990) and suggested by Roelofsen et al. (1965).

Duchesne and Daniel (1999) have reviewed various microscopic methods, applied in the study of wood fiber ultrastructure, including using AFM. The high degree of similarities between FE-SEM and other electron microscopy methods and observed fibrilar features on the pulp fiber surfaces by AFM is extraordinary. However, the applicability of the AFM in studying *in situ* pulp is questionable due to the fact that its ability to obtain good quality images is compromised by the condition of harsh physical–chemical environment, capable of altering the surface properties of both tip and substrate. Another paper by Niemi et al. (2002) specifically reviewed the applications of scanning probe microscopy in wood, paper, and fiber research. Both dry-air imaging and liquid imaging have been presented, and special attention has been given to the modification occurring at the nanoscale at the fiber surface in different stage of pulping.

6.3. AFM EXAMINATION OF PECTINS

Pectins are one of the main constituents of the plant cell walls, being highly branched polysaccharides whose primary, secondary, and tertiary structures are still little understood (Ridley et al., 2001). Moreover, the function of the pectins in the physiology of the plant cell, as well as the process of its biosynthesis, is only partially known (Ridley et al., 2001). One of the most important questions in plant physiology today is understanding the properties, functions, and biosynthesis of this class of polymers. The practical aspect of knowing more about the pectin structures comes from their role in food; that is, they are the main components of many plant-based foods for human consumption.

Because branched pectins are highly irregular structures, and are also intertwined with other plant cell wall constituents, it is virtually impossible to make them into crystalline form after their extraction, which makes them unsuitable for x-ray diffraction or neutron scattering studies. A classical method of chemical analysis introduces different artifacts, mainly in the forms of side-chain cleavage. Atomic force microscopy (AFM) allows direct observation, with sub-nanometer spatial resolution of extracted polysaccharides (Round et al., 1997, 2001), as well as of fragmented cell walls (Morris et al., 1997; Round et al., 1996).

In a study by Round et al. (1997), branching in isolated pectin polysaccharides was observed for the first time. One of the representative images of such a pectin-branched macromolecule is presented in Fig. 6.1. The large linear branch between 30 and 170 nm has been observed in that study. The pectin macromolecules were isolated from the cell walls of unripe tomatoes for the study. Extracted pectin solution (1–3 µg/ml) was applied to the surface

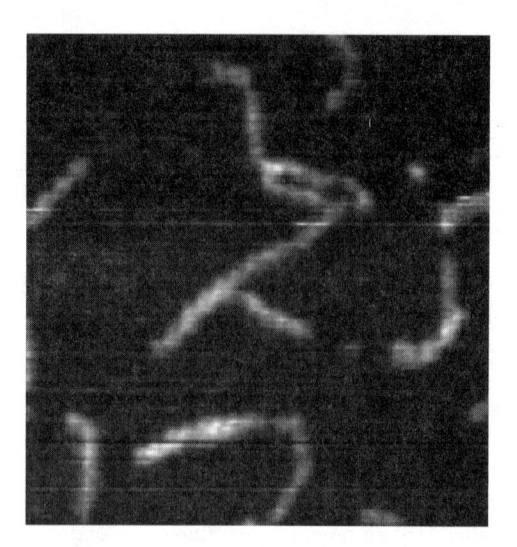

Figure 6.1. AFM image of pectins. Image size 250 × 250 nm. (Courtesy of Dr. A. N. Round, Institute of Food Research, Norwich, UK.)

of cleaved mica, and it was imaged by AFM in the contact mode in butanol. AFM micrographs of the extracts show aggregates and single molecules. To distinguish single pectin macromolecules from aggregates, the height measurement was used. According to Round et al. (1997), the observed lengths of the individual pectin chains are in the range between 30 and 390 nm, with lognormal distribution of chain lenght. Interestingly, the average molecular weight, calculated from the length of the macromolecular contours, are in the range of 50,000 Da, which is at least 200 times smaller than the estimated molecular weight of pectins. About 20% of the macromolecules shown in the AFM micrographs branched into side chains, and 30% of the branched macromolecules are multiply branched, forming a so-called "macromolecular brush" structure. The differentiation between covalent molecular branching and statistical overlapping of two independent chains has been performed, based on the height distribution. Because the average diameter of the polymer backbone is in the range from 0.5 to 0.8 nm, the height at the branching point should be in the same range. On the other hand, if the "node point" is higher, then it is to be assumed that the AFM image shows just the two overlapping polymer chains. Lengths of the side chains have been estimated to be in the range from 30 to 170 nm. One explanation is that the pectin branches are made of poly-galactouronic acids; while the alternative is that the branches are made of neutral sugar monomers, attached to the backbone of polymer via branched rhamnose. The later is less likely, since there is no confirmation of the existence of long neutral sugar chains, by methylation chemical analysis. Studies by the Round et al. (2001) shows that the distributions of the total amount of polymeric branches do not match the distribution of the neutral sugars in pectins. The authors in the study further suggest that the long chains are most likely polygalacturonic acid attached to the pectin macromolecular backbone, with occasional long neutral sugar branches. The majority of the neutral sugars are attached to the pectin backbone in the form of very short branched chains, a model which is in agreement with the results of sequencing analysis of pectins. The chemical analysis of pectins is performed by Semana hydrolysis followed by identification of monomers using gas chromatography (Blankey et al., 1983). It shows that there is no statistically significant relationship between the neutral sugars (arabinose and galactose) content and the branch lengths.

In a separate study by Morris et al. (1997), AFM was used to image fragments of the cell wall, pectin, and carrageenan. The polysaccharides were imaged *in situ* (i.e., embedded in their natural environment in the cell wall fragments) as well as extracted, from tomato and from the Chinese water chestnut. Both cell fragments and CDTA extracted polysaccharides were imaged on cleaved mica surface, under butanol, applying the differential contact mode ("error mode") AFM. The polysaccharide microfibers are easily observed as packed into the arrays, in laminar structure of the plant cell wall. However, it was impossible to distinguish between pectin and hemicelluloses fibers in the AFM micrographs generated.

6.4. USE OF AFM IN THE STUDY OF CELLULOSE

Native cellulose has been the most studied subject in polymer science. Cellulose microfibrils in a typical primary cell wall are elliptical in cross section with axes of 5–30 nm. They constitute 20–30% of the dry weight of the wall. Despite many reports, summarized in review articles (Bayer et al., 1998; O'Sullivan, 1997) and books (Kressig, 1993), the ultrastructure of cellulose has not been completely resolved. It is not yet clear how new cellulose chains are assembled into microfibrils at the cell surface, and our knowledge of the structure of the basic crystalline units is incomplete (Ha et al., 1997). Although the enzymatic apparatus for cellulose synthesis is poorly understood, it is apparently contained in particles observed in electron micrographs as ordered groups of rosettes in the plasmalemma (Hert and Hausser, 1984). The cellulose molecules extruded by a single rosette aggregate with those from adjacent rosettes to form the microfibrils, where cellulose molecules are packed in a parallel fashion (Varner and Lin, 1989). The microfibrils are then assembled into superstructures, such as the cell wall and fibers.

AFM is ideally suited to study cellulose surfaces at near-angstrom resolution. AFM images of *Valonia* cellulose microcrystals under propanol and water demonstrates a resolution approximating that of the best TEM micrographs (Baker et al., 1997). The AFM images revealed clear structural details consistent with the 0.54-nm repeat unit (glucose) along the cellulose chains. An intermolecular spacing of about 0.6 nm is also observed. AFM has also been used to evaluate the phase dimensions of cellulosic blends and composites. Although most of these studies employed primarily topographical imaging, there have also been studies involving phase imaging. For example, the distribution of lignin on the surface of a wood pulp fiber could be visualized on the phase image while it did not appear in the height images (Boras and Gatenholm, 1999). This is because the properties of lignin and its functions, such as adhesion, friction, and viscoelasticity, are different from those of cellulose and hemicelluloses. These studies suggest that AFM phase imaging can distinguish between the unmodified cellulose regions and the cellulose ester regions in the modified pulp fibers.

Digestion of cellulose by cellulolytic microorganisms has been the subject of intensive research efforts for decades, yet the exact course of the initial binding and disruption of the cellulose fibers remains unclear. Cellulolytic microorganisms, such as the fungus *Trichoderma reesei*, degrade crystalline cellulose by secreting mixtures of cellulases that have different but complementary modes of action on the substrate. Cellulases from both bacterial and fungal sources are responsible for the hydrolysis of the β-1,4-glucosidic bonds in cellulose. Since cellulose is resistant to microbial degradation, multiple enzyme systems are required to efficiently degrade its crystalline structure. Such systems comprise a collection of free cellulases and/or multicomponent complexes called cellulosomes (Bayer et al., 1998). The trend in recent years has been to determine the atomic structure and then to identify the catalytic residues of members of a given cellulase family. There have also been efforts to reveal the interaction of cellulase with cellulose and the mechanism of cellulase-catalyzed hydrolysis of cellulose (Chanzy and Henrissat, 1985; Hayashi et al., 1998). Studies on two cellulases from *Trichoderma reesei* on the cotton fibers provide the first physical evidence of (a) cellulase action at the surface of cotton fibers and (b) its movement along the fibers, (Lee et al., 2000). For over four decades there has been debate on whether a physical factor necessary for degrading crystalline cellulose is present in cellulases. AFM study of the interaction between catalytically active and inactive *Trichoderma reesei* cellulase components and cotton fibers have shown that inactivated major enzyme component (CBH I) still had the ability to bind to the cotton fiber. AFM also provided evidence that binding alone is not sufficient to disrupt the microfibril surface, but catalytic activity of CBH I is required for fiber disruption (Lee et al., 1996).

6.5. SCANNING PROBE MICROSCOPY ON LIGNIN

Although lignin has been extensively studied in different scientific disciplines for more than 100 years, much remains unknown about its formation, structure and bonding environment in the cell wall. Since lignin is highly interlinked with the other constituents of the cell wall, its isolation in unaltered form is impossible. There is opinion that lignin does not exist outside the cell wall. Therefore, knowledge on lignin structure has accumulated from the studies of lignin like polymers and isolated fragments of natural macromolecule. Experimental evidence has shown that enzymatic synthesis of lignin proceeds through free radical polymerization of phenolic alcohols (coniferyl, synapyl, and *p*-coumaryl alcohol — called monolignols), catalyzed by different peroxidases in the cell walls (Lewis and Yamamoto, 1990). Phenolic alcohols are synthesized in the cell starting from phenylalanine and tyrosine (Erdtman, 1957; Freudenberg and Niedercorn, 1958; Higuchi and Brown, 1963; Schubert and Nord, 1957). They are subsequently exported to the cell wall enclosed in vesicles, in the form of glucosides (Bardinskaya and Safonov, 1959; Schmid and Grisebach, 1982; Schmid et al., 1982). After fusion of the vesicles with the plasma membrane, glucosides are released by exocytosis (Pickett-Heaps, 1968) into the cell wall space. β-Glucosidase, localized in cell walls, hydrolyzes the glucosides (Marcinowski and Grisebach, 1977; Schmid et al., 1982), releasing phenolic alcohols, which serve as the substrate for the peroxidases in lignin synthesis. Hydrogen peroxide, necessary for the peroxidase reaction, is produced in the cell walls, but the mechanism responsible for its formation is still obscure. There is a lack of appropriate methodology that would allow us to directly monitor the processes of lignin synthesis and the activity of different peroxidases present in the cell wall (Lewis and Yamamoto, 1990). Most *in vivo* experiments are based on the histochemical methods of tissue dyeing in the course of lignification (Freudenberg et al., 1958; Harkin and Obst, 1973; Catesson et al., 1986; Chibbar and van Huystee, 1986), the use of the immunological probes (Bernardini et al., 1986; Conroy, 1986), or the electrophoretic identification of the isoenzymes (Gaspar et al., 1985; Imberty et al., 1985; Catesson et al., 1986). Monolignols have been rarely used as substrates in these *in vivo* experiments, due to the lack of suitable method for direct monitoring of their oxidation (Lewis and Yamamoto, 1990). The majority of experiments of such kind deal with the conversion of radioactively labeled lignin precursors, added to the medium where the plants were grown, into the lignin in the cell walls (Stone, 1953; Stewart et al., 1953; Marcinowski and Grisebach, 1977; Shann and Blum, 1987). *In vitro* experiments of phenolic alcohols polymerization in the presence of peroxidase have contributed to an understanding of the process of enzymic polymerization of phenolic alcohols in cell walls (Freudenberg et al., 1958; Nozu, 1967; Freudenberg and Toribio, 1969; Hwang, 1982).

As a response to external stress conditions, the amount of phenolic compounds increases in the cell walls, as well as production of hydrogen peroxide (Mehlhorn and Kunert, 1986). One of the first responses to the stress is the increase in activity of extracellular peroxidases, which use phenolic compounds and hydrogen peroxide as substrates. A parallel increase in lignin quantity occurs, as an indication of the stress intensity (Castillo, 1986; Pandolfini et al., 1992; Polle and Chakrabarti, 1994).

The structure and function of lignin have been extensively studied by various methods. The structural changes of lignin at the molecular level are in close connection with its protective function in the plant cell. Therefore, a study of the effect of external perturbations on the bonding pattern and molecular organization of lignin is important for understanding the relationship between mechanism of lignin formation and structure on one hand, and plant response to stress on the other.

Due to its complex structure, which is still puzzling even today, lignin is an interesting subject for ultramicroscopic studies. Elucidating the secondary and tertiary structure of lignin can be considered a holy grail in plant physiology. Besides the importance in understanding the role of lignin in higher plant cells, there is also a very practical aspect of this research. There has been a long history of microscopic investigation of the plant cells, starting with Antonvan Leeuwenhoek's first description of the cells in a corkwood. The modern lignin and plant cell wall microscopy can be traced back to the late 1970 with several investigations based on the transmission electron microscopy (Fengel et al., 1981; Falkehag, 1975) and scanning electron microscopy studies (Košikova et al., 1978; Zakutna et al., 1983; Karmanov et al., 1993) of plant cell wall and extracted lignins. Since then, the microscopic studies of lignin could be divided into the two major lines: one directed toward the fundamental investigation, and the other aimed at gaining knowledge necessary for improving practical applications such as pulp krafting and food processing. With regard to the fundamental aspect of lignin ultramicroscopy research, it is directed toward imaging either (a) primary and secondary cell walls or (b) lignin and lignin-model compounds.

Since isolation of lignin by various chemical procedures produces material of different complexity, there was a dilemma whether lignin formation in the cell wall is an uncontrolled process, the result of which would be a random polymer structure; or it is a process finely regulated in time and space (Lewis and Yamamoto, 1990). Scanning tunneling microscopic (STM) study of *in vitro* enzymic polymerization of coniferyl alcohol in time course studies (Fig. 6.2) show some of the intermediate products and the final polymer structure (Radotić et al., 1994). The STM images with molecular resolution (Figs. 6.2A and 6.2B) reveal individual aromatic monomers and many dimers and small oligomers. The generally accepted pathway of lignin synthesis starts from the coniferyl alcohol monomer from which a free radical is produced by hydrogen abstraction. The free radical, which has several resonance forms (Elder and Worley, 1984; Hwang, 1985), attacks at the β position of the propenyl group of a coniferyl alcohol monomer to form different dimer radical combinations. Subsequently, it undergoes various dehydrogenation, oxidation, hydration, and condensation reactions in which a variety of functional groups are formed on the aromatic ring and in the side chain. The structural pattern shown in the inset of Fig. 6.2B and proposed by Hwang (1985) on the basis of LCAO-MO calculations as one of the possible intermediates of lignin formation is compatible with the STM image in Fig. 6.2B. The images of final polymer, obtained 2 days after the start of the polymerization (Fig. 6.2C), show that the enzymatic polymer has a highly ordered structure even in *in vitro* conditions (Radotić et al., 1994). This study has also shown the modular organization of lignin polymer. An approximate molecular weight of $85,000 \pm 25,000$ for the observed structural unit of 5 ± 0.5 nm in diameter was obtained. This would mean that one structural unit, which we called supermodule, is composed of 500 monomers of coniferyl

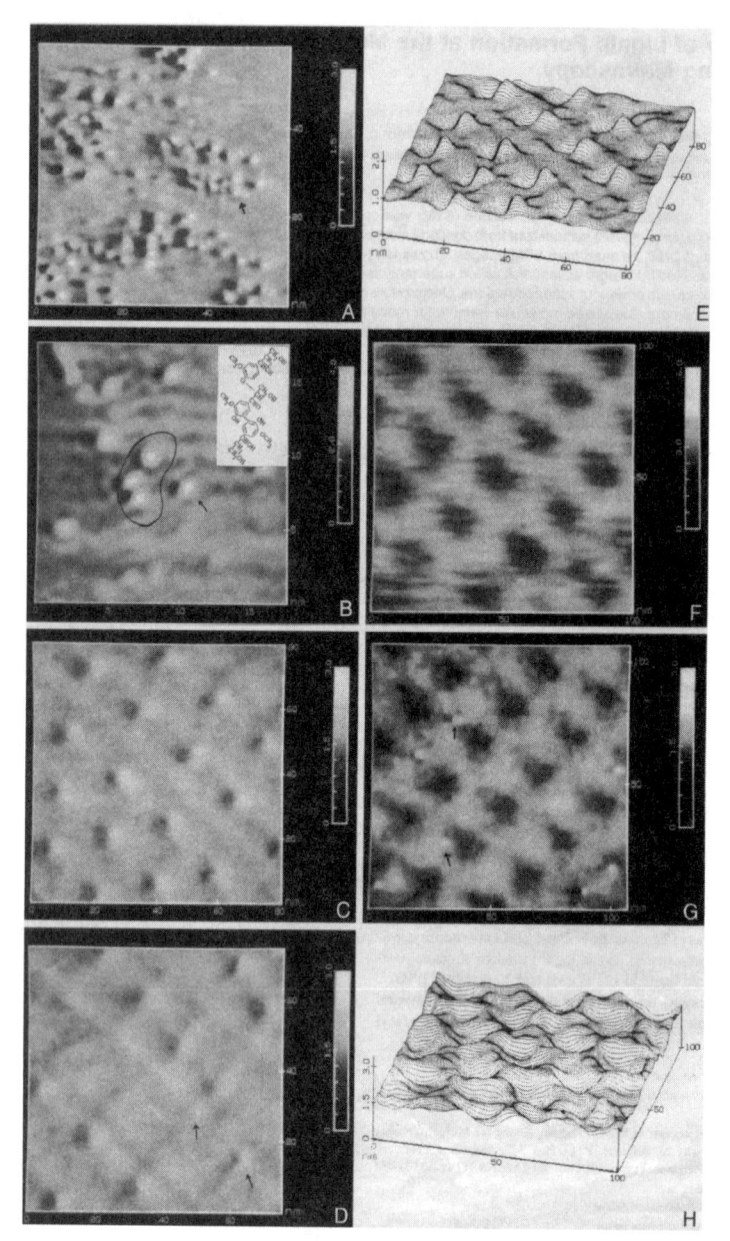

Figure 6.2. STM images of temporal formation and organization of DHP deposited onto HOPG substrate. (**A**) Two hours after mixing the components, ordered structure of DHP is not yet visible. CA molecules (*arrow*) are observable. $U = 199.9$ mV, $I_T = 0.50$ nA. (**B**) Detail from part A. CA molecules, 1.5 nm in diameter, are indicated by arrows. $U = 199$ mV, $I_T = 0.41$ nA. *Inset*: One of possible intermediates of lignin formation, proposed by Hwang, compatible with encircled pattern on the image. (**C**) DHP 2 days after mixing components, with the spherical regions. $U = 20.1$ mV, $I_T = 0.50$ nA. (**D**) The same as part C, seen at another place of the surface. Supermodules, 5 nm in diameter, are indicated by arrows. $U = 20$ mV, $I_T = 0.50$ nA. (**E**) Three-dimensional image of DHP presenting the layer of DHP with spherical regions. $U = 20.1$ mV, $I_T = 0.50$ nA. (**F**) DHP 2 days after mixing the components, without spherical regions. $U = 180.1$ mV, $I_T = 0.46$ nA. (**G**) The same as part F, showing another region of the sample. Supermodules, 5 nm in diameter, are indicated by arrows. $U = 180.1$ mV, $I_T = 0.60$ nA. (**H**) Three-dimensional image of DHP layer without spherical regions. $U = 180.1$ mV, $I_T = 0.46$ nA.

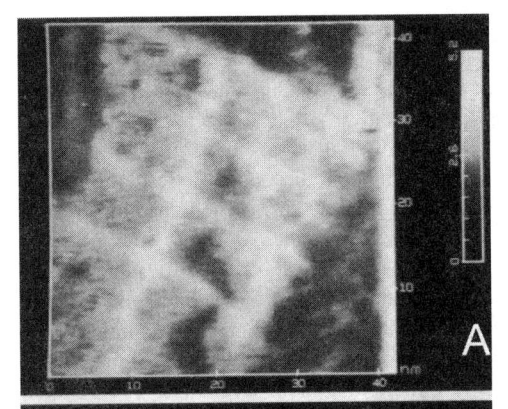

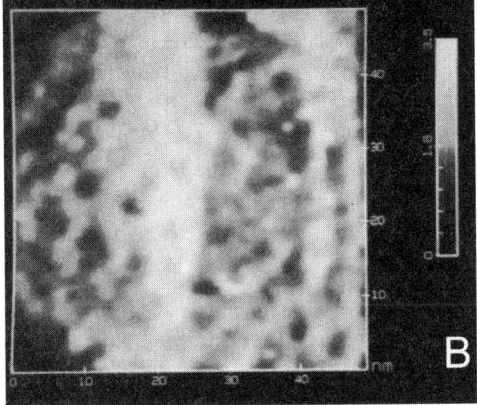

Figure 6.3. STM images of DHP deposited onto the gold substrate 2 days after mixing the components. **(A)** $U = 166.6$ mV, $I_T = 0.83$ nA. **(B)** Lattice-like structure of DHP. $U = 200$ mV, $I_T = 1.2$ nA.

alcohol. The substructure of supermodules is resolved in Fig. 6.3A. From the size of these subunits, the calculated approximate molecular weight is 6000 ± 4000. This would mean that a subunit, which is called a module, is made up of 20 monomers.

The calculated molecular weights are in good agreement with those obtained experimentally (Wayman and Obiaga, 1974). These results demonstrate that modules consist of 20 monomers and are formed first. They subsequently polymerize to form supermodules, which in turn aggregate to yield globules of different size, as schematically represented in Fig. 6.7. The bonds between the modules in a supermodule are weaker than the bonds between monomers within the modules. However, they are of covalent type, since both modules and supermodules exist as distinct molecular species in solution and are quite stable. On the other hand, the supermodules are interconnected by intermolecular forces within globules. These globules exist only within solid lignin and cannot be detected in solution as independent molecular species. Such a principle of polymer organization is summarized in Fig. 6.7, which highlights the major steps of polymerization and organization of enzymatic polymer. More detailed investigation of self-organization of lignin aggregates into higher structural levels, and influence of substrate has been examined by ESEM of synthetic lignin (Mićić et al., 2000b).

AFM study of polymer obtained using enzymic polymerization of coniferyl alcohol has confirmed the existence of strong intermolecular forces, which are responsible for holding lignin supermodules within the globule and also holding the globule together. Mićić et al. (2001a) have examined enzymatic polymer of coniferyl alcohol (DHP), using contact mode AFM. In the study, mechanical and rheological properties of individual globules of the lignin, as well as the forces of interaction between lignin globules, have been examined. The objective of this research was to answer if assembly of lignin semi-ordered structures, as observed by various imaging methods, is a result of random collision and packing of the lignin globules or is a consequence of intermolecular forces, such as hydrogen bonding and London force. Force spectroscopy studies have been performed using Si_3N cantilever tips, incubated with a lignin suspension for 12 h at $37°C$. The ESEM image of functionalized tip is shown in Fig. 6.4. Examination of mechanical and rheological properties of lignin globule was performed using a clean cantilever tip. Results show elastic behavior of the enzymatic DHP polymer. In force scans with clean tip there were no observed events which could be described as structure alteration such as

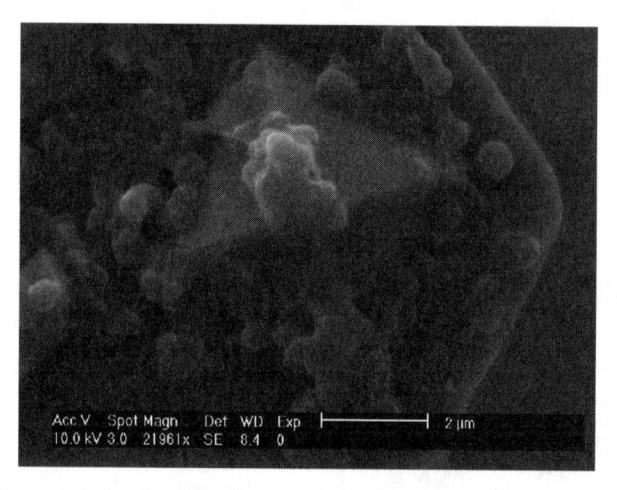

Figure 6.4. AFM tip functionalized with the lignin model compound (SEM micrography).

unwinding of lignin globules or filament pulling. Slight repulsive double-layer electrostatic force combined with a repulsive hydration force has also been observed. Measured adhesion forces were found to be in the range of 5–7 nN. Retraction curve shows adhesion due to van der Waals forces and indentation of the sample by the tip (Moy et al., 1994).

Much more interesting results are obtained with the use of a lignin functionalized tip. Some of the characteristic force scans are presented in Fig. 6.5. During the cantilever approach, which corresponds to compression, complex interactions are observed, resulting in nonlinear shape of the approaching curve. This shape indicates some kind of hydrophobic interaction, as there is a gradual pull-off. Also, some secondary structure unfolding and reorganization may contribute to the shape of the curve. The nonlinear event observed during the approach and retraction of the functionalized cantilever in the proximity of the substrate can be described as a fusion of two globules or globule secondary structure. However, these force curves clearly reveal the existence of three cohesion peaks during the cantilever retraction, within the distance corresponding to the diameter of individual lignin globules. Peaks with a force intensity range of 1–8 nN are reproducible, but their exact position and intensity vary. As lignin is the most entropic macromolecule in terms of the randomness of its structure as found in nature, it is believed that the discrepancy in position and intensity of peaks could be associated by the different ways that two lignin macromolecules or macromolecular assemblies can interact. The shape of the retraction peaks, which indicates a jump-off, expresses a strong hydrophobic van der Waals force and hydrogen bonding interactions.

Results of the AFM force scans, combined with the previous observation with ESEM, lead to the conclusion that the lignin globule consists of at least two-layered spherical, "onion-like" structure, which is composed of individual macromolecular subunits. An interior space between these shells may be filled with water or with water and air in order to accommodate hydrophobic and hydrophilic regions of lignin macromolecules. Layered structure of lignin has been finally confirmed by NSOM (Near Field Scanning Optical Microscopy) (Mićić et al., 2004). NSOM provides possibility to observe objects below diffraction limit. Schematic view of a lignin globule layered structure is presented on Fig. 7. Such organization of lignin globules may have physiological relevance, because it reduces the mass and amount of materials

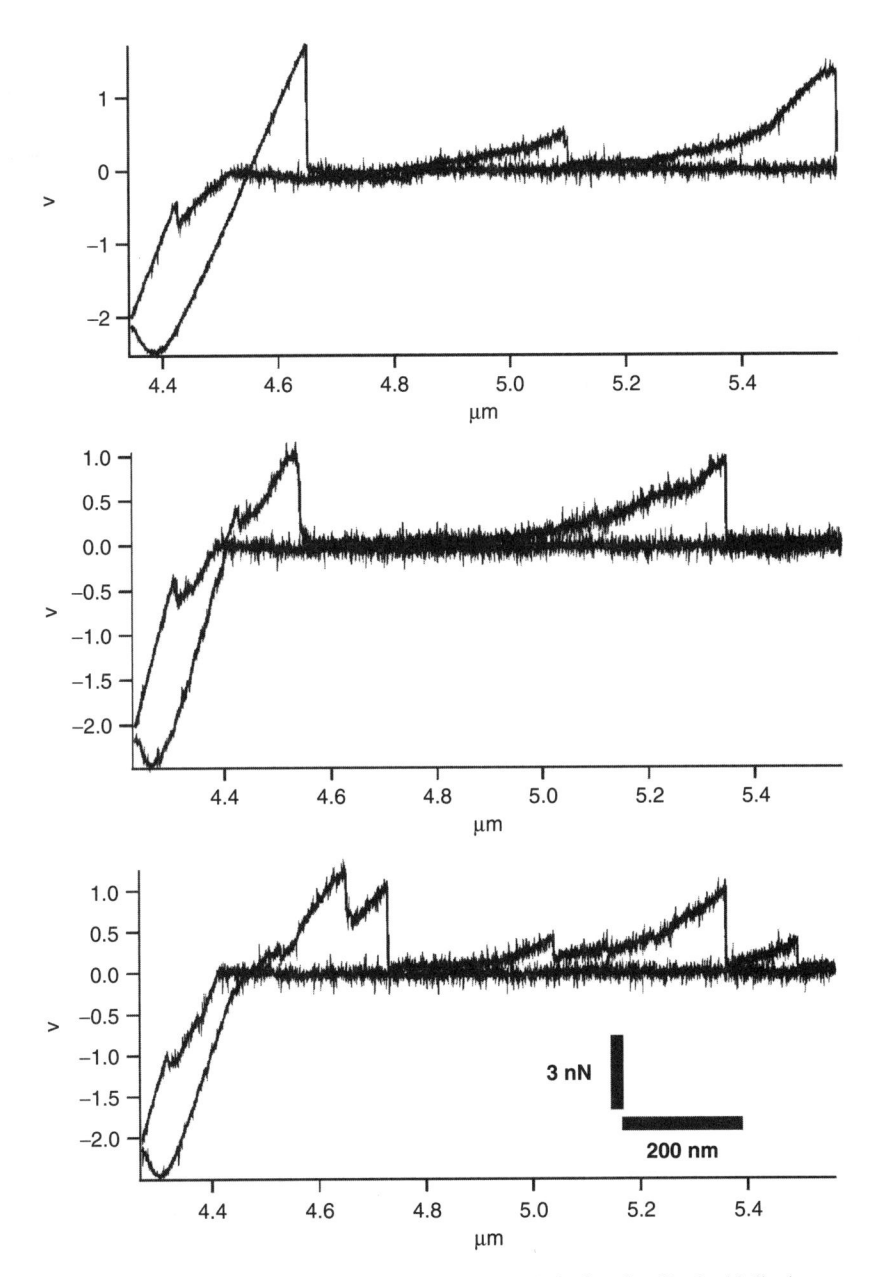

Figure 6.5. Atomic force scans obtained using AFM tip functionalized with lignin.

necessary to (a) create an individual lignin globule, (b) and make the globule act as a "shock absorber" in response to outside mechanical stress. Such structure also supports the concept of lignin-mediated water transport in plants (Laschimke, 1989). The structure is also in excellent agreement with electron spin resonance observations of the existence of areas of different mobility in lignin. The nonlinear shape of the peaks indicates complex interactions between lignin macromolecules in globular assemblies. The existence of intermolecular attractive forces between two lignin globules gives a clue that semi-ordered higher structures, such as channel and pore-like hexagonal structures, are not generated by the random collision of individual globules as earlier suggested by Falkehag (1975) and Jurasek (1995), but may be orchestrated by the actions of both hydrophilic and hydrophobic intermolecular attractive forces, such as $\pi-\pi$ interactions, hydrogen bonding, and van der Waals forces. The observed elastic property of lignin, along with its ability to reshape supramolecular structure under mutual interaction of globules, partially explains lignin's adaptability to grow directionally within a geometrically constrained cell matrix.

Light is one of the external factors affecting plant metabolism. There is a trend toward a permanent increase in low-wavelength UV radiation reaching the earth's surface (Bojkov, 1994), affecting plants as a form of stress. It has been found that short-wavelength UV light penetrates through the epidermal cell layer of the conifer needles into the mesophyll tissue (Delucia et al., 1992). Phenolic alcohols can be polymerized not only enzymatically, but also by photoirradiation to produce lignin-like polyer. However, their photochemical transformations have been less studied in the past. The cell wall is the first target of light action in the cell. The study of the mechanism of photochemical polymerization of phenolic alcohols to lignin-like polymer and the determination of structural differences of this polymer from the enzymatic polymer is a contribution to understanding how UV radiation affects plant tissue. Recent studies have shown that UV radiation induced polymerization of coniferyl alcohol proceeds through an ionic mechanism. Intermediate products of polymerization have been identified and characterized (Radotić et al., 1997). The structural differences between enzymatic and photochemical lignin can be well-discerned using STM (Fig. 6.2, 6.3, and 6.6). It appears that photochemical polymer contains the building units or modules, for larger assemblies such as supermodules.

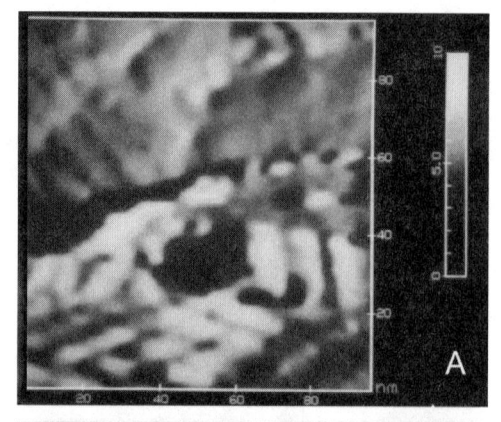

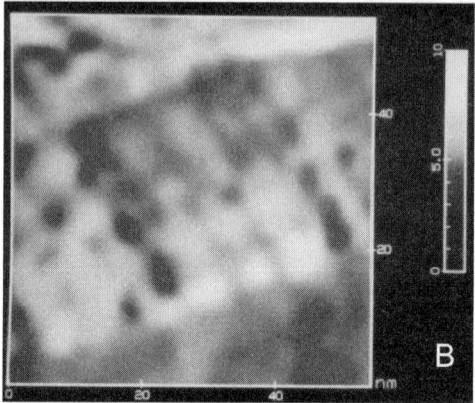

Figure 6.6. STM images of the photochemical polymer of coniferyl alcohol deposited onto a gold plate. (**A**) $U = 300$ mV, $I_T = 1$ nA, (**B**) $U = 300$ mV, $I_T = 1$ nA.

However, within the photochemical polymer there is a lower degree of order when compared with the enzymatic polymer (Radotić et al., 1998). The STM data are in agreement with the data of molecular mass distribution of the two polymers. The photochemical polymer contains, besides the two main fractions corresponding to the two fractions of the enzymatic polymer, two additional fractions that are probably the main reason for the observed structural differences. The overall features of the UV absorption spectrum, as well as ^{1}H NMR and Raman spectra of both polymers (Radotić, 1996; Radotić et al., 1998), show a general similarity, but also fine differences at the bond level.

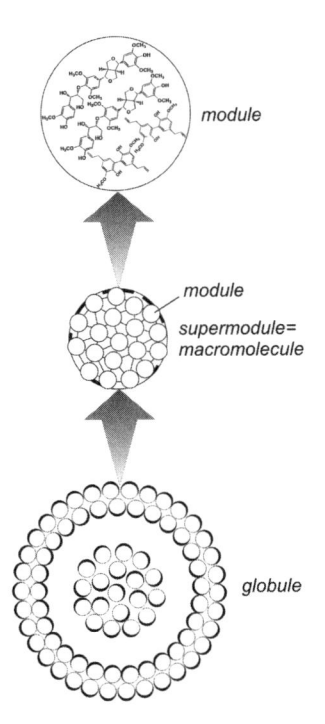

Figure 6.7. Schematic representation of lignin globule structure. The scheme is step-wise blown-up from globule to the assumed molecular structure of the first stage of polymerization (module). Thick and thin lines represent hydrophilic and hydrophobic regions, respectively.

AFM study of intermolecular forces in photochemical polymer confirms a higher randomness to the structure. Contrary to the enzymic polymer, photochemical polymers are nonfunctional (Mićić et al., 2001b). Results from this experiment provide evidence for the previously stated hypothesis that photochemical lignin polymerization may be one of a degrading effect of UV radiation on plant cells.

A comprehensive STM and AFM study of lignin model compounds and lignin has been done by Shevchenko et al. (1996). In the paper, the milled wood lignins from birch and spruce, dioxane extracted lignin from eucalyptus, DHP coniferyl alcohol-based polymeric model compound, and dimeric lignin model compound 1-(4-hydroxyphenyl)-2-(2-methodyphenoxy) ethanol were imaged. In the same paper, examples of humic acids were also examined, more specifically soil fulvic acid and river fulvic acid. All samples were deposited from the suspension as single droplets onto cleaved mice for AFM imaging, and STM imaging onto cleaved HOPG. The observations were justified by applying molecular modeling (SYBYL) to simulate observed motifs. Irregular aggregates of solid lignin compounds were observed using the AFM. The substrate conductivity and hydrophobicity significantly influence the lignin morphology at the single molecular level, in both lignin model compounds and in the extracted lignins. Molecular modeling of lignin forming the extended and flexible helices were also performed. The observation of such structures has been later confirmed by Mićić et al. (2000a,b) at the supramolecular level using the ESEM, at the 100 times higher orders of magnitude.

In order to avoid artifacts, AFM, in both the contact and tapping mode, has been employed by Simola-Gustafsson et al. (Simola et al., 2000; Simola-Gustafsson et al., 2001). Surface of the suspended colloids from pulp were analyzed in different stages of pulping process—that is, with and without the oxygen and alkali delignification. In these studies, there is a granular surface observed, like ones seen in studies of lignin model compounds (Mićić et al., 2000a,b). The smallest globular structures, measuring 20 nm in diameter, correspond to the expected molecular weight of a single lignin macromolecule—that is, a supermodule, which is 55,000 Da.

Since lignin is a constituent of the interfacial structure in a plant cell wall, there is the need to study lignins topographical features in a simulated environment. One way to achieve this is to making use of Langmuir–Blodgett films (Constantino et al., 2000; Pasquini et al., 2002). For example, Constantino et al. (2000) have examined Langmuir–Blodgett mono- and multiple-layer films of either (a) pure lignin extracted from sugar cane bagasse or (b) composite of lignin and amphiphillic salt (cadmium-stearate). Even given the poor stability of lignin monolayer, successful transfer to cleaved mica substrates have been possible. Langmuir–Blodgett film samples, ranging from monolayer to 21-fold layers, have been investigated using the AFM imaging in both contact and tapping mode. The Langmuir–Blodgett film of pure lignin express a smooth flat topography. In composite Langmuir–Blodgett films, comprised of cadmium-stearate amphiphile and lignin, there is phase separation, showing the clear separate domains of lignin and amphiphile. Lignin domains have more or less globular shape, with size similar to that of bulk lignin.

Pasquini et al. (2002) have also performed studies using Langmuir–Blodgett films of two different lignins, one cleaved of sugars and the other saccharified. The clean lignin has a flat surface, on L-B film morphology, while the saccharified forms irregular, rough surfaces. Based on the combined AFM and FTIR data, it is concluded that pure lignin arranges parallel to the substrate, forming compact film, while the saccharified lignin orients perpendicularly. Both lignins have random molecular organizations.

6.6. CONCLUSIONS

Scanning probe microscopies, especially AFM, is a powerful tool with unprecedented spatial resolutions and is finding its way into the different aspects of plant sciences, ranging from fundamental molecular biology and biophysics, to the applied fields of wood pulp, paper, and the food industry. The unique ability of the AFM to unveil details at the molecular or at least supramolecular level, without alternation of the sample, is providing a new opportunity for studies of fragile biological structures, even *in vivo*, which was not possible with previous high-resolution structural methods such as the electron microscopy. In this study, we have focused only on the plant cell wall structures, but nevertheless, AFM and other SPM techniques have been used and will continue to be used in much greater extent in studying other interesting aspects of plants, such as the photosynthetic center, structure of vacuoles, and the cell nuclei. In fundamental studies of the plant cell wall structures, we will witness in the near future significant advancements in our understanding of the complex structure of the plant cell wall. With further advent in design of the more user-friendly SPM setups, it is to be expected that in the near future SPM tools, especially tapping-mode AFM, will become a more common tool for the plant anatomist. Even more promising is the advancement of AFM techniques, which are still in their early stages of development but will enable simultaneous topographic and chemically specific imaging of structures at the nanoscale level. This will be mainly achieved by some variation of the near-field optical microscopy (NSOM), combined with the AFM, of which most promising candidates are a combination of AFM and surface-enhanced Raman spectroscopy (SERS) and fluorescence lifetime imaging (FLIM) (Mićić et al., 2002; Suh et al., 2002). It is to be envisioned that the applications of chemically specific imaging at the ultrastructural level *in vivo* will lead to a complete new understanding of the "metabolomics" of the plant cell. With the development of the high-throughput AFM microscope setups, it is expected that at some point AFM will probably be used as an analytical tool in the everyday production and quality control in the pulp and paper industry, as well as in food-related industries.

REFERENCES

Baker, A., Helbert, W., Sugiyama, J., and Miles, M. (1997). High-resolution atomic force microscopy of native *Valonia* cellulose I microcrystals. *J. Struct. Biol.* **119**:129–138.

Bardinskaya, M. S., and Safonov, V. I. (1959). Izuchenie biosinteza lignina v izolirivannih tkanjah morkovi. *Dokl. Akad. Nauk SSSR* **129**:205–208.

REFERENCES

Bayer, E. A., Chanzi, H., Yamed, R., and Shoham, Y. (1998). Cellulose, cellulases and cellulosomes. *Curr. Opin. Struct. Biol.* **8**:548–557.

Bernardini, N., Penel, C., and Grepin, H. (1986). Preliminary immunological studies on horseradish and spinach peroxidase isoenzymes. In: *Molecular and Physiological Aspects of Plant Peroxidases* (H. Greppin, C. Penel, and T. Gaspar, eds.), pp. 97–104, University of Geneva.

Blankey, A. B., Harris, P. J., Henry, R. J., and Stone, B. A. (1983). A simple and rapid preparation of alditol acetates for monosaccharide analysis. *Carbohydrate Res.* **113**:291–299.

Bojkov, R. D. (1994). The ozone layer—recent developments. *World Meteorol. Org. Bull.* **43**:113–116.

Boras, L., and Gatenholm, P. (1999). Surface properties of mechanical pulps prepared under various sulfonation conditions and preheating time. *Holzforschung* **53**:429–434.

Castillo, F. J. (1986). Extracellular peroxidases as markers of stress? In: *Molecular and Physiological Aspects of Plant Peroxidases* (H. Greppin, C. Penel, and T. Gaspar, eds.), pp. 419–426, University of Geneva.

Catesson, A. M., Imberty, A., Goldberg, R., and Czaninski, Y. (1986). Nature, localization and specificity of peroxidases involved in lignification processes. In: *Molecular and Physiological Aspects of Plant Peroxidases* (H. Greppin, C. Penel, and T. Gaspar, eds.), pp. 189–198, University of Geneva.

Chanzy, H., and Henrissat, B. (1985). Unidirectional degradation of Valonia cellulose microcrystals subjected to cellulase action. *FEBS Lett.* **184**:285–288.

Chibbar, R. N., and van Huystee, R. B. (1986). Immunochemical localization of peroxidase in cultured peanut cells. *J. Plant Physiol.* **123**:477–486.

Conroy, J. M. (1986). Immunological probes of relationships among plant peroxidases: polyclonal and monoclonal antibodies. In: *Molecular and Physiological Aspects of Plant Peroxidases* (H. Greppin, C. Penel, and T. Gaspar, eds.), pp. 85–96, University of Geneva.

Constantino, C. J. L., Dhanabalan, A., Cotta, M. A., Pereira-daSilva, M. A., Curvelo, A. A. S., and Oliveira, O. N., Jr. (2000). Atomic force microscopy (AFM) investigation of Langmuir–Blodgett (LB) films of sugar cane bagasse lignin. *Holzforschung* **54**:55.

Delucia, E. H., Day, T. A., and Vogelman, T. C. (1992). Ultraviolet-B and visible-light penetration into needles of two species of sub-alpine conifers during foliar development. *Plant Cell Environ.* **15**:921–929.

Duchesne, I., and Daniel, G. (1999). The ultrastructure of wood fibre surfaces as shown by a variety of microscopical methods—a review. *Nordic Pulp Paper Res. J.* **14**:129.

Elder, T. J., and Worley, S. D. (1984). The application of molecular orbital calculations to wood chemistry. *Wood Sci. Technol.* **18**:307–315.

Erdtman, H. (1957). Outstanding problems in lignin chemistry. *Ind. Eng. Chem.* **49**: 1385–1386.

Falkehag, S. I. (1975). Lignin in materials. *Appl. Polym. Symp.* **28**:247–257.

Fengel, V. D., Wegener, G., and Feckl, J. (1981). Beitrag zur Charakterisierung analytischer und technischer Lignine. *Holzforschung* **35**: 111–118.

Freudenberg, K., and Niedercorn, F. (1958). Umwandlung des phenylalanins in coniferin und fichtenlignin. *Chem. Ber.* **91**:591–597.

Freudenberg, K., and Toribio, P. (1969). Der abbau des dehydrierungsproduktes aus coniferylalcohol-($^{14}CH_2OH$) und des in gegenwart phenylalanin-($^{14}CO_2H$) gewarhesenen lignins. *Chem. Ber.* **102**:1312–1315.

Freudenberg, K., Harkin, J. M., Reichert, M., and Fukuzumi, T. (1958). Die an der verholzung beteiligten enzyme. Die dehydrierung des sinapinalkohola. *Chem. Ber.* **91**:581–590.

Gaspar, T., Penel, C., Castillo, F., and Greppin, H. (1985). A two-step control of basic and acidic peroxidases and its significance for growth and development. *Physiol. Plant.* **64**:418–423.

Goodwin, T. W., and Mercer, E. I. (1985). *Introduction to Plant Biochemistry*, Pergamon Press, Oxford.

Ha, M. A., Apperley, D. C., and Jarvis, M. C. (1997). Molecular rigidity in dry and hydrated onion cell walls. *Plant Physiol.* **115**:593–598.

Harkin, J. M., and Obst, J. R. (1973). Lignification in trees: Indication of exclusive peroxidase participation. *Science* **180**:296–297.

Hayashi, N., Sugiyama, J., Okano, T., and Ishihara, M. (1998). The enzymatic susceptibility of cellulose microfibrils of the algal-bacterial type and the cotton-ramie type. *Carbohydrate Res.* **305**:261–269.

Hert, W., and Hausser, I. (1984). Chitin and cellulose fibrillogenesis *in vivo* and their experimental alteration. In: (W. M. Dugger and S. Bartnicki-Garcia, eds.), *Structure, Function and Biosynthesis of Plant Cell Walls*, pp. 89–119, American Society of Plant Physiologists, Rockville, MD.

Higuchi, T., and Brown, S. A. (1963). Studies of lignin biosynthesis using isotopic carbon XI. Reactions relating to lignification in young wheat plants. *Canad. J. Biochem. Physiol.* **41**:65–76.

Hwang, R. H. (1982). An approach to lignification in plants. *Biochem. Biophys. Res. Commun.* **105**:509–514.

Hwang, R. H. (1985). A lignification mechanism. *J. Theor. Biol.* **116**:21–44.

Imberty, A., Goldberg, R., and Catesson, A. (1985). Isolation and characterization of Populus isoperoxidases involved in the last step of lignin formation. *Planta* **164**:221–226.

Jurasek, L. J. (1995). Toward a 3-dimensional model of lignin. *J. Pulp Paper Sci.* **21**:J274.

Karmanov, A. P., Fillipov, V. N., and Moskvicheva, T. V. (1993). Issledovanie poverhnostnoi morfologiceskoi strukturi lignina. *Khim. Drev.* **1–3**:123–127.

Košikova, B., Zakutna, L., and Joniak, D. (1978). Investigation of the lignin–saccharidic complex by electron microscopy. *Holzforschung* **32**:15–18.

Kressig, H. A. (ed.) (1993). *Cellulose. Structure, Accessibility and Reactivity*. Gordon and Breach Science, Yverdon.

Laschimke, R. (1989). Investigation of the weting behaviour of natural lignin—a contribution to the cohesion theory of water transport. *Thermochim. Acta* **151**:35–36.

Lee, I., Evans, B. R., Lane, L. M., and Woodward, J. (1996). Substrate–enzyme interactions in cellulase systems. *Biosci. Technol.* **58**:163–169.

Lee, I., Evans, B. R., and Woodward, J. (2000). The mechanism of cellulase action on cotton fibers: evidence from atomic force microscopy. *Ultramicroscopy* **82**:213–221.

Lewis, N. G., and Yamamoto, E. (1990). Lignin: Occurrence, biogenesis and biodegradation. *Annu. Rev. Plant Physiol. Plant Mol. Biol.* **41**:455–496.

Marcinowski, S., and Grisebach, H. (1977). Turnover of coniferin in pine seedlings. *Phytochemistry* **16**:1665–1667.

McCann, M. C., Wells, B., and Roberts, K. (1990). Direct visualization of cross-links in the primary plant cell wall. *J. Cell Sci.* **96**:323–334.

Mehlhorn, H., and Kunert, K. J. (1986). Ascorbic acid, phenolic compounds and plant peroxidases: A natural defense system against peroxidative stress in higher plants? In: *Molecular and Physiological Aspects of Plant Peroxidases* (H. Greppin, C. Penel, and T. Gaspar, eds.), pp. 437–449, University of Geneva.

Mićić, M., Jeremić, M., Radotić, K., and Leblanc, R. M. (2000a). A comparative study of enzymatically and photochemically polymerized artificial lignin supramolecular structures using environmental scanning electron microscopy. *J. Coll. Interface Sci.* **231**:190–194.

Mićić, M., Radotić, K., Jeremić, M., Mavers, M., and Leblanc, R. M. (2000b). Visualization of artificial lignin supramolecular structures. *Scanning* **22**:288–294.

Mićić, M., Benitez, I., Ruano, M., Mavers, M., Jeremić, M., Radotić, K., Moy, V., and Leblanc, R. M. (2001a). Probing the lignin nanomechanical properties and lignin–lignin interactions using the atomic force microscopy. *Chem. Phys. Lett.* **347**:41–45.

Mićić, M., Radotić, K., Benitez, I., Ruano, M., Jeremić, M., Moy, V., and Leblanc, R. M. (2001b). Topographical characterization and surface force spectroscopy of the photochemical lignin model compound. *Biophys. Chem.* **94**:257.

Mićić, M., Klymyshyn, N., Suh, Y. D., and Lu, H. P. (2003). Finite element methods simulation of the field distribution for the AFM metallic tip enhanced SERS microscopy, *J. Phys. Chem. B.* **107**:1574–1584..

Mićić, M., Radotić, K., Jeremić, M., Djikanović, D, Kämmer, S. (2004). Study of the lignin model compound supramolecular structure by combination of near-field scanning optical microscopy and atomic force microscopy. *Coll. Surf. B: Biointerfaces* **34**:33.

Morris, V. J., Gunning, A. P., Kirby, A. R., Round, A., Waldron, K., and Ng, A. (1997). Atomic force microscopy of plant cell walls, plant cell wall polysaccharides and gels. *Int. J. Biol. Macromol.* **21**:61–66.

Moy, V. T., Florin, E. L., and Gaub, H. E. (1994). Intermolecular forces and energies between ligands and receptors. *Science* **266**:257.

Niemi, H., Paulapuro, H., and Mahlberg, R. (2002). Application of scanning probe microscopy to wood, fiber and paper research. *Paperi, Ja. Puu-Paper and Timber* **84**:389–406.

Nozu, Y. (1967). Studies on the biosynthesis of lignin. *J. Biochem.* **62**:519–530.

O'Sullivan, A. C. (1997). Cellulose: The structure slowly unravels. *Cellulose* **4**:173–207.

Pandolfini, T., Gabrielli, R., and Comparini, C. (1992). Nickel toxicity and peroxidase activity in seedlings of *Triticum aestivum* L. *Plant Cell Environ.* **15**:719–725

Pasquini, D., Balogh, D. T., Antunes, P. A., Constantino, C. J. L., Curvelo, A. A. S., Aroca, R. F., and Oliviera, O. N. (2002). Surface morphology and molecular organization of lignins in Langmuir–Blodgett films. *Langmuir* **18**:6593.

Pickett-Heaps, J. D. (1968). Xylem wall deposition. Radioautographic investigations using lignin precursors. *Protoplasma* **3**:55–64.

Polle, A., and Chakrabarti, K. (1994). Effects of manganese deficiency on soluble apoplastic peroxidase activities and lignin content in needles of Norway spruce (*Picea abies*). *Tree Physiol.* **14**:1191–1200.

Radotić, K. (1994). Biophysical and biochemical characterization of the structure and function of the cell walls from maize (*Zea mays* L.). Doctorial thesis. University of Belgrade (in Serbian).

Radotić, K. (1996). Enzymatic and photochemical polymerization of coniferyl alcohol: A kinetic and structural study. *Polyphenol Commun.* **96**:497–498.

Radotić, K., Simi-Krsti, J., Jeremi, M., and Trifunovi, M. (1994). A study of lignin formation at the molecular level by scanning tunneling microscopy. *Biophys. J.* **66**:1763–1767.

Radotić, K., Zakrzewska, J., Sladi, D., and Jeremi, M. (1997). Study of photochemical reactions of coniferyl alcohol. I. Mechanism and intermediate products of UV radiation-induced polymerization of coniferyl alcohol. *Photochem. Photobiol.* **65**:284–291.

Radotić, K., Budinski-Simendic, J., and Jeremic, M. (1998). A comparative structural study of enzymic and photochemical lignin by scanning tunneling microscopy and molecular mass distribution. *Chem. Ind.* **52**:347–350.

Ridley, B. L., O'Neill, M. A., and Mohnen, D. A. (2001). Pectins: Structure, biosynthesis, and oligogalacturonide-related signaling. *Phytochemistry* **57**:929–967.

Roelofsen, P. A. (1965). Ultrastructure of the wall in growing cells and its relation to the direction of the growth. *Adv. Bot. Res.* **2**:69–149.

Round, A. N., Kirby, A. R., and Morris, V. J. (1996). Collection and processing of AFM images of plant cell walls. *Microscopy and Analysis*, **Sept**:33–35.

Round, A. N., McDougall, A. J., Ring, S. G., and Morris, V. J. (1997). Unexpected branching in pectin observed by atomic force microscopy. *Carbohydrate Res.* **303**:251–253.

Round, A. N., Rigby, N. M., MacDougall, A. J., Ring, S. G., and Morris, V. J. (2001). Investigating the nature of branching in pectin by atomic force microscopy and carbohydrate analysis. *Carbohydrate Res.* **331**:337–342.

Schmid, G., and Grisebach, H. (1982). Enzymic synthesis of lignin precursors. Purification and properties of UDPglucose: Coniferyl alcohol glucosyltransferase from cambial sap of spruce (*Picea abies* L.). *Eur. J. Biochem.* **123**:363–370.

Schmid, G., Hammer, D. K., Ritterbusch, A., and Grisebach, H. (1982). Appearance and immunohistochemical localization of UDP-glucose: coniferyl alcohol glucosyltranferase in spruce (*Picea abies* (L.) Karst) seedlings. *Planta* **156**:207–212.

Schubert, W. J., and Nord, F. F. (1957). Mechanism of lignification. *Ind. Eng. Chem.* **49**:1387–1390.

Shann, J. R., and Blum, U. (1987). The utilisation of exogenously supplied ferulic acid in lignin biosynthesis. *Phytochemistry* **26**:2977–2982.

Shevchenko, S. M., and Bailey, G. W. (1996). Nanoscale morphology of lignins and their chemical transformation products. *Tappi* **79**: 227.

Simola, J., Malkavaara, P., Alen, R., and Pletonen, J. (2000). Scanning probe microscopy of pine and birch kraft pulp fibers. *Polymer* **41**:21.

Simola-Gustafsson, J., Hortling, B., and Pletonen, J. (2001). Scanning probe microscopy and enhanced data analysis on lignin and elemental chlorine-free or oxygen delignified pine kraft pulp. *Coll. Polym. Sci.* **279**:221.

Stewart, A., Brown, K., Tanner, G., and Stone, J. E. (1953). Studies of lignin biosynthesis using isotopic carbon II. Short-term experiments with $C^{14}O_2$. *Canad. J. Chem.* **31**:755–760.

Stone, J. E. (1953). Studies of lignin biosynthesis using isotopic carbon I. Long-term experiment with $C^{14}O_2$. *Canad. J. Chem.* **31**:207–213.

Stout, G. H., and Jensen, L. H. (1968). *X-Ray Structure Determination*, Macmillan, New York.

Suh, Y. D., Schenter, G. K., Zhu, L., and Lu, H. P (2003). Probing nano-surface enhanced Raman scattering fluctuation dynamics using correlated AFM and confocal microscopy. *Ultramicroscopy* **97**:89–94.

Varner, J. E., and Lin, L.-S. (1989). Plant cell wall architecture. *Cell* **56**:231–239.

Wayman, M., and Obiaga, T. I. (1974). The modular structure of lignin. *Can. J. Chem.* **52**:2102–2111.

Zakutna, L., Košikova, B., and Eberingerova A. (1983). Partial alkaline delignification of spurce wood. IV Electron microscopy of the lignin and hemicellulose fractions isolated from pulp. *Dervarsky Vyskum* **28**:15.

CHAPTER 7

CELLULAR INTERACTIONS OF NANO DRUG DELIVERY SYSTEMS

RANGARAMANUJAM M. KANNAN

Department of Chemical Engineering and Material Science, and Biomedical Engineering, Wayne State University, Detroit, Michigan

OMATHANU PILLAI PERUMAL

Department of Pharmaceutical Sciences, South Dakota State University College of Pharmacy, Brookings, South Dakota

SUJATHA KANNAN

Department of Critical Care Medicine, Children's Hospital of Michigan, Detroit, Michigan

7.1. INTRODUCTION

The deciphering of the human genome has opened up multiple avenues for identifying new targets and developing novel therapies (Drews, 2000). With the availability of a plethora of genomic data and high-throughput screens, there is a compelling need to translate this "new knowledge" into successful therapies. Intelligent design of drug delivery systems (DDS) to meet this need requires a thorough understanding of the factors influencing the disease, drug, destination and delivery (Fig. 7.1). Drug delivery scientists have been successful in achieving spatial and temporal control of drug from DDS, but the path for transport of drug molecules from the delivery system to the site of action remains largely uncontrolled. Targeting and releasing the drug as close as possible to the site of action can fulfill this objective. The efficient functioning and regulation of the biological system are strictly controlled by structural hierarchy varying from centimeter to nanometer scales (Esfand and Tomalia, 2001). Many of the bioassemblies and endogenous molecules (such as enzymes, hormones and nucleic acids) exist and function at the nanoscale level. Majority of the therapeutic targets exist intracellularly or at the surface of the cells. Hence the goal of the drug delivery system is to deliver drugs to the intended target with minimal interference with normal cellular functioning and with the least systemic side effects.

With the development and use of force microscopy and advances in nanotechnology we have an opportunity to design and assemble DDS at nanoscale dimension to deliver the payload at the required target site at the appropriate time. By engineering nanoscale drug delivery vehicles to mimic the biological system, it is possible to achieve different degrees of targeting from organ to tissue and eventually at the cellular level.

Force Microscopy: Applications in Biology and Medicine, edited by Bhanu P. Jena and J.K. Heinrich Hörber.
Copyright © 2006 John Wiley & Sons, Inc.

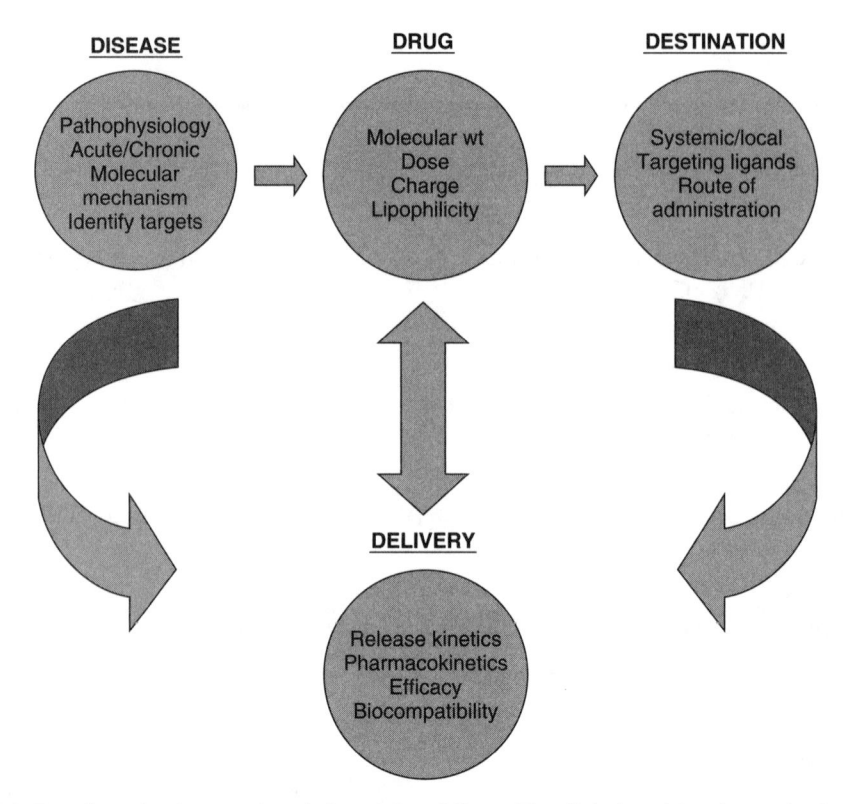

Figure 7.1. Four-dimensional approach to design of drug delivery. The clinical need based on pathophysiology of the disease is the major driving force for developing new drug delivery systems. "Smart" drug delivery systems can maximize the therapeutic efficacy of the drug by targeting it to the desired destination.

A better appreciation of the natural cell trafficking mechanisms in cells has helped in the design of nanoscale DDS targeted to specific transport pathways. Based on the location and function of cells, several specific transport mechanisms have been developed to meet the essential nutrient requirements of the intracellular organelles and also "kick out" unwanted foreign bodies and toxins to the extracellular medium (Fig. 7.2). Apart from simple diffusion of very small molecules, cells have evolved various mechanisms to internalize small molecules, macromolecules, and particles and target them to specific organelles within the cytoplasm. Collectively, these processes are referred to as "endocytosis" (De Duve, 1963), which includes (i) phagocytosis or cell eating, (ii) pinocytosis or cell drinking, (iii) clathrin-dependent receptor-mediated endocytosis, and (iv) clathrin-independent receptor-mediated endocytosis.

Phagocytosis, which involves the uptake of large particles (>300 nm) takes place in many cells and most importantly in specialized cells such as macrophages, where it plays an important role in cellular immunity. Macropinocytosis is a bulk internalization process and is seen in a number of phagocytic and nonphagocytic cells (Johannes and Lamaze, 2002; Amyere et al., 2002). Receptor-mediated endocytosis takes place in all nucleated vertebrate cells and plays an important role in many of the physiological processes. Clathrin-coated pits are ~150-nm invaginated structures of the plasma membrane varying from 0.4% of membrane surface area in adipocytes to 3.8%

INTRODUCTION 115

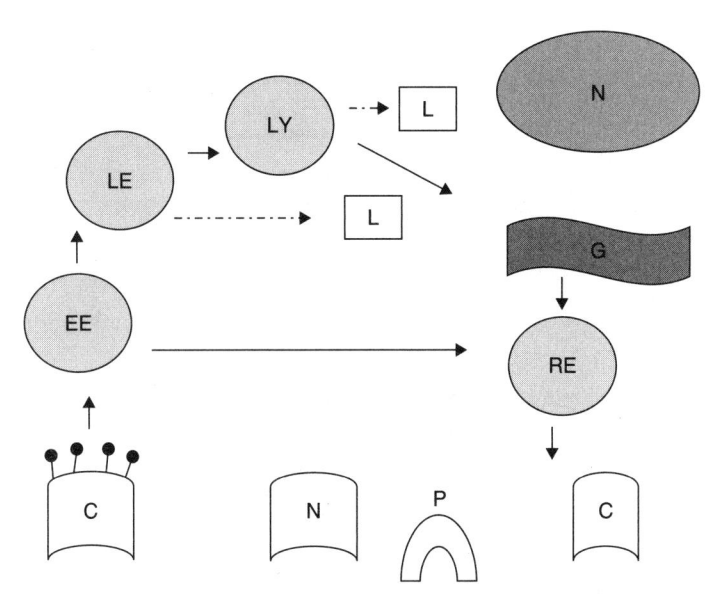

Figure 7.2. Various cell trafficking pathways. C, clathrin-coated pits; NC, non-clathrin pits; P, phagocytosis/macropinocytosis; CE, caveolae; EE, early endosome; LE, late endosome; LY, lysosome; L, ligand; N, nucleus; G, Golgi apparatus; RE, recycling endosomes. The pH in the endosomes is 5.9–6.0, while the lysosomal pH is 5.9–5.0. The broken arrows represent the escape of the ligand from the endososmes or from the lysosomes.

in some human fibroblasts (Goldberg et al., 1987; Nilsson et al., 1983). Some receptors are constitutively concentrated in the coated pits (low-density lipoprotein receptor), but others become concentrated upon ligand binding (epidermal growth factor). By electron microscopy, caveolae appear as ∼70-nm-diameter flask-shaped plasmalemmal invaginations formed through the oligomerization of structural coat protein caveolin (Peters et al., 1985). They are present on many cell types including fibroblasts, cardiomyocytes, and adipocytes mediating endocytosis to deliver a variety of molecules to intracellular compartments including endoplasmic reticulum, golgi apparatus, endosomes, and lysosomes (Schnitzer, 2001).

After internalization, the molecules are degraded and recycled back to the cell surface. Endoctyosis presents the ligand to the early endosomes and then to late endosomes, where the pH drops to 5.9–6.0. Then the molecules progress to lysosomes, where the pH further reduces from 6.0 to 5.0. In addition, the lysosomes contain enzymes to breakdown the cargo. Alternative fates such as trafficking to organelles like golgi apparatus or translocation into cytosol also occur. The receptors are recycled back to the cell surface by the recycling endosomes. Cellular trafficking is a complex process (Mukherjee et al., 1997) that depends on physicochemical properties of internalized molecules (ligands, pH optima of ligand-carrier dissociation) and the biological characteristics of the cell organelles that carry them (size, shape, luminal pH, lipids, and composition). Based on the intracellular targets and the unique functions of the target cells, nanoscale drug delivery devices that would exploit various cellular trafficking mechanisms can be designed such that the drug is delivered to the specific intracellular site.

Pathophysiological conditions such as cancer present certain unique characteristics that are not observed in normal tissue, including (i) extensive angiogenesis, (ii) defective (leaky) vascular architecture, (iii) impaired lymphatic drainage, and (iv) increased production/expression of permeability mediators on the cell surface (Brigger et al., 2002). These characteristics may also help in concentrating the nanoscale drug delivery systems at the site of the tumor.

By tagging suitable ligands to the nanovehicles, the drug cargo can be selectively targeted to the diseased cell. On the other hand, by using organelle-specific molecular probes and imaging techniques, the kinetics and location of nanovehicle can be followed to understand and validate the nanodrug delivery systems for intracellular drug delivery. This chapter discusses how the various nanovehicles can be engineered by manipulating the chemistry and the physicochemical properties of the vehicle to selectively and specifically interact with the biological system (Fig. 7.3), for achieving improved therapeutic efficacy and clinical outcome. The focus of this chapter is restricted to mainly lipid-based and polymer-based nanovehicles. To highlight the importance of size within the nanometer scale, nanovehicles varying in size from 1 nm (dendrimers) to 1000 nm (liposomes) have been discussed. However, the major emphasis is given on dendrimer-based drug delivery systems because these are the smallest of the nanoscale delivery systems and have been used as a case study with two model drugs (ibuprofen and methylprednisolone) to demonstrate the chemistry, cell entry dynamics, and biological activity of the nanovehicle (dendrimer) *per se* and in presence of the drug as a conjugate.

7.2. LIPID-BASED SYSTEMS

7.2.1. Conventional Liposomes

Liposomes are probably one of the earliest known nanosized drug delivery vehicles and are made of biodegradable lipid vesicle (phospholipids) with aqueous phase surrounded by a lipid bilayer. Liposomes have been used to entrap a wide variety of drug substances. Water-soluble drugs are entrapped in the aqueous compartment, while the lipophilic substances get incorporated in the lipid bilayer. Depending on the lipids, preparation method, and number and position of lipid lamellae, liposomes can vary widely in size (10–1000 nm) (Barenholz and Crommelin, 1994). They are rapidly taken up by the reticuloendothelial system from the circulation and hence can be used for passive targeting to the lymphatic system. The size of the liposomes is the main factor controlling the lymphatic uptake, and small

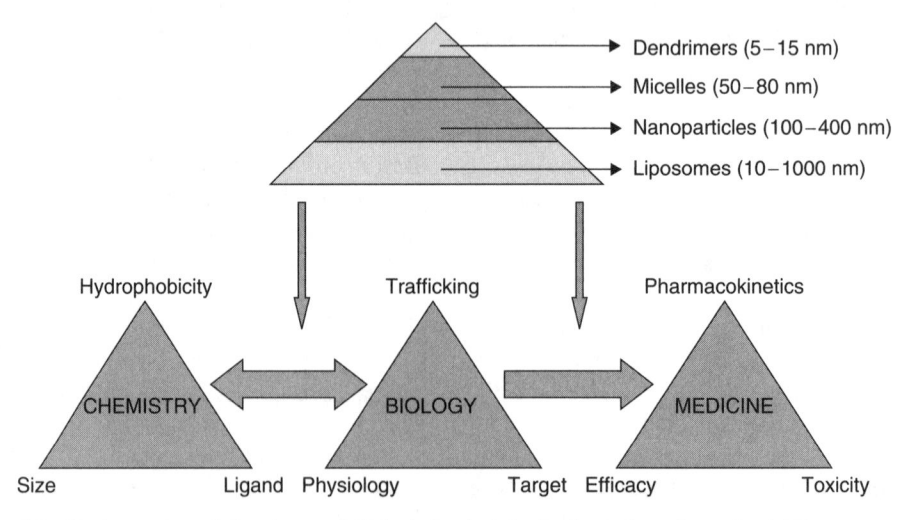

Figure 7.3. The interaction of chemistry and biological principles in the design of nanodrug delivery systems for medicine.

liposomes seem to achieve high lymphatic uptake (Oussoren and Storm, 1997; Oussoren et al., 1997). Similarly, surface charge and the route of administration also govern the lymphatic delivery of drugs (Kim and Han, 1995). Phagocytosis by macrophages plays an important role in the cellular uptake of liposomes. Alternate pathways of cellular entry of liposomes include fusion with the cell membrane, charge interaction of cationic liposomes with the membrane, endosomal uptake, and release of water-soluble contents into the cytoplasm (Felgner and Ringold, 1989).

Liposomes have been widely studied as a vehicle for targeting antitumor agents. The amount of liposomes accumulating in the tumor is dictated by (i) residence time in blood, (ii) transfer from blood to tumor interstitial spaces, (iii) local retention of liposomes in tumor tissue, and (iv) efflux from the tumor tissue to blood (Nagayasu et al., 1999). In tumor tissue, the permeability of capillary endothelium is enhanced, while the lymphatic system is poorly developed in comparison to normal tissues, and this phenomenon is termed as enhanced permeability and retention (EPR) effect (Fig. 7.4). This altered physiology has been exploited for passively targeting liposome-encapsulated drugs to the tumor (Jain, 1987). Smaller, long-circulating liposomes (~100 nm) have higher frequency of extravasation into tumor tissue through the leaky vasculature and are retained in the tumor longer due to the compromised lymphatic drainage in the rapidly dividing cells. The size of the liposomes also plays a role in dictating its blood residence time, as larger particles are rapidly cleared by macrophages by opsonization of plasma proteins. It has been reported that 100 nm represents an optimal size for blood to tumor transfer and for longer retention and biodistribution in tumor (Nagayasu et al., 1996). At the same time, the lipid composition is also critical for

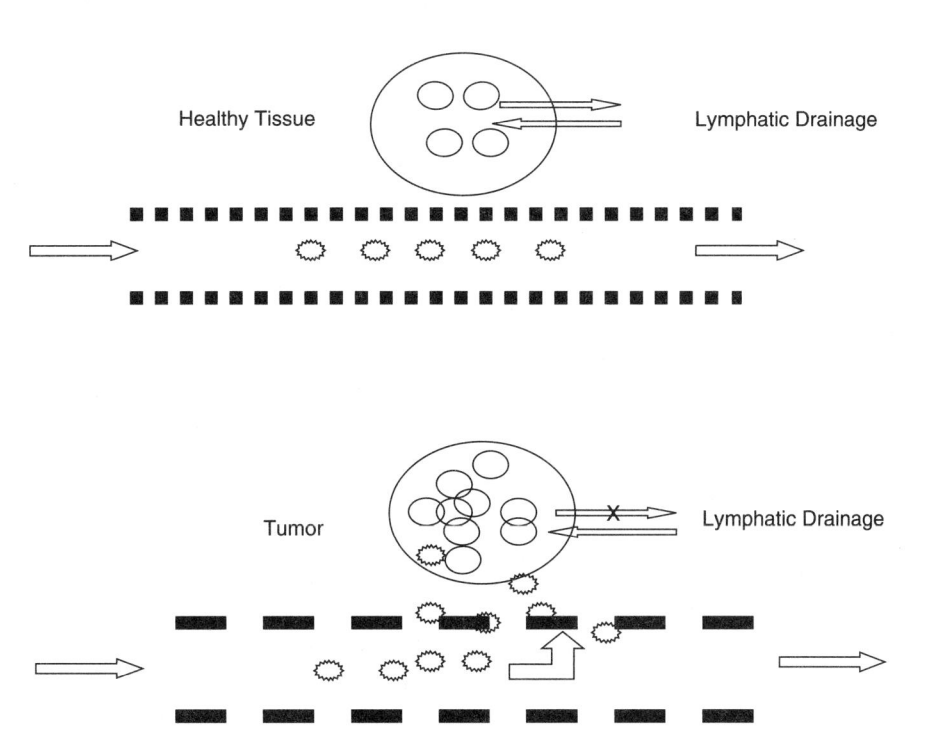

Figure 7.4. Schematic representation of enhanced permeability effect (EPR) in the tumor tissue in comparison to the healthy tissue.

controlling drug release in addition to prolonging the circulation half-life of liposomes.

7.2.2. Modified Liposomes

Drug delivery by conventional liposomes is severely limited by their rapid detection and blood clearance by the macrophages. This, in addition to reducing the efficacy of the drugs, also poses potential toxicity to the macrophages and causes bone marrow suppression (Senior, 1987; Daeman et al., 1997). Surface modification of liposomes with polyethylene glycol (PEG) or other selective hydrophilic polymers offers a number of advantages in this regard (Woodle, 1998); (i) Prolonged circulation is independent of liposomal bilayer, hence drug loading and release can be adjusted independent from blood circulation; (ii) pharmacokinetics and tissue distribution now becomes lipid independent, and therefore drug loading can be increased; (iii) ligands with other functionality can be attached to the polymer for active targeting. The modified liposomes are referred to as "stealth" or "sterically stabilized" liposomes, which reduces protein adsorption and consequently cellular interactions (Liu, 1997). This results in reduced blood clearance of liposomes and concomitant changes in tissue distribution. When conventional liposomes are used for intracellular delivery of macromolecules such as proteins and nucleic acids, the cargo is inactivated by the acidic environment and enzymes in lysosomes before reaching the site of action in the cytosol. Incorporation of lipids that undergo phase transformation in response to a pH stimuli have been used to deliver the contents intact to the cytosol (Hafez and Cullis, 2001). Dioleoyl phosphatidyl ethanolamine (DOPE) is a nonbilayer lipid that is used in the preparation of so-called "fusogenic liposomes." It derives its name due to its ability to undergo lamellar to hexagonal phase transformation in response to acidic pH stimulus (Felgner et al., 1987). Following endocytosis, pH-sensitive liposomes undergo destabilization and fuse with the endosomal membrane upon encountering acidic pH (5–6.5) in the endosomes. The membrane fusion helps in releasing the liposomal contents into the cytosol, thereby avoiding the degradation in the lysosomes (Fig. 7.5). Cationic lipid–DNA lipoplexes are commercially available as lipofectin, which contains a mixture of two cationic liposomes namely 2,3-bis(oleoyl)oxipropyl trimethyl ammonium chloride (DOTMA) and DOPE. This showed improved DNA transfection in several cells (de Lima et al., 2001). However, the cationic lipids cause formation of aggregates by interacting with the negatively charged serum proteins, hence affecting the biodistribution of liposomes in tissues (Zhu et al., 1993).

A novel fusogenic liposome known as "virosome" (Kaneda et al., 1999) was constructed using a fusogenic envelope derived from UV-inactivated heamaglutinating virus of Japan (HVJ; Sendai virus). As virosomes, liposomes acquire viral functions and can fuse with almost all cells except peripheral lymphocytes. Fusion is complete within 10–30 min at 37°C as opposed to 5–20 h with cationic liposomes. The unique characteristics of virosomes have been studied using confocal microscopy. Cytoplasm was found to be stained with FITC-dextran, when delivered by fusogenic liposomes, while aggregated fluorescence indicating endosomal presence in cells cultured with FITC-dextran-encapsulated conventional liposomes (Kunisawa et al., 2001). The endocytosis inhibitor (cytochalesin B) did not affect the delivery of virosomes, indicating that virosomes delivered their contents into the cytosol via an endocytosis-independent pathway by fusing directly with the cell membrane. Similarly, the delivery of nucleic acid by virosomes to the nucleus has been demonstrated using FITC-labeled oligonucleotide, where within 5 min fluorescence was observed in the nucleus and was stable for 72 h (Morishita et al., 1994).

Liposomes that respond to thermal stimulus can be designed by suitable choice of

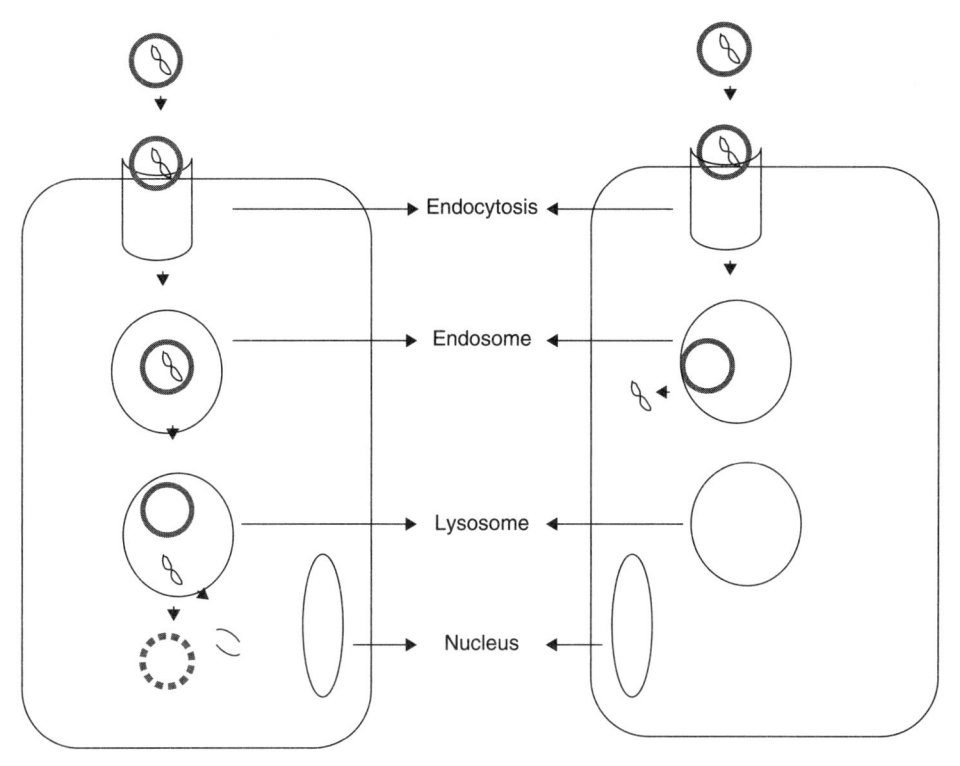

Figure 7.5. Schematic representation of the intracellular fate of conventional (left-hand panel; green) and fusogenic liposomes (right-hand panel; red). The fusogenic liposomes release the contents, when there is a changes in the lipid phase in the acidic pH environment of the endosome.

lipid that undergoes gel-to-sol transition in response to temperature. Alternatively, poly-(N-isopropyl acrylamide) [poly(NIPAM)], which has a lower consolute solution temperature, can be attached on the liposome surface. Such liposomes can be used to selectively release the drug contents in the tumor in conjunction with hyperthermia (Kono, 2001).

To achieve specific cellular targeting, various ligands have been attached to liposomes, including peptide-based ligands, sugar-based ligands, or glycoproteins. These ligand-decorated liposomes utilize the receptor-mediated endocytosis to get internalized into the cell. Galactosylated liposomes are taken up selectively by asialoglycoproteins receptor-positive liver parenchymal cells, while mannose residues resulted in specific delivery of genes to nonparenchymal liver cells (Hashida et al., 2001).

A very attractive tumor-selective ligand is folate receptor, which is absent in most normal tissues but elevated in over 90% of ovarian carcinomas and expressed at a high frequency in other human malignancies. As a low-molecular-weight (MW 441) ligand, folate offers numerous advantages over polypeptide based targeting ligands: (i) lack of immunogenicity, (ii) unlimited availability, (iii) functional stability, (iv) high affinity for the folate receptor (Kd $\sim 10^{-10}$ M), (v) defined conjugation chemistry, and (vi) favorable nondestructive cellular internalization pathway (Brzezinska et al., 2000; Michael and Curiel, 1994). It has been shown that folate-receptor-targeted liposomal oligonucleotide exhibited greater efficacy in suppressing cellular growth and inhibition of epidermal growth factor receptor expression compared to free oligonucleotide and nontargeted liposomal oligonucleotide (Wang and Low, 1998).

7.3. POLYMER-BASED SYSTEMS

7.3.1. Nanoparticles

Nanoparticles are submicron-sized (<1000 nm) systems, where the drug is dissolved, entrapped, or encapsulated and/or to which the drug is adsorbed or attached. According to the method of preparation, either nanospheres or nanocapsules can be obtained. Nanocapsules are vesicular systems in which the drug is confined to an oily or aqueous core surrounded by a unique polymeric membrane (Fig. 7.6). On the other hand, nanospheres are matrix systems in which the drug is uniformly dispersed (Fig. 7.6). Nanoparticles offer advantages over liposomes, because they are more stable and are more efficient drug carriers than liposomes (Fattal et al., 1991). Conventionally, nanoparticles are prepared mainly by two methods: (i) dispersion of preformed polymers and (ii) polymerization of monomers. Recently, supercritical fluids have become attractive alternatives to conventional methods, because the particles can be prepared in high purity without any trace organic solvents (Randolph et al., 1993). A variety of natural and synthetic polymers have been used for preparing nanoparticles. Synthetic polymers have the advantages of sustaining the release of encapsulated agent over a period of days to several weeks compared to natural polymers, which have a relatively short duration of drug release (Soppimath et al., 2001). Nanoparticles can be made from synthetic polymers such as polylactide–polyglycolide copolymers, polyacrylates, and polycaprolactones or natural polymers such as albumin, gelatin, alginate, collagen, and chitosan. Of these polymers, polylactides (PLA) and poly(DL-lactide-co-glycolide) (PLGA) have been the most extensively investigated for drug delivery. This polyester polymer undergoes hydrolysis in the body forming biologically compatible and metabolizable moieties (glycolic acid, lactic acid), which are eventually eliminated from the body by citric acid cycle.

Drug loading in the nanoparticles can be achieved by either incorporating drug at the time of nanoparticle production or by adsorbing drug to the preformed nanoparticles by incubating them in drug solution (Alenso et al., 1991; Ueda et al., 1998). The former method results in greater drug loading than the latter. Adsorption capacity is related to the hydrophobicity of the polymer and the specific surface area of the nanoparticles. In the case of the drug entrapment method, the monomer amount needs to be optimized for achieving high drug load (Radwan, 1995).

Surface active agents and stabilizers added during polymerization have an effect on drug loading apart from influencing particle size and molecular mass of nanoparticles (Egea et al., 1994). Yoo et al. (1999) have reported that the encapsulation efficiency can be improved by chemically conjugating to nanoparticles. For doxorubicin, an encapsulation efficiency of 97% was observed as opposed to 6.7% for unconjugated doxorubicin in PLGA nanoparticles.

Release rate of drugs from nanoparticles depends upon a number of factors including (i) desorption of surface bound/adsorbed drug; (ii) diffusion through the nanoparticle matrix; (iii) diffusion, in the case of a nanocapsule through the polymer wall;

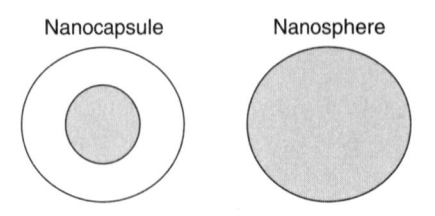

Figure 7.6. A comparative representation of nanocapsule and nanosphere. The blue layer in the nanocapsule represents encapsulated drug while drug is interdispersed with the polymer matrix in nanosphere.

(iv) nanoparticle matrix erosion; and (v) combined erosion/diffusion process. Thus diffusion and biodegradation govern the drug release from nanoparticles (Soppimath et al., 2001). The degradation of the polymer can be varied by altering the block copolymer composition and molecular weight, to tune the release of the therapeutic agent from days to months (Lin et al., 2000). Nanoparticles are efficiently internalized through an endocytic process, and the uptake is both concentration- and time-dependent (Panyam et al., 2002; Panyam and Labhasetwar, 2003a; Davda and Labhasetwar, 2002). In vascular smooth muscles, PLGA nanoparticle internalization has been shown in part to take place through fluid-phase pinocytosis and in part through clathrin-coated pits (Foster et al., 2001; Suh et al., 1998). Following their internalization into the cells, nanoparticles were observed to be transported to primary endosomes and then probably to sorting endosomes. The remaining fraction is transported to secondary endosomes, which then fuse with the lysosomes (Panyam and Labhasetwar, 2003b). Time-dependent studies have shown that nanoparticles escaped the endolysosomes within 10 min of incubation and entered the cytosol. From transmission electron microscopy, Panyam and Labhasetwar (2003a) found that PLGA nanoparticles undergo charge reversal (to cations) in the acid environment of the endosomes leading to localized destabilization of the membrane and escape of nanoparticles into the cytosol. This was further substantiated from studies with polystyrene, where no endolysosomal escape was observed, because these nanoparticles do not undergo charge reversal with change in pH. Furthermore, the authors observed exocytosis of PLGA nanoparticles, when the external concentration gradient of nanoparticles was removed. Probably, the albumin adsorbed onto nanoparticles interacted with the exocytic pathway leading to an increase in exocytosis (Tomoda et al., 1989). However, a fraction of the dexamethasone nanoparticle retained in the cell demonstrated sustained anti-proliferative action for 2 weeks as opposed to 3 days for drug in solution (Panyam and Labhasetwar, 2003a). Since PLGA nanoparticle can bypass the lysosomal compartment, it could serve as an attractive carrier for sustained gene expression (Labhasetwar et al., 1999). By varying the surface charge of nanoparticle, it is possible to target nanoparticles to specific intracellular cell organelles. Surface modification of biodegradable and long-acting polymeric nanoparticles can be achieved by (i) surface coating with hydrophilic polymers/surfactants and (ii) development of biodegradable copolymers with hydrophobic segments. Some of the surface coating materials include polyethylene glycol, polyethylene oxide, poloxamer, poloxamine, and polysorbate 80. Surface coating has been found to reduce opsonization by plasma proteins and improve the circulation time of nanoparticles (Soppimath et al., 2001). Polysorbate-coated poly(butyl cyanoacrylate) nanoparticle showed enhanced drug transport of anti-inflammatory drug dalargin (Alyautdin et al., 1995) into the brain by a number of mechanisms: (i) binding of nanoparticles to the inner endothelial lining of the brain capillaries and subsequently particles deliver drugs to the brain by providing a large concentration gradient, thus enhancing passive diffusion; and (ii) brain endothelial uptake by phagocytosis. Physicochemical properties of nanoparticles—namely, size, surface charge, hydrophobicity and release characteristics—can be varied by altering the composition or the method of preparation of nanoparticles to achieve targeted and sustained drug delivery to various tissues and cells.

7.3.2. Polymeric Micelles

Block copolymers with amphiphilic characters exhibiting a large solubility difference between hydrophilic and hydrophobic segments self-assemble in an aqueous milieu into polymeric micelles with nanoscopic dimensions (Tuzar and Kratochvil, 1976). These micelles have a fairly narrow size distribution and are characterized by their unique shell-core architecture, where hydrophobic segments are segregated from the aqueous exterior to form an inner core surrounded by a palisade of hydrophilic segments. Polymeric micelles show distinct advantages over surfactant-based micelles such as greater solubilization capacity, better kinetic stability, and lesser toxicity (Lavasanifar et al., 2002).

Polymeric micelles mimic aspects of biological transport systems (lipoproteins and viruses) in terms of structure and function. The hydrophilic shell helps them remain unrecognized by the blood circulation (Dunn et al., 1994). At the same time, the viral-like size (10–50 nm) prevents their uptake by reticulo-endothelial system (RES) and facilitates their extravasation at leaky capillaries for passive targeting to tumors. Poly(ethylene oxide) (PEO) is usually the hydrophilic shell, and others include poly(N-isopropyl acrylamide). A hydrophobic block includes poly(α-hydroxy acids) and poly(L-amino acids). Unlike the shell-forming block, the choice for the core-forming block is relatively diverse. Therefore, by suitable choice of the block copolymers the micelles can be constructed to respond to thermal or pH stimuli (Jeong et al., 1997; Tailefer et al., 2000). Targeting ligands may dot the surface of block copolymers for site-specific delivery.

Micellar cores serve as a nanoreservoir for loading and controlled release of hydrophobic molecules that are conjugated or complexed with the polymeric backbone or physically encapsulated in the core (Kwon, 1998; Allen et al., 1999; Kataoka et al., 2001). For drugs physically encapsulated in the core of the micelle, release is controlled by the rate of drug diffusion in the micellar core or breakup of the micelles. In the case of drug conjugation, the covalent bond between the drug and the polymer has to be cleaved for drug release. Water penetration and hydrolysis of the labile bonds in the micelles core (bulk erosion) followed by drug diffusion may occur in relatively hydrophobic liquid-like core structures. In the case of hydrophobic rigid cores, release is dependent on rate of micellar dissociation. Slow dissociation of micellar structure and further hydrolysis of labile bonds may result in a sustained drug release (Li and Kwan, 2000). The multifunctional nature of polymeric micelles makes it a versatile drug carrier capable of achieving selective and sustained drug delivery at different levels. Micelles based on PEO-b-poly(L-amino acids) (PLAA) block copolymers are unique among drug carriers owing to a tailor-made nonpolar core of PLAA, which can take up and protect water insoluble drug. The single greatest advantage of PEO-b-PLAA over other micelle forming block copolymers is its potential for attaching drugs and genes through the functional groups of the amino acid side chain (amine or carboxylic acid). Three types of nanodrug delivery systems based on these micelles have been investigated including (i) micelle forming block copolymer–drug conjugate, (ii) micellar nanocontainer, and (iii) polyion complex micelles. These strategies have been used to deliver antitumor agents and nucleic acids, where the block copolymer micelle protects the drug in the circulation and delivers it to the intracellular targets in a sustained manner (Kakizawa and Kataoka, 2002).

7.3.3. Dendrimers

7.3.3.1. Introduction. Dendrimers are highly branched, well-defined, polymeric nanostructures that have recently been explored as potential drug carriers (Liu and Frechet, 1999; Tomalia et al., 1985, 1986; Hawker and Frechet, 1990; Stiriba et al., 2002; Gebhart and Kabanov, 2001; Cloninger, 2002; Liu et al., 2000; Kolhe et al., 2003; Kono et al., 1999; Zhuo et al., 1999; Padilla De Jesus et al., 2002; Ihre et al., 2002; Malik et al., 2000; Wiwaattanapatapee et al., 2000; Kukowska-Latallo et al., 1996; Tajarobi et al., 2001; El-Sayed et al., 2002; Kannan et al., 2004; Kohle et al., 2004a,b; Khandare et al., 2004). The initial pioneering work by Tomalia and Newkome in the 1980s has paved the way for a number of research groups to contribute substantially in different methodologies for synthesis and specific applications of these dendrimers (Tomalia et al., 1985, 1986; Hawker et al., 1990). Due to their unique structure and properties, they have been evaluated for many applications, such as nanoscale catalysts, micelle mimics, encapsulation of drug molecules, biological recognition, chemical sensors, immunosensors in medical imaging, and drug/gene delivery (Fig. 7.7). Dendrimers are unimolecular micelles, which are much smaller (~10 nm) and more stable than liposomes. Dendritic polymers (size ~10 nm) are emerging as promising candidates for "smart" delivery nanodevices, because they *present a large density of tailorable functional groups at the periphery*. The nanoscale branching architecture provides them with several advantages over linear polymers, nanoparticles, and liposomes, such as rapid cellular entry, reduced macrophage uptake, and targetability. A range of polyamidoamine (PAMAM) dendrimers including partially hydrolyzed PAMAM dendrimers have been shown to be effective vectors for gene transfection (Stiriba et al., 2002; Gebhart and Kabanov, 2001; Cloninger, 2002).

One of the most exciting applications of dendrimers in the medical field has been the potential use of dendrimers for targeted drug delivery. These polymers have multiple functional end groups in their peripheries that can be easily modified to potentially attach a large number of drug molecules, ligands, or antibodies, making them ideal for use in drug targeting. Drugs can be either encapsulated within the dendrimer by electrostatic or hydrophobic interactions or complexed/conjugated to the end groups (Liu et al., 2000; Kohle et al., 2003).

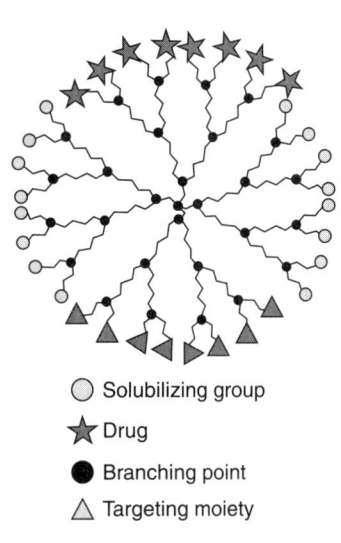

○ Solubilizing group
☆ Drug
● Branching point
△ Targeting moiety

Figure 7.7. The large number of surface functional groups within ~10 nm in diameter enables surface attachment of active molecules, ligands, imaging agents, and targeting moieties, leading to a multifunctional nanodevice that is much smaller than typical liposomes and nanoparticles.

Chemotherapeutic agents like methotrexate, folic acid, doxorubicin, and 5-fluorouracil have been conjugated to dendrimers and shown to successfully release the drug *in vitro* (Kono et al., 1999; Zhuo et al., 1999; Padilla De Jesus et al., 2002; Ihre et al., 2002). Dendritic unimolecular micelles have been synthesized with a hydrophobic core and a hydrophilic shell and have been used as a model for dendrimer-based drug delivery.

To take advantage of dendrimers for superior cellular delivery of drugs, it is essential to study and understand the transport of dendrimers into cells. Duncan and co-workers found that some generations of PAMAM dendrimers caused haemolysis and were cytotoxic even at lower concentrations. However, anionic dendrimers did not result in haemolysis and cytotoxicity (Malik et al., 2000). They also found that in an everted rat intestinal sac system, ^{125}I-labeled anionic dendrimers exhibit rapid serosal transfer rates and low tissue deposition. In contrast, cationic PAMAM dendrimers showed lower transport rates

TABLE 7.1. Properties of Dendrimers and Dendrimer–Drug Conjugates[a].

Dendrimers	Molecular Weight (Da)	Generation Number	Number of Surface Groups	Drug Molecules per Dendrimer (Payload)[c]
PAMAM-G4-NH$_2$ (G4)	14,215[b]	4	64	—
PAMAM-G4-OH (G4-OH)	14,279[b]	4	64	—
G4-OH-Ibuprofen	24,785[c]	4	64	51 (41%)
G4-OH-GA-Methyl prednisolone[d]	19,154[c]	3	64	13 (26%)

[a] Where imaging was used, these materials were FITC-labeled with 3–8 molecules of FITC per dendrimer
[b] Reported by Tomalia et al. (1990).
[c] Estimated from ^{1}H NMR.
[d] GA, glutaric acid spacer. OH represents hydroxyl terminated dendrimer

(Wiwaattanapatapee et al., 2000). Baker and co-workers studied the transfection ability of PAMAM dendrimers for DNA, across several cell lines and reported that PAMAM dendrimers can be used as an efficient transfection vehicle for DNA/dendrimer complexes (Kukowska-Latallo et al., 1996). This suggests that dendrimers can enter a variety of cells. Recently, Ghandehari and co-workers addressed the permeability of dendrimers across Madin–Darby canine kidney (MDCK) cells and caco-2 cell monolayers (Tajarobi et al., 2001; El-Sayed et al., 2002). They showed that permeability of the dendrimers across MDCK cell lines was in the order G4 ≫ G1 ≈ G0 > G3 > G2, where G represents the generation number of the dendrimer. Their results suggested that the permeability of the dendrimers appeared to be related to the generation number, concentration, incubation time, and probable modulation of the cell membrane by cationic dendrimers.

In recent studies by Kannan and co-workers, the intracellular trafficking of dendrimers, dendrimer–drug conjugates, and their cellular activities were investigated using a combination of flow cytometry, fluorescence and confocal microscopy, and UV/Vis spectroscopy (Kannan et al., 2004; Kohle et al., 2004a,b; Khandare et al., 2004). A key aspect of these studies was the establishment of time scales for entries and the "quantification" of the amount of dendrimer transported into cells. A variety of dendrimers and drug conjugates were investigated. Chemistry and cellular interactions of drug–dendrimer conjugates of anti-inflammatory drugs—namely, ibuprofen and methyl prednisolone—are highlighted.

7.3.3.2. Chemistry. Ibuprofen is a relatively small drug (molecular weight 206 Da), while methyl prednisolone (molecular weight 376 Da) is a relatively large steroidal drug. The drugs were conjugated to a neutral PAMAM-G4 hydroxyl-terminated dendrimer using N,N'-dicyclohexyl-carbodiimide (DCC) as a coupling agent and was followed by extensive purification before isolating the conjugate. For the larger methyl prednisolone, a highly reactive glutaric acid spacer molecule was used to overcome the steric hindrance of the aromatic ring in the drug. For the cell trafficking studies, these conjugates were further labeled with a fluorescent probe, fluorescein isothiocyanate (FITC). This resulted in dendrimer–drug–FITC conjugates. Care was taken to remove any trace amount of free FITC in the FITC-labeled conjugates. More details can be found elsewhere (Kannan et al., 2004; Kohle et al., 2004a,b; Khandare et al., 2004).

7.3.3.3. Cellular Interactions of Free Dendrimer. Rapid rate of cellular entry of the tagged G4-NH$_2$ is evident from the high degree of correlation in the absorbance loss in the supernatant (from UV analysis) and the simultaneous absorbance increase in the cell lysate (Fig. 7.8A) This correlates well with the flow cytometry data (Fig. 7.8B), which shows a significant (exponential) pickup in the FITC intensity, which is associated with the dendrimer uptake by the cells. These results were qualitatively consistent with fluorescence microscopy images on the FITC-labeled pure dendrimers (not shown). Approximately 90% of PAMAM dendrimers (amine-terminated generation 3 and 4) enter lung epithelial cells within 1 h. Surprisingly, a neutral PAMAM-G4-OH showed a comparable cellular entry profile to that of the cationic G4-NH$_2$. Epithelial cells are known to possess an overall anionic charge. Therefore the entry of cationic PAMAM dendrimers may be because of the electrostatic interaction with negatively charged epithelial cells probably via the fluid-phase pinocytic route. Since G4-OH does not have cationic amine surface groups, the cellular entry may be because

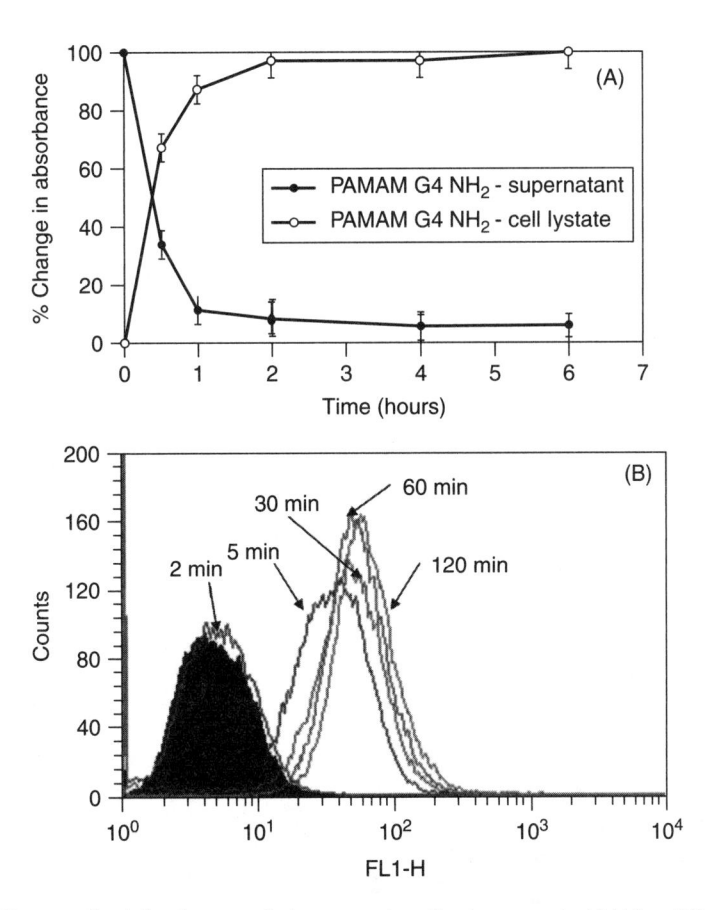

Figure 7.8. (A) The normalized absorbance evolution versus time. For the supernatant $[A(t) - A(\text{final})]/[A(\text{initial}) - A(\text{final})]$ is used. For the cell lysate, $[A(t)/A(\text{final})]$ is used. It can be seen that the increase in the normalized absorbance (A) of the FITC labeled PAMAM G4-NH$_2$ dendrimer with respect to time (t) in the cell lysate correlates very well with the decrease in absorbance in the cell supernatant. (B) Flow cytometry of the cell entry dynamics of pure PAMAM-G4-NH$_2$ in A549 lung epithelial cell line. The log of FITC absorption intensity (FL1-H x axis) is plotted against the number of cells (counts). The exponential increase in the cellular uptake of polymers within few minutes is evident. Based on this and the UV/Vis data (Part A), it is evident that dendrimers enter cells rapidly.

of the nonspecific adsorptive route, with some specific interactions due to the interior amine groups that would be cationic. Another important aspect in the rapid cellular entry in the A549 cells could be due to the higher permeation of cancer cells. Preliminary measurements on normal smooth muscle cells show that the entry is somewhat slower in normal cells (Kannan et al., 2004).

7.3.3.4. Cellular Interactions of Drug-Conjugated Dendrimer.
The transport of the drug carrying dendritic nanodevices into cells is vital to the ultimate delivery and the therapeutic effect. Dendrimer–drug conjugates with high drug payloads have been investigated. The neutral dendrimer PAMAM-G4-OH was conjugated with ibuprofen and methyl prednisolone. These studies are among the first to look at the transport and therapeutic efficiency of dendrimer-conjugated *noncancerous* drugs.

7.3.3.4.1. Dendrimer–Ibuprofen Nanodevices.
Using flow cytometry, it was found that the dendrimer–drug conjugate entered cells rapidly, with significant fluorescence intensity increase within 15 min, corresponding to a conjugate uptake of >30%. More than 50% of the G4-OH–ibuprofen conjugate entered the cells within 60 min. The cellular entry was also visualized by using phase contrast and fluorescence microscopy (Fig. 7.9A–D). It is evident that both the free and the dendrimer-conjugated ibuprofen entered the cells. As indicated by the microscopy images, the conjugates appear to be present mostly in the cytosol. For drugs such as ibuprofen and methyl prednisolone, the mechanism of action requires

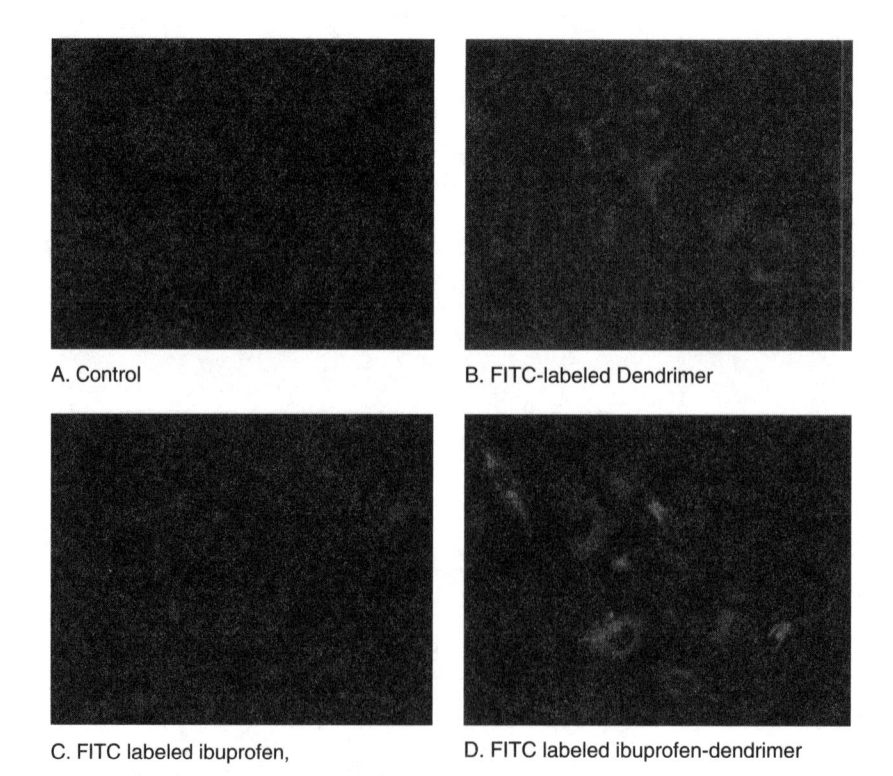

Figure 7.9. Confocal fluorescence images of A549 lung epithelial cells after treatment for 1 hour with (**A**) control, (**B**) FITC-labeled G4OH dendrimer, (**C**) FITC-labeled ibuprofen, and (**D**) FITC-labeled ibuprofen–dendrimer conjugate after 4 h.

them to be delivered to the cytoplasm. Therefore, the delivery of the drug to the cytoplasm is sufficient. On the other hand, for DNA delivery, the DNA complexed with the cationic dendrimer must be transported to the nucleus for it to be effective.

The mode of action of ibuprofen involves the acetylation of cyclooxygenase-2 (COX-2) which blocks access and egress to/from the active site, inhibiting the production of prostaglandin. The efficiency of the dendrimer–ibuprofen conjugates to suppress COX-2 was assessed by measuring the prostaglandin (PGE$_2$) in the cell supernatant. Free ibuprofen did not inhibit the prostaglandin release from A549 cells after 30 min of incubation, while the dendrimer–ibuprofen conjugate showed a significant inhibition within the same time period ($p < 0.05$). However, there was no significant difference ($p > 0.05$) in inhibition of the prostaglandin (PGE$_2$) release between free and conjugated ibuprofen at 60 and 360 min (Fig. 7.10). Neither blank dendrimer nor the solvent showed any suppression of PGE$_2$ synthesis. The above results imply that conjugates rapidly enter the cell and produce the desired pharmacological action at the target site in the cytosol. Although a moderate fraction of free ibuprofen enters the cells within 30 min, it may not achieve a sufficient concentration to elicit a rapid pharmacological response. On the other hand, the dendrimer–ibuprofen conjugate achieves a high local concentration in the cell due to its *high drug payload*. At this point, it is unclear whether the ibuprofen is released from the nanodevice inside the cell or if the drug is effective even in the conjugated form. However, once the device enters inside the cell, it is conceivable that the acidic pH and the enzymes in the endosomes would hydrolyze the ester bond in the conjugate, thereby releasing the free drug in the cytosol to suppress prostaglandin synthesis.

7.3.3.4.2. Dendrimer–Methyl Prednisolone Nanodevices.
Methyl prednisolone is a relatively hydrophobic steroid used to treat lung inflammation in asthma. Preliminary studies have indicated that dendrimers could increase the lung residence time in mice models for lung inflammation. To assess the cellular transport and activity of the steroid containing dendrimer conjugates, microscopy and prostaglandin assay were used. The A549 lung epithelial

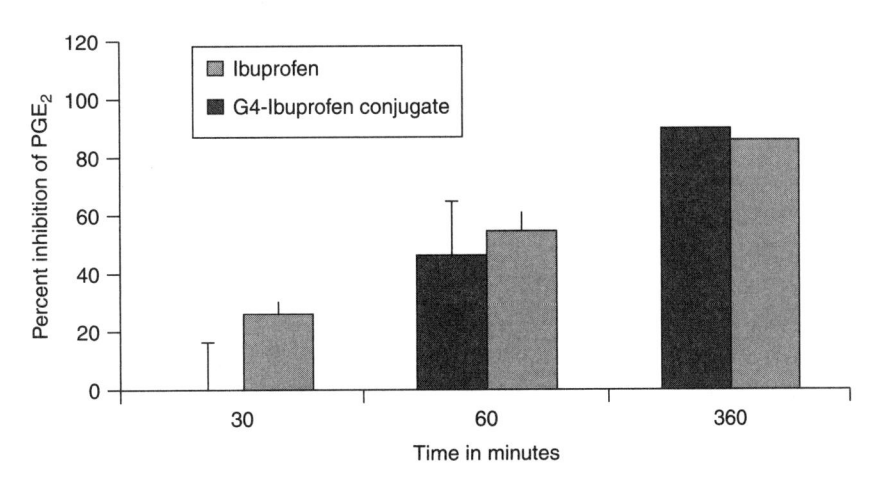

Figure 7.10. Percent inhibition of prostaglandin (PGE$_2$) for free ibuprofen and PAMAM-G4-OH–ibuprofen conjugate as a function of treatment time at 30, 60, and 360 min (average of four measurements with error bars) in A549 lung epithelial cell line. Blank dendrimer and solvent did not show any inhibition of prostaglandin release.

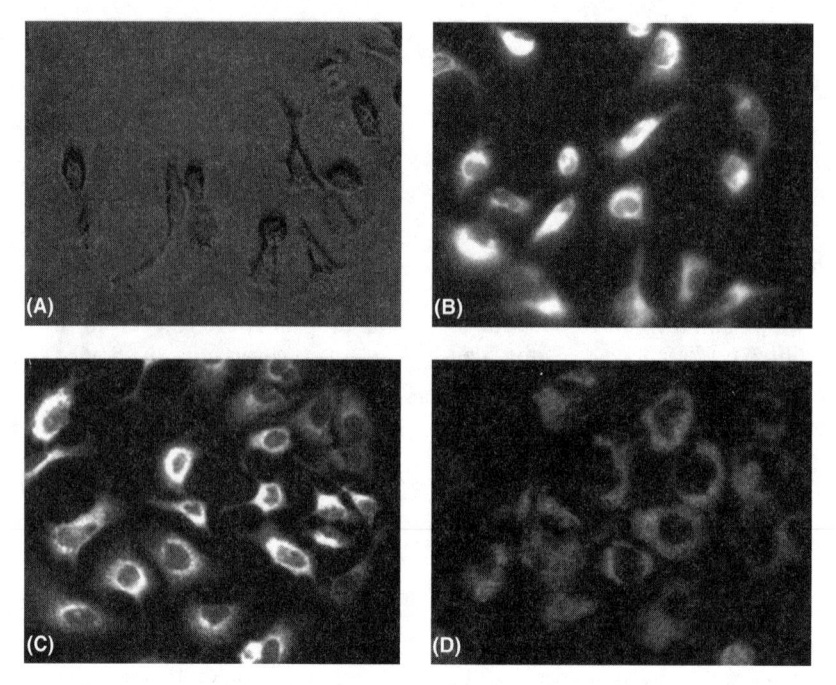

Figure 7.11. (A–C) Fluorescence microscopic images (magnification 40×) of A549 cells 4 h of treatment with (A) control (no treatment), (B) FITC labeled methylprednisolone, and (C) FITC-labeled dendrimer–methylprednisolone conjugate, (D) Confocal microscopy images of the FITC-labeled dendrimer–methylprednisolone conjugate inside the cell. The presence of the nanodevices in the cytoplasm is shown using both fluorescence and confocal microscopy.

cancer cells were treated with the G4-OH-methyl prednisolone conjugates. At the end of four hours, the cells were observed under fluorescence microscopy to identify the cellular localization of the free and drug conjugated dendrimer (Fig. 7.11A–C). Fluorescence microscopy images of treated A549 lung epithelial cells provide insights into the location and distribution of the drug and conjugates inside the cells. The cells without FITC show no fluorescence as expected (Fig. 7.11A). Both free and drug-conjugated dendrimers were mainly found to be in the cell cytoplasm. Furthermore, this was substantiated from confocal microscopic images in Figs. 7.11B and 7.11C, where the fluorescence is seen mostly in the cytoplasm. This suggests that the free methyl prednisolone and dendrimer–methyl prednisolone conjugate are mainly localized in the cytosol in the cell. Free methyl prednisolone entered the cells faster than the methyl prednisolone–dendrimer conjugate (data not shown). The efficiency of the endocytosis mechanism is influenced by the molecular size, and further studies are required to understand the cellular uptake mechanism. For methyl prednisolone, cytoplasmic delivery is sufficient, since it exerts pharmacological action by binding to the glucocorticoid receptors in the cytoplasmic compartment.

Dendrimer–methyl prednisolone conjugate inhibited prostaglandin synthesis to the same extent (more than 95% inhibition) as free methyl prednisolone (Fig. 7.12) after 4 h of treatment. The results indicate that the conjugate was able to enter the cell and produce the desired pharmacological action. It is interesting to note that even though only 40% of the conjugate was inside the cell as opposed to >50% of the free MP, it produced a comparable ($p > 0.05$) pharmacological action to free methyl prednisolone.

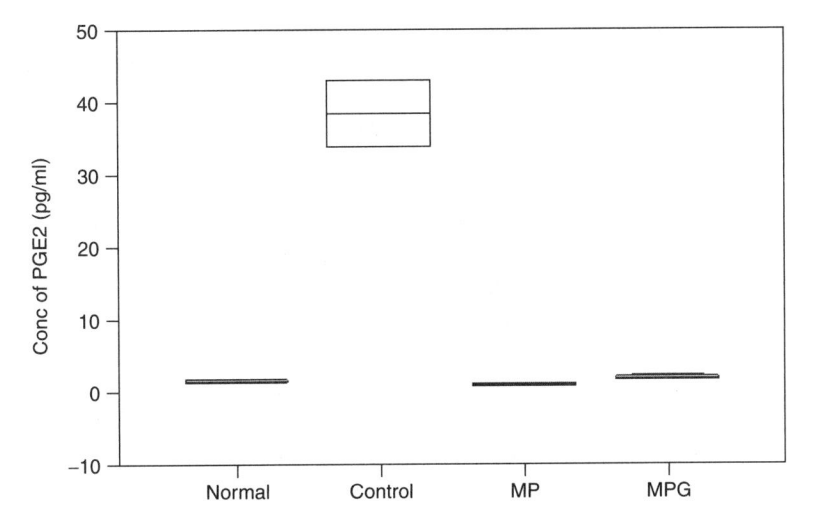

[Figure: Box plot showing PGE2 concentration (pg/ml) on y-axis (ranging from -10 to 50) for four conditions on x-axis: Normal, Control, MP, and MPG. Control shows elevated levels around 35-42 pg/ml; Normal, MP, and MPG show low levels near 0-2 pg/ml.]

Figure 7.12. Box plot showing prostaglandin (PGE$_2$) concentration in the cell supernatant for different treatment periods with A549 lung epithelial cell lines. Normal indicates the cells with no induction of prostaglandin synthesis. Control represents cells with prostaglandin induction but with no treatment. MP and MPG indicate the cells, which were induced to secrete prostaglandin and was treated with free methyl prednisolone and dendrimer conjugated methyl prednisolone respectively. The cells treated with MP/MP–dendrimer conjugate (4 h) suppressed prostaglandin secretion, and the PGE$_2$ levels were comparable to normal cells. Except normal and control treatment ($n = 2$).

From these results, it appears that high drug payload in the dendrimer conjugate produces a high local drug concentration inside the cell to elicit a significant therapeutic response. At this point, it is unclear whether the MP is released from the dendrimer conjugate inside the cell or if the drug is effective even in the conjugated form. However, once the device enters inside the cell, it is conceivable that the acidic pH and the enzymes in the endosomes would eventually hydrolyze the ester bond in the conjugate to release the free drug in the cytosol (McGraw et al., 1991). Based on previous results on dextran–methyl prednisolone conjugates, the hydrolysis process is likely to take much longer than the time frame used in current studies. Therefore, we may tentatively conclude that the conjugate may not have released the drug appreciably in 4 h. Nevertheless, the high payload drug conjugate shows significant therapeutic activity. It may be possible that the conjugate may slowly release the drug over a sustained period, thereby providing therapeutic effect over longer times. Further studies are required to investigate the stability of dendrimer–drug conjugate at various pHs and in the presence of enzymes to understand the intracellular hydrolysis of the conjugate.

7.3.3.5. Summary (Dendritic Nanodevices).

Dendritic polymers are a relatively new class of tree-like polymers that are typically much smaller and more stable than liposomes and nanoparticles. They have shown reduced macrophage uptake and decreased inflammatory response compared to liposomes. As more drug delivery approaches are developed using dendritic polymers (including hyperbranched polymers), understanding their cellular uptake is becoming critical. Fluorescence and confocal microscopy play a key rule in understanding the biological properties of these nanodevices. Most of the delivery efforts have focused on cancer drugs. Non-cancer drugs represent an interesting opportunity to study the cellular dynamics of these nanodevices. Based on the current understanding, they appear to enter cells very rapidly and transport the conjugated or complexed drugs effectively. A predominant quantity of the nanodevice appears to localize in the cytosol. The conjugated drugs are therapeutically effective, even if the drug is not released. As the drugs get released through the action of the enzymes on the conjugates to release the drug, these nanodevices may show sustained delivery also. These are subjects of current investigation.

7.4. CLINICAL ASPECTS AND FUTURE PERSPECTIVES

From a clinical viewpoint, an ideal drug delivery system is expected to (i) enhance the therapeutic efficacy of the drug and (ii) reduce the adverse effects of the drug on other organs. Cancer therapy for instance is severely limited by the toxicity caused by the antitumor agents on normal cells. Doxorubicin treatment is associated with severe cardiotoxicity and hepatotoxicity, but when it was administered using nanoparticles, neither of theses side effects were observed due to the localization of the nanoparticle formulation to the tumor (Kattan et al., 1992). It is critical that the drug delivery vehicle does not upset the normal physiology of the biological system, while exploiting the unique pathophysiological characteristics of diseased tissue or cell. Therefore, it is not sufficient to just selectively target the therapeutic agent to the cell, but is also essential to target to the specific compartment in the cell. Clinical relevance of specific drug targeting can be exemplified comparing the case of vaccine development for bacteria and viruses. Exogenous antigens (like bacteria) are usually taken up into cells by phagocytosis or endocytosis, which after degradation in the lysosomes are presented in a major histocompatibility complex (MHC)-class II restricted manner. Such antigen presentation induces antigen-specific antibody production but not cytotoxic T-lymphocyte response (CTL). On the contrary, endogenous antigens (cytoplasmic antigens such as viruses) are degraded by proteasomes in cytoplasm and presented with MHC class-I molecules, eventually leading to the induction of CTL response (Kunisawa et al., 2001). This antigenic response plays a key role against viral infectious diseases and cancer. Though conventional liposomes have met with some success for bacterial vaccines, they have achieved limited success for viral and tumor vaccines due to the lack of induction of CTL response (Sheikh et al., 2000). When diphtheria toxin A is delivered using conventional liposomes, no CTL response was detected because it is taken up by the endosomes and transported to the lysosomes. In sharp contrast, with fusogenic liposomes the CTL response was observed, because the toxin was released directly into the cytosol by the fusion of "fusogenic liposome" with the cell membrane (Nakanishi et al., 2000). Specific and selective targeting of nanovehicles can also be accomplished by attaching cell-specific ligands to the nanovehicles or by modifying the composition of the vehicle to respond to various chemical, physical, or biological stimuli. Targeting results in a high local drug concentration at the site of action with minimal systemic exposure, thus ultimately reducing the overall dose of the drug and consequently eliminating the dose related toxicity of the therapeutic agent. However, one of the challenges is to achieve a high drug payload in the nanovehicle. From this perspective, dendrimers offer tremendous potential as a high drug payload vehicle due to the presence of numerous functional groups on the surface (Kohle et al., 2004b). The nanosystems such as micelles and nanoparticles, in addition to transporting the therapeutic agent inside the cell, can also act as a sustained or controlled drug delivery system (Panyam and Labhasetwar, 2003b; Li and Kwon, 2000). Chemistry of the nanovehicle and the nature of the drug linkage to the carrier play an important role in modulating the release of the drug at the site of action (Joseph et al., 2004). By achieving sustained or controlled release of the drug from the vehicle, one can reduce the dosing frequency, which significantly improves the patient compliance to therapy.

Nanodelivery systems can positively alter the tissue distribution of the therapeutic agent by its ability to partition into specific tissue compartments, leading to increase in the plasma half-life and volume of distribution, while decreasing the plasma clearance of the drug. Pluronic block copolymers (PEO-b-PPO) or poloxamers have been shown to inhibit P-glycoprotein efflux which is partially responsible for multidrug resistance in cancer cells and reduced permeation across biomembranes (Batrakova et al., 1996, 1998). The polymer can form micelles which can carry drug in the nanocontainer of the micelle, thus bypassing the efflux system. Alternatively, the nanoparticles have been coated with poloxamer to improve the efficacy of the therapeutic agent (Soppimath et al., 2001).

Biocompatibility and safety of the vehicle is a critical component in clinical application of nano drug delivery systems. It has been shown that the biodistribution and biocompatibility of dendrimers depend on its surface charge and generation number (Malik et al., 2000). In general, studies have shown that balancing the surface hydrophobicity and hydrophilicity improves the biocompatibility of polymers. Neutral polymers and polyanions show less toxicity than polycations, while low-molecular-weight polymers have been found to undergo less protein adsorption and platelet adhesion (Wang et al., 2004). Therefore, understanding the structure–property relationships of vehicles holds the key for the rational design of nanodrug delivery systems.

A number of products based on nanodrug delivery systems are in various stages of clinical development (Duncan, 2003), and some of them are already available for clinical use such as Ambisome (amphotericn B liposome) and Daunoxome (Daunorubicin citrate liposomal formulation). The formulation issues (scaling up, sterilization, and lyophilization) and biological issues (biocompatibility, *in vitro–in vivo* correlations) remain to be resolved before other nanoscopic drug delivery systems can progress from the lab to the clinic. Nevertheless, the advances in nanotechnology has provided a platform for multidisciplinary approach to move closer to developing an ideal

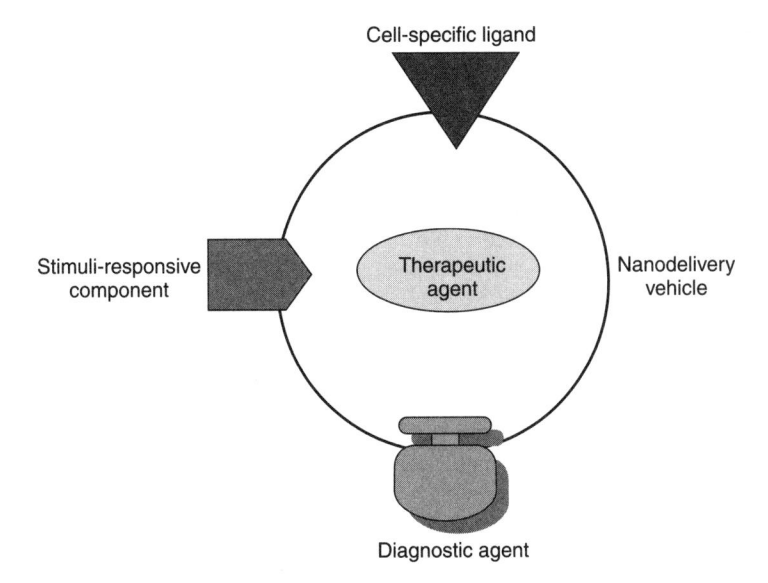

Figure 7.13. Schematic representation of an idealistic "smart NanoDrug delivery system."

"smart nanodrug delivery system," which can be decorated with a targeting ligand and a diagnostic or imaging ligand apart from the therapeutic agent of interest (Fig. 7.13). The availability of such "smart nanodrug delivery systems" in the near future would allow us to detect, diagnose, target, modulate delivery, and track the progression of therapy remotely and noninvasively.

REFERENCES

Alenso, M. J., Losa, C., Calvo, P., and Vila-Jato, J. L. (1991). Approaches to improve the association of amikacin sulfate to poly(cyanoacrylate) nanoparticles. *Int. J. Pharm.* **68**:69–76.

Allen, C., Maysinger, D., and Eisenberg, A. (1999). Nano-engineering block copolymer aggregates for drug delivery. *Colloids Surf. B: Biointerfaces* **16**:3–27.

Alyautdin, R., Gothier, D., Petrov, V., Kharkevich, D., and Kreuter, J. (1995). Analgesic activity of the hexapeptide dalargin adsorbed on the surface of polysorbate 80 coated poly(butylcyanoacrylate) nanoparticles. *Eur. J. Pharm. Biopharm.* **41**:44–48.

Amyere, M., Mettlen, M., Der Smissen, P. V., Platek, A., Paurastre, B., Veithan, A., and Courtoy, P. J. (2002). Origin, originality, functions, subversion and molecular signaling of macropinocytosis. *Int. J. Med. Microbiol.* **291**:487–494.

Barenholz, Y., and Crommelin, D. J. A. (1994). Liposomes as pharmaceutical dosage forms. In: *Encyclopedia of Pharmaceutical Technology* (J. Swarbrick, ed.), pp. 1–39, Marcel Dekker, New York.

Batrakova, E. V., Dorodnich, T. Y., Klinski, E. Y., Kliushenkova, E. N., Shemchukova, O. B., Goncharova, O. N., Arjakova, V. Y., Alakhov, V., and Kabanov, A. V. (1996). Anthracycline antibiotics noncovalently incorporated into the block copolymer micelles: *In vivo* evaluation of anticancer activity. *Br. J. Cancer* **74**:1545–1552.

Batrakova, E. V., Han, H. Y., Alakhov, V., Miler, D. W., and Kabanov, A. V. (1998). Effect of pluronic block copolymers on drug absorption in Caco2 cell monolayers. *Pharm. Res.* **15**:850–855.

Brigger, I., Dubernet, C., and Couveur, P. (2002). Nanoparticles in cancer therapy and diagnosis. *Adv. Drug Deliv. Rev.* **54**:631–651.

Brzezinska, D., Winska, P., and Balinksa, M., (2000). Cellular aspects of folate and antifolate membrane transport. *Acta Biochim. Pol.* **47**:735–749.

Cloninger, M. J. (2002). *Curr. Opin. Chem. Biol.* **6**(6):742.

Daeman, T., Regts, J., Meesters, M., Tenkate, M. T., Baker-Woudenberg, I. A. J. M., and Scherphof, G. L. (1997). Toxicity of doxorubicin entrapped within long-circulating liposomes. *J. Controlled Release* **44**:65–76.

Davda, J., and Labhasetwar, V. (2002). Characterization of nanoparticle uptake of endothelial cells. *Int. J. Pharm.* **233**:51–59.

De Duve, C. (1963). Lysosomes. In: *Ciba Foundation Symposium* (A. V. S. De Reuck and M. P. Cameron, eds.), pp. 411–412. Churchill, London.

de Lima, M. C. P., Simoes, S., Pires, P., Faneca, H., and Duzgunes, N. (2001). Cationic lipid–DNA complexes in gene delivery: from biophysics to biological applications. *Adv. Drug Deliv. Rev.* **47**:277–294.

Drews, J. (2000). Drug discovery: A historical perspective. *Science* **287**:1960–1964.

Duncan, R. (2003). The dawning era of polymer therapeutics. *Nature Rev.* **2**:347–360.

Dunn, S. E., Brindley, A., Davis, S. S., Davies, M. C., and Illum, S. (1994). Polystyrene-poly(ethyleneglycol) (PS-PEG-2000) particles as model systems for site specific drug delivery. 2. The effect of PEG surface density on the *in vitro* cell interaction and *in vivo* biodistribution. *Pharm. Res.* **11**:1016–1022.

Egea, M. A., Gamisani, F., Valero, J., Garcia, M. E., Garcia, M. L. (1994). Entrapment of cisplatin into biodegradable polyalkylcyanoacrylate nanoparticles. *Farmaco* **49**:211–217.

El-Sayed, M., Ginski, M., Rhodes, C., and Ghandehari, H. (2002). *J. Controlled Release* **81**(3):355.

Esfand, R., and Tomalia, D. A. (2001). Poly(amidoamine) (PAMAM) dendrimers: from biomimicry to drug delivery and biomedical applications. *Drug Discov. Today* **6**:427–436.

REFERENCES

Fattal, E., Rojas, J., Youssef, M., Couvuer, P., and Andremont, A. (1991). Liposome entrapped ampicillin in the treatment of experimental murin listeriosis and salmanellosis. *Antimicrob. Agents Chemother.* **35**:770–772.

Felgner, P. L., and Ringold, G. M. (1989). Cationic liposomes mediate transfection. *Nature* **337**:387–388.

Felgner, P. L., Gadek, T. R., Holm, M., Roman, R., Chan, H. W., Wenz, M., Northrop, J. P., Ringold, G. M., and Danielsen, M. (1987). Lipofectin: A highly efficient lipid-mediated DNA-transfection procedure. *Proc. Natl. Acad. Sci. USA.* **84**:7413–7417.

Foster, K. A., Yazdanian, M., and Audus, K. L. (2001). Microparticulate uptake: Mechanisms of *in vitro* cell culture models of the respiratory epithelium. *J. Pharm. Pharmacol.* **53**:57–66.

Gebhart, C. L., and Kabanov, A. V. (2001). *J. Controlled Release* **73**:401.

Goldberg, R. I., Smith, R. M., and Jarett, L. (1987). Insulin and α_2-macroglobulin-methylamine undergo endocytosis by different mechanisms in rat adipocytes. I. Comparison of cell surface events. *J. Cell. Physiol.* **133**:203–212.

Hafez, I. M., Cullis, P. R. (2001). Roles of lipid polymorphism in intracellular delivery. *Adv. Drug Deliv. Rev.* **47**:139–148.

Hashida, M., Nishikawa, M., Yamashita, F., and Takakura, Y. (2001). Cell specific delivery of genes with glycosylated carriers. *Adv. Drug Deliv. Rev.* **52**:187–196.

Hawker, C. J., and Frechet, J. M. J. (1990). Preparation of Polymers with Controlled Molecular Architecture. A New Convergent Approach to Dendritic Macromolecules. *J. Am. Chem. Soc.* **112**:7638.

Ihre, H. R., Padilla De Jesús, O. L., Szoka, F. C., Jr., and Frechet, J. M. J. Polyester (2002). Dendritic Systems for Drug Delivery Applications: Design, Synthesis, and Characterization. *Bioconjugate Chemistry.* **13**(3):443.

Jain, R. K. (1987). Transport of molecules across tumor vasculature. *Cancer Metastasis Rev.* **6**:559–593.

Jeong, B., Bae, Y. H., Lee, D. S., and Kim, S. W. (1997). Biodegradable block copolymer as injectable drug delivery systems. *Nature* **388**:860–862.

Johannes, L., and Lamaze, C. (2002). Clathrin-dependent or not: Is it still the question? *Traffic* **3**:443–451.

Joseph, A., Souza, M. D., and Topp, E. M. (2004). Release from polymeric prodrugs. Linkages and their degradation. *J. Pharm. Sci.* **93**:1962–1979.

Kakizawa, Y., and Kataoka, K. (2002). Block copolymer micelles for delivery of gene and related compounds. *Adv. Drug Deliv. Rev.* **54**:203–222.

Kaneda, Y., Saeki, Y., and Morishita, R. (1999). Gene therapy using HVJ-liposomes: The best of both worlds? *Mol. Med. Today* **5**:298–303.

Kannan, S., Kolhe, P., Kannan, R. M., Lieh-Lai, M., and Glibatec, M. (2004). *J. Biomater. Sci. Polymers* **15**(3):311.

Kataoka, K., Harada, A., and Nagasaki, Y. (2001). Block copolymer micelles for drug delivery: Design, characterization and biological significance. *Adv. Drug Deliv. Rev.* **47**:113–131.

Kattan, J., Droz, J. P., Couveur, P., Marino, J. P., Boutan-Laroze, A., Rougier, P., Brault, P., Vranckx, H., Grognet, J. M., Morge, X., and Sancho-Garrier, H. (1992). Phase I clinical trial and pharmacokinetics evaluation of doxorubicin carried by poly isohexylcyanoacrylate nanoparticles. *Invest. New Drugs* **10**:191–199.

Kim, C. K., and Han, J. H. (1995). Lymphatic delivery and pharmacokinetics of methotrexate after intramuscular injection of differently charged liposomes-entrapped methotrexate to rats. *J. Microencapsulation* **12**:437–446.

Kolhe, P., Khandare, J., Pillai, O., Kannan, S., Lieh-Lai, M., and Kannan, R. M. (2004). Hyperbranched polymer-drug conjugates with high drug payload for enhanced cellular delivery. *Pharm. Research.* **21**(12):2185.

Kolhe, P., Khandare, J., Pillai, O., Kannan, S., Lieh-Lai, M., and Kannan, R. M. (2005). Synthesis, cellular transport, and activity of polyamidoamine dendrimer–methylprednisolone conjugates. *Bioconjugate Chem.* **16**(4):1049.

Kolhe, P., Khandare, J., Pillai, O., Kannan, S., Lieh-Lai, M., and Kannan, R. M. (2006). Preparation, cellular transport, and activity of polyamindoamine-based dendritic nanodevices with a high drug payload. *Biomaterials.* **27**(4):660.

Kolhe, P., Misra, E., Kannan, R. M., Kannan, S., and Lieh-Lai, M. (2003). *Int. J. Pharmaceutics* **259**(1–2):143.

Kono, K. (2001). Thermosensitive polymer modified liposomes. *Adv. Drug Deliv. Rev.* **53**:307–319.

Kono, K., Liu, M., and Frechet, J. M. J. (1999). *Bioconjugate Chem.* **10**(6):1115.

Kukowska-Latallo, J. F., Bielinska, A. U., Johnson, J., Spindler, R., Tomalia, D. A., and Baker, J. R., Jr. (1996). *Proc. Natl. Acad. Sci. USA*, **93**, 4897.

Kunisawa, J., Nakagawa, S., and Mayumi, T. (2001). Pharmacotherapy by intracellular delivery of drugs using fusogenic liposomes: Application to vaccine development. *Adv. Drug Deliv. Rev.* **52**:177–186.

Kwon, G. S. (1998). Diblock copolymer nanoparticles for drug delivery. *Crit. Rev. Ther. Drug Carrier Syst.* **15**:481–512.

Labhasetwar, V., Bonadio, J., Goldstein, S. A., and Levy, R. J. (1999). Gene transfection using biodegradable nanospheres results in tissue culture and a rat osteotomy model. *Colloids Surf. B: Biointerfaces* **16**:281–290.

Lavasanifar, A., Samuel, J., and Kwon, G. S. (2002). Poly(ethylene oxide)-block–Poly(L-amino acid) micelles for drug delivery. *Adv. Drug Deliv. Rev.* **54**:169–190.

Lin, S. Y., Chen, K. S., Teng, H. H., and Li, M. J. (2000). in vitro degradation and dissolution behavior of microspheres prepared by low molecular weight polyesters. *J. Microencapsulation* **17**:577–586.

Liu, D. (1997). Biological factors involved in blood clearance of liposomes by liver. *Adv. Drug Deliv. Rev.* **24**:201–213.

Liu, M. J., and Frechet, J. M. J. (1999). *Pharmaceutical Sci. Technol. Today* **2**(10):393.

Liu, M., Kono, K., and Fréchet, J. M. J., (2000). *J. Controlled Release* **65**(1–2):121.

Li, Y., and Kwon, G. S. (2000). Methotrexate esters of poly(ethylene oxide) block-poly(2-hydroxyethyle-L-aspartamide), Part I: Effects of the level of methotrexate conjugation on the stability of micelles and on drug release. *Pharm. Res.* **17**:607–611.

Malik, N., Wiwattanakpatapee, R., Klopsch, R., Lorenz, K., Frey, H., Weener, J. W., Meijer, E. W., Paulus, W., and Duncan, R. (2000). Dendrimers: Relationships between structure and biocompatibility in vitro and preliminary studies on the biodistribution of ^{125}I labeled polyamidoamine dendrimers in vivo. *J. Controlled Release* **65**:133–148.

McGraw, T. E., Maxifield, F. R., Juliano, R. L. (1991). *Targeted Drug Delivery*, pp. 11–41. Springer Verlag, New York.

Michael, S. I., and Curiel, D. T. (1994). Strategies to achieve targeted gene delivery via the receptor mediated endocytosis pathway. *Gene Therapy* **1**:223–232.

Morishita, R., Gibbons, G. H., Kaneda, Y., Ogihara, T., and Dzau, V. J. (1994). Pharmacokinetics of antisense oligonucleotides (Cyclin B1 and C and C2 kinase) in the vessel wall: Enhanced therapeutic utility for restonsis by HVJ-liposome method. *Gene* **149**:3–9.

Mukherjee, S., Ghosh, R. N., and Maxfield, F. R. (1997). Endocytosis. *Physiol. Rev.* **77**:760–803.

Nagayasu, A., Uchiyama, K., and Kiwada, H. (1999). The size of liposomes: a factor which affects their targeting efficiency to tumors and therapeutic activity of liposomal antitumor drugs. *Adv. Drug Deliv. Rev.* **40**:75–87.

Nagayasu, A., Uchiyama, K., Nishida, T., Yamagiwa, Y., Kawai, Y., and Kiwada, H. (1996). Is control of distribution of liposomes between tumors and bone marrow possible? *Biochim. Biophys. Acta* **1278**:29–34.

Nakanishi, T., Hayashi, A., Kunisawa, J., Tsutsum, Y., Tanka, K., Yashiro-Ohtani, Y., Nakanishi, M., Fujiwara, H., Hamaoka, T., and Mayumi, T. (2000). Fusogenic liposomes efficiently deliver exogenous antigen through the cytoplasm into the MHC-class I processing pathway. *Eur. J. Immunol.* **30**:1740–1747.

Nilsson, J. J., Thyberg, J., Heldin, C. H., Westermark, B., and Wasteson, A. (1983). Surface binding and internalization of platelet derived growth factor in human fibroblasts. *Proc. Natl. Acad. Sci. USA* **80**:5592–5596.

Oussoren, C., and Storm, G. (1997). Lymphatic uptake and biodistribution of liposomes after subcutaneous injection. III. Influence of surface modification with poly(ethylene glycol). *Pharm. Res.* **14**:1479–1484.

Oussoren, C., Zuidema, J., Crommelin, D. J. A., and Storm, G. (1997). Lymphatic uptake and biodistribution of liposomes after subcutaneous injection. II. Influence of liposomal size, lipid composition and lipid dose. *Biochim. Biophys. Acta* **1328**:261–272.

Padilla De Jesus, O. L., Ihre, H. R., Gagne, L., Frechet, J. M. J., and Szoka, F. C., Jr. (2002). *Bioconjugate Chem.* **13**(3):453.

Panyam, J., and Labhasetwar, V. (2003b). Biodegradable nanoparticles for drug and gene delivery to cells and tissues. *Adv. Drug Deliv. Rev.* **55**:329–347.

Panyam, J., and Labhasetwar, V. Dynamics of endocytosis and exocytosis of poly (D,L-lactide-co-glycolide) nanoparticles in vascular smooth muscle cells. *Pharm. Res.* **20**:110–118. (2003a).

Panyam, J., Zhou, W. Z., Prabha, S., Sahoo, S. K., and Labhasetwar, V. Rapid endo-lysosomal escape of poly (D,L-lactide co-glycolide) nanoparticle. Implication for drug and gene delivery. *FASEB J.* **16**:1217–1226. (2002).

Peters, K. R., Carely, W. W., and Palade, G. E. (1985). Endothelial plasmalemmal vesicles have a characteristic stripped bipolar surface structure. *J. Cell. Biol.* **101**:2233–2338.

Radwan, M. A. (1995). In vitro evaluation of evaluation of poly isobutylcyanoacrylate nanoparticles as a controlled drug carrier for theophylline. *Drug Dev. Ind. Pharm.* **21**:2371–2375.

Randolph, T. W., Randolph, A. D., Meber, M., and Yeung, S. (1993). Sub-micron sized biodegradable particles of poly (L-lactic acid) via the gas antisolvent spray precipitation process. *Biotechnol. Prog.* **9**:429–435.

Schnitzer, J. E. (2001). Caveolae: from basic trafficking mechanisms to targeting transcytosis for tissue specific drug and gene delivery in vivo. *Adv. Drug Deliv. Rev.* **49**:265–280.

Senior, J. H. (1987). Fate and behavior of liposomes in vivo: A review of controlling factors. *Crit. Rev. Ther. Drug Carrier Syst.* **3**:123–193.

Sheikh, N. A., al- Shamisi, M., and Morrow, W. J. (2000). Delivery systems for molecular vaccination. *Curr. Opin. Mol. Ther.* **2**:37–54.

Soppimath, K. S., Aminabhavi, T. M., Kulkarni, A. R., and Rudzinski, W. E. (2001). Biodegradable polymeric nanoparticles as drug delivery devices. *J. Controlled Release* **70**:1–20.

Stiriba, S.-E., Frey, H., and Haag, R. (2002). *Angew. Chem. Int. Ed. Engl.* **41**:1329.

Suh, H., Jeong, B., Ralhi, R., and Kim, S. W. (1998). Regulation of smooth muscle cell proliferation using paclitaxel loaded poly(ethylene oxide)–poly(lactide-glycolide) nanospheres. *J. Biomed. Mater. Res.* **42**:331–338.

Tailefer, J., Jones, M. C., Brasseur, N., Van Liear, J. E., and Leroux, J. C. (2000). Preparation and characterization of pH responsive polymeric micelles for the delivery of photosensitizing anticancer drugs. *J. Pharm. Sci.* **89**:52–62.

Tajarobi, F., El-Sayed, M., Rege, B. D., Polli, J. E., and Ghandehari, H. (2001). *Int. J. Pharm.* **215**(1–2):263.

Tomalia, D. A., Baker, H., Dewald, J., Hall, M., Kallos, G., Martin, S., Roeck, J., Ryder, J., and Smith, P. (1985). A New Class of Polymers: Starburst–Dendrimers Macromolecules. *Polym. J.* **17**(1):117.

Tomalia, D. A., Baker, H., Dewald, J., Hall, M., Kallos, G., Martin, S., Roeck, J., Ryder, J., and Smith, P. (1985). Dendritic macromolecules: Synthesis of starburst dendrimers. *Macromolecules.* **19**:2466.

Tomalia, A., Naylor, A. M., Goddard, W. A., III., (1990). Starburst dendrimers: control of size, shape, surface chemistry, topology and flexibility in the conversion of atoms to macroscopic materials. *Angewandte. Chemie.* **102**(2):119.

Tomoda, H., Kishimoto, Y., and Lee, Y. C. (1989). Temperature effect of endocytosis and exocytosis by rabbit alveolar macrophages. *J. Biol. Chem.* **264**:15445–15450.

Tuzar, Z., and Kratochvil, P. (1976). Block and graft co-polymer micelles in solution. *Adv. Colloid Interface Sci.* **6**:201–232.

Ueda, M., Iwara, A., and Kreuter, J. (1998). Influence of the preparation methods on the drug release behavior of loperamide loaded nanoparticles. *J. Microencapsulation* **15**:361–372.

Wang, S., and Low, P. S. (1998). Folate mediated targeting of anti-neoplastic drugs, imaging agents and nucleic acids to cancer cells. *J. Controlled Release* **53**:39–48.

Wang, Y. Y., Robertson, J. L., Spillman, W. B., Jr., and Claus, R. O. (2004). Effect of the chemical structure and the surface properties of polymeric biomaterials on the biocompatibility. *Pharm. Res.* **21**:1362–1373.

Wiwaattanapatapee, R., Carreno-Gomez, B., Malik, N., and Duncan, R. (2000). *Pharm. Res.* **17**:991.

Woodle, M. C. (1998). Controlling liposome blood clearance by surface grafted polymers. *Adv. Drug Deliv. Rev.* **32**:139–152.

Yoo, H. S., Oh, J. E., Lee, K. H., and Park, T. G. (1999). Biodegradable nanoparticles containing doxorubicin–PLGA conjugate for sustained release. *Pharm. Res.* **6**:1114–1118.

Zhu, N., Liggitt, D., Liu, Y., and Debs, R. (1993). Systemic gene expression after intravenous DNA delivery into adult mice. *Science* **261**:209–211.

Zhuo, R. X., Du, B., and Lu, Z. R. (1999). *J. Controlled Release* **57**:249.

CHAPTER 8

ADAPTING AFM TECHNIQUES FOR STUDIES ON LIVING CELLS

J. K. HEINRICH HÖRBER

Department of Physics, University of Bristol, Bristol, United Kingdom

8.1. COMBINING AN AFM WITH OPTICAL MICROSCOPY AND ELECTROPHYSIOLOGICAL TECHNIQUES

The AFM (Binnig et al., 1986) was thought from the very beginning to be an ideal tool for biological research. Imaging living cells under physiological conditions and studying dynamic processes at the plasma membrane were envisioned (Drake et al., 1989), although it was clear that such experiments are quite difficult, as the AFM cantilever is by far more rigid than cellular structures. In this context the way cells are supported is quite important. Also a parallel optical observation is necessary to control the cantilever tip approach to distinct cellular features and to link the measurements to well-established optical microscopy techniques.

8.1.1. AFM on an Inverted Optical Microscope

In order to address these problems, we started in 1988 the development of a special AFM built into an inverted optical microscope, which finally provided the first reproducible images of the outer membrane of a living cell held by a pipette in its normal growth medium (Häberle et al., 1991, 1992; Hörber et al., 1992; Ohnesorge et al., 1997). A conventional piezo-tube scanner moved this pipette. The detection system for the cantilever movement used was in principle the normal optical detection scheme, but a glass fiber was used as a light source very close to the cantilever. This setup allowed a very fast scanning speed of up to a picture per second for imaging cells in the variable deflection mode. This is possible because the parts moving in the liquid are very small and, quite different from the imaging of cells adsorbed to a microscope slide, no significant excitation of disturbing waves or convection in the liquid occurs. It was possible to keep the cell alive and well for days within the liquid cell built. This made studies of live activities and kinematics possible, in addition to the application of other cell physiological measuring techniques. With this step in the development of scanning probe instruments, the capability of optical microscopy to investigate the dynamics of biological processes of cell membranes under physiological conditions could be extended with the help of the AFM into the nanometer range.

At high imaging rates of up to one frame per second, we finally could observe structures as small as 10 nm. This made it

---

Force Microscopy: Applications in Biology and Medicine, edited by Bhanu P. Jena and J.K. Heinrich Hörber.
Copyright © 2006 John Wiley & Sons, Inc.

possible to study processes such as the binding of labeled antibodies, endo- and exocytosis, pore formation, and the dynamics of surface and membrane cytoskeletal structures in general. The problem that still has to be solved, unfortunately for each cell individually, is to control the interaction between tip and plasma membrane structures, which are influenced quite strongly by the so-called extracellular matrix of cells containing a broad variety of sugars and other polymer structures.

With the integrated tip of the cantilever forces in the range of some 10 pN up to some 10 nN are applied to the investigated cell, the mechanical properties of cell surface structures therefore dominate the imaging process. On one hand, this fact mixes topographic and elastic properties of the sample in the images; on the other hand, it provides additional information about cell membranes and their dynamics in various situations during the life of the cell, if these two aspects can be separated in the AFM images. To separate the elastic and topographic properties, further information is needed, which can be provided either by additional topographic data from electron microscopy or by using AFM modulation techniques. The pipette–AFM concept is very well suited for such modulation measurements because, as mentioned, convection or excitation of waves in the solution due to the pipette movement is negligible and so the modulation can be done very fast. Nevertheless, for a thorough analysis of a cell wall elasticity map, one would have to record pixel-by-pixel a complete frequency spectrum of the cantilever response and derive image data in various frequency regimes. This would need too much time for a highly dynamic system like a living cell even with the speed possible with a cell at the end of a pipette. Experiments we have done on living cells showed in some cases a certain weak mechanical resonance in the regime of several kilohertz. Such resonance might be used to characterize cells and provide a kind of spectroscopic fingerprint for cellular processes or even drug effects, which may lead to new methods of medical diagnosis at the cellular level.

Figure 8.1 shows the schematic arrangement of the AFM above the objective of an inverted optical microscope. The sample area can be observed from below through a planar surface defined by a glass plate with a magnification of $600\times-1200\times$ and from above by a stereo microscope with a magnification of $40\times-200\times$. The illumination is from the top through the less well-defined surface of the aqueous solution. In order not to block the illumination, the manipulator for the optical fiber and for the micropipette point toward the focal plane at an angle of $45°$. The lever is mounted in a fixed position within the liquid slightly above the glass plate tilted by $45°$. The single-mode optical fiber is used to reduce the necessary optical components by bringing the end as close as possible to the lever. To do so, the last several millimeters of the fiber's protective jacket are removed. The minimum distance is determined by the diameter of the fiber cladding and the geometry of the lever. We normally use fibers for 633-nm light which have a nominal cladding diameter of 125 μm and levers which are 200 μm long. Holding the fiber at an angle of $45°$ with respect to the lever means that we can safely bring the fiber core as close as 150 μm to the lever. The 4-μm-diameter core has a numerical aperture of 0.1, and the light emerging from the fiber therefore expands with an apex angle of $6°$. For the geometry given above, the smallest spot size achievable is 50 μm, approximately the size of the triangular region at the end of the cantilever. Due to the construction of the mechanical pieces holding the pipette and the fiber positioner, the closest we can bring the position-sensitive quadrant photodetector to the lever is approximately 2 cm. This implies a minimum spot size of 2.1 mm, within the 3-mm × 3-mm borders of the detector. Using a 2-mW HeNe laser under normal operational conditions, the displacement

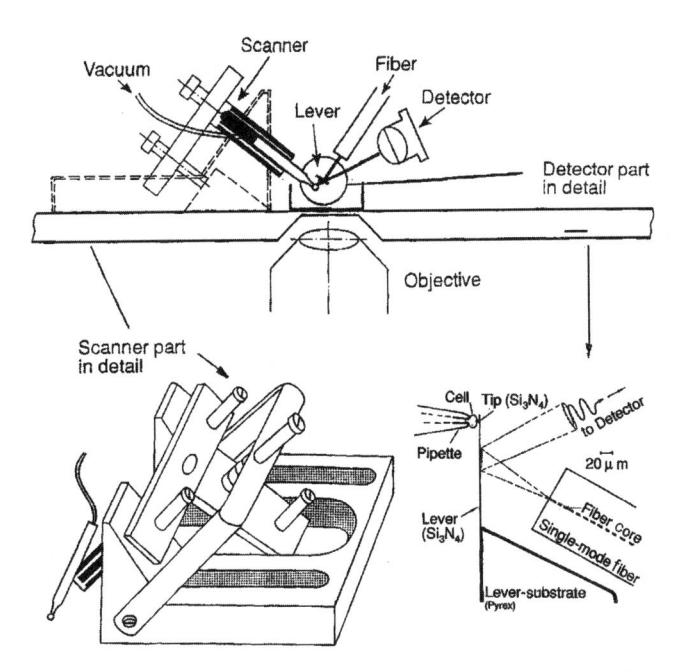

Figure 8.1. Schematic drawing of the AFM built onto an inverted optical microscope with a patch-clamp pipette as a sample holder. For the optical detection of the cantilever deflection, an optical fiber as a light source very close to the cantilever is used. The detection of the reflected light is done by a quadrant photodiode above the sample chamber.

sensitivity is below 0.01 nm with a signal-to-noise ratio of 10, comparable to the sensitivity of other optical detection techniques. The advantage of this method is that there are no lenses and there is no air–liquid or air–solid interface across which the incoming light beam must travel; only the outgoing light has to cross the water–air interface, because the detector cannot be immersed in the water.

The pipette used for holding the cell is made out of a 0.8-mm borosilicat glass capillary that is pulled to about (a) 2- to 4-μm opening in three pulling steps. It is mounted on the piezotube scanner and coupled to a fine and flexible Teflon tube, through which the pressure in the pipette can be adjusted by a piston or water pump. The pipette is fixed at an angle that allows imaging of the cell without the danger of touching the pipette with the lever. All these components are located in a container of 50-ml volume. The glass plate above the objective of the optical microscope forms the bottom of this container.

After adding several microliters of the cell suspension, a single cell can be sucked onto the pipette and fixed there by maintaining low pressure in the pipette. The other cells are removed through the pumping system. The fixed cell is placed close to the AFM lever by a rough approach with screws and finally positioned by the piezo scanner. When the cell is in close contact with the tip of the lever, scanning of the capillary with the cell attached leads to position-dependent deflections of the lever. The levers used are microfabricated silicon and silicon nitrite triangles with 200-μm length and with a spring constant of 0.12 N/m. The forces which can be applied with these levers, and the sensitivity of the detection system can be as low as 0.1 nN.

With such forces applied to the plasma membrane of a cell, the stiffness of the membrane structures dominates the images.

The scaffold of the cortical layer of actin filaments and actin-binding proteins that are cross-linked into a three-dimensional network and closely connected to the surface membrane dominate the images. These filaments have a structural role and are also involved in creating the changes seen in the AFM as the dynamics of the surface protrusions.

The relatively stable arrangements of the actin filaments are responsible for their relatively persistent structure. But these surface actin filaments are not permanent. During phagocytosis or cell movement, rapid changes of shape occur at the cell surface. These changes depend on the transient and regulated polymerization of cytoplasmic free actin or the depolarization during the breakdown of the actin filament complex. The time scale of cell surface changes on a larger scale observed with the AFM was about 1–2 h at room temperature. Except for small structures some 10 nm in size, everything is quite stable for 1–2 h. More and larger structures are rearranged only 10–15 min after this period has elapsed.

In a series of experiments, we have studied the membrane dynamics after infection of the cell by pox viruses. On the timescale of seconds to minutes after adding the virus solution to the chamber, the cell surface suddenly becomes smooth and so soft that the tip tends to penetrate the cell surface even with loading force far below 1 nN. This state of the cell might be related to the penetration of the viral particles through the cell membrane. After a few minutes the cell again becomes more rigid, and stable imaging is observed very similar to the situation before the infection. During the first hour after infection we usually did not detect any significant variation in the membrane structures imaged. It is known that within 4–6 h the first viruses are reproduced inside the cell and pass through the cell membrane via exocytosis (Stokes, 1976). However, approximately 2.5 h after infection, we already observed a series of processes occurring at the membrane. Single clear protrusions become visible and grow in size. The objects quickly disappear and the original structure of the membrane is restored. Such processes can occur several times in the same area and last about 90 s for small protrusions of about 20 nm and last up to 10 min for larger ones of up to 100-nm size. Each process proceeds distinctly and apparently independently of the others. They are never observed with uninfected cells and never until 2 h after infection.

The fact that the growing protrusions abruptly disappear after a while makes us believe that we observe the exocytosis of particles related to the starting virus reproduction, but not viruses. First progeny viruses are known to appear 5–8 h after infection, and they are clearly bigger than the structures described so far. It is known (Moss, 1986), however, that after 2–3 h, only the early stage of virus reproduction is finished and the final virus assembly has just begun. Since the protrusions are observed after this characteristic time span, we believe that they are related to the exocytosis of protein agglomerates originating from the virus assembly.

Significantly longer than 6 h after infection there are even more dramatic changes observed in the cell membrane. Large protrusions, with cross sections of 200–300 nm, grow out of the membrane in the vicinity of deep folds. These events occur much less frequently than those after 2 h. These protrusions also abruptly disappear, leaving behind small scars at the cell surface. Considering the timing and their size, we believe these protrusions are progeny viruses exiting the cell. Assuming that approximately 20–100 viruses exit the living cell and that roughly 1/40 of the cell surface is accessible to the AFM tip, one should be able to observe one or two of those events. We have actually observed 2 processes exhibiting the right size and timing during one 46-h experiment on a single infected cell, one after 19 h and one after 35 h. It is known from electron microscopy that individual viruses exit

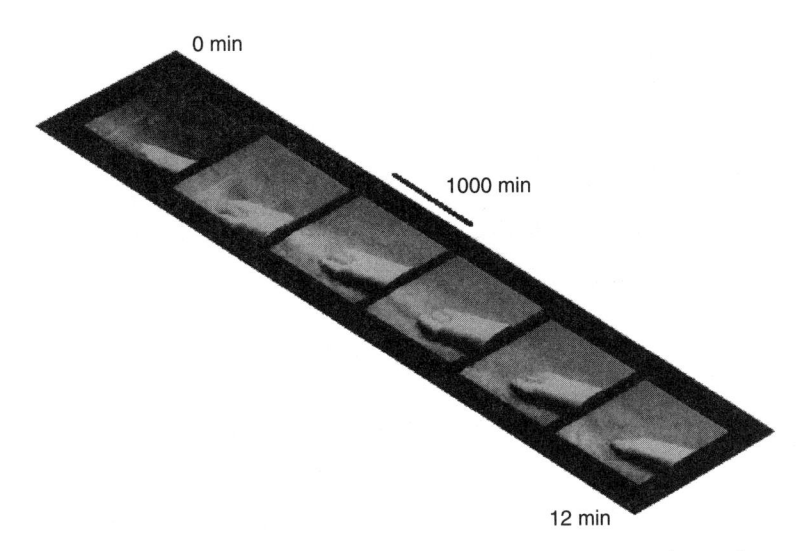

Figure 8.2. Exocytotic process seen about 19 h after the cell was infected by pox viruses. An unusually strong fingerlike structure is visible, and at one end a protrusion appears and disappears again after about 3 min. The size of the protrusion, the timing, and the evolution of the event are consistent with the interpretation that the release of a virus at the end of a microvillus structure is observed. There is a strong similarity to electron micrographs and fluorescence-labeled optical observation of such events.

sometimes the cell at the end of microvilli. Figure 8.2 actually shows such a protrusion and an exocytotic event occurring at the end. The release of another particle shown in Fig. 8.3 also occurs in a region where the cell membrane shows an array of microvilli. This striking similarity to results seen in electron and optical microscopy (Moss, 1986, (p. 40); Frischknecht et al., 1999) makes us believe that we indeed have imaged the exocytosis of a progeny virus through the membrane of a living infected cell.

8.1.2. Integrating the AFM into a Electrophysiological Setup

The method of fixing a cell to a pipette is flexible enough to allow integration of and combination with well-established cell physiological techniques of manipulation used in investigations of single cells. The structures observed can be partially related to known features of membranes, but more important than just imaging structures is the possibility provided by this technique to conduct dynamic studies of living organisms on this nanometer scale. In general, this technique makes studies of the evolution of cell membrane structures possible and provides information that brings us closer to understand not only the "being" of these structures, but also their "becoming."

An AFM with the cell fixed on end of a pipette, as described in the previous section, makes a combination of patch-clamp measurements on ion channels in the membrane of whole cells with force microscope studies possible. A logical step in the development of AFM technique combined with established cell biological techniques therefore was the development of a combined patch-clamp and AFM setup that can be used to investigate also excised membrane patches (Hörber et al., 1995) and in this way to study single ion channels in the membrane, which become activated by mechanical stress in the membrane. The importance of these ion channels is very obvious in our senses of touching and hearing. During developments that started in 1991, a new type of patch-clamp setup was built being much more stable to satisfy the needs of AFM

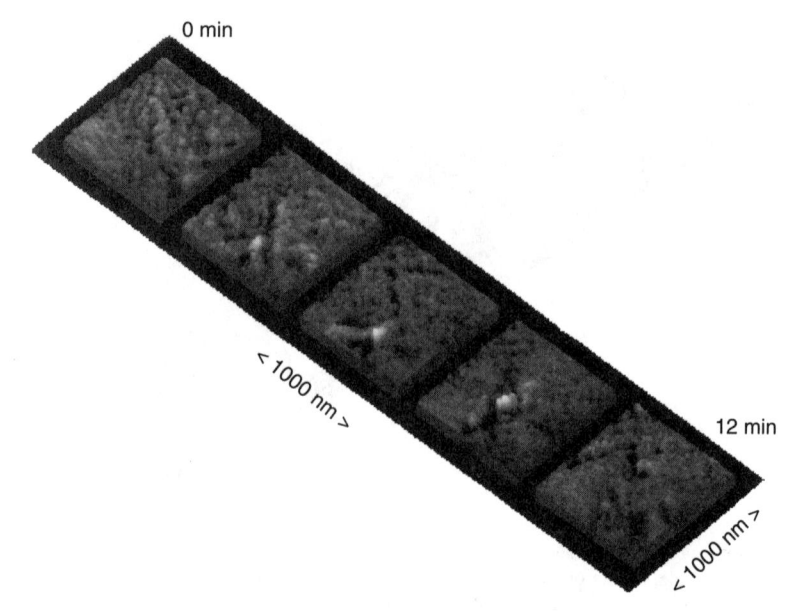

Figure 8.3. Exocytotic process seen about 35 h after viral infection of cultured monkey kidney cell. The protrusion in the lower part of the images over about 10 min slowly grow in size to about 200 nm in diameter. It disappears abruptly, leaving a scar on the cell membrane.

applications (Fig. 8.4) and able to accommodate the additional electrophysiological and optical components. The chamber, where a constant flow of buffer solution guarantees the right conditions for the experiments, consists of two optical transparent plastic plates; one at the top and one at the bottom, keeping the water inside just by its surface tension. In this way, a flow cell is created, which is accessible from two sides. The chamber is mounted on an *xyz*-stage together with the optical detection of the AFM lever movement and a double-barrel application pipette. The latter was integrated so that the setup can also be used for standard patch-clamp measurements with application of chemicals and markers to characterize the sample further. The patch-clamp pipette itself is mounted on a piezo-tube scanner fixed with respect to the objective of an inverse optical microscope that is necessary to control the approach of the pipette to the cell and to the AFM lever finally. In the experiments, small membrane pieces are excised from a cell inside the chamber containing none, one, or only a few of the mechano-sensitive ion channels. This enables studies of currents through a single channel, which opens and closes on the timescale of micro- to milliseconds resulting in currents in the pico-ampere range.

In such a setup with a membrane patch at the end of the pipette the AFM can image the tip of the pipette with the patch on top (Fig. 8.5). Furthermore, changes in the form of the patched membrane produced by changing pressure in the patch pipette can be monitored, along with the reaction of the membrane to the change of the electric potential across. In some images, cytoskeleton structures can be seen that are excised together membrane. On these stabilizing structures of the membrane and on the rim of the pipette, resolutions down to 10–20 nm can be achieved showing reproducible structures and changes that can be induced by the application of force.

Voltage-sensitive ion channels change shape in electrical fields, leading eventually to opening of the ion-permeable pore. To determine the size of this electromechanical

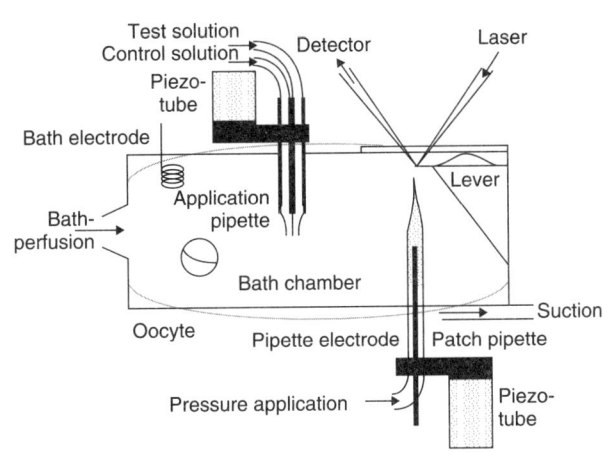

COMBINING AN AFM WITH OPTICAL MICROSCOPY AND ELECTROPHYSIOLOGICAL TECHNIQUES   **143**

Figure 8.4. Schematic top view of the patch-clamp/AFM setup within its bath chamber. The patch pipette is shown in front of the cantilever. Bath and pipette electrodes are used for electrophysiological recordings. A glass plate is positioned behind the cantilever to reduce fluctuations of laser beam direction by movement of the water–air interface. Different solutions can be applied to the excised patch by a two-barreled application pipette. If necessary, continuous perfusion of the path chamber can be stopped during AFM measurements.

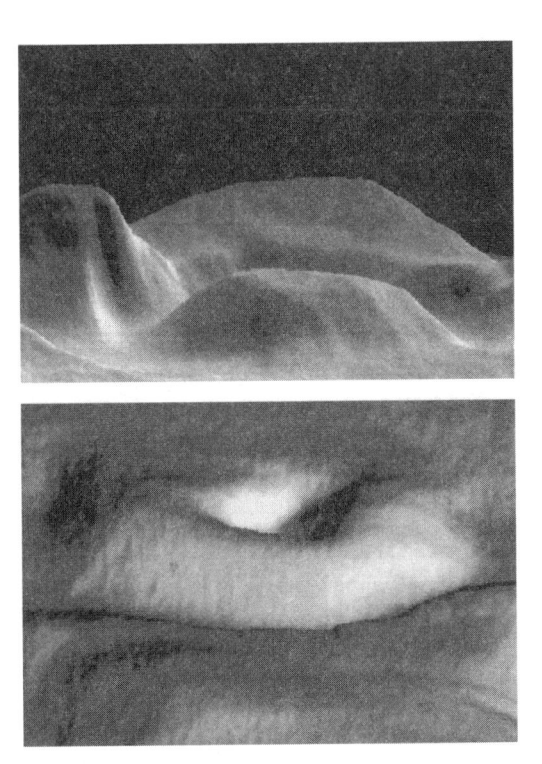

Figure 8.5. Three-dimensional representation of membrane patches at the top of a pipette pulled to 2 μm in diameter. In the upper image the higher part in the background corresponds to the glass rim and the central hill corresponds to the membrane patch spanning the central hole of the pipette. It is quite remarkable to see how this inside-out patch attaches to the glass being pulled first to the inside of the pipette, but then in the center extends outward again. In the image below, a slide suction is applied through the pipette. It does not pull the whole extending part of the membrane, which has a diameter of 1 μm, to the inside, but only pulls a small part in the middle. The height differences seen in the picture are about 200 nm.

transduction, we investigated with the AFM cells from a cancer cell line (HEK 293), which are kept at a certain membrane potential (voltage-clamped) (Mosbacher et al., 1998). We used either normal cells as controls, or cells transfected with *Shaker* K$^+$ ion channels. In control cells we found movements of 0.5 to 5 nm normal to the plane of the membrane. These movements tracked a ±10-mV peak-to-peak AC carrier stimulus to frequencies >1 kHz with a phase shift of 90–120°, as expected of a displacement current. The movement was outward with depolarization, and the holding potential was only weakly influenced by the amplitude of the movement. In contrast, cells transfected with a non-inactivating mutant of *Shaker* K$^+$ channels showed movements that were sensitive to the holding potential, decreasing with depolarization between −80 mV and 0 mV.

Further control experiments used open or sealed pipettes and cantilever placements just above the cells. The results suggested that the observed movement is produced by the cell membrane rather than by artificial movement of the patch pipette, acoustic or electrical interaction of the membrane, and the AFM tip. The large amplitude of the movements and the fact that they occurred also in control cells with a low density of voltage-sensitive ion channels imply the presence of multiple electromechanical motors. These experiments demonstrate that the AFM might be able to exploit the voltage-dependent movements as a source of contrast for imaging membrane proteins.

8.1.3. AFM Studies on Tissue Sections from the Inner Ear

Cochlear hair cells of the inner ear are responsible for the detection of sound. They encode the magnitude and time course of an acoustic stimulus as an electric receptor potential, which is generated by a still unknown interaction of cellular components. In the literature, different models for the mechanoelectrical transduction of hair cells are discussed (Corey and Hudspeth, 1983; Furness et al., 1996; Howard and Hudspeth, 1988; Hudspeth, 1983). All hypotheses have in common that a force applied to the so-called hair bundle (which consists of specialized stereocilia at the apical end of the cells) in the positive direction, toward the tallest stereocilia, opens transduction channels, whereas negative deflection closes them. For a better understanding of the transduction process, it is important to know what elements of the hair bundle contribute to the opening of transduction channels and to study their mechanical properties. The morphology of the hair cells is precisely described by scanning and transmission electron microscopy. Unfortunately, this method is restricted to fixed and dehydrated specimen. Therefore, we have extended our AFM development to enable the study of cochlear hair cells in physiological solution. To provide a clear view onto the tissue section, we built the combined AFM/patch-clamp setup onto an upright differential interference contrast (DIC) light microscope (Langer et al., 1997). A water immersion objective (40 × /0.75) provides a high resolution of about 0.5 µm even on organotypic cell cultures with thickness of about 300 µm. Using this setup (Fig. 8.6), it is possible to see ciliary bundles of inner and outer cochlear hair cells extracted from six- and eight-day-old rats and to approach these structures in a controlled way with the AFM cantilever tip and the patch pipette (Fig. 8.7). Images can be obtained by measuring the local force interaction between the AFM tip and the specimen surface while scanning the tip (Langer et al., 2000). The question whether morphological artifacts occurred at the hair bundles during AFM investigation was clarified by preparing the cell cultures for the electron microscope directly after the AFM measurements. Forces up to 1.5 nm applied in the direction of the stereocilia axis did not change the structure of the hair bundles.

The scan traces shown in Fig. 8.8 represent a typical example of force interaction

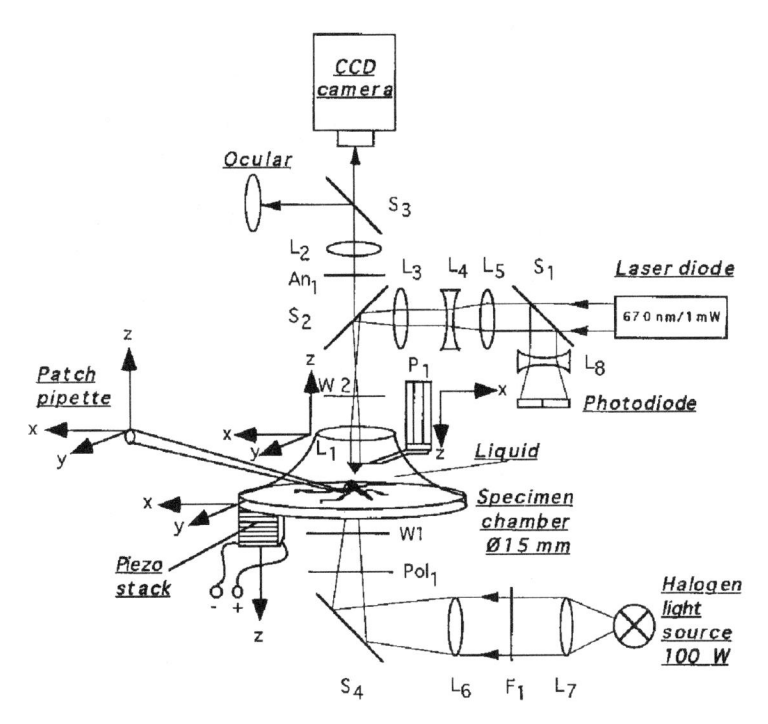

Figure 8.6. Schematic drawing of the setup built for whole cell tissue patch-clamp/AFM applications. A piezo-tube scanner P_1 moves the cantilever mounted on a titanium arm in xy and z direction. An xz-translation stage allows the lateral and vertical coarse positioning of the cantilever. Additionally, an xy-translation device moves the optical microscope. Thus, in combination with the vertical objective translation stage of the optical microscope, the laser beam is adjustable in all three orthogonal directions relative to the cantilever. The cantilever deflection is detected by the laser deflection method. A parallel beam of a laser-diode propagates through the beam-splitter cube S_1 and is reduced to a diameter of 1 mm by the convex lens L_5 and the concave lens L_4. The convex lens L_3 focuses the beam into the back-focal plane of the water immersion objective. Notch filter S_2 reflects the laser light to the objective lens L_1. It transmits light below 600 nm and above 700 nm, and it reflects with 98% in between. A Zeiss Achroplan 40 × /0.75-W water immersion objective lens creates a parallel laser beam, which is positioned on the cantilever. The reflected laser light propagates the same way back and is reflected by the beam-splitter cube S_1 through the concave lens L_8 onto a quadrant photodiode. With the help of this optical setup the shift of the reflected laser spot caused by a cantilever deflection or torsion is detected. For fine adjustment of the tip-sample distance the specimen chamber is moved vertically by a piezo-stack. A hydraulic micromanipulator, directly mounted on the specimen stage, is used to move the patch-clamp head stage together with the glass pipette in three orthogonal directions. The whole setup is built into an upright optical DIC-Microscope. For infrared DIC imaging the beam of a halogen light source propagates through an edge filter F_1 cutting below 700 nm. The light is focused by lenses L_6 and L_7 and reflected by mirror S_4 on the specimen chamber. The polarizer Pol_1, the Wollaston prisms W_1 and W_2, and the analyzer An_1 produce the DIC contrast. The imaging tubus lens L_2 creates an image which is detected by a CCD camera simultaneously viewed through the 10× eyepiece. It is possible to use all three units (optical microscope, AFM, and patch-clamp setup) simultaneously.

between the AFM tip and the stereocilia. For determination of the stereocilium position, all 200 scan lines recorded on each hair bundle were displayed as a greyscale image. The central stereocilium of a hair bundle was defined as "0" while adjacent stereocilia were labeled with numbers starting from "1" for the left half of the hair bundle and "−1" for the right half of the hair bundle. Traces shown represent the force interaction in excitatory direction. The stiffness of the stereocilia, when pushed, increased from 0.73 mN/m and reached a steady-state level at about 2 mN/m. The mean stiffness of 17

146 ADAPTING AFM TECHNIQUES FOR STUDIES ON LIVING CELLS

Figure 8.7. Principle of simultaneous AFM and patch-clamp recording composed from three separately taken SEM images. The central part represents the organ of Corti with V-shaped hair bundles on top. Above the front end of an AFM cantilever with its tip is placed and below the patch pipette contacting one of outer hair cells can be seen.

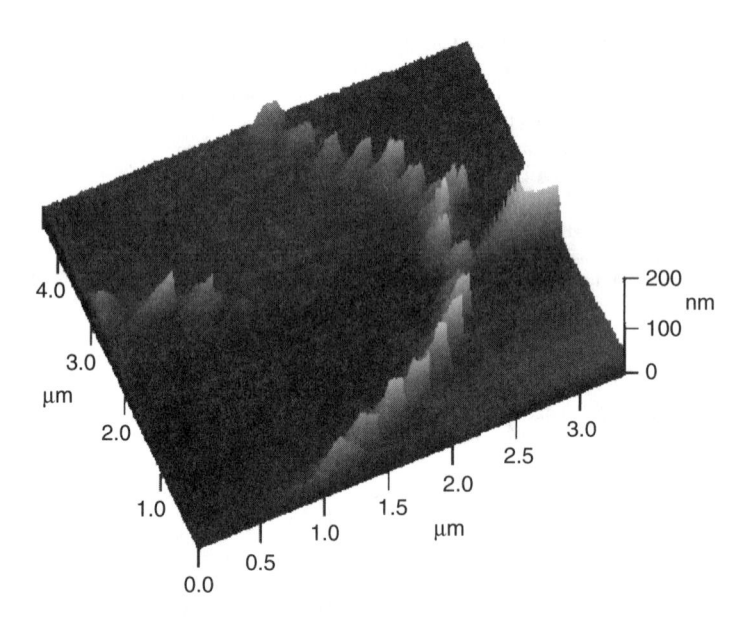

Figure 8.8. The overlayed saw-tooth patterns represent a typical example of force interaction between the AFM tip and the stereocilia. For determination of the stereocilium position, all 200 scan lines recorded on each hair bundle were displayed as a greyscale image. The central stereocilium of a hair bundle was defined as "0". At this position in the experiments with an additional glass capillary attached this capillary was positioned. Adjacent stereocilia were labeled with numbers starting from "1" for the left half of the hair bundle and "−1" for the right half of the hair bundle. Traces shown represent the force interaction in excitatory direction.

investigated hair bundles of outer hair cells can be calculated to be 2.5 m N/m in the excitatory direction and 3.1 m N/m in the inhibitory direction.

The average force constant showed only a weak dependence on the stereocilium position for the excitatory direction. Some stereocilia located at the center showed exceptionally high stiffness, whereas some located at the outer region (positions "−7" to "−11") revealed smaller stiffness.

In inhibitory direction, the mean force constant was slightly higher with a standard deviation being 2.5 times higher compared to the excitatory direction. The larger force constant might be explained by a direct contact between the taller and the adjacent shorter stereocilia while in excitatory direction shorter stereocilia are pulled by elastic tip links. These links connect the rows of taller and shorter stereocilia (Furness and Hackney, 1985; Lim, 1986; Furness et al., 1989; Zine and Romand, 1996). The high standard deviation might be an effect of the variation in spatial interaction between taller and shorter stereocilia. Depending on the angle between scan direction and the direction defined by the centers of adjacent taller and shorter stereocilia, mechanical compliance may vary in a wide range.

For the excitatory direction (Fig. 8.9) two distributions are distinguishable. One population of stiffness values is located around 2.2 mN/m, and a smaller second one is located around 3.1 mN/m. These stiffness data correspond to stereocilia located in the central region of investigated hair bundles. Comparing the stiffness plot in Fig. 8.9 and the corresponding greyscale image (Fig. 8.8), we can find a correlation between arrangement and stiffness of the stereocilia. The stereocilia standing a little apart from their neighbors show a significantly higher stiffness of 3.2 mN/m and 3.1 mN/m, respectively.

A possible hypothesis may be that lateral links connecting these stereocilia with

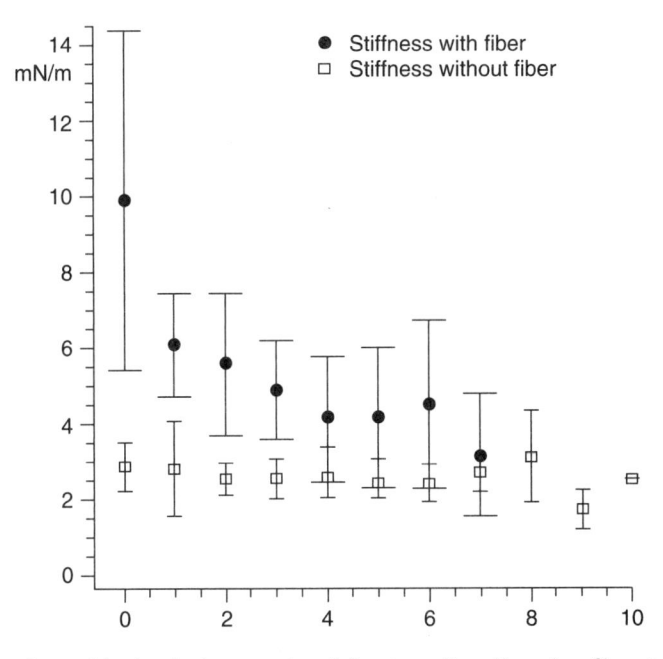

Figure 8.9. Changes observed in the elastic properties of the stereocilia with a glass fiber attached to the central stereocilium. Only direct neighbors are significantly affected, pointing to a small effect of side links to the observed stiffness.

their direct neighbors would be oriented in direction of the exerted force. This might allow transmission of additional force along these links to their neighbors.

This hypothesis implies that all other stereocilia would display much less interaction with their neighbors mediated by side links. This was directly addressed by a second AFM experiment. Force transmission between adjacent stereocilia was examined by AFM using a lock-in amplifier for detection of the transmitted forces together with a stimulation glass fiber touching an individual pair of stereocilia. The output of the AFM detector was connected to the input of a lock-in amplifier detecting the vertical oscillation of the AFM cantilever at 357 Hz. The lateral force F_L transmitted by lateral links was calculated from the output signal of the lock-in amplifier.

Force interaction between stereocilia and AFM tip led to displacements of stereocilia from zero to about 250 nm. Therefore, relative displacement between the directly stimulated stereocilium and the stereocilium displaced by the AFM tip is expected to result in continuous stretching of lateral links located in-between. This in principle allows detection of transmitted forces by lateral links for different states of stretching. Normalized forces rapidly decrease from the directly stimulated to the first adjacent stereocilium. Stereocilia located further away at positions "1" to "8" reveal only a slight decrease in relative force from 36% to 20%. Obviously, forces transmitted by lateral links rapidly decrease from a directly stimulated to adjacent stereocilia of rats at postnatal age day four. This result supports the hypothesis of a weak interaction between stereocilia by lateral links.

From these AFM measurements with the glass fiber attached also the stiffness of stereocilia can be determined as described before. We can use this information for detecting the mechanical effect of the touching glass fiber on stiffness of adjacent stereocilia. If lateral links contribute to the stiffness obtained at individual stereocilia, we would expect to see an increase in stiffness of adjacent stereocilia compared to data shown for stereocilia investigated without a fiber (Fig. 8.9). Stiffness data were calculated only for the excitatory direction, where the AFM tip displaces the stereocilia toward the fiber tip. Not only is the stiffness (filled circles) of the directly touched stereocilium at position "0" increased, but also the stiffness of stereocilia "1" to "4." Mean stiffness in excitatory direction is 4.8 mN/m. This is about 1.9 times higher compared to the mean stiffness of stereocilia not touched with a fiber (open squarres). For position "7" the stereocilium with and the stereocilium without a touching glass fiber show approximately identical stiffness.

These experiments proof that the AFM allows local stiffness measurement on the level of individual stereocilia with the measured stiffness representing the local elastical properties of the directly touched sterocilium and its nearest neighbors. The results show that stiffness depends on the orientation of links with regard to the direction of the stimulus and that a stimulating fiber had only little mechanical effect on adjacent stereocilia not in direct contact with the fiber.

For a partial decoupling of the tectorial membrane from hair bundles of outer hair cells, we would therefore suppose that only stereocilia still in contact with the tectorial membrane and their nearest neighbors are displaced by an incoming mechanical stimulus. Lateral links may not compensate loss in contact with the tallest stereocilia of outer hair cells. Decoupling of the tectorial membrane following exposure to pure tones at high sound pressure levels is supposed to protect hair cells and to avoid damage (Adler et al., 1993). A strong interaction between the tectorial membrane and hair bundles of outer hair cells seems to be essential for the efficacy of the cochlear amplifier and transduction of sound into an electrical signal.

Many micromechanical measurements have already been performed at entire stereociliary bundles of sensory hair cells using thin glass fibers directly attached to the bundle or fluid jets. The receptor potential or transduction current was measured in response to the displacement of stereocilia, but to study the kinetics of a single transduction channel over the whole range of its open probability requires a technique allowing stimulating a single stereocilium. As demonstrated already, AFM offers the opportunity to exert a force very locally to an individual stereocilium. In the experiments described in the following after supporting cells were removed using a cleaning pipette, a patch pipette filled with intracellular solution (concentrations in mM: KCl 135, $MgCl_2$ 3.5, $CaCl_2$ 0.1, EGTA 5, HEPES 5, Na_2ATP 2.5, pH 7.4) was attached to the lateral wall of one cell of the outermost row of outer hair cell. Thereby, the glass microelectrode forming a seal on the plasma membrane of an intact outer hair cell isolates a small patch (Fig. 8.7). This so-called "cell attached" configuration is the precursor of the whole-cell configuration where the microelectrode is in direct electrical contact with the inside of the cell. For low noise measurements of single ion channels the seal resistance should be typically in the range >1 GΩ. A pulse of suction applied to the pipette breaks the patch, thereby creating a hole in the plasma membrane, and provides access to the cell interior. During recording the electrical resistance between the inside of the pipette and the hair cell should be very small. Many voltage-activated K^+-ion channels are embedded in the lipid membrane of outer hair cells. Opening and closing of these channels increases the background noise level during transduction current measurements. Therefore, during transduction current measurements, the holding potential of the hair cell was set to -80 mV, corresponding to the reversal potential of K^+-ion channels. After forming a seal, the AFM tip was moved to the top of the corresponding hair bundle under light microscopic control. The AFM tip successively displaced each stereocilium within a hair bundle as described before. In contrast to force transmission measurements, a sinusoidal voltage was added to the normal AFM scan signal modulating the AFM tip in horizontal direction with 190 nm at 98 Hz. Thus, the hair bundle was slightly displaced several times while interacting with the lateral face of the AFM tip. The AFM tip repeatedly scanned in the same line while approaching the hair bundle toward the AFM tip. As expected, an inward current was not detected until displacement of a stereocilium in excitatory direction (Fig. 8.10).

In the experiments described before only a weak transmission of force from the directly stimulated stereocilium to adjacent stereocilia was observed implying that only few channels are opened. These results of elasticity measurement can be confirmed by electrophysiological findings, as shown in Fig. 8.10, where a set of transduction current measurements is displayed. The tip of an AFM cantilever repeatedly scanned across the same stereocilium of an outer hair cell. Applied horizontal forces of up to 0.8 nN (upper graphs) resulted in stereocilia displacements of about 350 nm in excitatory and 250 nm in inhibitory direction. The AFM tip displaces the stereocilium in excitatory direction between 89 and 131 ms, thereby opening transduction channels (middle graph). The current amplitude was determined for the period of interaction with the AFM tip, and the values for excitatory and inhibitory direction are put together in the histograms shown in Fig. 8.10 (lower graph). The distribution around 0 pA corresponds to the closed state of the channels, while the distribution around 19.1 pA corresponds to the maximal activation. Comparing this current amplitude to the maximum transduction current of about 462 pA measured for entire hair bundles of postnatal mice (Géléoc et al., 1997) confirms that the AFM is well-suited for local stimulation of individual stereocilia.

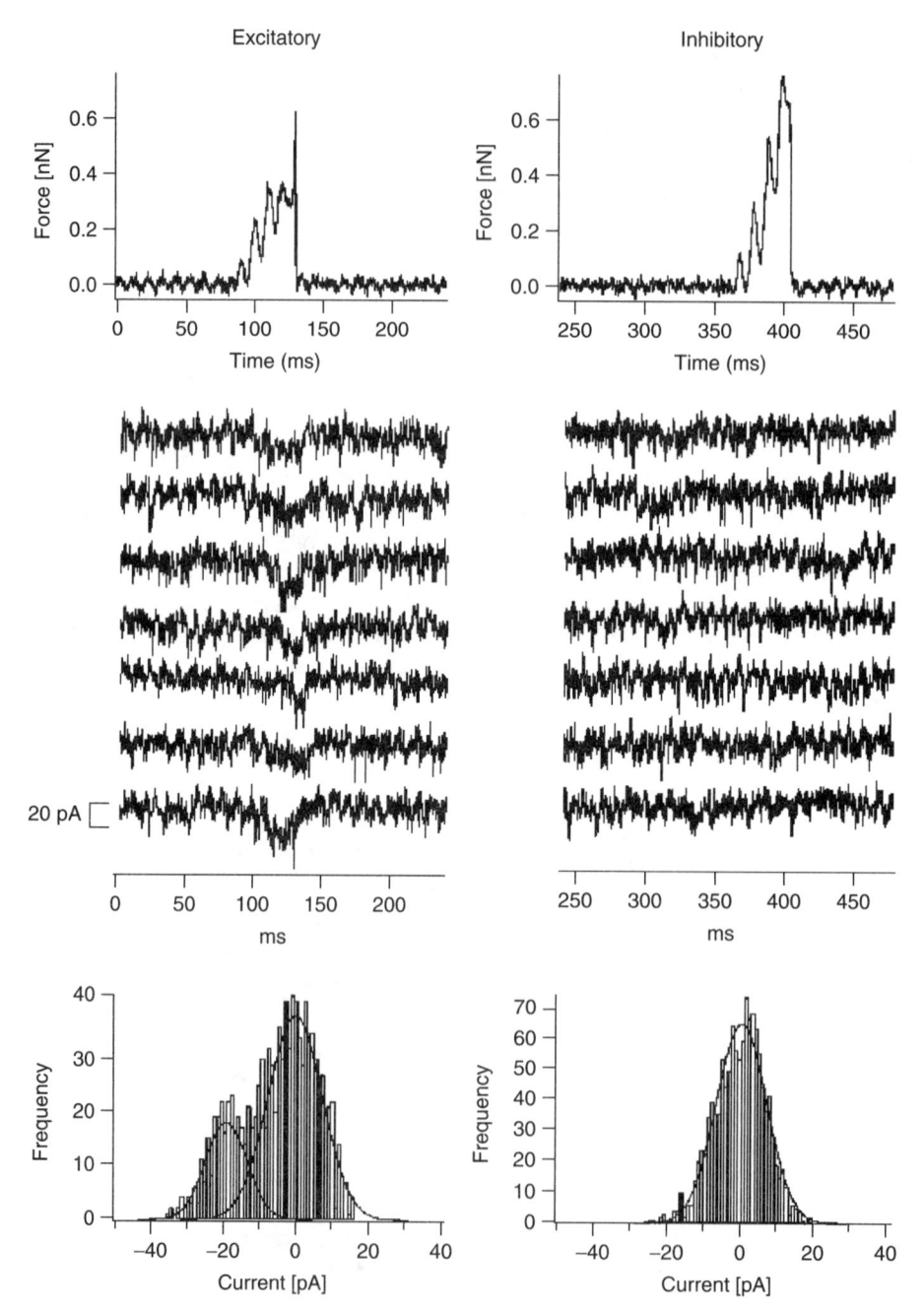

Figure 8.10. The upper two graphs show the force applied to a stereocilium while scanning either in excitatory (**left**) or inhibitory (**right**) direction. The observed current response is depicted in the traces in the middle of the picture. A reproducible opening of transduction channels between 89 ms and 131 ms can be seen in the left column. In contrast, force applications in the inhibitory direction did not result in the opening of transduction channels. In the lower section of the image, histograms of the currents measured at 89–131 ms and 366–407 ms are shown for both scanning directions. The distribution around zero represents the current for the closed state of the channel, while the second, smaller distribution seen only in excitatory direction around 19 pA represents the current for the open state of the transduction channel.

8.2. PHOTONIC FORCE MICROSCOPY (PFM) – REPLACING THE MECHANICAL AFM CANTILEVER BY AN OPTICAL TRAP

The AFM really turned out to be a very good tool for studies of live cells. However, cells have a rough surface, which often prevents the tip of a mechanical cantilever from following fine topographic details. Furthermore, forces of some 10 N are necessary for stable imaging, which is often strong enough to cause serious deformation or even destruction of soft cellular structures. Only stiff structures, such as the membrane cytoskeleton, are clearly visible in the pictures. Imaging inside cells is impossible due to the mechanical connection to the imaging tip. Therefore, a scanning probe microscope without a mechanical connection to the tip and working with extremely small loading forces is desirable. We developed such an instrument, the photonic force microscope (PFM) (Florin et al., 1997, 1998; Pralk et al., 1998, 1999).

In case of the PFM the mechanical cantilever of the AFM is replaced by the three-dimensional trapping potential of a laser focus (Fig. 8.11). The possibility to trap small particles with high stability in the focus of a laser was first described by Ashkin (1986). In the PFM a nano- to micrometer-sized particle (e.g., latex, glass, or metal bead) is used as a tip. The difference in the refractive index of the medium and the bead, the diameter of the bead, the laser intensity, and the intensity profile in the focal volume determine the strength of the trapping potential. Depending on the application, the potential is adjusted by changing the laser power. Usually the spring constant of the potential is two to three orders of magnitude softer than that of the softest commercially available cantilevers.

8.2.1. Setting Up a PFM

The PFM can make use of two position-sensing systems to determine the position of the trapped sphere relative to that of the potential minimum. One records the fluorescence intensity emitted by fluorophores inside a trapped latex or glass sphere, which are excited by the trapping laser via a two-photon process. The fluorescence intensity

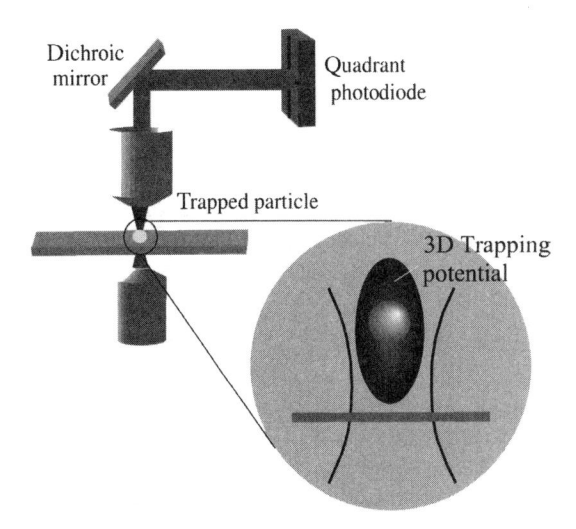

Figure 8.11. Schematic drawing of the photonic force microscope setup on top of an inverted optical microscope. The probe, a nanometer-sized spherical particle, is trapped by optical forces, and its position is measured with respect to laser focus using the interference of the laser light with the light scattered by the particle, which is detected by a quadrant photodiode in the back focal plan of the condenser.

provides an axially sensitive position signal with millisecond time and 10- to 20-nm spatial resolution. The other position detection system is based on the interference of the forward-scattered light from the trapped particle with the unscattered laser light at a quadrant photodiode. This provides a fast three-dimensional recording of the particle's position, and it is most sensitive perpendicular to the optical axis. The position of the bead can be measured with a spatial resolution of better than 1 nm at a temporal resolution of a microsecond or better. Dependent on the application, one or the other (or both) of these detection systems is used.

At room temperature, the thermal position-fluctuations of the trapped bead in weak trapping potentials reach several hundred nanometers. At first glance, these fluctuations seem to be disturbing noise that limits the resolution. However, due to the speed and resolution of the position sensor based on the forward-scattered light detection, the fluctuations of the bead can be tracked directly, opening new ways to analyze the interaction of the bead with its environment.

The PFM developed is based on an inverted optical microscope with a high numerical aperture objective lens providing a good optical control of the investigated structures. The laser light is coupled into the microscope using techniques known from laser scanning microscopy and is focused by the objective lens into the specimen plane. A 1064-nm Nd:YVO$_4$ laser is used since neither water nor biological material has significant absorption at this wavelength. A dichroic mirror behind the condenser deflects the laser light onto the quadrant photodiode. The difference between the left and right half of the diode provides the x position, the difference between upper and lower half provides the y position, and the sum signal provides the z-position. For displacements that are small relative to the focal dimensions, the signals change linearly with the position of the bead in the trap.

The trapped particle acts as a Brownian particle in a potential well, and its position distribution is therefore described by $E(r) = -kT * \ln p(r) + E_0$.

The so-called Botzmann distributions $p(r)$ is readily measured with the high spatial and temporal resolution available of the PFM. Therefore the three-dimensional trapping potential $E(r)$ can be determined just by knowing the temperature. A precision of one-tenth the thermal energy kT is achieved with a sufficient number of statistically independent position readings.

Scanning either the laser focus or the sample, the PFM can be used in a manner very similar to that of a conventional AFM to image surface topographies (Fig. 8.12). The applied forces range from a few piconewtons down to fractions of a pico-newton. The resolution is determined by the interaction area of the sphere used with the sample and by the thermal fluctuations of the bead. The PFM, like an AFM, can be used in different imaging modes. Besides the constant height mode, implementing a feedback loop for force control, also the constant force mode can be used. This mode provides, besides the force and the error signal, also the measurement of lateral friction forces. Furthermore, due to the large thermal fluctuations of the bead, a new imaging mode becomes feasible with the PFM by observing the spatial distribution of the thermal fluctuations of the bead, which avoids stationary objects. In this way, it is possible to image, for example, the three-dimensional (3-D) structure of polymer networks (Fig. 8.13), with a resolution now limited only by the precision of the detection system. Additionally, detailed information about the interaction potential between the bead and the surface of the objects imaged can be obtained. At present, PFM imaging of biological material is limited by nonspecific adhesion events between the bead and the sample, which can lead to the loss of the probe due to the weak forces of the laser focus. Although there seems to be no

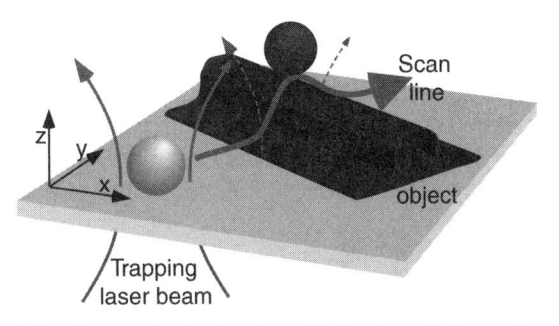

Figure 8.12. Constant height imaging with the photonic force microscope.

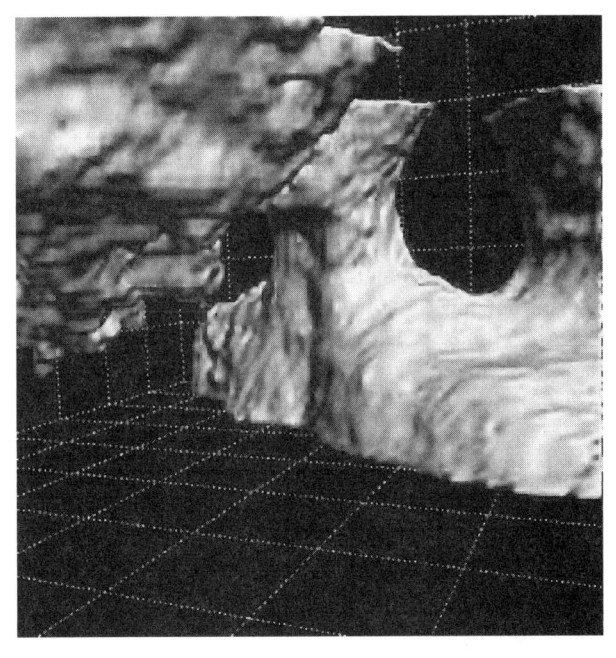

Figure 8.13. Three-dimensional thermal noise image of an agar network. The volume accessible to the bead fluctuations at a certain probability is shown as black space. The nonaccessible volume corresponding to the polymer fibers is rendered as a solid body. The minimal channels observed by the bead reflect its diameter of 216 nm.

general strategy to prevent nonspecific adhesion to complex biological material, it is possible to achieve specific binding, for instance, to membrane and other molecular structures and to track their motion.

8.2.2. Using Thermal Fluctuations to Measure Mechanical Properties of Single Molecules

Kinesins and kinesin like-proteins are of wide interest in biology because of their fundamental functions in the cell. They are responsible, for example, for targeting organelles through the cells and for setting up the mitotic spindle during cell division. The directed transport along the cytoskeletal filaments of microtubules is powered in an ATP-dependent way by these molecular motors. Conventional kinesin is a so-called plus-end-directed motor, able to transport organelles in a defined direction over a distance of up to several microns as a single molecule.

Structural and biochemical data of kinesin reveal a heavy chain folded as a globular N-terminal motor domain containing an ATP and a separate microtubule-binding site. The catalytic domain is followed by a neck region that consists of two short β-sheets and a coiled-coil helical structure probably responsible for dimerization. Behind this region, a hinge of variable length that is not predicted to form a coiled-coil is followed by the kinesin stalk that apparently is a α-helical coiled-coil interrupted by another hinge. The stalk finally connects to a poorly conserved tail structure that interacts with its light chain, binding to the cargo. Several studies on kinesins (e.g., Grummt et al., 1998) have shown that the neck and the first hinge region of the motor play an important role in kinesin directionality, velocity, and ATPase activity. Most of the molecules dimerize *in vivo* and therefore have two enzymatic head domains. The mechanism of movement and force generation required for intracellular transports is not yet fully understood. Several authors (e.g., Howard et al., 1989) suggest a hand-over-hand model where both heads would alternate to "walk" along the filaments. Having all its enzymatic and binding machinery in its heavy chain, the kinesin motor has the unique property of operating completely on its own, either as a dimer or even as a monomer. Earlier studies on single molecules have shown that the maximum force generated by the kinesin molecule to transport a microsphere along a microtubule is in the range of 5 pN (Svoboda and Block, 1994).

Obviously, characterizing mechanical properties of molecular motors is essential for a better understanding of how nano-machines, like kinesin, convert chemical energy into a mechanical movement. So far, attempts to explain in a satisfying way the function of kinesin using the knowledge gained by dynamical and structural studies have not been very successful. A major problem is that the high-resolution structural data are obtained on an ensemble of molecules in a rigid state and cannot resolve the dynamic of intermediate states required for the kinesin motility. Furthermore, the molecules used for structural analysis are truncated constructs missing the stalk. On the other hand, for *in vitro* motility assays, full-length constructs of the motors are used in an environment rather different from a protein crystal.

We focused our studies with the PFM on intrinsic mechanical properties of kinesin like its elasticity, comparing the mechanical behavior of two different full-length kinesin constructs by adsorbing them onto glass microspheres and letting them interact with a microtubule in the presence of different nucleotides. One type of motor we study is the wild type from drosophila with two head groups. Another type is a chimera missing one of the heads. To determine the mechanical properties of the kinesin molecule bound to a bead and to a microtubule in thermal equilibrium, the thermal position fluctuations of the bead are measured using the PFM. Boltzmann's equation can be used to analyze these fluctuations in 3-D and to determine the energy landscape defined by the kinesin as a molecular linker between bead and microtubule structure fixed to the surface. Finally, from the energy landscape measured (Fig. 8.14), the mechanical properties of single kinesin molecules in 3-D for different types of motors and their different nucleotide binding states can be extracted (Jeney et al., 2001). The measured stiffness along the tether axis can be then explained as elasticity of this structure. The rotational stiffness corresponds to restoring forces of the molecular hinges against the lateral bending due to thermal energy. The restoring forces the molecules develop against lateral bending are dependent of the microtubule orientation and are influenced by the microtubule's presence. A comparison between one- and two-headed motors shows that a two-headed motor behaves stiffer in all three directions. It points to the influence of the second head, which restricts the bead movement due to steric interactions with the microtubule and perhaps in cases

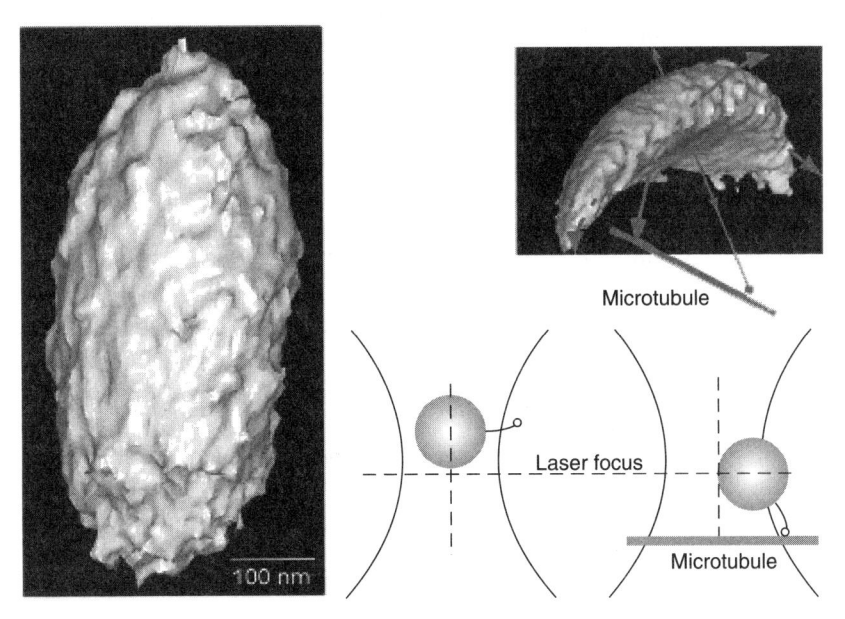

Figure 8.14. The isoenergy surfaces determined by the Boltzmann equation using the position probability data of a particle in the laser focus is shown in the left part of the figure. The schematic drawing shows the binding of this particle via a kinesin molecular motor to a microtubule structure fixed on the coverslip. In the upper right part of the figure, the change of the isoenergy surface after the binding event is shown in scale to the surface before binding.

of the observed high stiffness values in the ATP-free state even binds to it.

The stiffness of the motor bound with the nonhydrolyzable ATP analogue AMPPNP are much higher than the ones in the ADP bound states for one- and two-headed motors. This suggests that the binding strength between kinesin and tubulin influences the elasticity of the whole kinesin/bead/microtubule-system. It corresponds well with the finding that different nucleotides influence the position of the free domain relative to the bound one (Hirose et al., 1995; Arnal and Wade, 1998).

8.2.3. Using Thermal Fluctuations to Determine Local Membrane Viscosity

The motion of a thermally fluctuating particle in a harmonic potential like that of the laser trap can be characterized to a certain level by an exponentially decaying position autocorrelation function. This function has a characteristic autocorrelation time $\tau = \gamma/\kappa$, where γ denotes the viscous drag on the sphere and κ denotes the force constant of the optical trap. Thus, the local viscous drag and the diffusion coefficient $D = kT/\gamma$ of a sphere in a harmonic potential can be calculated from the measured autocorrelation time of the motion and the stiffness of the trapping potential. The stiffness of the trapping potential is determined as described before using Boltzmann's equation. To measure diffusion coefficient with a statistical error smaller than 10%, the observation interval must be about 1000 times longer than the autocorrelation time. Hence, the motion of the bead limits the temporal resolution of the viscosity measurement, and not the bandwidth of the detection system. Near surfaces (e.g., glass surfaces or cell membranes), the diffusion is reduced because of the spatial confinement (Happel and Brenner, 1965). Within an obstacle-free lipid bilayer, diffusion has been described by Saffman and Delbrück (1975) using a hydrodynamic model treating the bilayer as a continuum and assuming weak coupling to the surrounding liquid medium. The viscous drag γ_m on a cylindrical particle with radius r in a homogeneous lipid bilayer of thickness h is $\gamma_m = 4\pi\eta_m h/(\ln(\eta_m h/\eta_w r) - \varepsilon)$, where η_w denotes the viscosity of the surrounding fluid, η_m is the viscosity of the lipid bilayer, and ε is Euler's constant. This approximation is valid for proteins with radii large compared to the size of the lipid molecules and for $\eta_m \gg \eta_w$, which is true for cellular membranes. Thus, the membrane viscosity can be determined by binding a bead to a membrane structure of known diameter and measuring the total viscous drag on the bead as described above (Fig. 8.15). Such measurements can be performed with a temporal resolution of 0.3 s and within areas of 100 nm in diameter.

The measured viscous drag γ for the bead connected to a membrane component is the sum of the Stokes drag of the sphere $\gamma_s = 6\pi\eta_w r$ and the viscous drag γ_m of, for example, a single protein in the lipid bilayer. The Stokes drag of the sphere near the cell membrane increases due to the confinement and has to be taken into account for the correct calculation of γ_m. The observation of the strength of the lateral potential ensures that the observed membrane component diffuses freely.

Using the PFM technique on a single transmembrane protein of known size, it is possible to determine the membrane viscosity in living cells in areas about 100 nm in diameter, which translates into a diffusion coefficient of $D = 3.9 \times 10^{-8}$ cm^2/s at 36°C. The obtained local diffusion coefficients of proteins in the plasma membrane in intact cells agree for the first time well with the Saffman–Delbrück model validated by Peters and Cherry (1982) in dimyristoylphosphatidylcholine (DMPC) bilayers using FRAP to measure the diffusion coefficient of bacteriorhodopsin to be between 0.15×10^{-8} cm^2/s and 3.4×10^{-8} cm^2/s, depending on the protein/lipid ratio.

Once the membrane viscosity has been obtained, the technique allows us to determine the diameter of other membrane structures of unknown size and to monitor continuously their diffusion characteristics. We performed such experiments to determine the size of membrane structures called "rafts" (Pralle et al., 2000). Lipid rafts are involved in the polarized sorting of proteins and cellular signaling (Simons and Ikonen, 1997; Keller and Simons, 1998). According to our study in the plasma membrane of fibroblast-like cells, these structures are indeed quite stable on the timescale of minutes, justifying the model of "rafts" as lipid–protein complexes floating raft-like in the membrane. Their diameter was determined to be about 50 nm. The broad distribution of viscous drag values observed for the rafts studied indicates that even one raft type in one cell might have a distribution of sizes and that the size of a single raft might be dynamic.

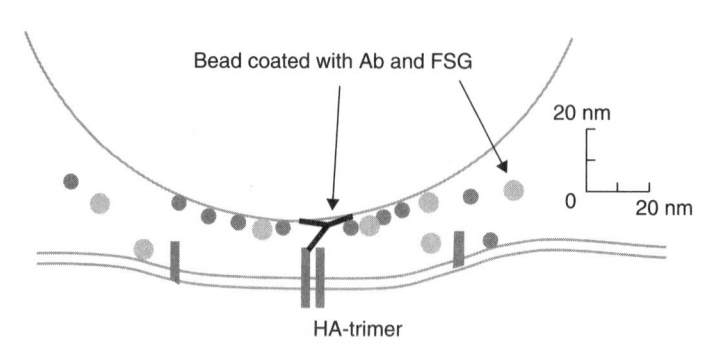

Figure 8.15. Scaled model of the experimental situation, when a sphere ($r = 108$ nm) is bound via an adsorbed antibody to a transmembrane protein. The lipid bilayer is symbolized by the gray double line.

Transferring these values to other cell types and non-GPI–anchored proteins might not be possible since raft size and stability are most likely dependent on the lipid and protein constituents. Indeed, recent data suggest that rafts in the apical membrane of MDCK behave differently from rafts in fibroblasts (Verkade et al., 2000).

According to our study, rafts in the plasma membrane of fibroblast-like cells diffuse as a rather stable platform with an average area of 2100 nm^2. The size estimate allows an assessment of the maximal contents of one raft. If these were composed purely of lipid molecules, having a radius comparable to phosphoethanolamine ($r = 0.44$ nm), one raft would contain almost 3500 lipid molecules. How many proteins a raft contains depend on how densely packed the proteins would be. If they were as densely packed as rhodopsin molecules are in frog rods (Blasie and Worthington, 1969), or as densely packed as the spikes in the envelope of Semliki Forest virus (Cheng et al., 1995), a raft would contain 55–65 proteins, respectively. Clearly, the packing density in a lipid raft in a mammalian plasma membrane would be lower.

The consequences of a small raft size and stable association is that proteins within rafts are restricted in their interactions with other proteins. To use rafts as platforms in membrane trafficking (e.g., in apical transport from the Golgi complex) would imply that these rafts have to be clustered together by an apical sorting machinery to form a container comprising several rafts. A growing body of evidence implicates lipid rafts in signal transduction. Well-studied examples include IgE receptor signaling and T- and B-cell activation. If rafts normally contain only a limited set of proteins, then clustering of rafts would be necessary to achieve the concentration of interacting molecules required to elicit a signal above the activating threshold. There is ample indication that clustering is an essential feature of signal transduction processes involving rafts.

In these experiments that finally led to a physical proof of the existence of rafts, the PFM has proven to be a powerful tool to study biophysical properties of the plasma membrane as well as interaction of single molecules or complexes of molecules within the membrane. Future applications of the PFM will cover a wide range from 3-D *in vivo* imaging of biological samples to a detailed investigation of single-molecule properties.

REFERENCES

Adler, H. J., Poje, C. P., and Saunders, J. C. (1993). *Hearing Res.* **71**:214.

Arnal, I., and Wade, R. H. (1998). *Structure* **6**:33.

Ashkin, A., (1986). *Opt. Lett.* **11**:288.

Binnig, G., Quate, C. F., and Gerber, C. (1986). *Phys. Rev. Lett.* **56**:930.

Blasie, J. K., and Worthington, C. R. (1969). *J. Mol. Biol.* **39**:407.

Cheng, R. H., Kuhn, R. J., Olson, N. H., Rossman, M. G., Choi, H.-K., and Baker, T. S. (1995). *Cell.* **80**:621.

Corey, D. P., and Hudspeth, A. J., (1983). *J. Neurosci.*, **3**:962.

Drake, B., Prater, C. B., Weisenhorn, A. L., Gould, S. A. C., Albrecht, T. R., Quate, C. F., Cannell, D. S., Hansma, H. G., and Hansma, P. K. (1989). *Science* **243**:1586.

Florin, E.-L., Pralle, A., Hörber, J. K. H., and Stelzer, E. H. K. (1997). *J. Struct. Biol.* **119**:202.

Florin, E.-L., Pralle, A., Stelzer, E. H. K., and Hörber, J. K. H. (1998). *Applied Phys. A*, **66**:S75.

Frischknecht, F., Moreau, V., Röttger, S., Gonfloni, S., Reckmann, I, Superti-Furga, G., and Way M. (1999). *Nature* **401**:926.

Furness, D. N., and Hackney, C. M. (1985). *Hearing Res.* **18**:177.

Furness, D. N., Richardson, G. P., and Russell, I. J. (1989). *Hearing Res.* **38**(1–2):95.

Furness, D. N., Hackney, C. M., and Benos, D. J. (1996). *Hearing Res.* **93**:136.

Géléoc, G. S. G., Lennan, G. W. T., Richardson, G. P., and Kros, C. J. (1997). *Proc. R. Soc. Lond.* **B 264**:611.

Grummt, M., Woehlke, G., Henningsen, U., Fuchs, S., Schleicher, M., and Schliwa, M. (1998). *EMBO J.* **17**:5536.

Häberle, W., Hörber, J. K. H., and Binnig, G. (1991). *J. Vac. Sci. Technol.* **B9**:1210.

Häberle, W., Hörber, J. K. H., Ohnesorge, F., Smith, D. P. E., and Binnig, G. (1992). *Ultramicroscopy* **42–44**:1161.

Happel, J., and Brenner H. (1965). Low Reynolds Number Hydrodynamics, Prentice Hall, Englewood Cliffs, NJ.

Hirose, K., Lockhart, A., Cross, R. A., and Amos L. A. (1995). *Nature* **376**:277.

Hörber, J. K. H., Häberle, W., Ohnesorge, F., Binnig, G., Liebich, H. G., Czerny, C. P., Mahnel, H., and Mayr, A. (1992). *Scanning Microscopy* **6**:919.

Hörber, J. K. H., Mosbacher, J., Häberle, W., Ruppersberg, P., and Sakmann, B. (1995). *Biophys. J.* **68**:1687.

Howard, J., and Hudspeth, A. J. (1988). *Neuron*, **1**:189.

Howard, J., Hudspeth, A. J., and Vale, R. D. (1989). *Nature* **342**:154.

Hudspeth, A. J. (1983). *Annu. Rev. Neurosci.* **6**:187.

Jeney, S., Florin, E.-L., and Hörber, J. K. H. (2001). Methods in Molecular Biology; Kinesin Protocols, (I. Vernos, ed.), Humana Press, Totowa, NJ.

Keller, P., and Simons, K. (1998). *J. Cell Biol.* **140**:1357.

Langer, M. G., Öffner, W., Wittmann, H., Flösser, H., Schaar, H., Häberle, W., Pralle, A., Ruppersberg, J. P., and Hörber, J. K. H. (1997). *Rev. Sci. Instrum.* **68**:2583.

Langer, M. G., Koitschev, A., Haase, H., Rexhausen, U., Hörber, J. K. H., and Ruppersberg, J. P. (2000). *Ultramicroscopy* **82**:269.

Lim, D. J. (1986). *Hearing Res.* **22**:117.

Mosbacher, J., Langer, M., Hörber, J. K. H., and Sachs, F. (1998). *J. Gen. Physiol.* **111**:65–74.

Moss, B. (1986). Replication of poxviruses. In: Fundamental Virology (B. N. Fields and D. M. Knape, eds.), Raven Press, New York.

Ohnesorge, F. M., Hörber, J. K. H., Häberle, W., Czerny, C.-P., Smith, D. P. E., and Binnig, G. (1997). *Biophys. J.* **73**:2183.

Peters, R., and Cherry, R. J. (1982). *Proc. Natl. Acad. Sci. USA* **79**:4317.

Pralle, A., Florin, E.-L., Stelzer, E. H. K., and Hörber, J. K. H. (1998). *Applied Phys. A* **66**:S71.

Pralle, A., Prummer, M., Florin, E.-L., Stelzer, E. H. K., and Hörber, J. K. H. (1999). *Microsc. Res. Technique* **44**:378.

Pralle, A., Keller, P., Florin, E.-L., Simons, K., and Hörber, J. K. H. (2000). *J. Cell Biol.*, **148**:997.

Saffman, P. G., and Delbrück, M. (1975). *Proc. Natl. Acad. Sci. USA* **72**:3111.

Simons, K., and Ikonen, E. (1997). *Nature* **387**:569.

Stokes, G. V. (1976). *J. Virol.* **18**(2):636.

Svoboda, K., and Block, S. M. (1994). *Cell* **77**:773.

Verkade, P., Harder, T., Lafont, F., and Simons, K. (2000). *J. Cell Biol.* **148**:727.

Zine, A., and Romand., R. (1996). *Brain Res.* **721**:49.

CHAPTER 9

INTERMOLECULAR FORCES OF LEUKOCYTE ADHESION MOLECULES

XIAOHUI ZHANG AND VINCENT T. MOY
Department of Physiology and Biophysics, University of Miami School of Medicine
Miami, Florida

9.1. INTRODUCTION

Leukocytes (i.e., white blood cells) are found in the systemic circulation. Their primary roles are to aid in tissue repair and host defense against invading pathogens. However, when left unchecked, the accumulation of leukocytes may damage the tissue as in the case of inflammatory diseases such as rheumatoid arthritis, asthma, and atherosclerosis. In order for leukocytes to fulfill their functions, they must leave the circulation and migrate through the tissue toward their targets. This process, termed *extravasation*, is mediated by leukocyte adhesion molecules (Springer, 1990).

In order to leave the bloodstream, leukocytes must first attach to and subsequently permeate through the wall of blood vessels. Leukocyte extravasation involves four stages: tethering or rolling, cell activation, firm adhesion, and finally transmigration (Fig. 9.1). Each stage engages different sets of adhesion molecules (Springer, 1994; Kubes and Kerfoot, 2001). Leukocyte rolling is mediated by selectin molecules that are expressed on vascular endothelial cells and by their adhesive partners (e.g., P-selectin glycoprotein ligand-1 (PSGL-1)) on leukocytes. Adhesion mediated by selectins is relatively weak and transient. These bonds are not strong enough to stop leukocytes from the blood flow. Rather, a continuous cycle of the formation and dissociation of selectins/ligand complexes permits the leukocyte to roll along the endothelium. Rolling enables leukocytes to survey the endothelial surface for chemotactic signals. When appropriate signals are encountered, these signals activate a second class of leukocyte adhesion molecule, the integrins. The activated integrins bind tightly to their adhesive partners, which are molecules of the immunoglobulin superfamily (IgSF) that are expressed on the endothelial surface. These interactions arrest the rolling leukocytes. Integrins important for leukocyte firm adhesion include leukocyte function-associated antigen-1 (LFA-1) and very late antigen-4 (VLA-4) (Kubes and Kerfoot, 2001). Their major endothelial ligands are intercellular adhesion molecule-1 (ICAM-1) and vascular cell adhesion molecule-1 (VCAM-1), respectively (Fig. 9.1).

Leukocyte adhesion has been extensively studied for more than two decades. Sequence and structure of many leukocyte adhesion molecules have been determined. However,

---

Force Microscopy: Applications in Biology and Medicine, edited by Bhanu P. Jena and J.K. Heinrich Hörber.
Copyright © 2006 John Wiley & Sons, Inc.

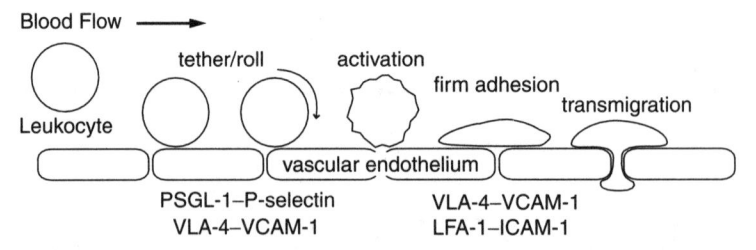

Figure 9.1. Recruitment of leukocytes in inflamed tissues. Adhesion molecules including selectins, immunoglobulin superfamily members, and integrins, mediate the initial tethering/rolling, firm adhesion, and transmigration of leukocytes. Leukocytes are captured from the circulating blood by transient VLA-4–VCAM-1 or selectin-glycan-based interactions. Following activation by chemokines, the leukocytes adhere firmly to the endothelial cells via integrin-mediated (i.e., VLA-4 or LFA-1) adhesion. The selectins (P-, E-, and L-selectins) are a family of membrane-bound glycoproteins that bind a carbohydrate structure called sialyl Lewis X presented on their ligands. Integrins are αβ heterodimeric adhesion molecules that bind the immunoglobulin superfamily cell adhesion molecules (CAMs).

it is not known how these molecular structures give rise to the mechanical strength of the leukocyte adhesion complexes. Recently, we applied the atomic force microscopy (AFM) (Binnig et al., 1986) to determine the effects of a pulling force on selectin and ingetrin–ligand complexes. Similar approaches, including the biomembrane force probe (BFP) (Evans, 2001), have been used to study the unbinding of the streptavidin–biotin complex (Florin et al., 1994; Merkel et al., 1999), complementary strands of DNA (Lee et al., 1994), and other adhesion systems (Merkel, 2001). These direct force measurements have provided the means to probe the dissociation pathway of biomolecular complexes. In this review, we discuss the application of AFM on the unbinding of cell adhesion complexes.

9.2. SINGLE-MOLECULE UNBINDING: THEORY AND APPLICATIONS

9.2.1. Theory

The theoretical framework for understanding how a pulling force affects the dissociation of an adhesion complex was first formulated by Bell (1978) and later expanded on by Evans and other researchers (Evans and Ritchie, 1997). The Bell model describes the influence of an external force on the rate of bond dissociation. This model is based on the conventional transition state theory, in which a molecular complex needs to overcome an activation energy barrier before final separation. If only a single barrier dominates the dissociation process, the dissociation rate of this interaction is given by

$$k_{\text{off}} = \alpha \frac{k_B T}{h} \exp\left\{\frac{-\Delta G^*}{k_B T}\right\} \quad (9.1)$$

where ΔG^* is the activation energy, T is absolute temperature, k_B is Boltzmann's constant, h is Planck's constant, and α is a prefactor that characterizes the potential well. When the complex is exposed to a pulling force, the applied force adds a $-fx$ term to the potential of the system. If the potential barrier is steep, adding this term to the free energy reduces the activation barrier by approximately $f\gamma$, where γ is the distance between the bound state and the transition state along the reaction coordinate. Thus, the force dependent dissociation rate of the complex is given by

$$k_{\text{off}}(f) = \alpha \frac{k_B T}{h} \exp\left\{\frac{-(\Delta G^* - f\gamma)}{k_B T}\right\}$$
$$= k^0 \exp\left\{\frac{f\gamma}{k_B T}\right\} \quad (9.2)$$

where k^0 is the unstressed dissociation rate. Hence, the Bell model predicts that the dissociation rate of the complex increases exponentially with a pulling force. The two parameters k^0 and γ are often referred to as the Bell model parameters. These two parameters characterize the energy potential of the protein–ligand complex. k^0 characterizes the depth of the potential, and γ characterizes the width of the potential and dictates the force resistance of a molecular complex. If a complex has a small γ (i.e., the activation potential is narrow), then an external force will have less effect on its force dependent dissociation rate $k_{\text{off}}(f)$. On the other hand, if the activation potential is wide, the complex will be sensitive to an external force since $f\gamma$ adds a larger term to the intermolecular potential.

Equation (9.1) describes how bond dissociation is changed by a constant pulling force. However, a constant pulling force is difficult to maintain in an AFM experiment. Instead, a dynamic force approach is generally used to characterize the forced dissociation of ligand–receptor complexes (Evans and Ritchie, 1997). Under conditions of a constant loading rate r_f, the probability density function for the forced unbinding of an adhesion complex is given by

$$P(f) = k^0 \exp\left\{\frac{\gamma f}{k_B T}\right\} \exp\left\{\frac{k^0 k_B T}{\gamma r_f}\right.$$
$$\left. \times \left[1 - \exp\left(\frac{\gamma f}{k_B T}\right)\right]\right\}. \quad (9.3)$$

From Evans and Ritchie (1997), the most probable unbinding force f^* (i.e., the maximum of the distribution $\partial P(f)/\partial f = 0$) is

$$f^* = \frac{k_B T}{\gamma} \ln\left\{\frac{\gamma}{k^0 k_B T}\right\} + \frac{k_B T}{\gamma} \ln\left\{r_f\right\} \quad (9.4)$$

Equation (9.4) shows that f^* is a linear function of the logarithm of the loading rate. The Bell model parameters are obtained from the plot of f^* versus $\ln(r_f)$, the dynamic force spectrum (DFS) of the complex (Evans and Ritchie, 1997).

The energy landscape of a complex may consist of multiple sharp activation barriers (Fig. 9.2A) (Evans and Ritchie, 1997). In this case, the DFS is predicted to have multiple linear regimes with ascending slopes, as shown in Fig. 9.2B. The increase in slope from one regime to the next indicates that an outer barrier has been suppressed

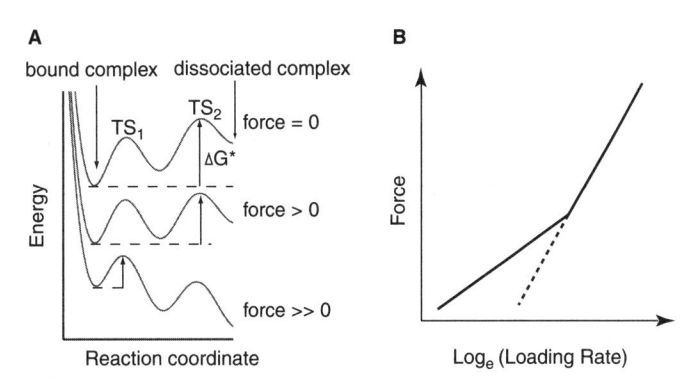

Figure 9.2. Dynamic force spectroscopy. **(A)** Effects of an applied force on a protein–ligand interaction potential consisting of two transition states: TS_1 and TS_2. In the absence of an applied force (top trace), the dissociation kinetics of the complex are determined by the outer energy barrier (i.e., TS_2). An external force tilts the energy potential and suppresses the outer barrier (middle trace). Further increase in force results in a potential that is governed by the properties of the inner energy barrier (i.e., TS_1) (bottom trace). **(B)** Two linear regimes are predicted for a cascade of two sharp energy barriers. The increase of slope indicates that the outer barrier has been suppressed and that the inner barrier has become the dominant kinetic impedance to detachment.

by force and that an inner barrier dominates the dynamic response of the complex (Fig. 9.2A). This theory is supported by recent experiments from different groups (Merkel et al., 1999; Evans et al., 2001; Zhang et al., 2002; Li et al., 2003). Multiple activation barriers were found in a number of ligand–receptor systems, including the (strep)avidin/biotin complexes and all tested integrin–ligand complexes. A partial list of these studies is tabulated in Table 9.1.

9.2.2. Force Unbinding of Individual Selectin/sLeX Complexes

DFS has been employed by us and others to characterize the mechanical properties of selectin/ligand complexes (Evans et al., 2001; Zhang et al., 2002, 2004; Li ct al., 2003). A series of representative AFM force scans (i.e., force versus piezo displacement) are shown in Fig. 9.3A. In these measurements, the two interacting surfaces were the AFM tip, functionalized with sialyl Lewis X, a carbohydrate ligand for all selectins, and a substrate coated with P-selectin. Figure 9.3B presents the DFS of the P-selectin–sLeX complex (Zhang et al., 2004). The force spectrum can be divided into two loading regimes: The unbinding force increased exponentially with loading rate from 70 to 9000 pN/s. Beyond 9000 pN/s, a faster exponential increase is clearly evident. Fitting Eq. (9.4) to both the slow and fast loading regimes yielded the Bell model parameters for the outer and inner activation barriers of the complex, respectively. For the P-selectin–sLeX complex, the barrier widths γ for the outer and inner barriers were 4.5 Å, and 0.80 Å, respectively. Similar results were reported for the forced unbinding of L-selectin–PSGL-1 complex (Evans et al., 2001).

9.2.3. Probing Integrin Activation on Live Leukocytes

The regulation of integrin affinity plays an important role in leukocyte extravasation. The strength of the integrin–ligand bond is drastically increased when the integrin molecule is activated by intracellular signals. In order to access the bond strength of individual integrin–ligand complexes in different affinity states, it is best to maintain the integrins in their native environments—that is, on the surface of live cells. To achieve this goal, we have developed an experimental protocol to couple a live leukocyte onto the AFM cantilever (Figs. 9.4A and 9.4B) (Zhang et al., 2002; Li et al., 2003) based on an earlier work by Benoit et al. (2000). This approach allows us to measure the unbinding forces on the surface of live leukocytes.

As shown in Figs. 9.4A and 9.4B, the AFM force measurements were carried out with an LFA-1 expressing T-cell hybridoma (i.e., 3A9) coupled to the AFM cantilever. Purified ICAM-1 was immobilized on a tissue culture dish. High-affinity state of LFA-1 was induced by the addition of 5 mM Mg^{2+} and 1 mM ethylene glycol bis(β-aminoethylether)-N,N,N',N'-tetraacetic acid (EGTA). Figure 9.4C plots the most probable unbinding force of individual LFA-1–ICAM-1 complexes formed between 3A9 and immobilized ICAM-1 over three orders of magnitude change in loading rate. As shown, the LFA-1–ICAM-1 force spectrum can be divided into two loading regimes. Induction of high-affinity LFA-1 by Mg^{2+}/EGTA resulted in stronger LFA-1–ICAM-1 unbinding forces in the slow loading regime (<10,000 pN/s). However, the unbinding forces of resting and Mg^{2+}–EGTA-activated complexes were nearly identical in the fast loading regime (>10,000 pN/s).

An analysis of Fig. 9.4C revealed that the unbinding of the LFA-1–ICAM-1 complex involves overcoming at least two activation barriers. Fitting Eq. (9.4) to the slow and fast regimes gives the Bell model parameters for the outer and inner barriers of the complex, respectively. The γ values for the inner and outer barriers are approximately 0.2 and 2 Å, respectively. Moreover, our analysis revealed that the observed increase in unbinding

TABLE 9.1. Summary of Reported AFM Unbinding Studies.

Ligand–Receptor Pair	Loading Rate (pN/s)	Rupture Forces (pN)	Bell Parameter γ (Å)	References
Leukocyte Adhesion Complexes				
LFA-1–ICAM-1	50–50,000	20–320	0.2, 2	Zhang et al. (2002)
VLA-4–VCAM-1	30–60,000	15–130	1, 5.5	Zhang et al. (2003)
P-selectin–sLeX	70–100,000	20–220	0.8, 4.5	Zhang et al. (2004)
E-selectin–sLeX	200–100,000	40–160	0.9, 5	Zhang et al. (2004)
L-selectin–sLeX	300–100,000	20–140	0.8, 4.5	Zhang et al. (2004)
P-selectin–PSGL-1	100–10,000	30–220	1.4	Hanley et al. (2003)
P-selectin–PSGL-1	—[a]	110–170	2.5	Fritz et al. (1998)
L-selectin–sLeX[b]	10–100,000	10–180	0.6, 4	Evans et al. (2001)
L-selectin–PSGL-1[b]	10–100,000	20–160	0.6, 4	Evans et al. (2001)
Other Ligand–Receptor Complexes				
$\alpha_5\beta_1$–fibronectin	50–50,000	40–170	0.9, 4	Li et al. (2003)
Streptavidin–biotin	0.05–60,000	5–170	1.2, 5	Merkel et al. (1999)
Avidin–biotin	0.05–60,000	5–170	1.2, 3, 30	Merkel et al. (1999)
Con A–D-mannose	400–5000	80–125	2.7	Chen and Moy (2000)
Plant lectin–asialofetuin	100–30,000	37–65	4–6	Dettmann et al. (2003)
GTPase Rap–impβ[c]	300–80,000	40–90, 75–160[d]	N/A	Nevo et al. (2003)
VE-cadherin pair	—[a]	30–50	5.9	Baumgartner et al. (2000)
Anti-γ-GT–γ-GT[e]	N/A	131 ± 44	N/A	Wielert-Badt et al. (2002)
Anti-ferritin–ferritin	N/A	49 ± 10	N/A	Allen et al. (1997)
Anti-HSA–HSA[f]	N/A	240 ± 48	N/A	Hinterdorfer et al. (1996)
Anti-βhCG–βhCG[g]	N/A	100 ± 47	N/A	Allen et al. (1999)
Glycoprotein csA pair	N/A	23 ± 8	N/A	Benoit et al. (2000)
GroEL[h] pair	N/A	420 ± 100	N/A	Vinckier et al. (1998)
ICAM-1–anti-ICAM-1	N/A	100 ± 50	N/A	Willemsen et al. (1998)
Insulin pair	N/A	1340–1350	N/A	Yip et al. (1998)
Meromyosin–actin	N/A	15–25	N/A	Nakajima et al. (1997)
Ocular mucin pair	N/A	100–4000	N/A	Berry et al. (2001)
Proteoglycan pair	N/A	40 ± 15	N/A	Dammer et al. (1995)
$\alpha_5\beta_1$–GRGDSP peptide	N/A	32 ± 2	N/A	Lehenkari and Horton (1999)
$\alpha_v\beta_3$–echistatin	N/A	97 ± 15	N/A	Lehenkari and Horton (1999)
$\alpha_v\beta_3$–GRGDSP peptide	N/A	42 ± 4	N/A	Lehenkari and Horton (1999)
$\alpha_v\beta_3$–osteopontin	N/A	50 ± 2	N/A	Lehenkari and Horton (1999)

[a] Authors reported unbinding forces versus different pulling velocities; we were unable to convert the velocity to loading rates.
[b] Studies using biomembrane force probe.
[c] impβ: nuclear import receptor importin β1.
[d] Authors reported two populations of unbinding forces, reflecting the existence of two conformation states in the Rap–impβ complexes.
[e] γ-GT: γ-glutamyl-transpeptidase.
[f] HSA: Human serum albumin.
[g] βhCG: β subunit of human chorionic gonadotrophin.
[h] GroEL: Chaperonin GroEL from *E. coli*.

force following LFA-1 activation in the slow loading regime stemmed from an increase in the outer activation energy barrier, which is manifested in a lowering of the dissociation rate constant from 4.6/s to 0.15/s. The inner activation barrier was unaffected by the integrin activation, since no significant change was found in the dynamic strength of the complex in the fast loading regime following LFA-1 activation.

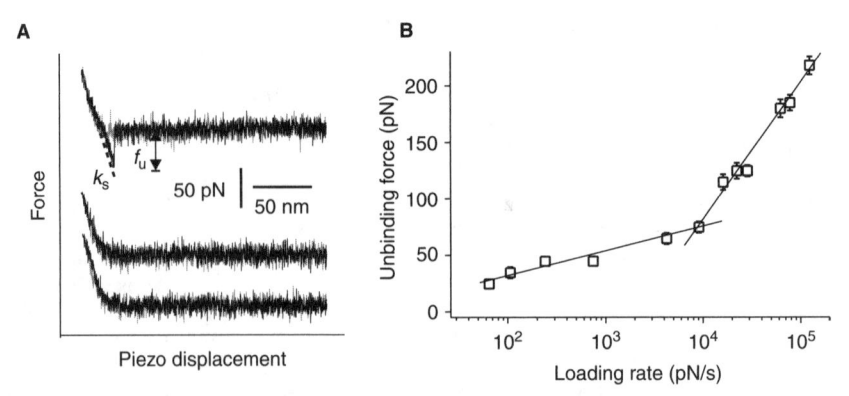

Figure 9.3. Dynamic force spectra of P-selectin–sLeX interactions. (**A**) A series of three consecutive AFM force traces of the P-selectin–sLeX interaction. Gray and black traces were recorded during the approach and withdrawal of the cantilever, respectively. In the top trace, the force f_u is the unbinding force. The loading rate of the unbinding force was derived from the system spring constant k_s. To study individual bonds, proteins were immobilized in low surface densities; and the interaction between tip and substrate was further reduced by lowering contact duration. Under these conditions, some force scans showed no adhesion (middle and bottom traces). The specificity of the P-selectin–sLeX interactions was confirmed by a reduction in the frequency of adhesion following the addition of a function blocking antibody against P-selectin, as well as by the low adhesion frequency measured between the sLeX functionalized cantilever and immobilized bovine serum albumin. (**B**) Dynamic force spectra of the P-selectin–sLeX interactions. The best-fit curves (solid lines) were obtained using Eq. (9.4) applied for each of the two loading regimes. Error bars are the standard error of the mean (SEM). Some error bars are smaller than the symbols.

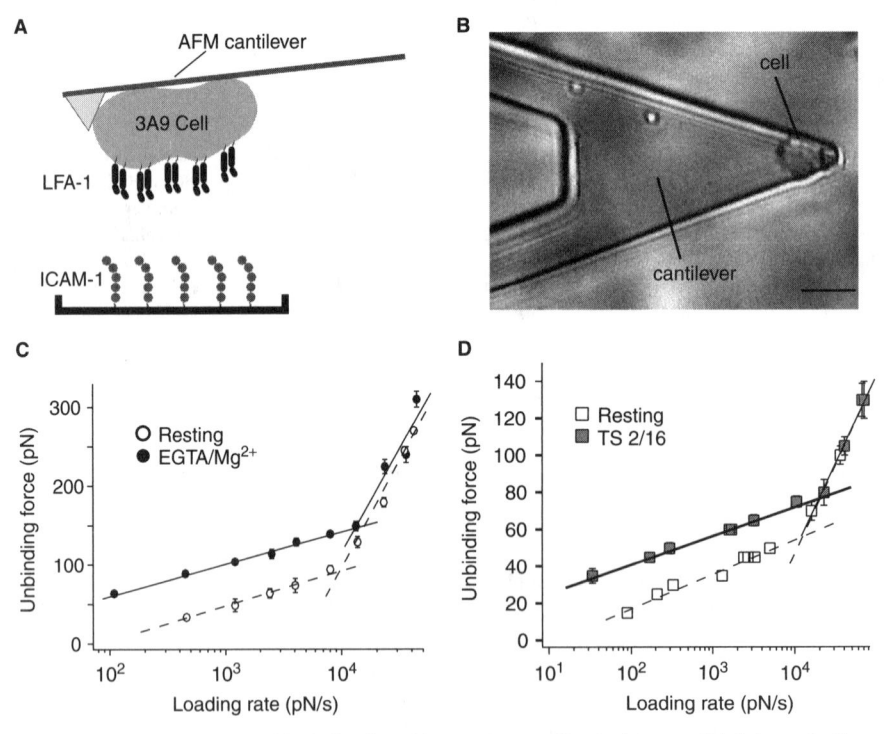

Figure 9.4. Dynamic force spectra of integrin–ligand interactions on live leukocytes. (**A**) Schematic diagram of the AFM experiments using live 3A9 cells. (**B**) Micrograph of a live leukocyte attached to the AFM cantilever. The bar is 20 μm. (**C**) DFS of the low- and high-affinity LFA-1–ICAM-1 complexes. (**D**) DFS of the low- and high-affinity VLA-4–VCAM-1 complexes on live U937 cells.

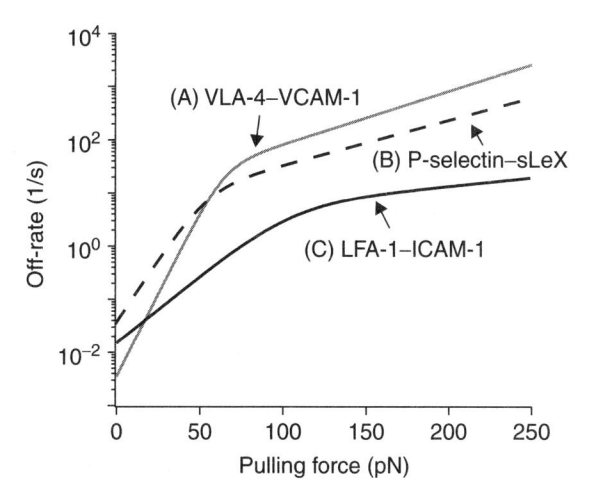

Figure 9.5. Kinetic profiles of the **(A)** VLA-4–VCAM-1, **(B)** P-selectin–sLeX, and **(C)** LFA-1–ICAM-1 complexes.

Similarly, the DFS of the VLA-4–VCAM-1 complex was measured using live U937 cells (Fig. 9.4D). The high-affinity state of VLA-4 was induced by the addition of an activation antibody TS2/16. After activation, the unbinding forces of the VLA-4–VCAM-1 complex were elevated over the range of loading rates between 100 and 20,000 pN/s, but did not change the dynamic response of the complex at loading rates greater than 20,000 pN/s. The dissociation rates of the outer barrier for low- and high-affinity VLA-4–VCAM-1 complexes are 1.4/s and 0.035/s, respectively. However, no significant difference was found in dissociation rates (~80/s) for the inner barrier of both affinity states. Thus, the VLA-4–VCAM-1 system shows similarities to the LFA-1–ICAM-1 system. In both integrin systems, induction of high-affinity states elevated the energy potential of their outer activation barriers, but had minimal effect on the inner barriers.

For a molecular complex with two activation barriers, the force-dependent dissociation rate is given by

$$k_{\text{off}} = 1/\{k_1^{0^{-1}} \exp[-f\gamma_1/k_B T] + k_2^{0^{-1}} \exp[-f\gamma_2/k_B T]\} \quad (9.5)$$

where the superscripts 1 and 2 refer to the inner and outer activation energy barriers, respectively (Evans et al., 2001). Figure 9.5 presents the kinetic profiles of the P-selectin–sLeX, high-affinity VLA-4–VCAM-1, and high-affinity LFA-1–ICAM-1 interactions. The kinetic profiles revealed the profound impact of a pulling force on the dissociation rate of these complexes. As shown, the dissociation rate constants are sensitive to weak pulling forces, but are less responsive to stronger pulling forces (>~50 pN for both P-selectin and VLA-4; >~100 pN for LFA-1). This resistance to stronger pulling force can be explained by the Bell–Evans model. At weak pulling forces, the dissociation rates of these complexes are determined by the properties of the outer barriers, which are relatively wide and, hence, more sensitive to the applied forces. However, at stronger pulling forces, the outer barriers are suppressed, and the dissociation rates are governed by the properties of the inner barrier, which is steep and less responsive to a pulling force.

It is interesting to compare the dynamic responses of these three interactions and relate them to their functions. Previous studies revealed that the P-selectin–sLeX and

LFA-1–ICAM-1 interactions mediate leukocyte rolling and firm adhesion, respectively, and that the VLA-4–VCAM-1 interaction can mediate both leukocyte rolling and firm adhesion (Kubes, 2002). A comparison of kinetic profiles of these complexes revealed that both the P-selectin–sLeX complex and the VLA-4–VCAM-1 complexes are more sensitive to a pulling force than the LFA-1–ICAM-1 complex (Fig. 9.5). This finding suggests that the P-selectin–sLeX and VLA-4–VCAM-1 interactions are more suited for cell rolling, because during cell rolling, the adhesive complexes need to dissociate readily by shear force (Orsello et al., 2001). The more stable complex, the LFA-1–ICAM-1, is more suitable for mediating firm adhesion, in line with observations by Lawrence et al. (1991). In their flow chamber experiments, leukocytes rolled only on P-selectin but not on ICAM-1-coated surfaces.

9.3. SUMMARY

In this review, we have presented examples on how to use the AFM to probe the intermolecular forces of leukocyte adhesion molecules. AFM studies on three major leukocyte adhesion complexes revealed that a common characteristic of these complexes is the presence of at least two activation barriers. The inner barriers are crucial for these complexes to resist strong pulling forces. In addition, experiments using live leukocytes revealed that integrin activation enhances only the energy of their outer activation barriers.

The methods described herein are not limited to the chosen system, and has been used to study a broad range of protein–ligand interactions (see Table 9.1 for a partial list). With the completion of the human genome project, there is a compelling need to characterize the biophysical and functional features of a tremendous amount of protein-protein interactions. It is predictable that in the near future, the AFM will play an important role in such studies.

ACKNOWLEDGMENTS

We thank A. Chen, E. Wojcikiewicz, F. Li, and D. Bogorin for insightful discussions and thank C. Freites for technical support. This work was sponsored by grants from the AHA, the NSF-BITC and the NIH (GM55611). XZ was a Predoctoral Fellow (0215139B) of the AHA.

REFERENCES

Allen, S., Chen, X., Davies, J., Davies, M. C., Dawkes, A. C., Edwards, J. C., et al. (1997). Detection of antigen–antibody binding events with the atomic force microscope. *Biochemistry* **36**(24):7457–7463.

Allen, S., Davies, J., Davies, M. C., Dawkes, A. C., Roberts, C. J., Tendler, S. J., et al. (1999). The influence of epitope availability on atomic-force microscope studies of antigen–antibody interactions. *Biochem. J.* **341**(Pt 1):173–178.

Baumgartner, W., Hinterdorfer, P., Ness, W., Raab, A., Vestweber, D., Schindler, H., et al. (2000). Cadherin interaction probed by atomic force microscopy. *Proc. Nat. Acad. Sci. USA* **97**(8):4005–4010.

Bell, G. I. (1978). Models for the specific adhesion of cells to cells. *Science* **200**:618–627.

Benoit, M., Gabriel, D., Gerisch, G., and Gaub, H. E. (2000). Discrete interactions in cell adhesion measured by single-molecule force spectroscopy. *Nature Cell Biol.* **2**(6):313–317.

Berry, M., McMaster, T. J., Corfield, A. P., and Miles, M. J. (2001). Exploring the molecular adhesion of ocular mucins. *Biomacromolecules* **2**(2):498–503.

Binnig, G., Quate, C. F., and Gerber, C. (1986). Atomic force microscope. *Phys. Rev. Lett.* **56**:930–933.

Chen, A., and Moy, V. T. (2000). Cross-linking of cell surface receptors enhances cooperativity of molecular adhesion. *Biophys. J.* **78**(6):2814–2820.

Dammer, U., Popescu, O., Wagner, P., Anselmetti, D., Guntherodt, H. J., and Misevic, G. N. (1995). Binding strength between cell adhesion proteoglycans measured by atomic force microscopy. *Science* **267**(5201):1173–1175.

Dettmann, W., Grandbois, M., Andre, S., Benoit, M., Wehle, A. K., Kaltner, H., et al. (2003). Differences in zero-force and force-driven kinetics of ligand dissociation from beta-galactoside-specific proteins (plant and animal lectins, immunoglobulin G) monitored by plasmon resonance and dynamic single molecule force microscopy. *Arch. Biochem. Biophys.* **383**(2):157–170.

Evans, E. (2001). Probing the relation between force–lifetime–and chemistry in single molecular bonds. *Annu. Rev. Biophys. Biomol. Struct.* **30**:105–128.

Evans, E., and Ritchie, K. (1997). Dynamic strength of molecular adhesion bonds. *Biophys. J.* **72**(4):1541–1555.

Evans, E., Leung, A., Hammer, D., and Simon, S. (2001). Chemically distinct transition states govern rapid dissociation of single L-selectin bonds under force. *Proc. Natl. Acad. Sci. USA* **98**(7):3784–3789.

Florin, E. L., Moy, V. T., and Gaub, H. E. (1994). Adhesion forces between individual ligand–receptor pairs. *Science* **264**(5157):415–417.

Fritz, J., Katopodis, A. G., Kolbinger, F., and Anselmetti, D. (1998). Force-mediated kinetics of single P-selectin/ligand complexes observed by atomic force microscopy. *Proc. Natl. Acad. Sci. USA* **95**(21):12283–12288.

Hanley, W., McCarty, O., Jadhav, S., Tseng, Y., Wirtz, D., and Konstantopoulos, K. (2003). Single molecule characterization of P-selectin/ligand binding. *J. Biol. Chem.* **278**(12):10556–10561.

Hinterdorfer, P., Baumgartner, W., Gruber, H. J., Schilcher, K., and Schindler, H. (1996). Detection and localization of individual antibody–antigen recognition events by atomic force microscopy. *Proc. Nat. Acad. Sci. USA* **93**(8):3477–3481.

Kubes, P. (2002). The complexities of leukocyte recruitment. *Semin. Immunol.* **14**(2):65–72.

Kubes, P., and Kerfoot, S. M. (2001). Leukocyte recruitment in the microcirculation: the rolling paradigm revisited. *News Physiol. Sci.* **16**:76–80.

Lawrence, M. B., and Springer, T. A. (1991). Leukocytes roll on a selectin at physiologic flow rates: Distinction from and prerequisite for adhesion through integrins. *Cell* **65**(5):859–873.

Lee, G. U., Chrisey, L. A., and Colton, R. J. (1994). Direct measurement of the forces between complementary strands of DNA. *Science* **266**(5186):771–773.

Lehenkari, P. P., and Horton, M. A. (1999). Single integrin molecule adhesion forces in intact cells measured by atomic force microscopy. *Biochem. Biophys. Res. Commun.* **259**(3):645–650.

Li, F., Redick, S. D., Erickson, H. P., and Moy, V. T. (2003). Force measurements of the alpha(5)beta(1) integrin–fibronectin interaction. *Biophys. J.* **84**(2):1252–1262.

Merkel, R. (2001). Force spectroscopy on single passive biomolecules and single biomolecular bonds. *Phys. Rep. Rev. Sect. Phys. Lett.* **346**(5):344–385.

Merkel, R., Nassoy, P., Leung, A., Ritchie, K., and Evans, E. (1999). Energy landscapes of receptor–ligand bonds explored with dynamic force spectroscopy. *Nature* **397**(6714):50–53.

Nakajima, H., Kunioka, Y., Nakano, K., Shimizu, K., Seto, M., and Ando, T. (1997). Scanning force microscopy of the interaction events between a single molecule of heavy meromyosin and actin. *Biochem. Biophys. Res. Commun.* **234**(1):178–182.

Nevo, R., Stroh, C., Kienberger, F., Kaftan, D., Brumfeld, V., Elbaum, M., et al. (2003). A molecular switch between alternative conformational states in the complex of Ran and importin beta1. *Nature Struct. Biol.* **10**(7):553–557.

Orsello, C. E., Lauffenburger, D. A., and Hammer, D. A. (2001). Molecular properties in cell adhesion: A physical and engineering perspective. *Trends Biotechnol.* **19**(8):310–316.

Springer, T. A. (1990). Adhesion receptors of the immune system. *Nature* **346**:425–434.

Springer, T. A. (1994). Traffic signals for lymphocyte recirculation and leukocyte emigration: The multistep paradigm. *Cell* **76**(2):301–314.

Vinckier, A., Gervasoni, P., Zaugg, F., Ziegler, U., Lindner, P., Groscurth, P., et al. (1998). Atomic force microscopy detects changes in the interaction forces between GroEL and substrate proteins. *Biophys. J.* **74**(6):3256–3263.

Wielert-Badt, S., Hinterdorfer, P., Gruber, H. J., Lin, J. T., Badt, D., Wimmer, B., et al. (2002). Single molecule recognition of protein binding epitopes in brush border membranes by force microscopy. *Biophys. J.* **82**(5):2767–2774.

Willemsen, O. H., Snel, M. M., van der Werf, K. O., de Grooth, B. G., Greve, J., Hinterdorfer, P., et al. (1998). Simultaneous height and adhesion imaging of antibody–antigen interactions by atomic force microscopy. [comment]. *Biophys. J.* **75**(5):2220–2228.

Yip, C. M., Yip, C. C., and Ward, M. D. (1998). Direct force measurements of insulin monomer–monomer interactions. *Biochemistry* **37**(16):5439–5449.

Zhang, X., Wojcikiewicz, E., and Moy, V. T. (2002). Force spectroscopy of the leukocyte function-associated antigen-1/intercellular adhesion molecule-1 interaction. *Biophys. J.* **83**(4):2270–2279.

Zhang, X., Craig, S. E., Humphries, M. J., and Moy, V. T. (2003). Unpublished data.

Zhang, X., Bogorin, D. F., and Moy, V. T. (2004). Molecular basis of the dynamic strength of the sialyl Lewis X–selectin interaction. *Chem. Phys. Chem.* **4**:100–107.

CHAPTER 10

MECHANISMS OF AVIDITY MODULATION IN LEUKOCYTE ADHESION STUDIED BY AFM

EWA P. WOJCIKIEWICZ AND VINCENT T. MOY

Department of Physiology and Biophysics, University of Miami Miller School of Medicine, Miami, Florida

10.1. INTRODUCTION

Integrins play an important role in the regulation of leukocyte adhesion to target cells. Integrins are αβ heterodimeric glycoproteins derived from the noncovalent association of one of 18 α chains and one of 8 β chains (Hynes, 1992). In leukocytes, α_L combines with β_2 to form leukocyte function-associated antigen-1 (LFA-1), the major leukocyte integrin (Sanchez-Madrid et al., 1983). LFA-1 facilitates leukocyte adhesion by binding to its ligands, which include intercellular adhesion molecule-1 (ICAM-1), a member of the immunoglobulin superfamily, expressed on the target cell (Marlin and Springer, 1983; Springer, 1990).

Integrins possess the ability to modulate the adhesive states of cells, which is a critical process necessary for proper immune function (Dustin and Springer, 1991). LFA-1 is expressed in its inactive form on resting lymphocytes and binds to ICAM-1 with low affinity. *In vivo*, LFA-1 activation takes place following the engagement of the T lymphocyte by an antigen presenting cell (APC). This process results in the strengthening of the transient interaction between the two cells that lasts long enough for the T lymphocyte to become activated before the cells separate.

Both receptor affinity modulation and receptor clustering (avidity modulation) contribute to the regulation of integrin-mediated leukocyte adhesion (Lollo et al., 1993). The mechanism for affinity modulation involves the cytoplasmic engagement of the LFA-1 β subunit. This "inside-out" signal, transmitted though the β subunit, crosses over to the α subunit via the β-propeller–β I-like domain interface. The β propeller of the α subunit then activates the I domain (Lu et al., 2001) by inducing a downward displacement of the I domain α7 helix and the subsequent opening of the ICAM-1 binding site. This resulting conformational change quickly stabilizes the LFA-1–ICAM-1 complex (Hughes et al., 1996; Shimaoka et al., 2001; Lupher et al., 2001). The need for an inside-out signal for the induction of high-affinity LFA-1 can be circumvented by high concentration of extracellular Mg^{2+} and by certain antibodies directed against the β chain (Shimaoka et al., 2001; Lupher et al., 2001; McDowall et al., 1998).

The usage of the term "avidity modulation" most frequently refers to redistribution or clustering of receptors that results

---

Force Microscopy: Applications in Biology and Medicine, edited by Bhanu P. Jena and J.K. Heinrich Hörber.
Copyright © 2006 John Wiley & Sons, Inc.

in augmented cell adhesion. The mechanism of avidity modulation remains ill-defined and may include polarization of receptors to the zone of cell–cell contact, clustering of receptors to form focal adhesion sites, and dimerization of receptors. The dimerization of receptors may result in the formation of a dimeric complex that functions as a cooperative unit that ruptures simultaneously. Although the relative contributions of these underlying mechanisms to adhesion still needs to be resolved, it is well-established that the initial event following the engagement of the antigen receptor of T lymphocytes in the avidity modulation of leukocytes is the release of LFA-1 from cytoskeletal constraints (Stewart et al., 1998). Recent studies have shown the mobilization of LFA-1 on T lymphocytes in contact with APCs. Once released from the cytoskeleton, the receptor molecules form ring-like structures termed *supramolecular activation clusters* (SMACs). The SMACs consist of antigen receptors localized at the center of cell–cell contact and LFA-1–ICAM-1 complexes in the periphery (Monks et al., 1998).

Normally avidity modulation in the leukocyte is brought about as a result of T-cell receptor signaling. Experimentally, changes in avidity can be induced by pharmacological agents that exert their effects on intracellular pathways, bypassing the requirement of surface receptor engagement (Berry and Nishizuka, 1990). One such agent that promotes the mobilization of LFA-1 and enhanced adhesion is phorbol myristate acetate (PMA) (Rothlein and Springer, 1986). PMA is a potent activator of protein kinase C (PKC). The activation of PKC leads to an increase of intracellular Ca^{2+} concentrations, $Ins(1,4,5)P_3$- kinase activity, and phosphorylation of Rack1, MacMARCKS (macrophage-enriched myristoylated alanine-rich C kinase substrate), and L-plastin (van Kooyk and Figdor, 2000; Zhou and Li, 2000; Jones et al., 1992). Rack1 functions as a scaffold protein that recruits PKC to the site of action (van Kooyk and Figdor, 2000). MacMARCKS and L-plastin are PKC substrates and are involved in releasing LFA-1 from cytoskeletal constraints. This takes place through the action of the Ca^{2+}-dependent protease, calpain, which is activated due to the increased levels of intracellular Ca^{2+} and releases LFA-1 from the cytoskeleton (Stewart et al., 1998). PMA also acts indirectly on the activity of cytohesin-1, resulting in its association with LFA-1. Cytohesin-1 has been shown to induce cytoskeletal reorganization and subsequently promote cell spreading (Kolanus et al., 1996).

Avidity changes can also be brought about experimentally through the use of the Ca^{2+} ionophore ionomycin and Ca^{2+} mobilizers such as thapsigargin, which increase intracellular Ca^{2+} more directly than PMA (Thastrup et al., 1990; Thomas and Hanley, 1994). Thapsigargin functions by inhibiting the ATPase pumps of Ca^{2+} storage organelles. These pumps normally pump Ca^{2+} from the cytosol into Ca^{2+} stores. When they are inhibited, there is a resulting influx of intracellular Ca^{2+} that not only comes from the internal Ca^{2+} stores but also occurs as a result of capacitative fluxing through the plasma membrane (Stewart et al., 1998). Both Ca^{2+} mobilizers have been shown to function in a similar way as PMA. Clustering has been shown as a result of both ionomycin and thapsigargin application, and their mechanism of action had been shown to depend on the calpain protease. The use of calpeptin, a calpain protease blocker, resulted in maximal inhibition of T-cell adhesion (Tsujinaka et al., 1988). No affinity modulation had been reported following the use of Ca^{2+} mobilizers using more traditional cell adhesion studies (Stewart et al., 1998).

Our previous work showed that cell spreading was part of the activation mechanism of PMA (Wojcikiewicz et al., 2003). Using the AFM, we were able to quantify direct force measurements of PMA-stimulated 3A9 cells, which are murine hybridoma cells that express the LFA-1 integrin but no other receptor for ICAM-1 (Lollo et al., 1993). LFA-1 of 3A9 cells is constitutively inactive but can be activated to a high avidity state by PMA. Our AFM data revealed that PMA-stimulated cells did not exhibit a strengthening of individual bonds between LFA-1/ICAM-1, but instead exhibit an increase in the number of bonds that were formed as well as an increase in the work of de-adhesion (Wojcikiewicz et al., 2003). This is thought to be largely a result of cell spreading, which we were able to visualize through reflection interference contrast microscopy (unpublished data). The cells were found to become more compliant following PMA stimulation as well, which was shown in our AFM compliance measurements. The combination of cell spreading and increased cell compliance appears to be a multiplicative effect leading to a 10-fold increase of leukocyte adhesion (Wojcikiewicz et al., 2003)

The first part of this report details work that was done in order to separate the contribution of cell spreading from that of LFA-1 redistribution or clustering toward enhanced leukocyte adhesion to ICAM-1. This was accomplished by restricting the lateral diffusion of LFA-1 by cross-linking the cell surface and then stimulating the cells by PMA. Under these conditions, we are able to investigate the role of cell spreading in cell adhesion without contribution from receptor redistribution.

The second part of this report focuses on integrin activation through a direct increase of intracellular Ca^{2+}. This was accomplished by stimulating the 3A9 T cells with thapsigargin and ionomycin. Thapsigargin acts by inhibiting the ATPase pumps on the Ca^{2+} storage organelles and ionomycin is a Ca^{2+} ionophore. Both methods increased leukocyte adhesion to ICAM-1 at levels that were comparable to the increases previously reported with PMA. However, unlike PMA, which only acts through avidity modulation, other mechanisms may be involved when intracellular Ca^{2+} is increased directly.

10.2. METHODS

10.2.1. Cells and Reagents

The 3A9 cell line was maintained in continuous culture in RPMI 1640 medium supplemented with 10% heat-inactivated fetal calf serum (Irvine Scientific, Santa Ana, CA), penicillin (50 U/ml, Gibco BRL, Grand Island, NY), and streptomycin (50 μg/ml, Gibco BRL) and were expanded on a 3-day cycle (Kuhlman et al., 1991).

ICAM-1/Fc chimera consisted of all five extracellular domains of murine ICAM-1 (Met 1-Ala 16), and the Fc fragment of human IgG_1 and was purchased from R & D Systems, Inc. (Minneapolis, MN). The ability of this protein to bind LFA-1 was confirmed using ELISA and adhesion assays. From these experiments, we were able to conclude that the LFA-1 binding epitope D1 of ICAM-1 is available for binding. Antibodies against LFA-1 (i.e., M17/4.2 and FD441.8) and against ICAM-1 (i.e., BE29G1) were purified from culture supernatant by protein G affinity chromatography (Sanchez-Madrid et al., 1983; Kuhlman et al., 1991). Stock solutions of PMA (10,000X) (Sigma, St. Louis, MO) were prepared at 1 mM in DMSO. Stimulation of cells was carried out at 37°C for 5 min. Alternatively, thapsigargin (Calbiochem, La Jolla, CA) and ionomycin, calcium salt, *S. congobatus* (Calbiochem) were used for cell stimulation at 5 μM and 0.7 μM, respectively (Stewart et al., 1998). Stimulation of cells was carried out at 37°C for 30 min.

10.2.2. Cross-Linking of 3A9 Cells

The 3A9 cells were exposed to either disuccinimidyl suberate (DSS) or bis(sulfosuccinimidyl)suberate (BS^3), (Pierce, Rockford, IL). Both cross-linkers were prepared at 20 mM, DSS in DMSO and BS^3 in water. The cross-linking reaction was prepared at room temperature in PBS. One-tenth of a milligram of 3A9 cells was incubated for 30 min in a final concentration of 2 mM of either DSS or BS^3.

The reaction mixture was quenched with 1 M Tris, pH 7.5, added to a final concentration of 10 mM for 15 min.

10.2.3. Protein Immobilization

Twenty-five microliters of ICAM-1/Fc at 50 μg/ml in 0.1 M NaHCO₃ (pH 8.6) was adsorbed overnight at 4°C on the center of a 35-mm tissue culture dish (Falcon 353001, Becton Dickinson Labware, Franklin Lakes, NJ). Unbound ICAM-1/Fc was removed and bovine albumin (Sigma) at 100 μg/ml in PBS was used to block the exposed surface of the dish. A similar protocol was used to immobilize the anti-LFA-1 antibodies (i.e., FD441.8 and M17/4.2).

10.2.4. AFM Measurements of Adhesive Forces

The AFM used in our laboratory was a homemade modification of the standard AFM and was designed to be operated in the force spectroscopy mode (Fig. 10.1A) (Benoit et al., 2000; Heinz and Hoh, 1999; Willemsen et al., 2000; Wojcikiewicz et al., 2004). 3A9 cells were attached to the AFM cantilever by concanavalin A (con A)-mediated linkages (Zhang et al., 2002). To prepare the con A-functionalized cantilever, the cantilevers were soaked in acetone for 5 min, UV irradiated for 30 min, and incubated in biotinamidocaproyl-labeled bovine serum albumin (biotin-BSA, 0.5 mg/ml in 100 mM NaHCO₃, pH 8.6; Sigma) overnight at 37°C. The cantilevers were then rinsed three times with phosphate buffered saline (PBS, 10 mM PO₄³⁻, 150 mM NaCl, pH 7.3) and incubated in streptavidin (0.5 mg/ml in PBS; Pierce; Rockford, IL) for 10 min at room temperature. Following the removal of unbound streptavidin, the cantilevers were incubated in biotinylated Con A (0.2 mg/ml in PBS; Sigma) and then rinsed with PBS.

To attach the 3A9 cell to the cantilever, the end of the Con A-functionalized cantilever was positioned above the center of a cell and carefully lowered onto the cell for approximately 1 s. When attached, the cell is positioned right behind the AFM cantilever tip. To obtain an estimate of the strength of the cell-cantilever linkage, we allowed the attached cell to interact with a substrate coated with Con A for 1 min. Upon retraction of the cantilever, separation always ($N > 20$) occurred between the cell and the Con A-coated surface. The average force needed to induce separation was greater than 2 nN. Thus, these measurements

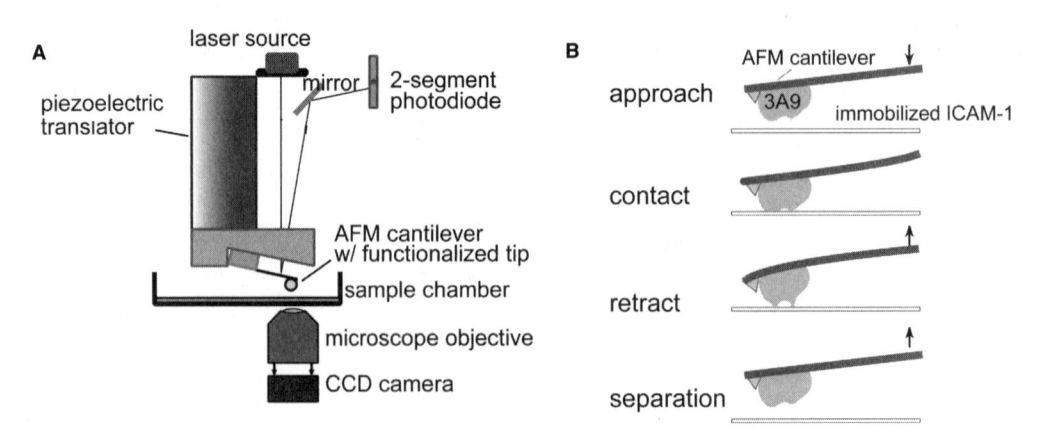

Figure 10.1. (A) Complete schematic diagram of the AFM. (B) Schematics of AFM force measurements. The principal events of cell–substrate interaction during an AFM force versus displacement measurement are: approach of the 3A9 cell onto a surface coated with ICAM-1, contact between cell and substrate, retraction of the 3A9 cell, and separation of the cell from the substrate. Arrows indicate the direction of cantilever movement.

revealed that the linkages supporting cell attachment to the cantilever is greater than 2 nN and much larger than the detachment force required to separate the bound 3A9 cell from immobilized ICAM-1 (Zhang et al., 2002).

A piezoelectric translator was used to lower the cantilever/cell onto the sample. During the acquisition of a force scan, the cantilever was bent (Fig. 10.1B), causing the beam of a 3-mW diode laser (Oz Optics; emission 635 nm) that was focused on top of the cantilever to be deflected, which was measured by a position-sensitive 2-segment photodiode detector. AFM cantilevers were purchased from TM Microscopes (Sunnyvale, CA). The largest triangular cantilever (320 μm long and 22 μm wide) from a set of five on the cantilever chip was used in our measurements. These cantilevers were calibrated by analysis of their thermally induced fluctuation to determine their spring constant (Hutter and Bechhoefer, 1993). The experimentally determined spring constants were consistent with the nominal value of 10 mN/m given by the manufacturer (Wojcikiewicz et al., 2003).

10.2.5. AFM Measurements of Cell Elasticity

In this report, the AFM also served as a microindenter that probes the mechanical properties of the cell. The bare AFM tip is lowered onto the cell surface at a set rate, typically 2 μm/s, to obtain the cell elasticity measurements. After contact, the AFM tip exerts a force against the cell that is proportional to the deflection of the cantilever. The deflection of the cantilever was recorded as a function of the piezoelectric translator position during the approach and withdrawal of the AFM tip. The force–indentation curves of the cells were derived from these records using the surface of the tissue culture dish to calibrate the deflection of the cantilever. Estimates of Young's modulus were made on the assumptions that the cell is an isotropic elastic solid and the AFM tip is a rigid cone (Wu et al., 1998; Hoh and Schoenenberger, 1994; Radmacher et al., 1996). According to this model, initially proposed by Hertz, the force (F)–indentation (α) relation is a function of Young's modulus of the cell, K, and the angle formed by the indenter and the plane of the surface, θ, as follows:

$$F = \frac{K}{2(1-\nu^2)} \frac{4}{\pi \tan \theta} \alpha^2 \qquad (10.1)$$

Young's modulus was obtained by least-square analysis of the force–indentation curve using routines in the Igor Pro (WaveMetrics, Inc., Lake Oswego, OR) software package. The indenter angle, θ, and Poisson ratio, ν, were assumed to be 55° and 0.5, respectively.

Measurements of cell adhesion and elasticity were carried out at 25°C in fresh tissue culture medium supplemented with 10 mM HEPES buffer. Cells were stimulated by 5 mM $MgCl_2$ plus 1 mM EGTA or 100 nM PMA. The activation of 3A9 by Mg^{2+} was immediate. 3A9 cells were exposed to PMA for ~5 min at 37°C prior to the start of the experiments. All experiments involved making contact with the same cell up to 50 times. There was no dependence on previous contacts observed in either the elasticity or adhesion AFM studies (Wojcikiewicz et al., 2003).

10.3. RESULTS

10.3.1. Contribution of Integrin Lateral Redistribution in Leukocyte Adhesion

We have carried out studies aimed at resolving the contribution of cell spreading from LFA-1 redistribution on the cell surface (clustering). This was accomplished through the use of both an external and an internal cross-linking agent, which allowed us to examine leukocyte adhesion to ICAM-1 under conditions where the movement of receptors on the cell surface is retarded and cell spreading is prevented. 3A9 T cells were exposed to bis(sulfosuccinimidyl) suberate (BS^3), a water-soluble, membrane-impermeable cross-linker that restricts the lateral diffusion of LFA-1 but should not cross-link the cytoskeleton. The BS^3 cross-linked cells were then activated by PMA, which induced the cell to spread, while the lateral diffusion of membrane receptors remained retarded. Under these conditions, we were able to investigate the role of cell spreading in cell adhesion while limiting the contribution from receptor redistribution.

AFM force measurements were carried out to determine the adhesive properties of the cross-linked cells. As illustrated in Fig. 10.1B, a 3A9 cell, coupled to the end of an AFM cantilever, was lowered onto a substrate coated with ICAM-1. The work of de-adhesion was obtained by integrating the adhesive force over the distance traveled by the cantilever upon its retraction from the substrate. The AFM adhesion data revealed that both BS^3 cross-linked and untreated 3A9 cells had similar levels of adhesion. On average, the work of de-adhesion to detach resting 3A9 cells from immobilized ICAM-1 was 3.16×10^{-16} J for untreated cells and 2.89×10^{-16} J for the BS^3 cross-linked (Fig. 10.2). This is to be expected, because clustering is not thought to occur until the cell is stimulated. Following PMA stimulation, both the untreated and the BS^3 cross-linked cells showed a significant increase in adhesion (Fig. 10.2). The work of de-adhesion of the BS^3 cross-linked cells following PMA stimulation was ~20% smaller on average than that of the untreated cells. In contrast, 3A9 cells that were internally cross-linked with disuccinimidyl suberate (DSS) exhibited diminished adhesion to immobilized ICAM-1, even following PMA stimulation. These observations are attributed to the cross-linked cell's inability to spread or induce LFA-1 clustering.

To establish that the cytoskeleton of the cell is not cross-linked by BS^3, we

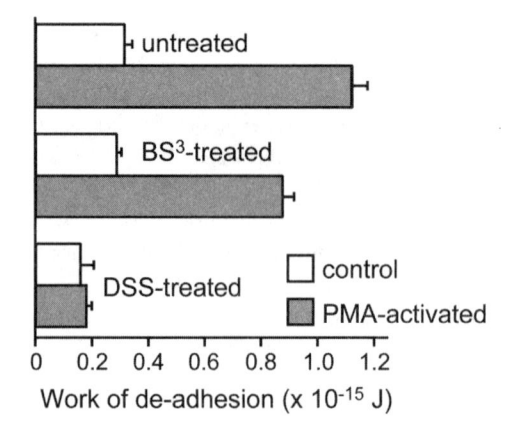

Figure 10.2. Work of de-adhesion of resting (white bars) and PMA-stimulated (gray bars) 3A9 cells. The cells compared were either not cross-linked or cross-linked with the water-soluble, membrane-impermeable cross-inker (BS3) or the lipid-soluble, membrane-permeable cross-linker (DSS). The calculated areas of de-adhesion were acquired with a compression force of 200 pN, 5-s contact, and a cantilever retraction speed of 2 μm/s. The error bar is the standard error.

measured the elasticity of the cross-linked cells by AFM. The Young's modulus values for both the unstimulated and PMA-stimulated cells that have been cross-linked with BS^3 are similar to those of cells that have not undergone cross-linking. As shown in Fig. 10.3, the Young's modulus values of the cross-linked cells were 1.6 ± 0.08 kPa for unstimulated cells (1.4 ± 0.4 kPa, without cross-linking) and 0.3 ± 0.02 kPa for PMA-stimulated cells (0.3 ± 0.01 kPa, without cross-linking). Elasticity measurements conducted on the DSS cross-linked cells showed Young's modulus values of 5.1 ± 0.3 kPa for the unstimulated cells and 4.6 ± 0.5 kPa PMA for the PMA stimulated cells. These values are considerably higher than both the untreated and BS^3 cross-linked cells and are indicative of stiffer, less compliant cells that are also less likely to undergo spreading following PMA-stimulation (Fig. 10.3). The ability of both groups of cross-linked cells to spread was further verified using reflection interference contrast microscopy, where spreading was only observed for the PMA-stimulated cells

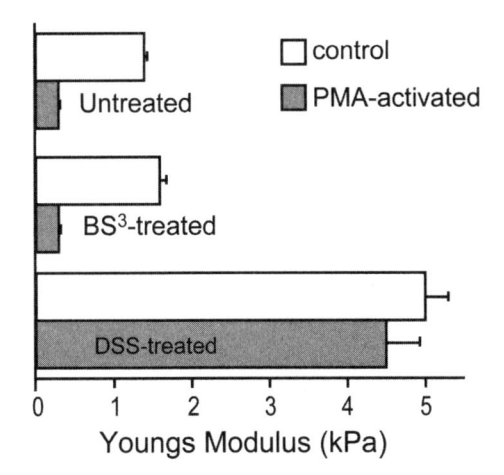

Figure 10.3. Young's modulus of resting (white bars) and PMA-stimulated (gray bars) 3A9 cells. The cells compared were either not cross-linked or cross-linked with the water-soluble, membrane-impermeable cross-linker (BS3) or the lipid-soluble, membrane-permeable cross-linker (DSS). The error bar is the standard error.

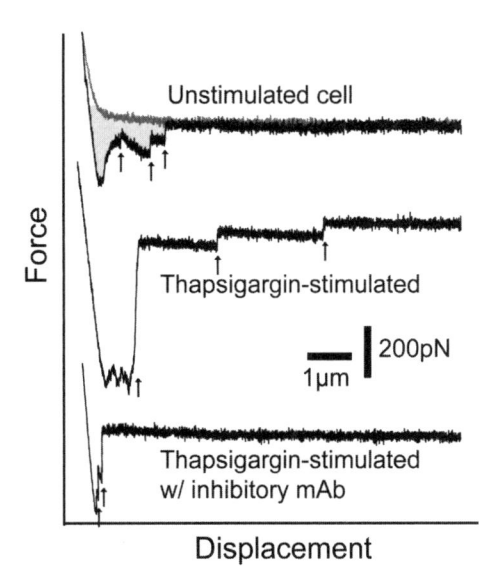

Figure 10.4. Force versus displacement traces of the interaction between 3A9 cells and immobilized ICAM-1. The measurements were carried out with a resting cell (first trace) and a thapsigargin-stimulated cell (second trace). The measurements were acquired with a compression force of 200 pN, 5-s contact, and a cantilever retraction speed of 2 μm/s. The third trace corresponds to a measurement acquired from a thapsigargin-stimulated cell in the presence of an LFA-1 (20 μg/ml FD441.8) function-blocking antibody. The shaded area estimates the work of de-adhesion. Arrows point to breakage of LFA-1–ICAM-1 bond(s).

that had been cross-linked with BS^3 (data not shown).

10.3.2. Effects of Ionomycin and Thapsigargin Stimulation on Leukocyte Adhesion

The next series of experiments focused on the effects of direct intracellular Ca^{2+} increase on the adhesion of the 3A9 cells to ICAM-1. We examined the effects of two agents, ionomycin and thapsigargin, which have been shown to enhance leukocyte adhesion through increases in [Ca$_i^{2+}$] (Stewart et al., 1998). Figure 10.4 presents a series of AFM force measurements carried out with resting and thapsigargin-stimulated 3A9 cells. The adhesive interaction of the cantilever bound cell and the ICAM-1-coated substrate was determined by the work done by the cantilever to stretch the cell and to detach it from immobilized ICAM-1. The work of de-adhesion was derived by integrating the adhesive force over the distance traveled by the cantilever. The work of de-adhesion, as can be seen in the second force scan in Fig. 10.4, is much greater following stimulation of the cell with thapsigargin. The work of de-adhesion increased from 2.82 × 10^{-16} J in the resting state to 1.31 × 10^{-15} J following thapsigargin stimulation (Fig. 10.6). Treatment of the thapsigargin stimulated cells with an antibody against LFA-1 (FD441.8) resulted in a work of de-adhesion that was greatly reduced (5.95 × 10^{-17} J), indicating the specificity of this interaction.

Stimulation of the 3A9 cells with 0.7 μM of ionomycin also resulted in a significant increase of the work of de-adhesion. Figure 10.5 shows a series of measurements obtained with ionomycin-stimulated cells. The work of de-adhesion following ionomycin stimulation is significantly larger than that of the resting cell as can be seen in Fig. 10.5. The work of de-adhesion

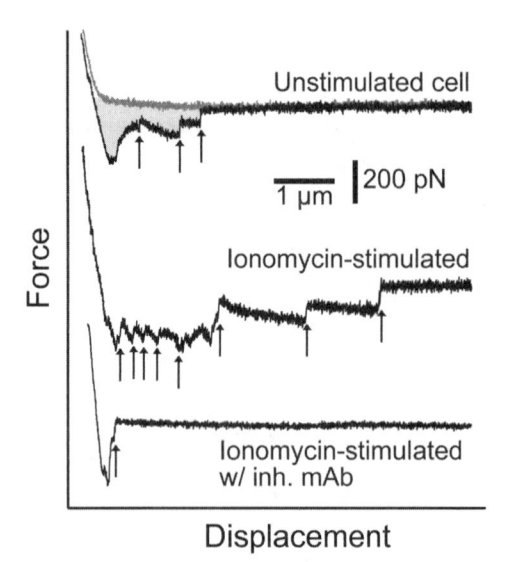

Figure 10.5. Force versus displacement traces of the interaction between 3A9 cells and immobilized ICAM-1. The measurements were carried out with a resting cell (first trace) and an ionomycin-stimulated cell (second trace). The measurements were acquired with a compression force of 200 pN, 5 s contact and a cantilever retraction speed of 2 μm/s. The third trace corresponds to a measurement acquired from an ionomycin-stimulated cell in the presence of an LFA-1 (20 μg/ml FD441.8) function-blocking antibody. The shaded area estimates the work of de-adhesion. Arrows point to breakage of LFA-1–ICAM-1 bond(s).

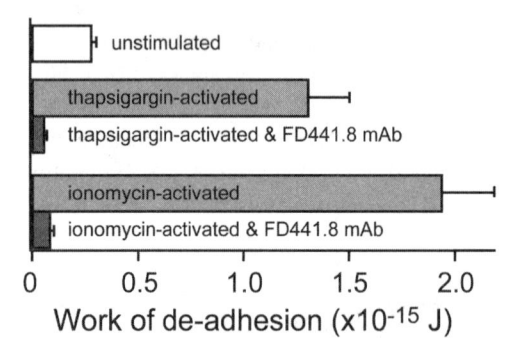

Figure 10.6. Work of de-adhesion of resting and stimulated 3A9 cells bound to immobilized ICAM-1/Fc. Cells were stimulated with 5 μM thapsigargin or 0.7 μM ionomycin. The inhibitory monoclonal antibody used was FD441.8 (anti-LFA-1; 20 μg/ml). The numbers given here were derived from measurements that were acquired with a compression force of 200 pN, 5-s contact, and a cantilever retraction speed of 2 μm/s. The error bar is the standard error.

increased from 2.82×10^{-16} J in the resting state to 1.94×10^{-15} J following ionomycin stimulation (Fig. 10.6). The ionomycin-stimulated cells that were treated with antibody against LFA-1 had a greatly reduced area of de-adhesion of 8.64×10^{-17} J (Fig 10.6).

10.4. DISCUSSION

The main goal of these studies was to elaborate on the contribution of integrin clustering to the enhanced adhesion of 3A9 cells to ICAM-1. As was reported earlier in Wojcikiewicz et al. (2003), our force data did not suggest the formation of large clusters of LFA-1 on the cell surface. This would have been observed as irregularities in the adhesion force measurements following PMA-stimulation due to areas on the cell surface exhibiting high and low receptor densities (Wojcikiewicz et al., 2003). It is still conceivable that microclustering is taking place; and although no cooperativity of bonds was observed in this or the previously published data, clustering may still aid in enhancing adhesion by distributing forces among clustered LFA-1–ICAM-1 complexes.

Retarding the mobility of LFA-1 receptors on the cell surface through BS^3 cross-linking resulted in a ~20% decrease of the area of de-adhesion following PMA stimulation as compared to the untreated cells (Fig. 10.2). Both RICM and elasticity studies revealed that the cells were still spreading following PMA stimulation. We attribute the drop in the area of de-adhesion following BS^3 cross-linking to be due to the absence of clustering. This indicates that the rearrangement of LFA-1 on the cell surface is an important part of PMA-stimulation of 3A9 T cells, but it is not the major mechanism responsible for the observed enhanced adhesion.

The 3A9 cells that were exposed to the lipid-soluble cross-linker, DSS, exhibited low levels of adhesion. This cross-linker was expected to prevent receptor mobility on the cell surface as well as any cytoskeletal changes leading to cell spreading due to its internal cross-linking capacity. Eliminating cell spreading in addition to clustering resulted in a very significant drop in the level of adhesion. Even following stimulation with PMA, these cells had lower areas of de-adhesion than the resting, untreated 3A9 T cells (Fig. 10.2).

Our previous study highlighted the importance of cell spreading as a key mechanism for enhanced adhesion of T cells to ICAM-1 following PMA stimulation (Wojcikiewicz et al., 2003). The cross-linking studies reinforce the importance of cell spreading. The acquired elasticity measurements can be used to estimate the area of contact for the 3A9 cells. The Young's modulus of the unstimulated BS^3 cross-linked cells was ∼1.6 kPa (Fig. 10.3). An estimate of contact area A_c for a given compression force F, cell radius R, and Young's modulus K is given by the Hertz model, that is, $A_c = \pi \times \sqrt[3]{(RF/K)^2}$ (Israelachvili, 1992). For $F = 200$ pN, $R = 5$ μm, and $K = 1600$ Pa, the estimated contact area is ∼2.3 μm^2. When K is reduced to 0.3 kPa following PMA stimulation, the estimated area of contact is ∼7.0 μm^2. These data suggest that cell spreading can contribute to a 300% increase in cell adhesion for the BS^3 cross-linked cells. The area of contact for the DSS cross-linked cells changed little and remained smaller than for resting cells that had not been cross-linked. The estimated contact area was ∼1.06 μm^2 for the DSS cross-linked cells (for $K = 5100$ Pa) and ∼1.14 μm^2 for the PMA-stimulated cells (for $K = 4600$ Pa). These small contact areas for this group of cells largely contributed to their low adhesion.

Cell clustering has been examined in many studies and multiple systems. It is an important mechanism that has been implicated in adhesion enhancement. Our studies focused on examining the contribution of this mechanism through the use of PMA stimulation that has been shown to induce receptor clustering. Based on this methodology, we can conclude in these studies that cell compliance changes following PMA stimulation provide a major contribution to the increase in the area of de-adhesion. Clustering of receptors on the cell surface appears to contribute to the adhesion process most likely through the re-distribution of force among the receptor micro-clusters and could be responsible for up to 20% of the enhanced adhesion effect that is observed following PMA stimulation of 3A9 cells.

The latter part of this study focused on two other methods of cell stimulation, both of which involved increasing the levels of $[Ca_i^{2+}]$. The first involved stimulating the 3A9 T cells with 5 μM thapsigargin, which resulted in an ∼5-fold increase in the work of de-adhesion. Stimulation of the cells with 0.7 μM ionomycin resulted in an ∼7-fold increase in the work of de-adhesion. These changes are similar to those observed in our earlier studies using PMA-stimulation, where an ∼5-fold increase in the work of de-adhesion was observed following stimulation. In all three cases, the enhancement in adhesion of the 3A9 cells to ICAM-1 was similar in magnitude, with ionomycin stimulation having the greatest effect in these experiments.

The mechanism of PMA stimulation is better defined than the mechanisms of ionomycin and thapsigargin stimulation. In addition to the intracellular signaling changes taking place following PMA stimulation, the resultant Ca^{2+} fluxes following treatment of cells with ionomycin or thapsigargin have also been shown to activate actin-binding proteins, causing actin dissociation and cytoskeletal rearrangements (Stossel, 1989). Both compounds have been shown to induce LFA-1 clustering on T cells through a mechanism involving protease calpain (Stewart et al., 1998).

As in stimulation with PMA, both cell clustering and cell spreading most likely play a role in the observed enhancement of adhesion following stimulation with thapsigargin and ionomycin. However, other mechanisms of enhanced adhesion may also be involved. Upon closer examination of the ionomycin data, it becomes clear that they differ from the PMA data. The individual LFA-1–ICAM-1 ruptures are greater in magnitude than those observed following PMA stimulation and resemble those following treatment with Mg^{2+}/EGTA (Fig. 10.5) (Wojcikiewicz et al., 2003; Zhang et al., 2002). This may be due to affinity modulation taking place alongside of cell spreading and receptor clustering. Affinity changes lead to the strengthening of the individual LFA-1–ICAM-1 bonds. It is also possible that the LFA-1–ICAM-1 bonds are breaking in a cooperative manner—that is, simultaneously. The mechanism that allows for the enhancement of adhesion following thapsigargin stimulation may also be different from what takes place following PMA stimulation. The strength of the individual LFA-1–ICAM-1 rupture forces is not greater than following PMA stimulation. However, we did observe large steps during the initial unbinding of the cell from ICAM-1, as can be seen in the second force scan of Fig. 10.4. This could be an indication of receptor cooperativity. The observed differences between PMA, thapsigargin, and ionomycin stimulation most likely stem from the differences in the activation process that results following their application: PMA enhances adhesion through the activation of PKC, while both thapsigargin and ionomycin do so by increasing $[Ca^{2+}_i]$ directly. There still remain many unanswered questions regarding these mechanisms of activation.

ACKNOWLEDGMENTS

This work was supported by grants from the AHA (Florida/Puerto Rico Affiliate), NSF-BITC and the NIH (GM55611).

REFERENCES

Benoit, M., Gabriel, D., Gerisch, G., and Gaub, H. E. (2000). Discrete interactions in cell adhesion measured by single-molecule force spectroscopy. *Nature Cell Biol.* **2**:313–317.

Berry, N., and Nishizuka, Y. (1990). Protein kinase C and T cell activation. *Eur. J. Biochem.* **189**:205–214.

Dustin, M. L., and Springer, T. A. (1991). Role of lymphocyte adhesion receptors in transient interactions and cell locomotion. *Annu. Rev. Immunol.* **9**:27–66.

Heinz, W. F., and Hoh, J. H. (1999). Relative surface charge density mapping with the atomic force microscope. *Biophys. J.* **76**:528–538.

Hoh, J. H., and Schoenenberger, C. A. (1994). Surface morphology and mechanical properties of MDCK monolayers by atomic force microscopy. *J. Cell Sci.* **107**:1105–1114.

Hughes, P. E., Diaz-Gonzalez, F., Leong, L., Wu, C., McDonald, J. A., Shattil, S. J., and Ginsberg, M. H. (1996). Breaking the integrin hinge. A defined structural constraint regulates integrin signaling. *J. Biol. Chem.* **271**:6571–6574.

Hutter, J. L., and Bechhoefer, J. (1993). Calibration of atomic-force microscope tips. *Rev. Sci. Instrum.* **64**:1868–1873.

Hynes, R. O. (1992). Integrins: Versatility, modulation, and signaling in cell adhesion. *Cell* **69**:11–25.

Israelachvili, J. N. (1992). Intermolecular and Surface Forces, Academic Press, London.

Jones, S. L., Wang, J., Turck, C. W., and Brown, E. J. (1992). A role for the actin-bundling protein L-plastin in the regulation of leukocyte integrin function. *Proc. Natl. Acad. Sci. USA* **95**:9331–9336.

Kolanus, W., Nagel, W., Schiller, B., Zeitlmann, L., Godar, S., Stockinger, H., and Seed, B.

(1996). Alpha L beta 2 integrin/LFA-1 binding to ICAM-1 induced by cytohesin-1, a cytoplasmic regulatory molecule. *Cell* **86**:233–242.

Kuhlman, P., Moy, V. T., Lollo, B. A., and Brian, A. A. (1991). The accessory function of murine intercellular adhesion molecule-1 in T lymphocyte activation. Contributions of adhesion and co-activation. *J. Immunol.* **146**:1773–1782.

Lollo, B. A., Chan, K. W., Hanson, E. M., Moy, V. T., and Brian, A. A. (1993). Direct evidence for two affinity states for lymphocyte function-associated antigen 1 on activated T cells. *J. Biol. Chem.* **268**:21693–21700.

Lu, C., Shimaoka, M., Zang, Q., Takagi, J., and Springer, T. A. (2001). Locking in alternate conformations of the integrin alpha Lbeta 2 I domain with disulfide bonds reveals functional relationships among integrin domains. *Proc. Natl. Acad. Sci. USA* **98**:2393–2398.

Lupher, M. L., Jr., Harris, E. A., Beals, C. R., Sui, L. M., Liddington, R. C., and Staunton, D. E. (2001). Cellular activation of leukocyte function-associated antigen-1 and its affinity are regulated at the I domain allosteric site. *J. Immunol.* **167**:1431–1439.

Marlin, S. D., and Springer, T. A. (1987). Purified intercellular adhesion molecule-1 (ICAM-1) is a ligand for lymphocyte function-associated antigen 1 (LFA-1). *Cell* **51**:813–819.

McDowall, A., Leitinger, B., Stanley, P., Bates, P. A., Randi, A. M., and Hogg, N. (1998). The I domain of integrin leukocyte function-associated antigen-1 is involved in a conformational change leading to high affinity binding to ligand intercellular adhesion molecule 1 (ICAM-1). *J. Biol. Chem.* **273**:27396–27403.

Monks, C. R., Freiberg, B. A., Kupfer, H., Sciaky, N., and Kupfer, A. (1998). Three-dimensional segregation of supramolecular activation clusters in T cells. *Nature* **395**:82–86.

Radmacher, M., Fritz, M., Kacher, C. M., Cleveland, J. P., and Hansma, P. K. (1996). Measuring the viscoelastic properties of human platelets with the atomic force microscope. *Biophys. J.* **70**:556–567.

Rothlein, R., and Springer, T. A. (1986). The requirement for lymphocyte function-associated antigen 1 in homotypic leukocyte adhesion stimulated by phorbol ester. *J. Exp. Med.* **163**:1132–1149.

Sanchez-Madrid, F., Simon, P., Thompson, S., and Springer, T. A. (1983). Mapping of antigenic and functional epitopes on the alpha- and beta-subunits of two related mouse glycoproteins involved in cell interactions, LFA-1 and Mac-1. *J. Exp. Med.* **158**:586–602.

Shimaoka, M., Lu, C., Palframan, R. T., von Andrian, U. H., McCormack, A., Takagi, J., and Springer, T. A. (2001). Reversibly locking a protein fold in an active conformation with a disulfide bond: Integrin alpha L I domains with high affinity and antagonist activity *in vivo*. *Proc. Natl. Acad. Sci. USA* **98**:6009–6014.

Stewart, M. P., McDowall, A., and Hogg, N. (1998). LFA-1-mediated adhesion is regulated by cytoskeletal restraint and by a Ca^{2+}-dependent protease, calpain. *J. Cell Biol.* **140**:699–707.

Springer, T. A. (1990). Adhesion receptors of the immune system. *Nature* **346**:425–434.

Stossel, T. P. (1989). From signal to pseudopod. How cells control cytoplasmic actin assembly. *J. Biol. Chem.* **264**:18261–18264.

Thastrup, O., Cullen, P. J., Drobak, B. K., Hanley, M. R., and Dawson, A. P. (1990). Thapsigargin, a tumor promoter, discharges intracellular Ca^{2+} stores by specific inhibition of the endoplasmic reticulum Ca^{2+}-ATPase. *Proc. Natl. Acad. Sci. USA* **87**(7):2466–2470.

Thomas, D., and Hanley, M. R. (1994). Pharmacological tools for perturbing intracellular calcium storage. *Methods Cell Biol.* **40**:65–89.

Tsujinaka, T., Kajiwara, Y., Kambayashi, J., Sakon, M., Higuchi, N., Tanaka, T., and Mori, T. (1988). Synthesis of a new cell penetrating calpain inhibitor (calpeptin). *Biochem. Biophys. Res. Commun.* **153**(3):1201–1208.

van Kooyk, Y., and Figdor, C. G. (2000). Avidity regulation of integrins: The driving force in leukocyte adhesion. *Curr. Opin. Cell Biol.* **12**:542–547.

Willemsen, O. H., Snel, M. M., Cambi, A., Greve, J., De Grooth, B. G., and Figdor, C. G. (2000). Biomolecular interactions measured by atomic force microscopy. *Biophys. J.* **79**:3267–3281.

Wojcikiewicz, E. P., Zhang, X., Chen, A., and Moy, V. T. (2003). Contributions of molecular binding events and cellular compliance to the modulation of leukocyte adhesion. *J. Cell Sci.* **116**:2531–2539.

Wojcikiewicz, E. P., Zhang, X., and Moy, V. T. (2004). Force and compliance measurements on living cells using atomic force microscopy (AFM). *Biol. Proc. Online* **6**:1–9.

Wu, H. W., Kuhn, T., and Moy, V. T. (1998). Mechanical properties of L929 cells measured by atomic force microscopy: Effects of anticytoskeletal drugs and membrane crosslinking. *Scanning* **20**:389–397.

Zhang, X., Wojcikiewicz, E., and Moy, V. T. (2002). Force spectroscopy of the leukocyte function-associated antigen-1/intercellular adhesion molecule-1 interaction. *Biophys. J.* **83**:2270–2279.

Zhou, X., and Li, J. (2000). Macrophage-enriched myristoylated alanine-rich C kinase substrate and its phosphorylation is required for the phorbol ester-stimulated diffusion of beta 2 integrin molecules. *J. Biol. Chem.* **275**:20217–20222.

CHAPTER 11

RESOLVING THE THICKNESS AND MICROMECHANICAL PROPERTIES OF LIPID BILAYERS AND VESICLES USING AFM

GUANGZHAO MAO AND XUEMEI LIANG
Department of Chemical Engineering and Materials Science, Wayne State University, 5050 Anthony Wayne Drive, Detroit, Michigan

11.1. INTRODUCTION

Lipid bilayers and vesicles serve as models for biomembranes and cells (Blumenthal et al., 2003) and show increasing applications in medical and nonmedical fields (Barenholz, 2001; Lasic, 1998). A number of methods have been established to make two-dimensional (2-D) lipid bilayers (Cooper, 2004): (1) supported lipid bilayers by vesicle fusion or Langmuir–Blodgett (LB) dip coating, (2) tethered bilayer membranes by hydrophilic linkers or biotinylated receptors, (3) polymer-supported lipid layers, (4) micro-arrayed lipid layers, and (5) black lipid membranes spanning the holes of porous supports. Small unilamellar vesicles can be prepared from multilamellar vesicle dispersions by either sonication (Huang, 1969) or extrusion (Hope et al., 1985). The study of structure, morphology, and stability of lipid bilayers and vesicles is important to the understanding of membrane fusion as well as in drug delivery, gene therapy, and biosensor design.

AFM images a surface by scanning a sharp tip attached to a cantilever at a close distance to the surface (Quate, 1994). AFM is one of the newest and most important tools for biomembrane analysis because it provides unrivaled molecular-level understanding of structure, stability, and layer interactions. AFM provides surface topographical images with spatial resolution close to 1 Å and force–distance curves with detection limit close to 10^{-12} N. AFM has become the preferred method for imaging soft materials such as molecular crystals, proteins, and living cells (Radmacher et al., 1992). The ideal way to image lipid layers and biomembranes is to conduct scanning in the natural solution environment of the constituents. Minimizing image force to below a threshold breakthrough force further ensures minimum disturbance of the surface structure. Force measurement has become an indispensable part of AFM analysis of biomembranes and biomaterials. AFM force curves provide information on the structure and surface charges, stability, and surface interactions of lipid layers in addition to the bilayer thickness. AFM imaging and force measurement are combined to provide a micromechanical map of the biomembrane or cell

---

Force Microscopy: Applications in Biology and Medicine, edited by Bhanu P. Jena and J.K. Heinrich Hörber.
Copyright © 2006 John Wiley & Sons, Inc.

181

surface in force mapping or force–volume imaging (A-Hassan et al., 1998; Heinz and Hoh, 1999).

This chapter describes various experimental methods used to determine the structure of lipid layers, specifically the bilayer thickness and elastic constants using EggPC as an example.

11.2. SUPPORTED LIPID BILAYER (SLB) THICKNESS

The structure and stability of SLBs with or without biological or synthetic additives have been studied extensively by AFM (Dufrêne and Lee, 2000). The ideal way to image fluid-like lipid bilayers with the least disturbance is the soft contact or tapping imaging mode, which uses the standard silicon nitride (Si_3N_4) tips in the natural solution environment. After a solid substrate, such as mica, graphite, glass, silicon wafer with or without surface modification, is brought to a close distance to the AFM tip, the lipid vesicle solution is injected directly into an AFM liquid cell through inlet tubing. The liquid cell is often sealed by an O-ring. Sometimes an open-cell configuration is preferred. SLBs are often formed *in situ* as a result of vesicle rupture and fusion after sufficient incubation time. Excess vesicles can be removed by flushing the liquid cell with buffer solutions. A different pH or salt solution can be introduced during measurement by solution exchange through the inlet/outlet tubing.

Bilayer thickness can be determined by the step height at the bilayer domain edges. An example of bilayer thickness determination is given in Fig. 11.1 of egg yolk phosphatidylcholine or EggPC (Mao et al., 2004; Liang et al., 2004a). Multilamellar EggPC vesicle solution is extruded through a polycarbonate membrane with an average pore size of 200 nm using a LiposoFast extruder from Avestin. The extrusion method produces larger vesicles that rupture into bilayers upon adsorption to produce the EggPC SLB. AFM imaging and force measurement are conducted using a Nanoscope IIIa AFM (Digital Instruments) and an E scanner (maximum scan area = 14.2 × 14.2 μm^2). AFM sectional height analysis of the SLB step edges in water yields a value of 6.3 ± 0.6 nm. The bilayer film thickness agrees with the hydrated bilayer thickness measured by x-ray diffraction (McIntosh et al., 1987). The value is larger than the headgroup-to-headgroup bilayer distance (~4 nm) due to hydration of the headgroup. Sackmann (1996) points out that freely supported lipid–protein bilayers

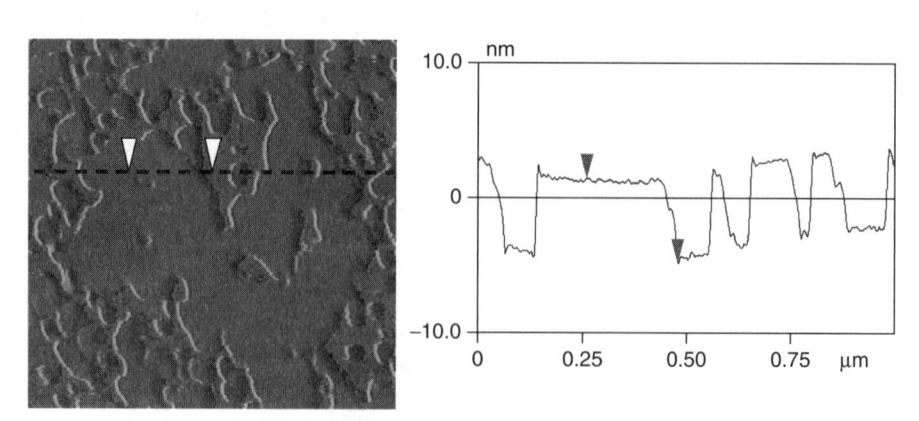

Figure 11.1. The AFM amplitude image and sectional height analysis of the corresponding height image of EggPC SLB patches on mica in water. The thickness of the EggPC bilayer patches is 6.27 ± 0.57 nm, which agrees with 6.58 ± 0.37 nm from force-curve analysis.

are separated from the substrate by a 1-nm-thick water layer. ^{1}H-NMR estimates that the average thickness of the water layer between the single bilayer and the glass bead surface is 1.7 ± 0.5 nm (Bayerl and Bloom, 1990).

In addition to three-dimensional (3-D) topographical imaging, AFM enables direct force measurements between the tip and surface by moving an AFM tip up and down at one point on the sample surface. The force–distance curves, or short force curves, yield not only bilayer thickness but also its micromechanical properties. It has been shown that film thickness from force curves matches precisely the value from sectional height analysis of bilayer step edges if similar image force is used in both cases. AFM force calibration plots are converted to force curves by defining the point of zero force and the point of zero separation (Prater et al., 1995). Zero force is determined by identifying the region at a large separation, where the deflection is constant. Zero separation is determined from the constant compliance region at high force where deflection is linear with the expansion of the piezoelectric crystal or by the end of the jump-in process. Discontinuity, called the *jump-in point*, has been a typical feature in force curves measured on lipid bilayers and adsorbed surfactant films above a threshold force (Patrick et al., 1997; Dong and Mao, 2000; Dufrêne et al., 1998; Loi et al., 2002; Franz et al., 2002). This threshold force has been used as the upper limit of the image force in the soft-contact AFM imaging mode most useful for organic, polymeric, and biological samples (Manne et al., 1994). The jump-in process during tip approach is due to spring instability. The process corresponds to the rupture and removal of the bilayer portion from the tip/substrate gap, most probably by a lateral push-out mechanism. The threshold force (or maximum steric barrier) is reported to be proportional to the surface excess of the film, and it can be used to compare packing density within similar adsorption class (Eskilsson et al., 1999).

Figure 11.2 is a representative force curve measured on the EggPC SLB. Only force values obtained with the same AFM tip are compared. The radius of the contact tip (= 33.2 ± 6.6 nm) is calibrated by imaging the TGT01 gratings (Mikro-Masch) (Villarrubia,

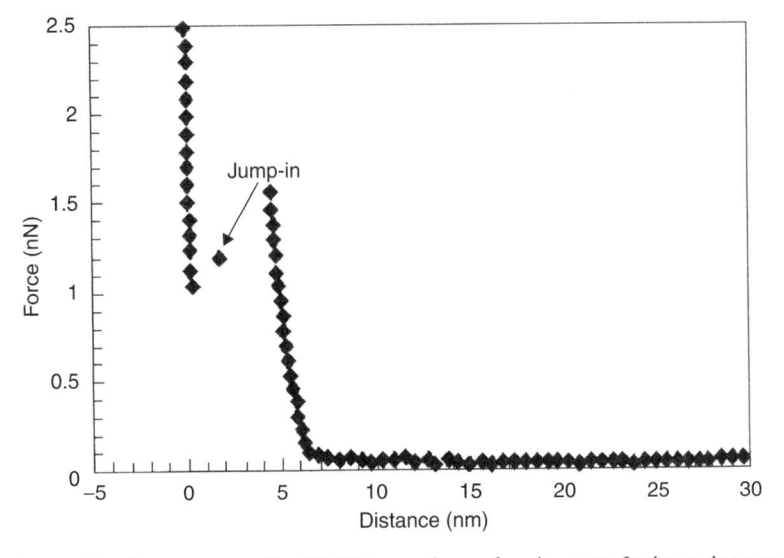

Figure 11.2. Approaching force curve on EggPC SLB on mica surface in water. It shows that a repulsive force starts at 6.58 ± 0.37 nm and the jump-in distance near 4.57 ± 0.27 nm on the EggPC bilayer patch.

1997). The spring constant of the cantilever is calibrated using the deflection method against a reference cantilever (Park Scientific Instruments) of known spring constant (0.157 N/m) (Tortonese, 1997). The calibrated spring constant (0.17 ± 0.05 N/m) is used. Force curves are obtained in liquid contact mode only. Multiple force curves are obtained on the same SLB. In Fig. 11.2, the repulsion starts at 6.6 ± 0.4 nm and the jump-in occurs at 4.6 ± 0.3 nm with a maximum force = 1.6 ± 0.3 nN. The repulsion up to the jump-in point can be described as a combination of hydration, steric force, and mechanical deformation for bilayer films. The force curves measured on the lower, flat background in both extruded and sonicated vesicle adsorption cases show characteristics of tip–mica interactions in water. It can be concluded that mica is not fully covered by the SLB in this case.

Next let's look at the SLB film thickness variation due to the incorporation of a macromolecular additive as studied by AFM force measurement (Liang et al., 2004b). Four Pluronic copolymers (L81, L121, P85, F87) (BASF) are used to prepare Pluronics-modified EggPC SLBs. The Pluronics are incorporated into EggPC vesicles according to literature procedures (Kostarelos et al., 1995, 1999). The concentration of Pluronics is kept at 0.02% w/w well below the critical micelle concentration (CMC). The nominal structures and the number of EO (hydrophilic) and PO (hydrophobic) units are listed in Table 11.1. Pluronic copolymers are known to enhance the stability of lipid vesicles (Woodle et al., 1995; Kostarelos et al., 1998; Johnsson et al., 1999).

TABLE 11.1. Molecular Structure and Properties of Pluronic® Copolymers Provided by BASF.

Pluronic®	Composition	Molecular Weight
L81	$(PEO)_2(PPO)_{40}(PEO)_2$	~2666
L121	$(PEO)_4(PPO)_{60}(PEO)_4$	~4000
P85	$(PEO)_{26}(PPO)_{39.5}(PEO)_{26}$	4600
F87	$(PEO)_{61.1}(PPO)_{39.7}(PEO)_{61.1}$	7700

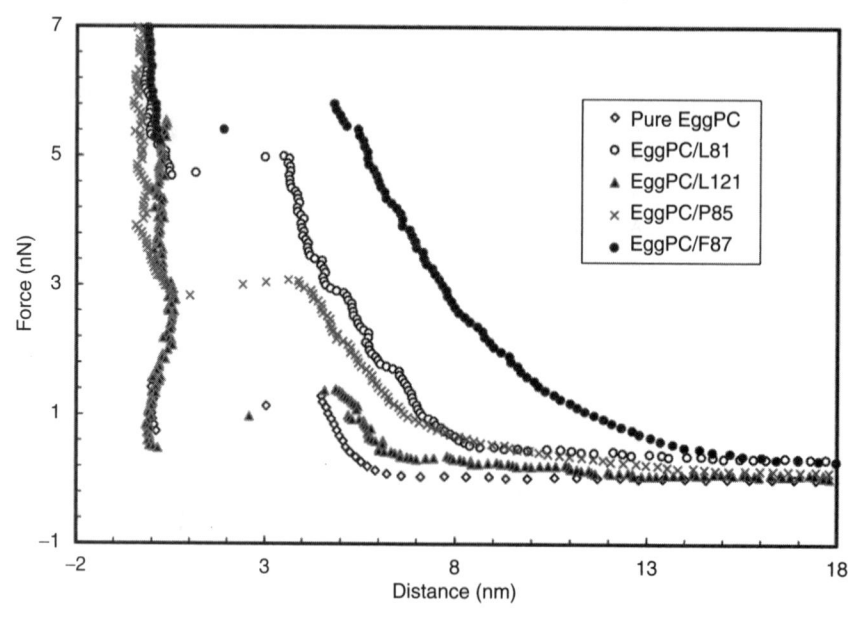

Figure 11.3. Force curves on EggPC, EggPC/L81, EggPC/L121, EggPC/P85, and EggPC/F87 bilayer. The force curves on the bilayer systems are characterized by one repulsive regime and one jump-in point.

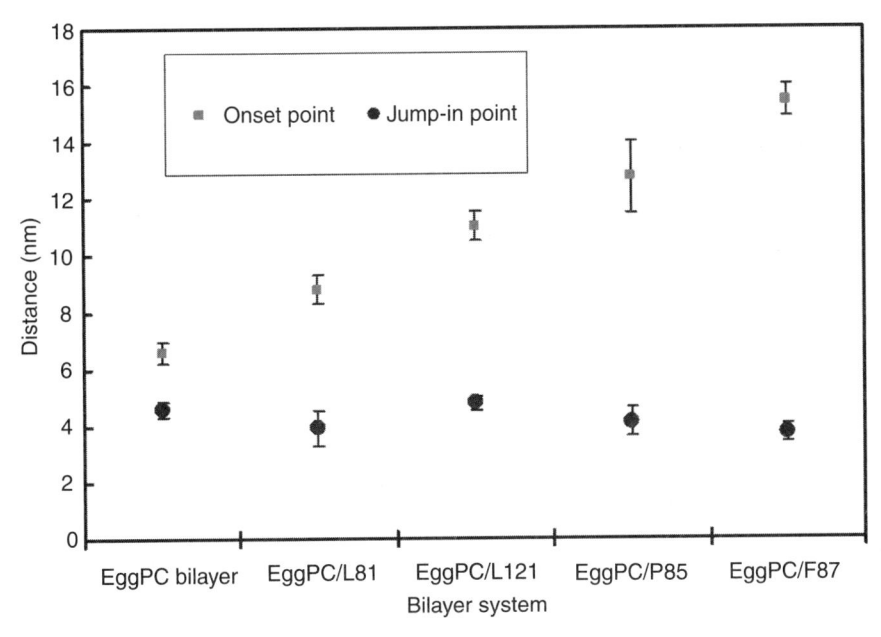

Figure 11.4. Comparison of onset point (square) and jump-in point (circle) distances for EggPC, EggPC/L81, EggPC/L121, EggPC/P85, and EggPC/F87 bilayer films. The error bar is calculated based on the standard deviation of jump-in distance and onset point from a number of force curves.

Fused bilayers are observed for Pluronics with short PEO chain length. Fig. 11.3 compares the force curve of EggPC SLB with those treated with Pluronics (EggPC/L81, EggPC/L121, EggPC/P85, and EggPC/F87). The approaching force curves show typical features of SLBs including the steric repulsion up to a threshold followed by the jump-in to contact. The distances of the repulsive force onset and jump-in points are plotted in Fig. 11.4. The force onset distance corresponds to an unperturbed bilayer thickness at a small contact force (~0.2 nN). Fig. 11.4 shows that the onset distance increases with increasing PEO chain length, which suggests that Pluronic copolymers have been integrated into the SLB with PEO chain protruding into the solution. Similar chain-length dependence of adsorbed Pluronic copolymer layers on a self-assembled monolayer (SAM) of n-octadecyltrichlorosilane has been found by Wang et al. (2002). A likely structure of the Pluronics-modified SLB is drawn in Fig. 11.5. On the other hand, the jump-in distances remain unaffected by the PEO chain length. This observation is consistent with the interpretation of the force gap as the tip jumping across the interior of the bilayer.

The PEO chains sticking out of the EggPC lipid bilayer can be either random coils or extended chains. Fig. 11.6 compares our experimental data of the force onset distances with calculations based on random coil and fully extended chain configurations, respectively. The following equations are used. Fully extended chain length is $L_f = a \times N$. Random coil chain length is $L_r = a \times N^{3/5}$, where a is the EO monomer chain length (0.35 nm) and N is the number of EO units. Pluronics-modified EggPC bilayer thickness is $L_f^{bilayer} = L_0 + 2L_f$ (nm) if the chain is fully extended and $L_r^{bilayer} = L_0 + 2L_r$ (nm) if the chain is a random coil. L_0 is the undisturbed EggPC bilayer thickness. Fig. 11.6 shows that experimental data are in good agreements with calculated values based on the random coil chain.

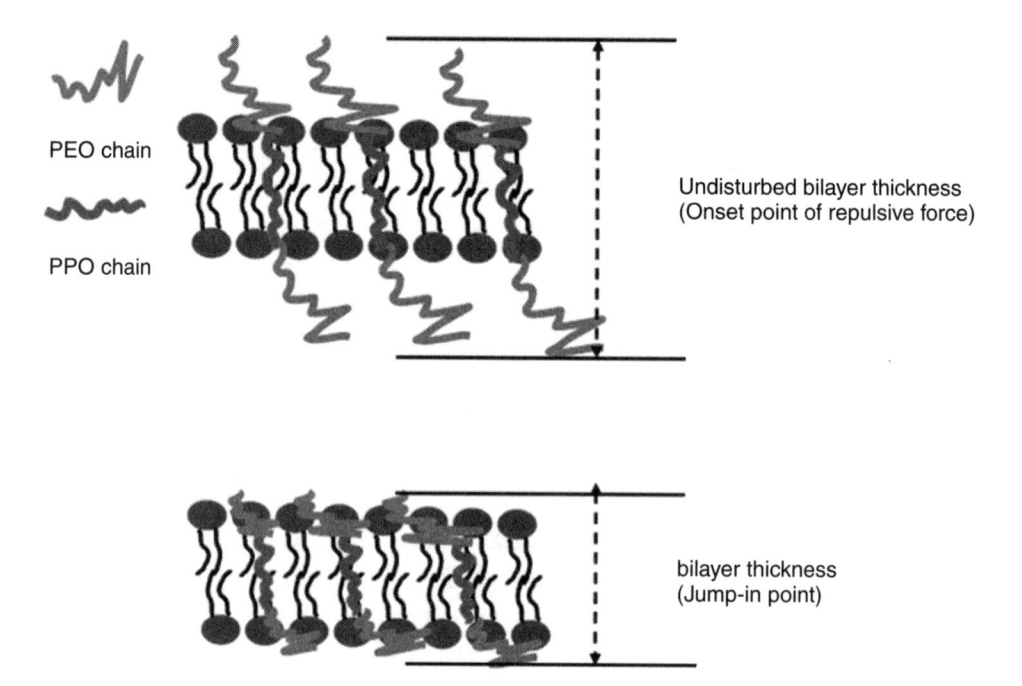

Figure 11.5. (Top) Scheme of onset point of repulsive force on bilayer (undisturbed bilayer thickness). (Bottom) Scheme of jump-in point (bilayer thickness). The onset point of the repulsive force increased with increasing PEO chain length and the jump-in point is around 4–5 nm.

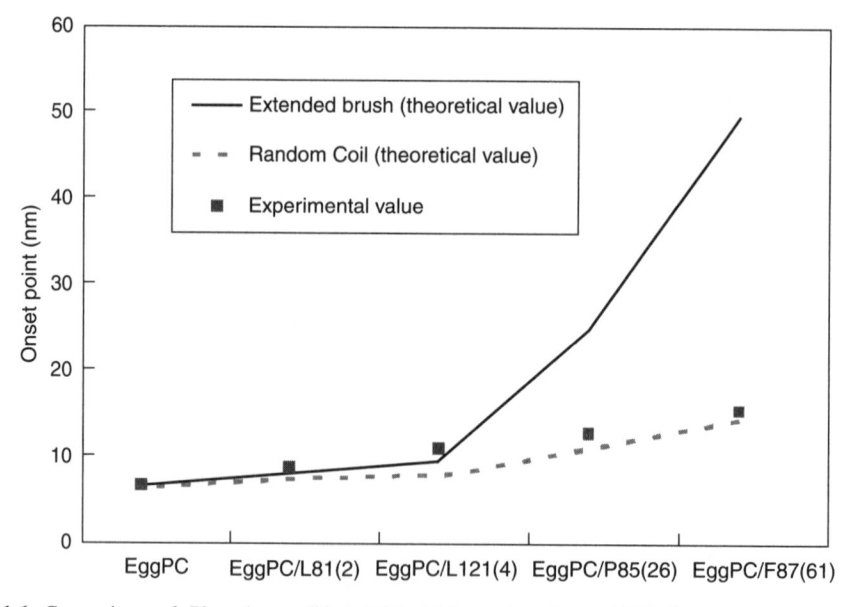

Figure 11.6. Comparison of Pluronics-modified SLB thickness based on AFM force curves and theoretical calculations. The number in the parentheses in x axis is the PEO chain length.

11.3. SONICATED UNILAMELLAR VESICLE (SUV) BILAYER THICKNESS AND MORPHOLOGY

Compared to other micromechanical methods such as shape fluctuation method (Servuss et al., 1976; Schneider et al., 1984a, b), magnetic-field-induced orientation (Sakurai and Kawamura, 1983), and micropipette aspiration method (Evans and Rawicz, 1990; Waugh et al., 1992), AFM can provide information on (1) bilayer thickness, (2) adsorbed liposomes of small sizes between 20 and 100 nm, (3) mechanical property variation at nanoscale, (4) local deformation and rupture events induced by the tip, and (5) surface and adhesive forces. AFM measurement focuses on the properties of vesicle population with the smallest sizes because large vesicles rupture into bilayers upon adsorption. Small vesicles with size between 50 and 150 nm have been found to passively target several different tumors because this size range is a compromise between loading efficiency (which increases with increasing size), stability (which decreases with increasing size), and ability to extravasate in tissues with enhanced permeability (which decreases with increasing size) (Lasic, 1998). Size control may be a simple way for targeted drug delivery because vesicle distribution among different organs and tissues is related to its size.

Compared to studies on SLBs, much fewer articles address the structure of adsorbed yet unruptured vesicles. Shape instability, size, and softness of vesicles often prevent unambiguous AFM imaging. Perhaps the earliest images of intact liposomes are obtained by Shibata-Seki et al. (1996) from dipalmitoyl phosphatidylcholine (DPPC) and cholesterol mixture. Others subsequently confirm the characteristic features of adsorbed vesicles as first reported. The features include (1) flattened spheres with length to width ratio less than 1 (about 0.4 in their case) and (2) image quality dependence on contact pressure either by changing image force or using tips of different curvature.

Egawa and Furusawa (1999) reported a conical relief image of sonicated phosphatidylethanolamine (PE) vesicles with diameter around 100 nm adsorbed on its own bilayer. The conical image has been attributed to unruptured yet flattened vesicle with height to diameter ratio of 0.175. Thomson et al. (2000) imaged liposomes 70 nm in diameter made from p-ethyldimyristoyl phosphatidylcholine (EDMPC) and cholesterol mixture on aminopropylsilane modified mica. Closely packed liposomes have been imaged but can be easily disturbed by the scanning process. The article by Raviakine and Brisson (2000) provides many details about vesicle adsorption and fusion by AFM imaging. In the article, supported vesicular layer (SVL) is used to differentiate adsorbed and intact vesicles from adsorbed and single-bilayer disks as in SLBs. It is found that vesicles of all sizes adsorb on mica, but only vesicles with sizes below a critical rupture radius remain intact. The bilayer disks exhibit constant height of 5 nm, while the intact vesicles exhibit height variation from 10 to 40 nm. The shape of liposomes are found to change both with applied image force and adhesion between biotin and streptavidin that are incorporated into vesicle and substrate layer, respectively (Pignataro et al., 2000). Kumar and Hoh (2000) reported intact vesicles of phospholipid and cholesterol mixture that exhibited saucerlike structure in addition to the usual rounded protrusion (domelike structure). The saucerlike structure is attributed to wetting and fusion near the edge of the vesicle. A force curve is obtained on the intact vesicle that shows monotonic repulsion with onset at 15 nm. The flattening from the outer edges toward the center has been studied as a function of time by Jass et al. (2000) and is described as the first step in a multistep processes leading toward the formation of SLBs.

Here we describe an AFM study of EggPC SUVs with diameter less than 50 nm, adsorbed on mica (Mao et al., 2004; Liang et al., 2004). AFM tip is moved on top of individual vesicles so that mechanical properties of the smallest vesicles can be measured at the nanoscale. The EggPC vesicle undergoes reversible shape changes from convex to flattened and concave shape with increasing image force. In addition to the monotonic repulsion due to vesicle resistance to the AFM tip advancement and compression, there exist several characteristic breaks in the force versus distance curves. Hertzian analysis of the slope of the repulsion gives a measure of the vesicle elastic properties. The breaks in the force curve are interpreted as the tip jumping across the bilayer and allow the determination of bilayer thickness.

A well-established recipe is used to prepare EggPC SUVs (Huang, 1969). Multilamellar vesicle (MLV) solution is obtained by dissolving appropriate amounts of EggPC lipids in chloroform/methanol (2:1 v/v) and evaporating the solvent with nitrogen. After drying in a desiccator connected to a rotary vacuum pump for 30 min, the lipids are resuspended by stirring them in an aqueous buffer solution (20 mM NaCl) at a concentration of 0.5 mg/ml. SUVs are produced from the MLV suspension by sonication to clarity (about 1 h) in a sonicator bath (Branson 2200). The suspension is kept in an ice bath during the sonication process. Sonicated samples are centrifuged for 1 h at 16,000 rpm to remove large lipid fragments by Sorvall OTD70B Ultraspeed Centrifuge.

Images of EggPC vesicles are obtained in both liquid contact mode and liquid tapping mode. Fig. 11.7 shows the amplitude and deflection images for tapping and contact, respectively, and the cross-sectional height profiles from the corresponding height images. The images consist of spherical objects on a flat background. The lateral width or diameter of the spheres is measured to be 48.6 ± 11.4 nm and 69.3 ± 12.8 nm on the sectional height profiles of tapping and contact, respectively. The diameter is taken from the width of the peak at the baseline in sectional height profiles. The lateral width values are higher than the vesicle diameter in solution 37.0 ± 7.9 nm by dynamic light scattering. The discrepancy is generally attributed to the flattening of vesicles on surface and tip casting its shadow over the object, so-called tip convolution effect. The height of vesicles from sectional height

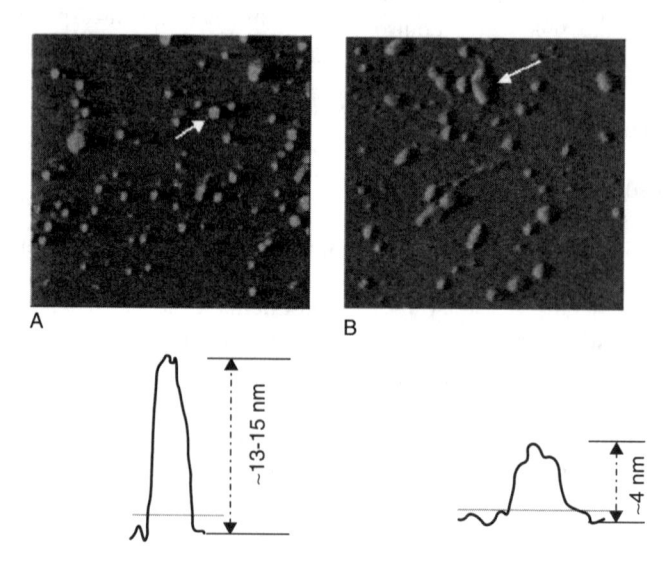

Figure 11.7. AFM images and cross-section profiles of EggPC vesicles on mica by tapping mode (**A**) and contact mode (**B**). The image size is 1 × 1 μm². The height profile belongs to the particle pointed by the arrow obtained in height image. (**A**) Amplitude image with Z scale = 20 nm. (**B**) Deflection image with Z scale = 5 nm.

profiles in Tapping and Contact Mode is determined to be 13.9 ± 2.2 nm and 3.9 ± 0.4 nm, respectively. The height value from tapping mode is about 40% to 50% of the vesicle diameter in solution. Causes for the unreasonably low height values from contact mode include vesicle movement during contact mode imaging and the vesicles being severely compressed by the tip. The intermittent contact motion of the tapping tip is known to reduce frictional and adhesive forces on the sample surface. By minimizing image force, it is possible to obtain stable images of intact vesicles in either tapping or contact mode. Tapping causes less deformation of the vesicle, and it can maintain a smaller image force than contact mode.

With increasing image force, EggPC vesicles exhibit three distinctive morphologies: convex-shaped vesicles where the highest point is at the center as shown in Fig. 11.8A, disk-shaped vesicles with uniform height across much of the vesicle as shown in Fig. 11.8B, and concave-shaped vesicles where the center is depressed as shown in Fig. 11.8C. The lateral diameter of the vesicle increases from 48.6 ± 11.4 nm for convex shape to 73.7 ± 11.5 nm for disk shape and to 89.8 ± 6.6 nm for concave shape. The morphological change is reversible under varying image force. The convex-shaped vesicles are similar to the conical relief and dome-like structure in other studies (Egawa and Furusawa, 1999; Kumar and Hoh, 2000). The disk-shaped vesicles are similar to the saucer-like vesicles (Kumar and Hoh, 2000; Jass et al., 2000). While the saucer-like vesicles are cited as an intermediate state of vesicle fusion due to adhesion between vesicle and substrate, the morphological changes observed here can only be attributed to the compressive tip. The small size of the SUVs means that the adhesion alone is not enough to collapse the vesicle edge to form saucer-like vesicles (Seifert and Lipowsky, 1996). The morphological change due to tip compression is schematically illustrated in Fig. 11.9. A convex shape is obtained in the lowest indentation region of the force curve; a planar shape is obtained at an intermediate indentation, and a concave shape is obtained at the highest indentation. For the concave shape, when the tip moves across the vesicle with an applied force held more or less constant, the amount of indentation is largest at the center of the vesicle. Several factors may contribute to maximum depression at the center:

1. It is known that the mechanical response of thin films couples to

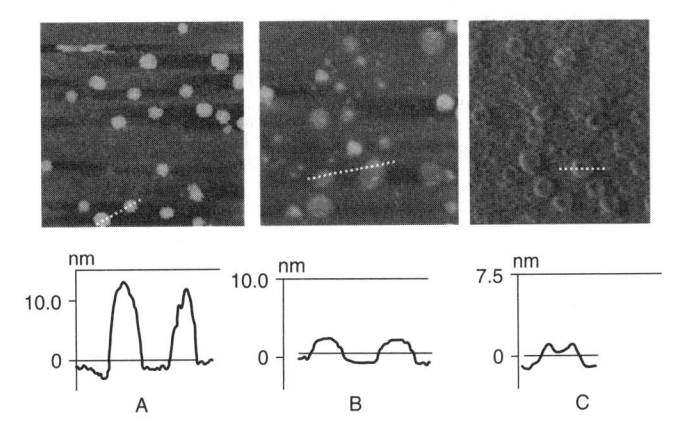

Figure 11.8. AFM height images obtained by tapping mode with different shapes. The image size is 0.75×0.75 μm^2. Z scale = 20 nm. The height profile across the dotted line is given. (**A**) Convex vesicles. (**B**) Planar vesicles. (**C**) Concave vesicles.

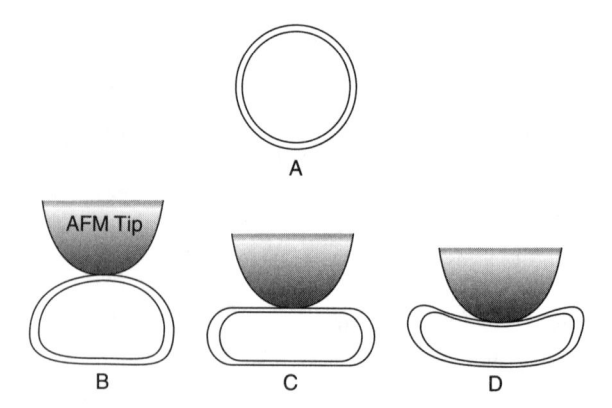

Figure 11.9. Tip compression scheme on vesicle morphology: **(A)** Vesicle in solution. **(B)** Convex vesicle imaged with minimum image force. **(C)** Planar vesicle imaged with intermediate indentation. **(D)** Concave vesicle imaged with highest indentation.

the substrate, which results in an increase (50%) of the apparent modulus (Domke and Radmacher, 1998). The coupling increases with decreasing film thickness. Therefore, the tip indents more in the central region of a sessile droplet.

2. A high volume percentage of material is bound by the surface near the edge of a droplet and spreads less upon compression of the tip. Contrarily, at the center of the vesicle, less resistance exists against squeezing of the trapped liquid portion.

Figure 11.10 represents a typical force curve captured on EggPC vesicles. While it is impossible to place the tip exactly at the center of the vesicle, the force curve with the largest onset repulsion distance was selected from a set of force curves obtained in the vicinity of a vesicle, and it is used to represent the interaction between the apexes of AFM tip and vesicle. The onset force distance of 32.2 nm in Fig. 11.10 falls in

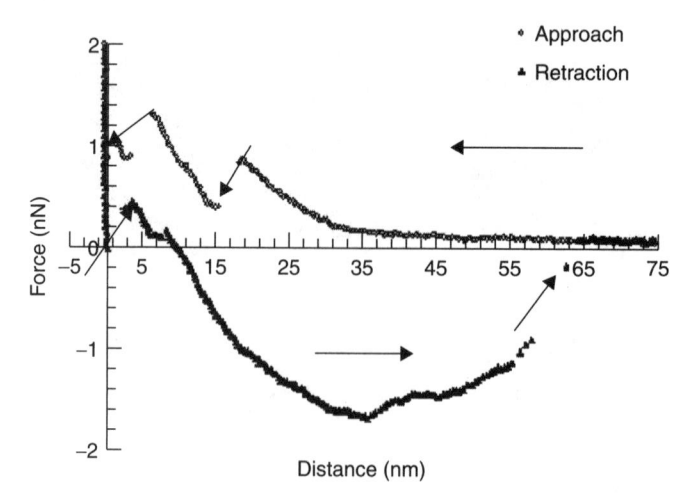

Figure 11.10. Force versus distance curves on the EggPC vesicle. Two jump-in events during approach and two jump-out events during retraction are marked by arrows.

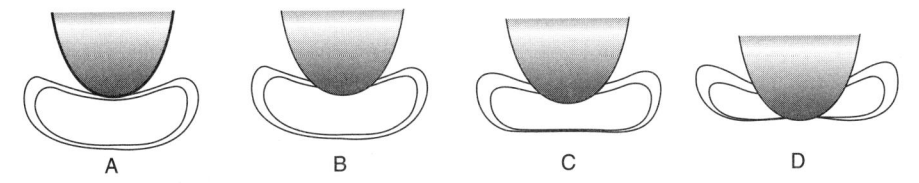

Figure 11.11. Scheme of tip effect on vesicle during approach force measurement. (**A**) Elastic compression of vesicle top fraction. (**B**) Tip penetration of the upper bilayer. (**C**) Further compression with tip bridging the gap. (**D**) Tip penetration of the lower bilayer.

the range of measured vesicles size between 30 and 40 nm. The repulsion is not continuous, but displays characteristic breaks. During approach, one or two breaks are observed, around 19 and 7 nm respectively. While the exact locations of the breaks vary somewhat, the gap distance between the beginning and the end point of the jump-in is constant at 4.8 ± 0.4 nm. During retraction, similar jump-out gap is also observed though apparently not as well defined. This unique gap distance coincides with lipid bilayer thickness.

We hypothesize that during approach, the bilayer portion at the top of vesicle gives away to the pressing tip and slides aside, followed by continuous compression of the tip on the vesicle, until tip approaches and breaks through the bottom portion of the bilayer enclosure. It is surprising that vesicle maintains its enclosed shape while the tip bridges across different parts of the bilayer shell. The stepwise deformation during approach is shown in Fig. 11.11: (1) compression of vesicle, (2) breakthrough of top portion, (3) further compression of vesicle with tip bridging top portion, and (4) breakthrough of bottom portion with tip spanning the whole vesicle. The process is reversed during retraction except that the vesicle is stretched to 150% of its intrinsic diameter before the final detachment of tip from vesicle. This adhesive interaction may also contribute to the shape stability of vesicle bridged by the AFM tip. Vesicles are known to self-heal after perforation due to high line tension.

11.4. SUV MICROMECHANICAL PROPERTIES

The mechanical properties of phospholipid membranes can be related to many vesicle behaviors, such as their formation, stability, size, shape, fusion, and budding processes. Bending rigidity is a fundamental and characteristic mechanical property of vesicles and is related to the stability and strength of bilayer (Svetina and Zeks, 1996).

Force curves can be used to extract quantitative micromechanical constants of SLBs and SUVs. The force curves can be fitted to the Hertzian model $\delta = AF^b$ (δ is the indentation on the vesicle and F is the load force) by assuming a spherical shape for the tip. The indentation, δ, from the difference between the cantilever distance $z - z_0$ and cantilever deflection $d - d_0$ is described in Eq. (11.1):

$$|z - z_0| - (d - d_0) = \delta = A(d - d_0)^{2/3}$$
$$= 0.825 \left[\frac{k^2 (R_{tip} + R_{ves})(1 - \upsilon_{ves}^2)^2}{E_{ves}^2 R_{tip} R_{ves}} \right]^{1/3}$$
$$\times (d - d_0)^{2/3} \quad (11.1)$$

E_{ves} is the Young's modulus of the vesicle, R_{tip} and R_{ves} are the radii of the tip and vesicle, respectively, υ_{ves} is the Poisson's ratio of the vesicle, and k is the cantilever spring constant. υ_{ves} is assumed to be 0.5 (Laney et al., 1997; Radmacher et al., 1996). The spring constant and tip radius are calibrated to be 0.17 N/m and 33 nm, respectively. R_{ves} is taken to be equal to $z_0/2$. Experimental data can be fitted to

Eq. (11.1) with two fitting parameters A and z_0. In general, the fitted z_0 value is found to be consistent with the visually examined contact point. Thus, z_0 in our experiment is determined by the onset point of the repulsive force. E_{ves} is then computed from A by least-square fitting.

Bending modulus k_c is deduced from Young's modulus based on Eq. (11.2) (Evans, 1974):

$$k_c = \frac{E_{ves} h^3}{12(1 - \upsilon_{ves}^2)} \quad (11.2)$$

h is the bilayer thickness.

Laney et al. (1997) have calculated elastic properties from averaged data of approach and retraction force curves, and they discard the force plots with discontinuities. The approach part of the force curves is suitable for the indentation calculation because significant adhesive forces in retraction can affect the measurement of indentation (Vinckier and Semenza, 1998). Generally, the force curves on vesicles are characterized by two repulsive regimes.

Figure 11.12A shows a typical deflection versus z position plot on a vesicle containing cholesterol (EggPC 80:cholesterol 20) (Liang et al., 2004c). Fig. 11.12B is a force curve converted from Fig. 11.12A by defining points of zero force and zero separation. The force curve can be divided into four regions as labeled. In region I, the noncontact region, the tip is far away from vesicle and the force between the tip and the vesicle is zero. Region II illustrates the elastic deformation of the vesicle under tip compression and therefore can be used to calculate Young's modulus. Region III corresponds to further tip compression after the tip penetrates the vesicle's top bilayer. Region IV reflects the cantilever deflection when it is in contact with the hard mica substrate after penetrating through the vesicle's bottom bilayer. Theoretically, the slope of region IV in Fig. 11.12A should be 1.0 because the deflection of the cantilever is identical to the z direction movement of sample on hard surface (Weisenhorn et al., 1993). Based on the fitted data, a slope of 0.9967 ± 0.0036 is obtained.

By fitting our experimental data (region II and region III) to the Hertzian model we obtain exponent b of 0.6632 (region II) and 0.9227 (region III), respectively. The poor fit on the high loading force (region III) ($b = 0.9227$) suggests that the Hertzian model severely fails in region III. The exponent $b = 0.6632$, which is close to the 2/3 in region II, is remarkable. Although the Hertzian model describes the contact between two solid bodies on the basis of continuum elasticity theory without adhesion force (Radmacher et al., 1994), the good fit in region II suggests that Hertzian model may also be applicable to describing the elastic deformation between the tip and the vesicle within the limit of small indentation. Region II illustrates the elastic deformation of the vesicle under tip compression, and it is used in subsequent calculations for determining the micromechanical properties of the vesicle. Fig. 11.12C is the indentation versus load force converted from data in Fig. 11.12A, region II. It shows that in the beginning of the compression (indentation), the Hertz model can simulate the experimental findings very well. A deviation from the model is observed at high load force (larger indentation, the second repulsive regime), illustrating the limitation of the model.

Young's modulus is calculated using $R_{tip} = 33$ nm, $R_{ves} = z_0/2$ (the onset point of repulsive force is regarded as the size of the vesicle), $\upsilon_{ves} = 0.5$, and $k = 0.17$ N/m. Bending modulus is a characteristic property of the vesicles, which is closely related to the activities of liposomes and the gel–liquid phase transition of liposome's bilayer membrane. According to solid-state mechanics, Young's modulus is related to bending modulus as $E_{ves} = k_c / I$ where I is the cross-sectional moment. The value I for a

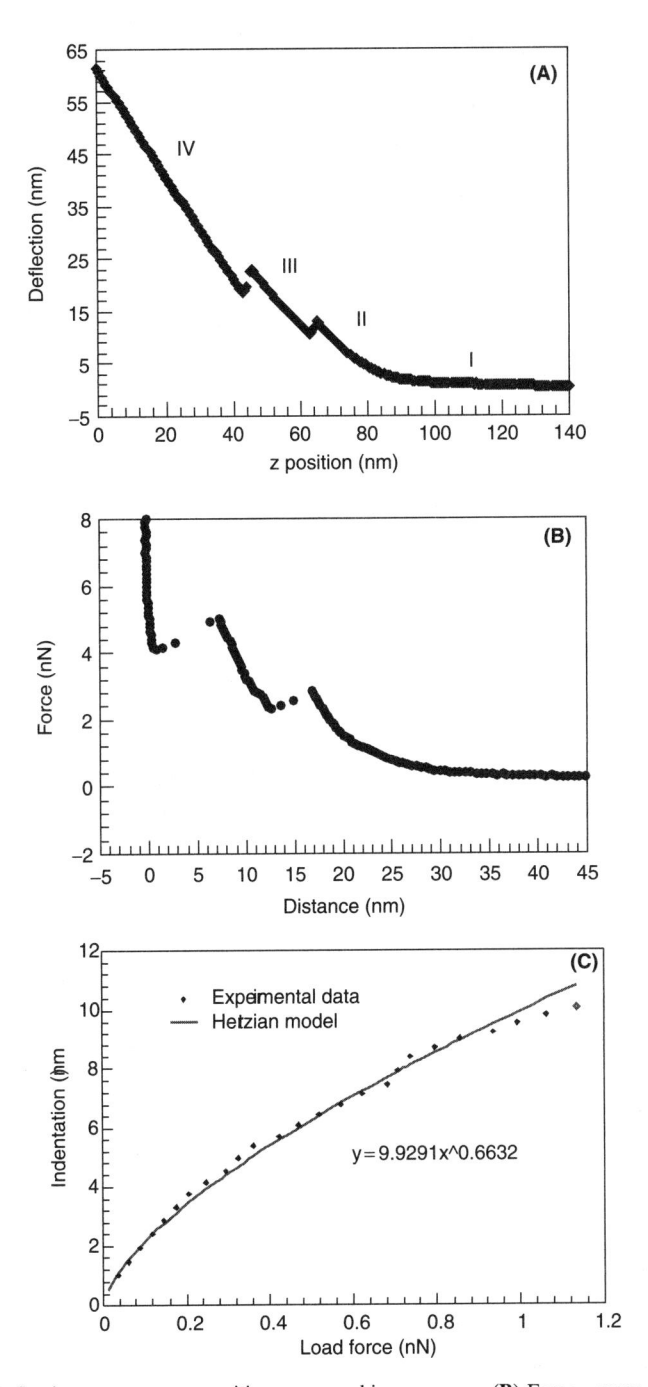

Figure 11.12. (**A**) Deflection versus z position approaching curve. (**B**) Force curve converted Part A. (**C**) Force-curve data fit with Hertzian model for region II in Part A. The squares are experimental data. Experimental data can be described by power equation $\delta = 9.9291 F^{0.6632}$. The solid line is the Hertzian model $\delta = AF^{2/3}$ with $A = 9.9291$. The measured compression (indentation) versus loading force agrees with the Hertzian model in the case of the first repulsive force region.

three-dimensional, isotropic planar surface is $h^3/[12(1 - \upsilon^2)]$, where h is the thickness of the bilayer and υ is Poisson's ratio (Laney et al., 1997; Radmacher et al., 1996; Evans, 1974). The bending modulus k_c is calculated using Eq. (11.2) with $h = 4.57 \times 10^{-9}$ m and $\upsilon = 0.5$. The Young's modulus and the bending modulus for pure EggPC are found to be 1.97 ± 0.75 MPa and $(0.21 \pm 0.08) \times 10^{-19}$ J, respectively. The calculated values are compared with literature values in Tables 11.2 and 11.3. The Young's modulus of the EggPC vesicle from force curves is one order of magnitude smaller than the reported value ($\sim 10^7$ MPa) (Hantz et al., 1986). The discrepancy is probably due to different measurement environment (Mao et al., 2004).

TABLE 11.2. Young's Modulus of Biological Samples.

Material	Young's Modulus E (MPa)	Method	Remarks	References
Synaptic vesicles	0.2–1.3	Force mapping	Size: 90–150 nm adsorbed on mica	38a Laney et al. (1997)
DMPC vesicles	15	Osmotic swelling	Size: 160–180 nm in solution	44 Hantz et al. (1986)
DOPC vesicles				
EggPC vesicles	1.97 ± 0.75	Force plot	Size: < 60 nm adsorbed on mica	Mao et al. (2004), Liang et al., (2004a)

[a] DMPC, dimyristoylphosphatidylcholine; DOPC, dioleoylphosphatidylcholine.

TABLE 11.3. Comparison of Bending Modulus of Egg Yolk Phosphatidylcholine.

Method	Size/Shape	Bending Modulus $k_c(\times 10^{-19}$ J$)$	T (°C)	References
Phase contrast microscopy	Long unilamellar tubular vesicle (11 μm < L <34 μm, 17 < L/r < 83)	2.3 ± 0.3	22.0	Servuss et al. (1976)
	Cylindrical vesicle (>10 μm)	1–2	25	Schneider et al. (1984a)
	Quasi-spherical vesicle (>10 μm)	1–2		Schneider et al. (1984b)
	Spherical (>10 μm)	0.4–0.5		46a, b
Magnetic-field-induced orientation	Cylindrical rods (5 to 30-μm diameter, <200 μm long)	0.4	25	Sakurai and Kawamura (1983)
AC electric field	Spherical vesicle (diameter >20 μm)	0.247		Kummrow and Helfrich (1991)
	Spherical vesicle (~15 to 70-μm diameter)	$0.66 \pm 0.06, 0.45 \pm 0.05$		Angelova et al. (1992)
AFM force curve	Spherical vesicle (diameter < 60 nm) on mica substrate	0.21 ± 0.08	22 ± 1	Mao et al. (2004), Liang et al. (2004a)

The calculated bending modulus is in the same range as that reported in literature between 10^{-19} and 10^{-20} J (Table 11.3). Variations in bending modulus measured have been reported, and the reason for the discrepancy still needs further analysis (Kummrow and Helfrich, 1991; Angelova et al., 1992; Niggemann et al., 1995).

Table 11.4 lists the bending moduli of pure, cholesterol-modified, and Pluronics-modified EggPC vesicles. The data show that EggPC vesicles are stiffened by adding cholesterol and Pluronic copolymers.

There is a significant increase in bending modulus when cholesterol is incorporated into EggPC vesicles. It has been reported that the phospholipid bilayer packing geometrical structures are changed by cholesterol insertion and thus by changing fluidity and intravesicle interaction (Liu et al., 2000). After the cholesterol is incorporated into phospholipid bilayers, the small hydrophilic 3β-hydroxyl headgroup of cholesterol is located in the vicinity of the lipid ester carbonyl groups, and the hydrophobic steroid ring orients itself parallel to the acyl chains of the lipid (Ladbrooke et al., 1968). Thus, the movement of the acyl chains of the phospholipid bilayer has been restricted. Below the gel phase transition temperature (T_m) of lipids, cholesterol addition increases the fluidity of lipids, while above T_m the mobility and fluidity of the lipid chains are restricted (T_m of EggPC = $-15°C$) (New, 1990). Moreover, hydrogen bonding between cholesterol's β-OH and the carbonyl groups of the lipid enhances the stability of the bilayer (Presti et al., 1982). The interaction between the cholesterol and phospholipid bilayer results in an increase in membrane cohesion, as shown by increases in the mechanical stiffness of the membranes.

The slope of the long-range repulsive force regime in Fig. 11.3 can be used to calculate the elastic properties of the Pluronics-modified EggPC SUVs based on Eqs. (11.1) and (11.2). The slopes of EggPC/F88 and EggPC/F127 are found to be lower than that of EggPC/F108. The slope of EggPC/F127 is close, but slightly higher than that of EggPC/F88. Both F127 and F88 have similar PEO size, while the PPO chain of F127 is longer. Table 11.3 lists the calculated bending modulus values. The bending modulus of Pluronics-modified EggPC SUVs increases by an order of magnitude from that of pure EggPC. The magnitude of the increase is comparable with cholesterol/EggPC vesicles. These data show that the bending moduli of Pluronics-modified EggPC SUVs are a function of PEO and PPO chain lengths. The dramatic increase can be attributed to PPO block incorporation and PEO chain on the surface. For Pluronics-modified SUVs, the physical stability is improved by the PEO steric repulsion (Kostarelos et al., 1998, 1999). The bending modulus obtained on F127 vesicles is $(1.64 \pm 0.28) \times 10^{-19}$ J, slightly larger than that of F88 ($(1.50 \pm 0.40) \times 10^{-19}$ J).

TABLE 11.4. Bending Modulii of Pure, Cholesterol-Modified, and Pluronics-Modified EggPC Vesicles.

Sample	Bending Modulus	References
Pure EggPC	0.27 ± 0.10	Mao et al. (2004), Liang et al. (2004a)
EggPC:Chol (85:15)	1.68 ± 0.21	Liang et al. (2004c)
EggPC:Chol (80:20)	1.49 ± 0.09	
EggPC:Chol (70:30)	1.44 ± 0.56	
EggPC:Chol (50:50)	1.81 ± 0.41	
EggPC/F88	1.50 ± 0.40 (103 EO, 40 PO)	Liang et al. (2004b)
EggPC/F127	1.64 ± 0.28 (100 EO, 65 PO)	
EggPC/F108	2.06 ± 0.38 (133 EO, 50 PO)	

The increase can be attributed to longer PPO chain length of F127. While comparing bending moduli of F127 and F108, the circumstances are somewhat complicated since both have different PPO and PEO chain lengths. The bending modulus of F108 is higher than that of F127, although the PPO chain length of F108 is smaller. However, F108 has longer PEO chain, and stiffer shell-like "coating" on the vesicles seems to contribute to bending modulus increase.

It is well known that the stability of polymer-coating liposome is improved in terms of steric repulsive force between the polymer chains dangling outside of the liposome. In earlier studies, the steric stabilization effect has been described by aggregation and fusion among vesicles in solution (Kostarelos et al., 1998, 1999). Kostarelos et al. (1998) draw a similar conclusion by comparing prolonged vesicle durability against flocculation and ζ-potential change. It has been reported that the bending and mechanical properties of vesicles from diblock polymer exceeds those of phospholipid liposomes by a factor of 5 or more (Discher et al., 1999). Thus, the significant bending modulus improvement can be attributed to PEO (incorporated) and PPO chains (dangling outside of EggPC vesicles). Fig. 11.13 shows a schematic of the steric stabilization effect of copolymer on liposome. A block copolymer may aggregate to form micelles (core/shell structure) consisting of a swollen core of insoluble blocks surrounded by a flexible fringe of soluble block (Tuzar et al., 1993). As an analogy to the polymer micelle, the PEO chains dangling outside the vesicle surface may exhibit a shell-like effect to provide steric stabilization to the vesicle. The shell-like structure also serves as a kind of "net" or "coating" of vesicles (Ringsdorf et al., 1988). With longer PEO chain length (>19 units), a finite steric barrier is formed (Wang et al., 2002). The PPO chain incorporated into the lipid bilayer acts in a similar fashion as cholesterol to restrict the lipid molecular movement and fluidity. For copolymer with longer PEO chain length (F88, F108, and F127),

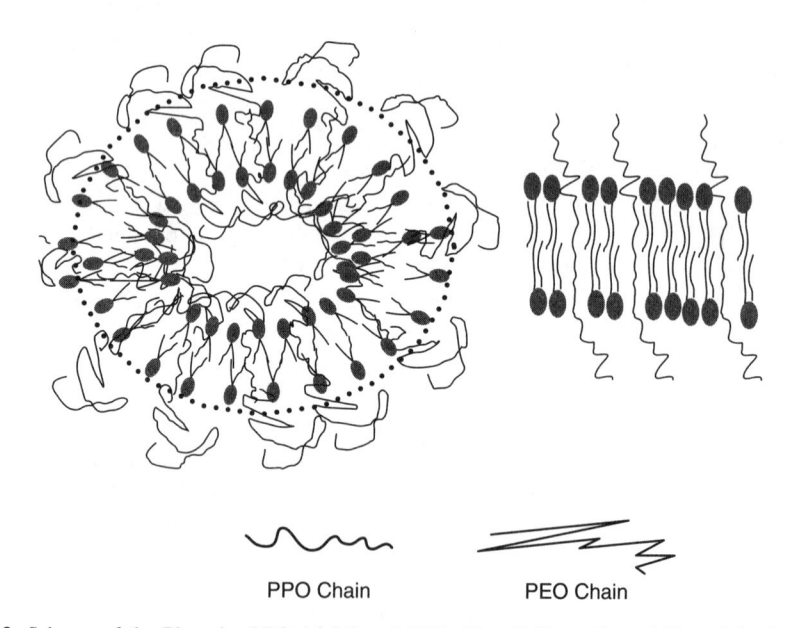

Figure 11.13. Scheme of the Pluronics PEO (**right**) and PPO effect (**left**) on the stability of the EggPC vesicle. (**Left**) SUVs with long attached PEO chains can form shell-like structure (black dotted circle) due to sufficient steric repulsion provided by the dangling PEO chains away from the vesicle surface. (**Right**) The PPO chain incorporated inside the lipid bilayer restricts the bilayer fluidity and movement, and makes the lipid bilayer membrane more rigid.

shell-like wall structure is strong enough to keep all EggPC vesicles in an intact form upon adsorption on mica (Liang et al., 2005).

11.5. SUMMARY

This chapter describes experimental methods to determine the thickness and micromechanical properties of supported bilayers and vesicles using AFM imaging and force measurement. Fused bilayers and sonicated unilamellar vesicles of EggPC are used as examples to illustrate the various approaches for film thickness determination. The AFM imaging of lipid layers is ideally conducted in natural biological solutions using tapping or contact mode at below a threshold force. The adsorbed vesicles of EggPC below 50 nm in size can be imaged in an intact state using the soft contact mechanism. The bilayer thickness can be accurately determined by the step height at bilayer domain edges. Alternatively, the jump-in distances can be used to obtain the bilayer thickness. The jump-in distances change when biological or polymeric additives are integrated into the lipid bilayer. The distance increase can be matched to the hydrophilic block chain length when nonionic Pluronic copolymers are adsorbed onto the EggPC bilayer. Jump-in points are also observed in supported intact vesicles, with one corresponding to the upper bilayer and another corresponding to the lower, surface-attached bilayer. The two gaps are identical in value, which is the lipid bilayer thickness. These gaps can be interpreted as abrupt jumps of the AFM tip across the top and bottom portion of the bilayer enclosure during approach and retraction. AFM can be also used to monitor the shape changes at different image forces. The indentation curve converted from the force curve can be fitted to the Hertz spherical contact model in order to extract the Young's modulus of the vesicle. The calculated Young's modulus and bending modulus of the EggPC SUVs are $(1.97 \pm 0.75) \times 10^6$ Pa and $(0.21 \pm 0.08) \times 10^{-19}$ J, respectively. The elastic constants are compared to literature values from various micromechanical tests of liposomes. AFM methods are the most direct way to determine bilayer film thickness and offer new insights into adsorption, spreading, fusion, self-healing, and mechanical properties of biomembranes and liposomes at the nanoscale.

REFERENCES

A-Hassan, E., Heinz, W. F., Antonik, M. D., D'Costa, N. P., Nageswaran, S., Schoenenberger, C.-A., and Hoh, J. H. (1998). Relative microelastic mapping of living cells by atomic force microscopy. *Biophys. J.* **74**:1564.

Angelova, M. I., Soleau, S., Meleard, P., Faucon, J. F., and Bothorel, P. (1992). Preparation of giant vesicles by external AC electric fields, kinetic and applications. *Progr. Colloid Polym. Sci.* **89**:127.

Barenholz, Y. (2001). Liposome application: Problems and prospects, *Curr. Opin. Colloid Interface Sci.* **6**:66.

Bayerl, T. M., and Bloom, M. (1990). Physical properties of single phospholipid bilayers adsorbed to micro glass beads. A new vesicular model system studied by 2H-nuclear magnetic resonance. *Biophy. J.* **58**:357.

Blumenthal, R., Clague, M. J., Durell, S. R., and Epand, R. M. (2003). Membrane fusion. *Chem. Rev.* **103**:53.

Cooper, M. A. (2004). Advances in membrane receptor screening and analysis. *J. Mol. Recognit.* **17**:286.

Discher, B. M., Won, Y., Ege, D. S., Lee, J. C., Bates, F. S., Discher, D. E., and Hammer, D. A. (1999). Polymersomes: Tough vesicles made from diblock copolymers. *Science* **284**:1143.

Domke, J., and Radmacher, M. (1998). Measuring the elastic properties of thin polymer films with the atomic force microscope. *Langmuir* **14**:3320.

Dong, J., and Mao, G. (2000). Direct study of $C_{12}E_5$ aggregation on mica by AFM imaging and force measurements. *Langmuir* **16**:6641.

Dufrêne, Y. F., and Lee, G. U. (2000). Advances in the characterization of supported lipid films

with the atomic force microscope. *Biochim. Biophys. Acta* **1509**:14.
Dufrêne, Y. F., Boland, T., Schneider, J. W., Barger, W. R., and Lee, G. U. (1998). Characterization of the physical properties of model membranes at the nanometer scale with the atomic force microscope. *Faraday Discuss.* **111**:79.
Egawa, H., and Furusawa, K. (1999). Liposome adhesion on mica surface studied by atomic force microscopy. *Langmuir* **15**:1660.
Engelhardt, H. P., Duwe, H., and Sackmann, E. Bilayer bending elasticity measured by Fourier analysis of thermally excited surface undulations of flaccid vesicles. *J. Phys. Lett. France* **46**:L395.
Eskilsson, K., Ninham, B. W., Tiberg, F., and Yaminsky, V. V. (1999) Effects of adsorption of low-molecular-weight triblock copolymers on interactions between hydrophobic surfaces in water. *Langmuir* **15**:3242.
Evans, E. A. (1974). Bending resistance and chemically induced moments in membrane bilayers. *Biophys. J.* **14**:923.
Evans, E., and Rawicz, W. (1990). Entropy-driven tension and bending elasticity in condensed-fluid membranes. *Phys. Rev. Lett.* **64**:2094.
Faucon, J. F., Mitov, M. D., Méléard, P., Bivas, I., and Bothorel, P. (1989). Bending elasticity and thermal fluctuations of lipid membranes. Theoretical and experimental requirements, *J. Phys. France* **50**:2389.
Franz, V., Loi, S., Müller, H., Bamberg, E., and Butt, H. J. (2002). Tip penetration through lipid bilayers in atomic force microscopy. *Colloids Surf. B* **23**:191.
Hantz, E., Cao, A., Escaig, J., and Taillandier, E. (1986). The osmotic response of large unilamellar vesicles studied by quasielastic light scattering. *Biochim. Biophys. Acta* **862**:379.
Heinz, W. F., and Hoh, J. H. (1999). Spatially resolved force spectroscopy of biological surfaces using the atomic force microscope. *Tibtech* **17**:143.
Hope, M. J., Bally, M. B., Webb G., and Cullis, P. R. (1985). Production of large unilamellar vesicles by a rapid extrusion procedure. Characterization of size distribution, trapped volume and ability to maintain a membrane potential. *Biochim. Biophys. Acta* **821**:55.

Huang, C.-H. (1969). Studies on phosphatidylcholine vesicles formation and physical characteristics. *Biochemistry* **8**:344.
Jass, J., Tjärnhage, T., and Puu, G. (2000). From Liposomes to Supported, planar bilayer structures on hydrophilic and hydrophobic surfaces: An atomic force microscopy study. *Biophys. J.* **79**:3153.
Johnsson, M., Silvander, M., Karlsson, G., and Edwards, K. (1999). Effect of PEO-PPO-PEO triblock copolymers on structure and stability of phosphatidylcholine liposomes. *Langmuir* **15**:6314.
Kostarelos, K., Luckham, P. F., and Tadros, T. F. (1995). Addition of block copolymers to liposomes prepared using soybean lecithin. Effects on formation, stability and the specific localization of the incorporated surfactants investigated. *J. Liposome Res.* **5**:117.
Kostarelos, K., Luckham, P. F., and Tadros, T. F. (1998). Steric stabilization of phospholipid vesicles by block copolymers—vesicle flocculation and osmotic swelling caused by monovalent and divalent cations. *J. Chem. Soc. Faraday Trans.* **94**:2159.
Kostarelos, K., Tadros, T. F., and Luckham, P. F. (1999). Physical conjugation of (tri-) block copolymers to liposomes toward the construction of sterically stabilized vesicle systems. *Langmuir* **15**:369.
Kumar, S., and Hoh, J. H. (2000). Direct visualization of vesicle–bilayer complexes by atomic force microscopy. *Langmuir* **16**:9936.
Kummrow, M., and Helfrich, W. (1991). Deformation of giant lipid vesicles by electric fields. *Phys. Rev.* **44**:8356.
Ladbrooke, B. D., Williams, R. M., and Chapan, D. (1968). Studies on lecithin–cholesterol–water interactions by differential scanning calorimetry and x-ray diffraction. *Biochim. Biophys. Acta* **150**:333.
Laney, D. E., Garcia, R. A., Parsons, S. M., and Hansma, H. G. (1997). Changes in the elastic properties of cholinergic synaptic vesicles as measured by atomic force microscopy. *Biophys. J.* **72**:806.
Lasic, D. D. (1998). Novel applications of liposomes. *Tibtech* **16**:307.
Liang, X., Mao, G., and Ng, K. Y. S. (2004a). Probing small unilamellar EggPC vesicles on

mica surface by atomic force microscopy, *Colloids Surf. B: Biointerfaces* **34**:41.

Liang, X., Mao, G., and Ng, K. Y. S. (2005). Effect of chain lengths of PEO-PPO-PEO on small unilamellar liposome morphology and stability: An AFM investigation, *J. Colloid Interface Sci.*, **285**:360–372.

Liang, X., Mao, G., and Ng, K. Y. S. (2004c). Mechanical properties and stability measurement of cholesterol-containing liposome on mica by atomic force microscopy. *J. Colloid Interface Sci.* **278**:53.

Liu, D., Chen, W., Tsai, L., and Yang, S. (2000). Effects of cholesterol on the release of free lipids and the physical stability of lecithin liposomes. *J. Chin. Inst. Chem. Engrs.* **31**:269.

Loi, S., Sun, G., Franz, V., and Butt, H.-J. (2002). Rupture of molecular thin films observed in atomic force microscopy. II. Experiment. *Phys. Rev. E* **66**:031602.

Manne, S., Cleveland J. P., Gaub, H. E., Stucky G. D., and Hansma, P. K. (1994). Direct visualization of surfactant hemimicelles by force microscopy of the electrical double layer. *Langmuir* **10**:4409.

Mao, G., Liang X., and Ng, K. Y. S. (2004). Direct force measurement of liposomes by atomic force microscopy. In: The Dekker Encyclopedia of Nanoscience and Nanotechnology (J. A. Schwarz, C. I. Contescu, and K. Putyera, eds.), pp. 923–932, Marcel Dekker, New York.

McIntosh, T. J., Magid, A. D., and Simon, S. A. (1987). Steric repulsion between phosphatidylcholine bilayers. *Biochemistry* **26**:7325.

New, R. R. C. (1990). In: Liposomes: A Practical Approach (R. R. C. New, Ed.), Oxford University Press, New York.

Niggemann, G., Kummrow, M., and Helfrich, W. (1995). The bending rigidity of phosphatidylcholine bilayers: Dependences on experimental method, sample cell sealing and temperature. *J. Phys. II France* **5**:413.

Patrick, H. N., Warr, G. G., Manne, S., and Aksay, I. A. (1997). Self-assembly structures of nonionic surfactants at graphite/solution interfaces. *Langmuir* **13**:4349.

Pignataro, B., Steinem, C., Galla, H-J., Fuchs, H., and Janshoff, A. Specific adhesions of vesicles monitored by scanning force microscopy and quartz crystal microbalance. *Biophys. J.* **78**:487.

Prater, C. B., Maivald, P. G., Kjoller, K. J., and Heaton, M. G. (1995) *Probing Nano-scale Forces with the Atomic Force Microscope.* Application Note No. 8.

Presti, F. T., Pace, R. J., and Chen, S. I. (1982). Cholesterol–phospholipid Interaction in membranes. 2. Stoichiometry and molecular packing of cholesterol-rich domains. *Biochemistry* **21**:3831.

Quate, C. F. (1994). The AFM as a tool for surface imaging. *Surf. Sci.* **299**/300:980.

Radmacher, M., Tillmann, R. W., Fritz, M., and Gaub, H. E. (1992). From molecules to cells: Imaging soft samples with the atomic force microscope. *Science* **257**:1900.

Radmacher, M., Fritz, M., Kacher, C. M., Walters, D. A., and Hansma, P. K. (1994). Imaging adhesion forces and elasticity of lysozyme adsorbed on mica With the atomic force microscope. *Langmuir* **10**:3809.

Radmacher, M., Fritz, M., Kacher, C. M., Cleveland, J. P., and Hansma, P. K. (1996). Measuring the viscoelastic properties of human platelets with the atomic force microscope. *Biophys. J.* **70**:556.

Reviakine, I., and Brisson, A. (2000). Formation of supported phospholipid bilayers from unilamellar vesicles investigated by atomic force microscopy. *Langmuir* **16**:1806.

Ringsdorf, H., and Schlarb, B. Venzmer, J. (1988). Molecular architecture and function of polymeric oriented systems. *J. Angew. Chem. Int. Ed. Engl.* **27**:113.

Sackmann, E. (1996). Supported membranes: Scientific and practical applications. *Science* **271**:43.

Sakurai, I., and Kawamura, Y. (1983). Magnetic-field-induced orientation and bending of the myelin figures of phosphatidylcholine. *Biochim. Biophys. Acta* **735**:189.

Schneider, M. B., Jenkins, J. T., and Webb, W. W. (1984a). Thermal fluctuations of large cylindrical phospholipid vesicles. *Biophys. J.* **45**:891.

Schneider, M. B., Jenkins, J. T., and Webb, W. W. (1984b). Thermal fluctuations of large quasi-spherical bimolecular phospholipid vesicles. *J. Phys. France* **45**:1457.

Seifert, U., and Lipowsky, R. (1996). Shapes of fluid vesicles. In Handbook of Non-medical Applications of Liposomes—Theory and Basic Sciences, Vol. I, (D. D. Lasic and

Y. Barenholz, eds.), pp. 43–85, CRC Press, Boca Raton, FL.

Servuss, R. M., Harbich, W., and Helfrich, W. (1976). Measurement of the curvature-elastic modulus of egg lecithin bilayers. *Biochim. Biophys. Acta.* **436**:900.

Shibata-Seki, T., Masai, J., Tagawa, T., Sorin, T., and Kondo, S. (1996). In-situ atomic force microscopy study of lipid vesicles adsorbed on a substrate. *Thin Solid Film* **273**:297.

Svetina, S., and Zeks, B. In: (D. D. Lasic and Y. Barenholz, eds.), Handbook of Nonmedical Applications of Liposomes, Chapter 1, CRC Press, Boca Raton, FL.

Thomson, N. H., Collin, I., Davies, M. C., Palin, K., Parkins, D., Roberts, C. J., Tendler, S. J. B., and Williams, P. M. (2000). Atomic force microscopy of cationic liposomes. *Langmuir* **16**:4813.

Tortonese, M., and Kirk, M. (1997). Characterization of application specific probes for SPMs. *SPIE* **3009**:53.

Tuzar, Z., and Kratochvil, P. (1993). In: Surface and Colloid Science, Vol. 15 (E. Matijevic, ed.), Plenum Press, New York.

Villarrubia, J. S. (1997). Algorithms for scanned probe microscope image simulation, surface reconstruction, and tip estimation. *J. Res. Natl. Inst. Stand. Technol.* **102**:425.

Wang, A., Jiang, L., Mao, G., and Liu, Y. (2002). Direct force measurement of silicone- and hydrocarbon-based ABA triblock surfactants in alcoholic media by atomic force microscopy. *J. Colloid Interface Sci.* **256**:331.

Waugh, R. E., Song, J., Svetina, S., and Žekš, B. (1992). Local and nonlocal curvature elasticity in bilayer membranes by tether formation from lecithin vesicles. *Biophys. J.* **61**:974.

Weisenhorn, A. L., Khorsandi, M., Kasas, S., Gotzos, V., and Butt, and H.-J. Deformation and height anomaly of soft surfaces studied with an AFM. *Nanotechnology* **4**:106.

Woodle, M. C., Newman, M. S., and Working, P. K. (1995). In: Stealth Liposomes (D. Lasic and D. F. Martin, eds), pp. 103–118, CRC Press. Boca Raton, FL.

Vinckier, A., and Semenza, G. (1998). Measuring elasticity of biological materials by atomic force microscopy. *FEBS Lett.* **430**:12.

CHAPTER 12

IMAGING SOFT SURFACES BY SFM

ANDREAS JANKE
Leibriz-Institut fur Polymerforschung Dresden eV., Hohe Str. 6, 01005 Dresden, Germany

TILO POMPE
Leibriz-Institut fur Polymerforschung Dresden eV., Hohe Str. 6, 01005 Dresden, Germany

12.1. INTRODUCTION

Since its invention in 1986 (Binning et al., 1986), the atomic force microscope has become a scientific tool with steadily growing use. Its open modular construction allow for the inclusion of various input parameters and applications. Various sensing mechanisms and improvements in the sensing tools permit the microscope's application in material research and physics as well as imaging techniques. Nowadays, scanning probe microscopy (SPM) is a tool for microscopy, force spectroscopy, and nanometer manipulation in many fields of industry and research. With the scanning probe microscope, imaging of crystal structures with atomic resolution became as readily possible as (a) the measuring and mapping of the forces of specific chemical bonds on surfaces or (b) writing nanometer-sized structures of mono-molecular layers onto surfaces.

Most recently, scanning force techniques have found their way into life science applications. The major reason for this may be found in SPM's own characteristics: the high resolution of the scanning probe microscope in the nanometer range, the great sensitivity of the probe techniques, and the microscope's open construction. This last characteristic allows the flexible adaptation of scanning force microscopy (SFM) techniques to the environmental conditions, which is absolutely necessary for investigating living systems. This adaptability is one major advantage using SPM and its available techniques over high-resolution techniques such as (a) scanning electron microscopy or (b) its improvement for higher-pressure conditions, environmental scanning electron microscopy. On the other hand, SPM optical techniques, which permit application under environmental conditions, are limited to resolutions of only about 0.5 μm. However, SPM offers the opportunity to image, investigate, and manipulate soft structures with nanometer resolution under several controlled environmental conditions, (namely temperature, humidity, gas content) and directly in fluids. Furthermore, SPM readily allows measurement of properties of living tissues, such as elasticity of cell structure elements and binding forces of antibodies.

Despite its usefulness as a research tool in life science, the use of the SPM requires certain precautions and special conditions

---

Force Microscopy: Applications in Biology and Medicine, edited by Bhanu P. Jena and J.K. Heinrich Hörber.
Copyright © 2006 John Wiley & Sons, Inc.

because of the softness and high sensitivity of living structures. This often results in deviation from standard SFM applications. The goal of this chapter is to provide an overview of the basic principles of SPM techniques in SFM, along with its applications and problems, with special focus on the use of these techniques in living structures. Various probing tools for these techniques are described, and the chapter concludes with the presentation of examples illustrating several of the special features, problems, and artifacts that can arise using SPM in the imaging of soft surfaces.

12.2. IMAGING SOFT SURFACES: PHYSICS AND METHODS

In this section, the setup of typical scanning force microscopes, along with methods and probes used for investigation, is explained. The section cannot provide a detailed application description for certain experiments. For that the reader is referred to other chapters of this book. However, some special features generally important for imaging soft surfaces are outlined.

12.2.1. Experimental Setup

The first section will deal with the general setup of a scanning force microscope. It consists schematically of three major parts. Figure 12.1 illustrates the setup described next. The central part is the detection unit, which is a small tip mounted on a cantilever. The cantilever itself is fixed at a stage, which can be moved by a piezoelectric crystal. This piezoelectric crystal belongs to the second major part, which is the scanning unit, consisting of piezoelectric crystals. Two different setups are available. In one, the sample is moved in x, y, and z direction on a stage and the cantilever stage remains fixed. In the second setup, the piezoelectric crystals and the detection unit are integrated in one stage and the sample remains fixed. The scanning

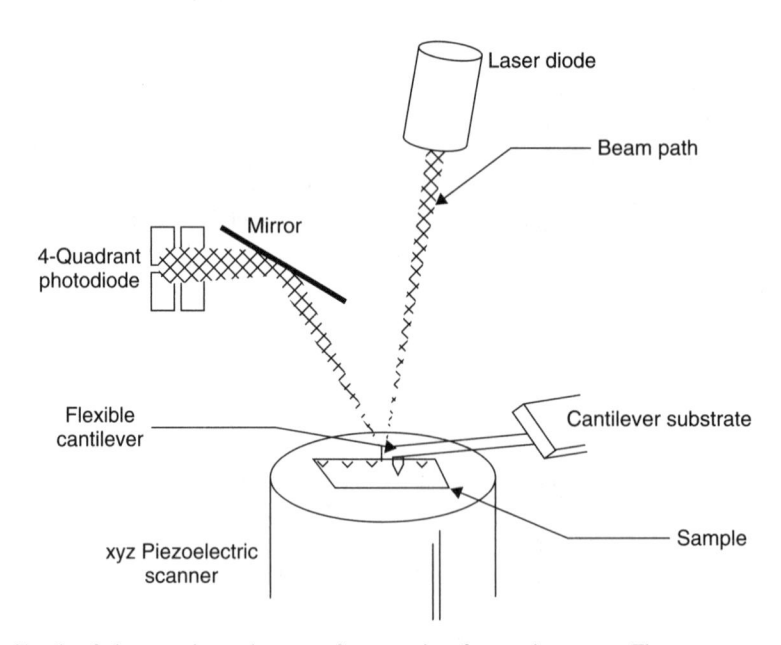

Figure 12.1. Sketch of the experimental setup of a scanning force microscope. The *xyz*-scanner provides the movement of the mounted sample in the three directions. The beam deflection principle is illustrated. The laser beam hits the cantilever. By reflection of the laser beam, distortion and bending of the cantilever are detected with very high precision on the four-quadrant photodiode.

range of piezoelectric crystals is limited, so appropriate scanners must be used to ascertain the dimensions of the range of interest (e.g., atomic resolution—small-range scanner; imaging of whole cells—large-range scanner, up to 200 μm). The movement of the piezoelectric crystals is controlled by an electronic and computer unit, which is the third major part.

How is the detection of surface forces or structures performed? The main principle is the beam deflection. A beam of a laser diode is focused on the back of the very end of a cantilever, on which a tip is mounted on the other side. The reflected laser beam is detected by a four-quadrant photodiode. When the cantilever is bent or distorted due to interactions between the tip and the surface by any kind of force, the bending or distortion gives rise to an amplified change of the laser beam position on the photodiode. In this way, changes of the tip position down to the angstrom range can be detected. The signal of the photodiode is used to give a feedback to the controlling unit, so that the piezoelectric crystals can be adjusted in the z direction to keep the bending on a constant level while scanning in $x-y$ direction over the surface. This mode is called *imaging mode* or *constant force mode*, meaning that a constant force between tip and surface leads to a constant bending of the cantilever. In this mode the z-axis position of the cantilever is plotted over the $x-y$ plane leading to a constant force map, which is used as the topography image.

In the spectroscopy mode the feedback is disabled and the tip is moved in the z direction only. The distortion of the cantilever is measured in dependence of the distance of the tip from the surface giving the interaction force of the tip with the sample surface in dependence of distance.

With this general setup, SFM imaging started its development to a powerful tool in a wide variety of applications. Many of them used special modes of imaging and spectroscopy and special probing tools to differentiate between different forces. The next two sections will focus on these modes and tools.

12.2.2. Modes

The operation modes of SPM are determined by the type of tip scanning.

In *contact mode* the tip is in a repulsive contact with the surface and is dragged with a constant velocity over the sample. Scanning is possible in two different variations of this mode: *constant force mode*, already mentioned above, and *constant height mode*. The constant force mode is useful for imaging rough surfaces (minimizing artifacts due to variable load) and for friction force microscopy (a constant normal load is a requirement for good friction imaging). A disadvantage of this mode, especially for imaging very rough samples, is the limited scan velocity by the time constants of the feedback loop; typical scan frequencies are in the range of 1–6 Hz per scan line. In the deflection mode the piezo-tube extension (in z direction) is constant and the cantilever deflection is recorded. This mode is useful for flat samples, and in this case the load is only nearly constant and high scan speeds are allowed because the feedback gains has been set to zero. Typical scan frequencies range up to 60 Hz. A high scan speed can be useful for scanning small areas below 300 nm to minimize the influence of thermal drift of sample and sample holder.

Under ambient conditions (humid air), only the contact mode provides the highest lateral resolution of 0.1–0.2 nm down to the true atomic resolution under appropriate conditions (Ohnesorge and Binnig, 1993). However, scanning under ambient conditions has disadvantages for the contact mode: high local pressure and shear stresses on the surface. Each surface in air is covered with fluid, in most cases water; the thickness of this layer depends on the hydrophobicity of the surface. In contact mode the fluid layers form a meniscus between surface and tip, and the resulting capillary forces

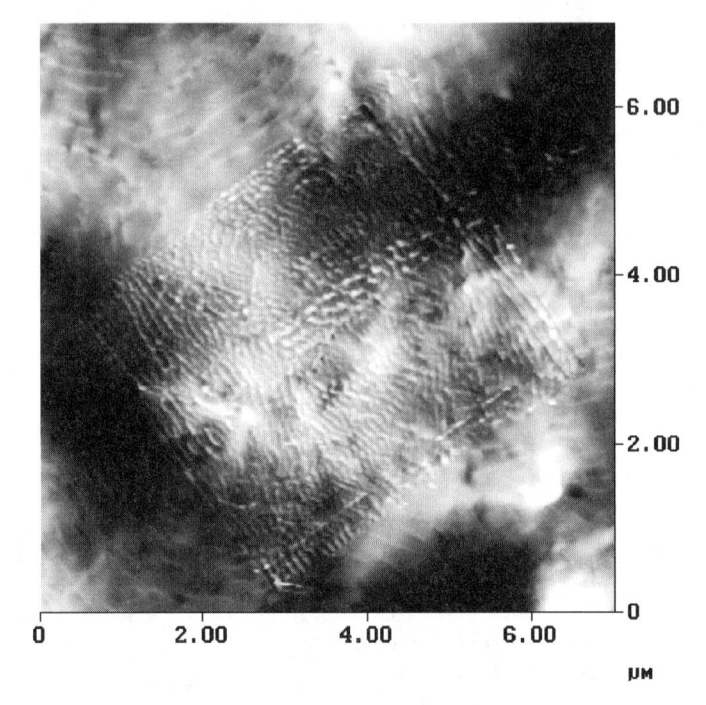

Figure 12.2. Contact mode image (7 μm × 7 μm) of the surface of a ternary copolyester. The inner region shows the result of scanning a 4-μm × 4-μm image (scan angle 45°) with the same normal load and with the same scan frequency, but the smaller image size results in lower tip velocity; the tip is now able to move a certain amount of material and can generate an artificial structure.

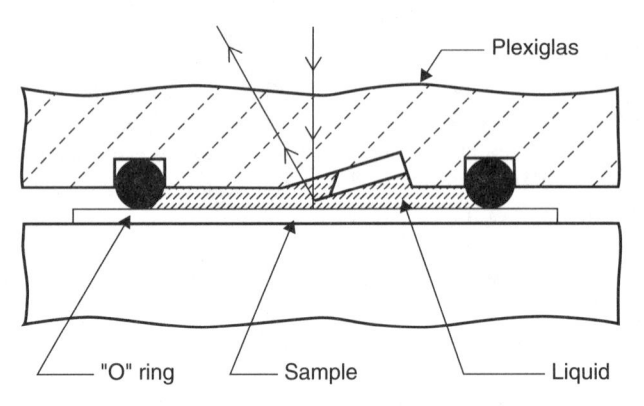

Figure 12.3. Sketch of a liquid cell for a "scanning sample" setup.

are the main contribution to the contact forces in a range from 5 to 1000 nN. With the small contact area (10–100 nm^2), even very low normal forces in the nanonewton-range result in very high normal pressure and shear stresses in the range of 0.1–100 GPa (Tsukruk, 1997). In most cases, such high local stresses damage soft surfaces, an example of which is shown in Fig. 12.2.

Imaging in fluids (water, buffer solution) can circumvent this drawback. In this case, both sample surface and cantilever are immersed in the liquid (Fig. 12.3), resulting in no capillary forces. The remaining contact forces are in the range of a few hundred pico-newtons.

For topographic imaging, the fast scan direction is normally parallel to the cantilever symmetry axis and the photodiode signal generated by the vertical movement of the cantilever is used for the feedback system of a z-scanner and for topography information. With the fast scan direction perpendicular to the cantilever symmetry axis, another form of SPM in contact mode can be performed: friction force microscopy (FFM), also called lateral force microscopy (LFM). The signal at the four-quadranted photodiode is generated by two kinds of cantilever motion (Fig. 12.4), the vertical bending provides information about the topography (= (A + C) − (B + D)), and the torsion and the horizontal bending of the cantilever are a measure of the friction force (= (A + B) − (C + D)). Topography and friction images are taken simultaneously, making it possible to investigate material contrast alongside surface morphology. Small height steps that are hardly visible in topography images can be detected in friction images. Figures 12.5 and 12.6 show the principles and an example of FFM.

To overcome surface damage problems due to the high local pressure in contact mode, the **dynamic mode** (often called "tapping mode®" or "intermittent contact mode") has been developed (Zhong et al., 1993; Spatz et al., 1995; Winkler et al., 1996; Tamayo and Garcia, 1996). In this mode, the instant tip-surface contact is replaced with short contacts of an oscillating tip (Fig. 12.7). The cantilever with the tip is mounted on a piezo stack in a special cantilever holder that drives the cantilever near or at its resonance frequency (typically in the range of 70–300 kHz). The oscillation is detected as RMS signal (amplitude) at the photodiode. In the vicinity of the surface, weak interactions between tip and surface can significantly reduce the amplitude of cantilever oscillation: The tip slightly "taps" on the sample surface during scanning, contacting the surface at the bottom of its oscillation turning point. The amplitude of the oscillation is reduced by two mechanisms: (1) shifting of the effective resonance frequency of the cantilever-interacting surface system and (2) damping due to energy dissipation. The feedback loop of the controller maintains the reduced amplitude constant by variation of the z distance of the scanner at each (x, y) data point (like the constant force mode in contact mode AFM), while also providing information about the topography. If the free amplitude (with the tip far away from the surface) is high enough (i.e., greater than 20 nm), the absorbed fluid layer on the surface will be penetrated without getting stuck. At sufficient low free amplitudes (<5 nm) the fluid layer can be imaged.

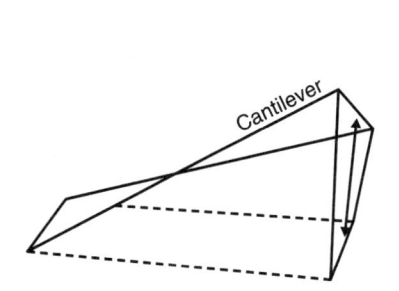

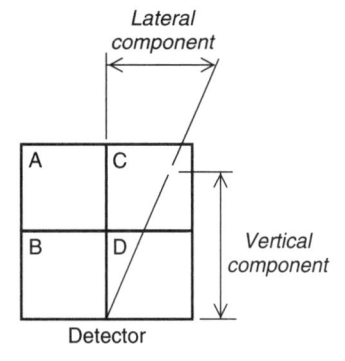

Figure 12.4. Cantilever motion in contact mode. The vertical component (signal (A + C) − (B + D)) gives information about topography, and the lateral component (signal (A + B) − (C + D)) results from friction forces.

206 IMAGING SOFT SURFACES BY SFM

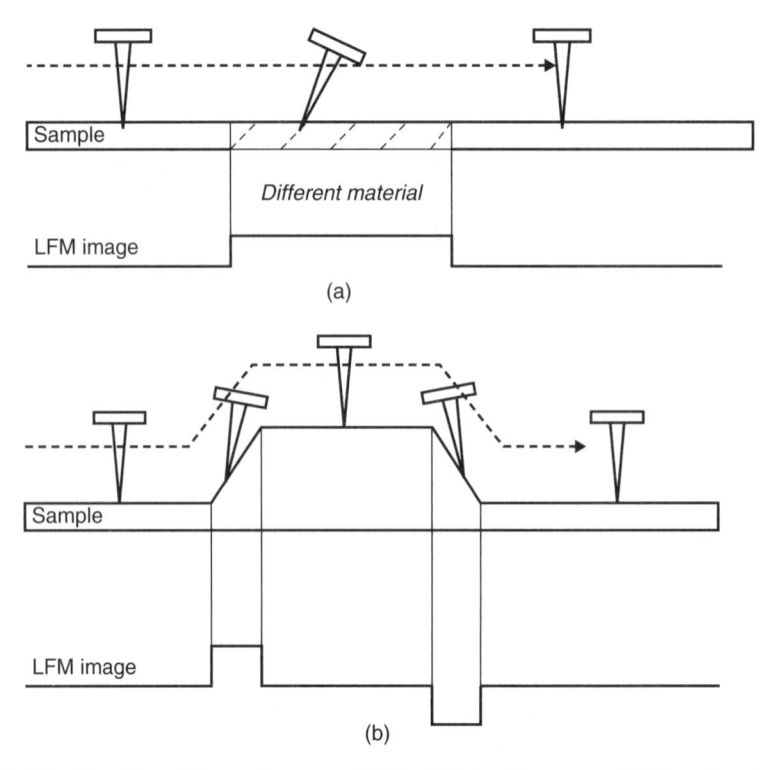

Figure 12.5. Principle of lateral force microscopy (LFM). (**A**) Material contrast. (**B**) Detection of small edges.

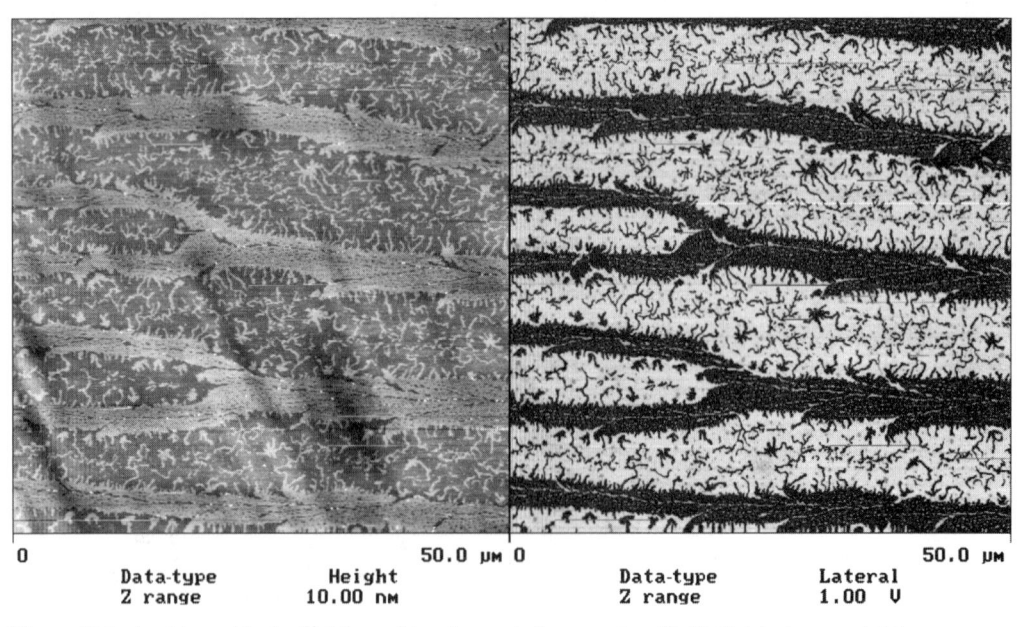

Figure 12.6. Arachine acid adsorbed from chloroform solution on mica. (**Left**) Height image—brighter parts are higher regions. (**Right**) Simultaneously taken friction image (dark—low friction; bright—high friction). The arachine acid shows a significantly lower friction than mica, and the coverage is more clearly seen as in the height image.

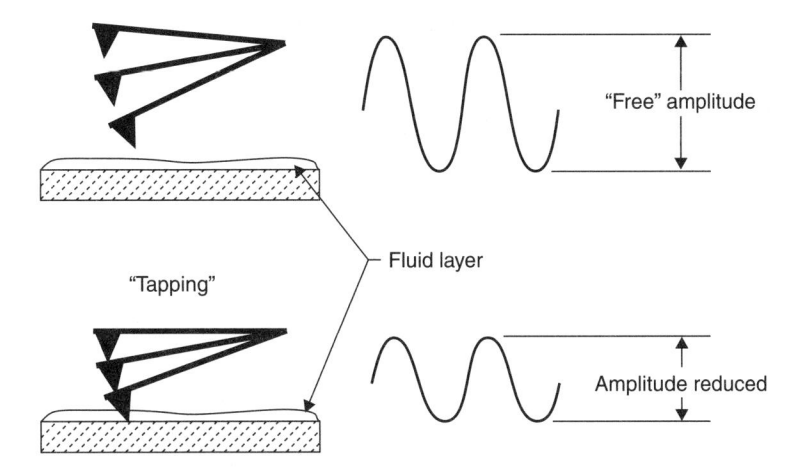

Figure 12.7. Principle of tapping mode. The reduced amplitude will be maintained during scanning.

Maintaining a constant reduced amplitude means also a constant tip–sample interaction is maintained during scanning. Compared to contact mode, the typical tip–surface interaction forces are reduced by at least one order of magnitude (like the forces in fluid cell operation in contact mode). Only compression forces—and most soft samples are insensitive to such forces—have any effect on the sample surface; shear forces are absent. Therefore, dynamic mode is the first choice for imaging soft sample surfaces under ambient conditions. The scan velocity in the dynamic mode is slightly lower than in contact mode; scan frequencies in tapping mode should not exceed 1.5 Hz per scan line (Digital Instruments, 1993). The lateral resolution on most samples (depending on scan size and tip radius) is in the range of 1–5 nm.

The next technique of force reduction between tip and sample surface involves the dynamic mode in fluids. In this way, forces down to tens of pico-newton can be reached. Furthermore, active electronic de-damping of the cantilever oscillation (called Q control) can enhance the sensitivity of the dynamic mode by orders of magnitude due to the enhancement of the Q factor. The quality factor Q determines how fast an oscillating system (cantilever + tip) can respond to a change of the interaction forces acting on the tip. The quality factor is defined as $Q = \Delta\omega/\omega_0$, where ω_0 is resonance frequency and $\Delta\omega$ is half of full maximum width of the resonance curve. At low Q (Fig. 12.8A), a relatively strong interaction (change in frequency) is needed to reach the chosen reduced amplitude. At higher Q (Fig. 12.8B), the system is much more sensitive: The reduced amplitude can be reached by tiny changes of the tip–surface interaction. The quality factor can be changed by Q control (Anczykowski et al., 1998; Humphris et al., 2000), an extra feedback circuit with a phase shifter and an adjustable amplifier added to the standard SFM setup. This circuit applies an additional force via the standard piezo, which is used to drive the cantilever base. The frequency range of modern Q-control devices (5 kHz to 1 MHz) allows applications from liquid dynamic mode (low frequency) to high frequency dynamic mode in air. An example of Q control is shown in Fig. 12.9: tapping mode images of the surface of a poly(hydroxybutyrate) film taken in air with a used, broken tip; without Q control (left) an artifact appears due to a "double" tip, with Q control (right), only the nearest tip touches the surface resulting in an artifact free image.

Apart from amplitude detection, the dynamic mode also allows the detection

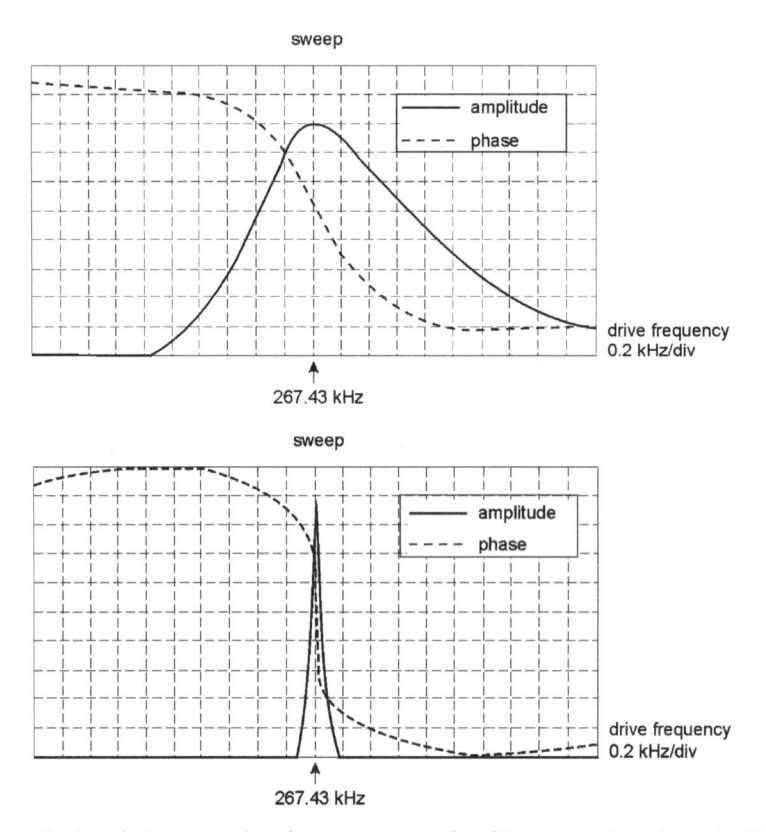

Figure 12.8. Amplitude and phase curve in a frequency sweep of a 125-μm-long dynamic mode silicon cantilever (Nanosensors, Germany) in air. **(A)** Without Q control, $Q = 195$. **(B)** With Q control, $Q = 2542$.

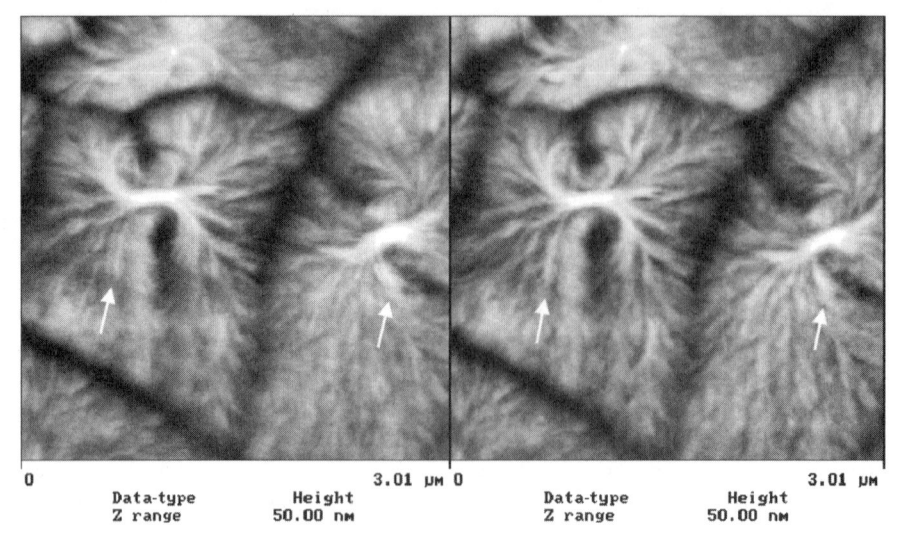

Figure 12.9. An example of Q control: Surface of poly(hydroxybutyrate) film on silicon wafer imaged by tapping mode in air with a used broken tip. **(Left)** Without Q control, artifacts resulting from a "double" tip are clearly seen (arrows). **(Right)** With Q control, only the nearest tip of the "double" tip touches the surface, the tip artifacts are vanished.

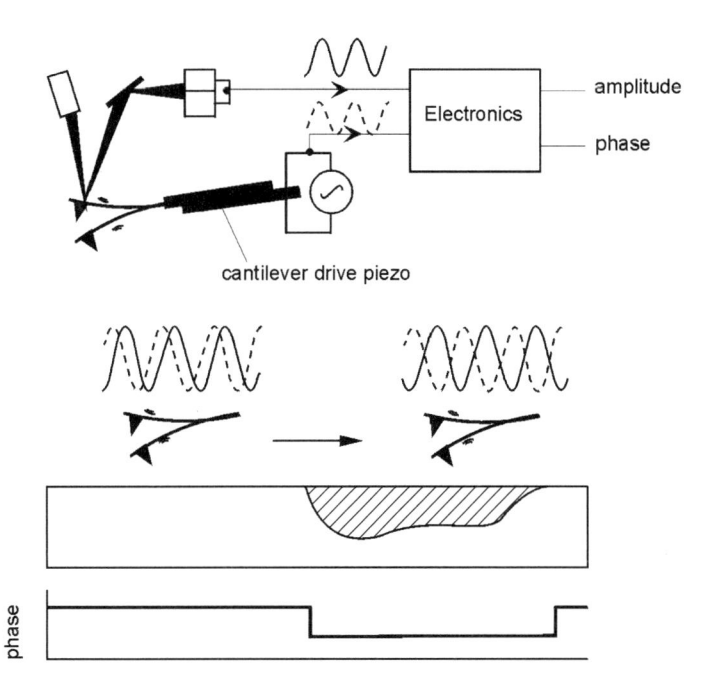

Figure 12.10. Principle of phase shift detection. The electronics calculate the phase shift between the phase signal of the cantilever drive voltage and the phase signal of the real cantilever oscillation detected at the photodiode.

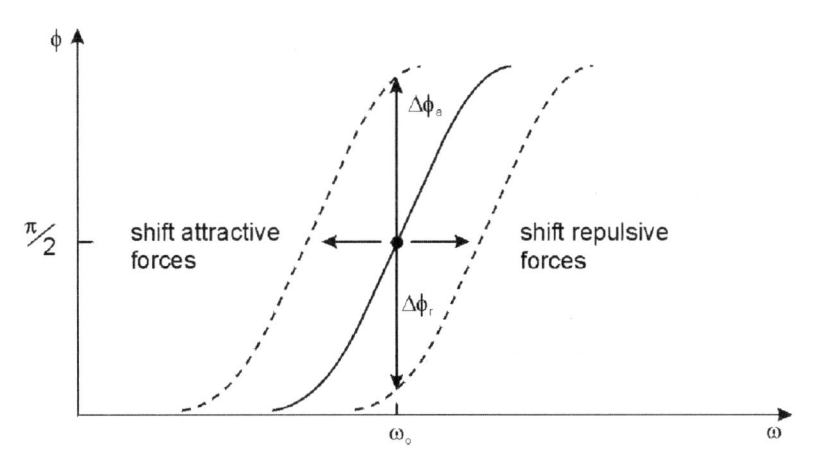

Figure 12.11. Phase shift depending on the kind of tip–surface interaction (explanation in the text).

of phase shift between the oscillating voltage at the cantilever drive piezo and the real cantilever oscillation detected at the photodiode (Fig. 12.10). Interactions of the tip with the surface can reduce the amplitude, and the phase shift can be changed by the influence of different material properties—for example, viscoelasticity, stiffness, and electric and magnetic fields. The mechanism of phase imaging is depicted in Fig. 12.11: The phase of the free vibrating cantilever at its resonance frequency ω_0 is $\Phi = 90°$; under the influence of attractive forces, the effective spring constant decreases, leading to lower resonance frequency, but the cantilever will be further driven at ω_0, with the result of a phase shift $\Delta\Phi_a$. This behavior is reversed in the case of repulsive interaction, leading to a phase shift $\Delta\Phi_r$. For practical phase imaging it is important to ensure scan conditions for either repulsive interaction (stiffness or viscoelasticity contrast) or attractive interaction (adhesion contrast). According to Magonov et al. (1997a, b), the scan conditions for repulsive interaction are a free amplitude greater than 100 nm and a setpoint ratio (reduced amplitude/free amplitude) of 0.5, the scan conditions for attractive interaction are a free amplitude lower than 20 nm and in most cases a setpoint ratio of 0.9; furthermore, the phase shift depends on the cantilever's Q factor and spring constant k: $\Delta\Phi \sim Q$ and $1/k$.

The above-mentioned phase shift, which is due to the influence of long-range forces (electric, magnetic fields), can be best detected in the so-called **lift mode** (Digital Instruments, 1993). In this mode, one line will be scanned twice (Fig. 12.12). After the first pass, which provides the topography, the tip will be lifted a certain value, and then in the second pass it will follow the topography in a constant distance, with only the long-range forces acting to the tip. For detecting electric fields, it is possible to apply voltage to the tip; for detecting magnetic fields, special magnetic tips are necessary. The second pass is performed without true material contact, like a noncontact regime. Under ambient conditions this **noncontact mode** can only be performed for very long-ranging forces (>50 nm), because otherwise near to the surface the fluid layer will form material contact by liquid bridge formation including action of strong attractive capillary forces to the tip.

12.2.3. Cantilevers and Probes

As mentioned at the beginning of the chapter, a very fine tip mounted on a cantilever is the central part of the SPM. It embodies the sensing tool for imaging and spectroscopy. Because of that importance a wide variety of cantilevers and probes exist. For many experiments, however, the choice of cantilever and tip is absolutely crucial for the success of the experiments.

Most commercially available cantilevers are made of silicon nitride or silicon. Generally, silicon nitride cantilevers are softer compared with those made of silicon. Furthermore, V-shaped and beam-shaped cantilevers are the two major constructions (Fig. 12.13). The material properties and the measures of the cantilever provide two essential parameters for SPM: (1) The force constant k and (2) the Q factor (quality factor).

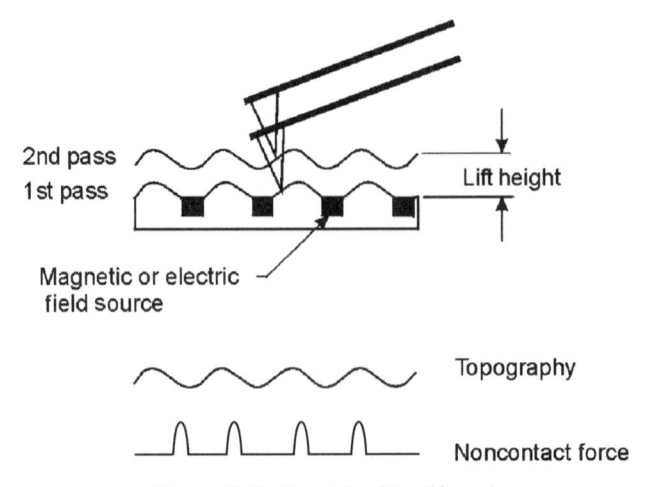

Figure 12.12. Principle of the lift mode.

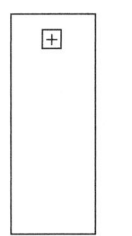

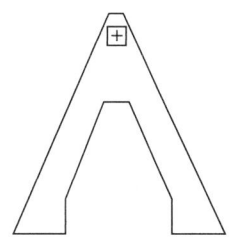

Figure 12.13. Sketch of the two different basic construction forms of cantilevers: (**Left**) Beam-shaped and (**right**) V-shaped. The tip is placed in the center of the turning point of the V-shaped cantilever or near the edge of the beam-shaped cantilever.

The force constant is defined by Hooke's law as the ratio between the applied force F and the cantilever deflection Δd:

$$k = \frac{F}{\Delta d}$$

When low forces should be applied for imaging soft surfaces, cantilevers with low force constants should be used. For dynamic mode the resonance frequency (which is related to the force constant) and the Q factor of the resonance are important measures. The Q factor is a measure of the sharpness of the resonance of the cantilever, thus providing a indication about the sensitivity to changes in oscillation of the cantilever due to external forces.

In addition, cantilevers with a one-sided magnetic coating are available, which are used for dynamic mode with a magnetic drive, in contrast to the standard drive by a piezoelectric crystal.

The position of the tip on the cantilever is a further difference between the available cantilevers. Mostly the tip is placed near the end of the beam-shaped cantilever or in the center of the turning point of the V-shaped cantilever (Fig. 12.13). However, there are also cantilevers available where the tip is at the very end of the cantilever. That shape allows for optical control at the exact point of imaging during the imaging process from above, and it allows an easier way to achieve exact positioning, too.

To differentiate between various tips, one has to look at the aspect ratio and the tip radius that are necessary for the experiments. Standard tips are available with tip radii from 10 nm to 50 nm and cone angles around 40°. However, many procedures may be used to alter these tip characteristics. A tip can be oxide-sharpened, and a fine carbon tip can be grown on it by electron beam deposition. Furthermore, carbon nanotubes can be attached to a tip, rendering it very fine indeed (Wong et al., 1998). Larger microspheres or beads can be glued to the tip, to create a more blunt sensing tool.

Once the geometry of the tip is known or established, its surface properties can be further altered to sense different signals from the sample surface. An electric field can be applied to a conducting tip to measure varying electric surface properties. Capacitance, magnetic, and thermal gradients can be measured by modified tips. Chemical modification of the tip is often used to attach certain molecules and special chemical groups to the tip, or just alter the surface tension of the tip.

Before an experiment and while imaging, one should take care about the shape of the tip. Contamination can result in double tip images or reduction in lateral resolution. There are a few good ways to clean tips; mostly by oxidation of the organic layers on the tip by UV light, or by using oxidizing acid solutions. Other cleaning methods include the use of oxygen or argon plasma (Luginbuhl et al., 2000) or carbon dioxide snow (Hills, 1995).

For biological applications such as the imaging of biomolecules, cells, tissues, and biomaterials, the choice of cantilevers and probes may be summarized as follows. Silicon tips are often used in dynamic mode in air. Silicon nitride tips or very soft silicon tips are used in contact and dynamic mode in liquids. For imaging biomolecules, mostly sharp tips are used, while for imaging living cells blunt tips should chosen.

12.3. EXAMPLES AND SPECIAL PROBLEMS

12.3.1. Liquid Droplets

This section provides an example showing some basic principles that are important for imaging soft liquid-like surfaces or surfaces with a liquid layer on top. Its purpose is to show the high sensitivity of the scanning force microscope, as well as describe certain precautions necessary in this special application. The principles explained in this section are also important when imaging biological structures in air, because these are mostly very soft and either contain a lot of water or have a water layer on top.

In 1996, imaging of high-viscosity liquids like glycerin (Tamayo and Garcia, 1996) and dendrimers (Sheiko et al., 1996) was demonstrated. Soon thereafter, Herminghaus et al. demonstrated the imaging of low-viscosity liquid surfaces with nanometer resolution (Herminghaus et al., 1997). Now, even water structures could be imaged in investigations on wetting properties on the nanometer scale (Pompe and Herminghaus, 2000). The basic mechanism of this imaging technique may perhaps best be explained in the model by Fery et al. (1999) since it provides useful insights in the imaging of liquid and liquid-like surfaces (Fig. 12.14).

For imaging liquid surfaces, the intermittent contact mode is used, meaning that a oscillating tip is scanning the surface. The damping of the oscillation provides the feedback signal. The interaction between tip and liquid surface occurs via the intermittent formation of a very small liquid neck, which is illustrated in Fig. 12.14. During every oscillation cycle (at a frequency of approximately 300 kHz), the intermittent formation and disruption of the small liquid neck occur. At the turning point of the oscillation cycle the tip just touches the surface. By moving away from the surface, a nanometer-sized liquid neck is formed for only a short period of time in the oscillation cycle. During this time an attractive capillary force is acting on the tip. When the tip moves further away, the liquid neck becomes energetically unstable and disruption occurs. The difference in surface energy of the formed liquid neck is thought to be the amount of dissipated energy. Thus the creation of the small liquid neck during every oscillation cycle leads to an energy dissipation of the oscillating tip–cantilever system. Hence, the cantilever oscillation is damped and the damping can be detected from the feedback system.

The force of the nanometer-sized liquid neck is very small and thus the feedback mechanism has to be very sensitive to detect

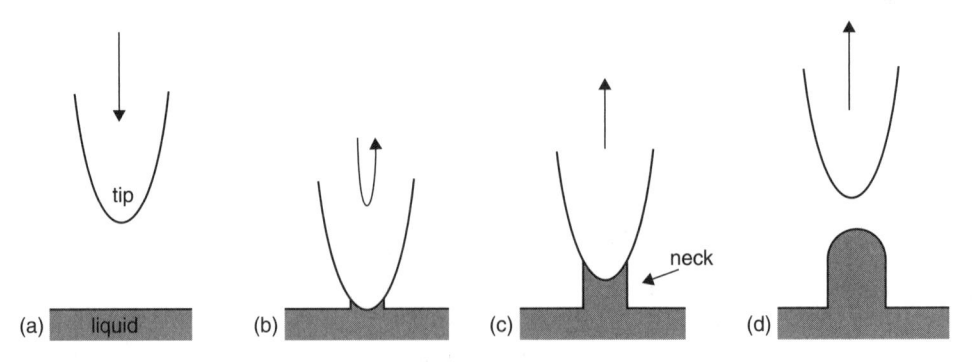

Figure 12.14. Sketch of liquid neck formation and disruption. (A) The tip moves toward the surface. (B) At the turning point of oscillation, it just touches the surface. (C) By moving upward, a nanometer-sized liquid neck is formed. (d) At a certain distance of the tip from the surface, the liquid neck becomes energetically unstable and disrupts from the tip.

the small damping of the cantilever oscillation. From an analysis of the oscillation signals, the power dissipation was determined by 10^{-12} W. The model suggested that the liquid neck has a radius of only 4 nm and exhibits an average attractive force of only 10^{-10} N to the tip. From this it is clear that usually only 2–5% damping of the free oscillation amplitude is allowed for the feedback circuit. When a stronger damping value is set in the feedback system, the tip actually goes into the liquid and is damped by viscous friction. Strong oscillations of the feedback system occur. Imaging of the liquid surface with nanometer resolution is no longer possible. The result of these perturbations is the change of the profile of a liquid droplet as shown in Fig. 12.15.

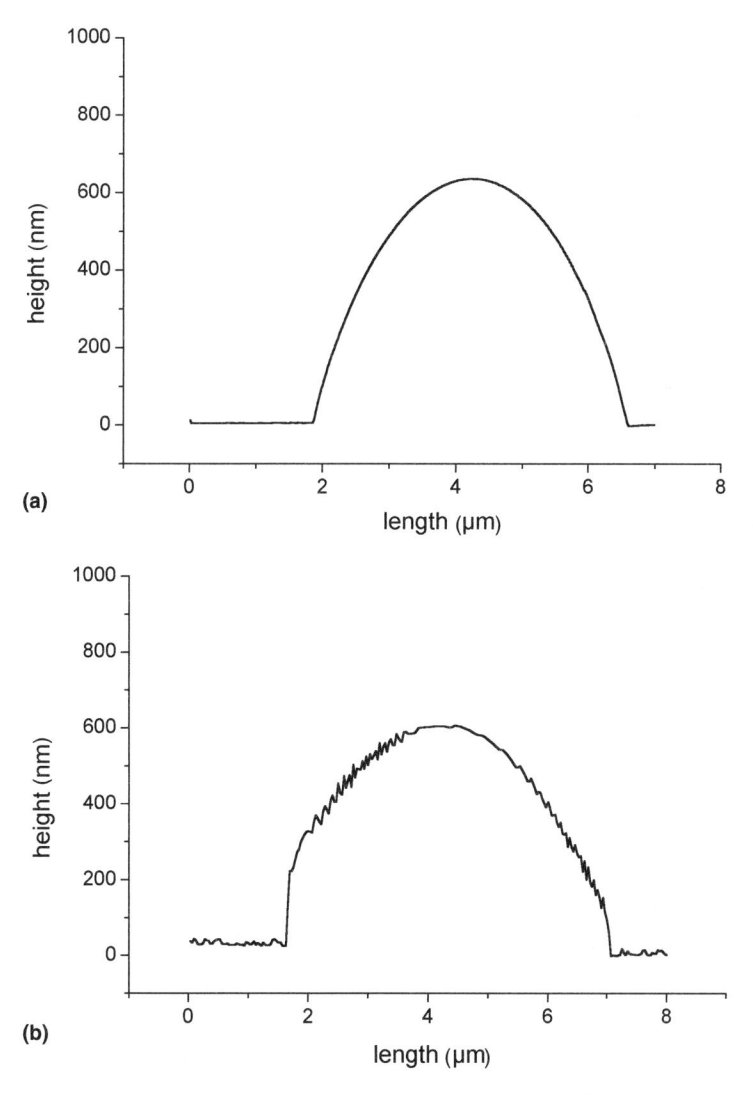

Figure 12.15. (**A**) Section of an SFM topography image of a hexaethylene glycol droplet. (**B**) Another profile of a droplet imaged with stronger damping conditions set in the feedback circuit. For the increased damping of the cantilever the tip has to crash into the liquid. The imaging with high resolution is no longer possible. Strong oscillations of the feedback circuit occur.

The very small dissipation energy of the cantilever oscillation is the reason that only small deviations from the optimal imaging conditions can disturb the whole imaging process. Therefore, the wetting properties of the tip and its geometry are crucial imaging parameters. The model for the imaging process (Fery et al., 1999) verified this importance. It was shown that there is a dependence of the size of the liquid neck formed on the tip radius and on the assumed contact angle of the measured liquid on the tip surface. For large tip radii and low contact angles, the liquid neck becomes quite large, and this leads to an overly strong force on the tip, with the tip bent into the liquid surface. Imaging of liquid surfaces would be not possible under such conditions. On the other hand, very small tip radii and very high contact angles lead to very small liquid necks. However, a very small liquid neck cannot provide enough energy dissipation to the tip oscillation. In this case, the damping could not be detected by the feedback system and the tip would crash into the liquid surface, making imaging impossible.

We therefore see that imaging of liquid surface can only be successfully performed with certain tip radii and liquids of particular surface tension. In the experiments considered here, droplets of salt solutions, hexaethylene glycol, and water were investigated. A typical image of such a micrometer-sized droplet is shown in Fig. 12.16.

Recently it has become possible to increase the Q factor of the cantilever oscillation in the dynamic mode, as was mentioned in the previous section concerning modes for imaging soft surfaces. This method is thought to increase the opportunities for and stability of imaging liquid surfaces and liquid-like structures. The improved control of imaging in the context of purely attractive forces and the higher sensitivity should allow for better imaging in applications at the very end of the sensitivity range of feedback systems in conventional configurations.

12.3.2. Swollen Polymer Surfaces

This section presents examples demonstrating the influence of scan conditions on the image of soft sample surfaces. AFM in liquids is the first choice for imaging soft samples because low contact force is essential. Polymers are excellent materials on which to practice AFM in liquids, since polymers show a swelling behavior in water.

At first a blend of 10% poly(acrylic acid) (PAA) and 90% polyamide 6 (PA6) prepared as thin film (thickness 20 nm) on silicon wafer (Steinert, 2000) is examined. PAA is the actual component that swells in water. Figure 12.17A shows a 3D contact mode image of the surface scanned in air by a contact mode silicon cantilever; note that the surface is relatively flat and less structured. The associated force–distance curve is shown in Fig. 12.17B. In the above-mentioned spectroscopy mode, the tip moves only in the z direction and the deflection of the cantilever as a function of Δz is measured. During the approach (extending),

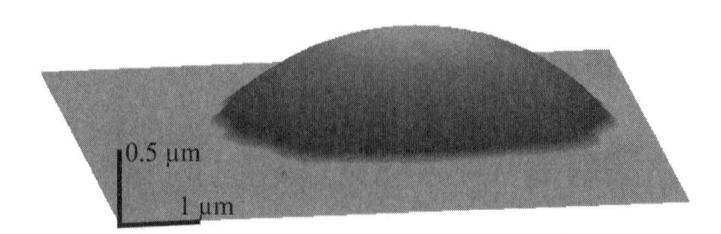

Figure 12.16. Image of a hexaethylene glycol droplet on a silicon wafer substrate. Due to the artificially patterned wettability of the substrate, the droplet exhibits a distinct curved form especially near the contact line, demonstrating the high resolution of the imaging technique.

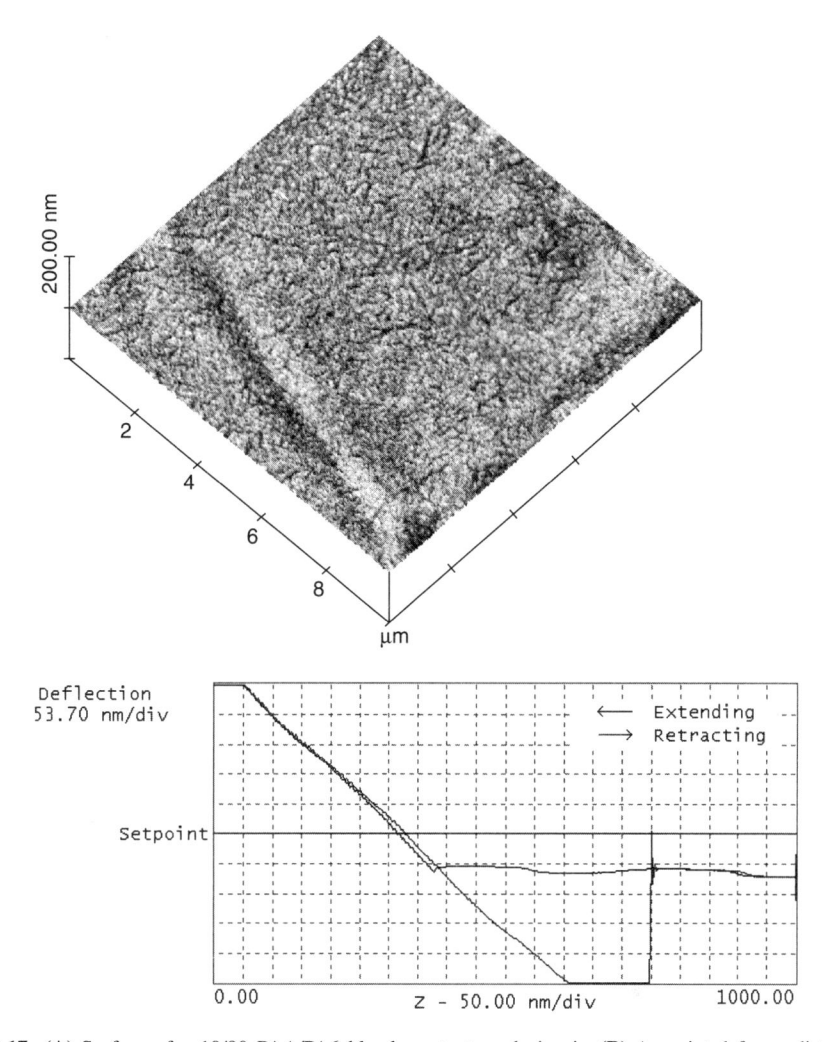

Figure 12.17. (A) Surface of a 10/90 PAA/PA6 blend, contact mode in air. (B) Associated force–distance curve (explanation in the text).

the contact with the surface is marked by the beginning of the slope at the left side. During the retracting of the tip, adhesive forces hold the tip down on the surface beyond the contact point until the spring force ($F = k\Delta z$) equals the adhesion force; hence, the tip jumps out. This adhesion force results from the capillary forces between the fluid layers on the surface and on the tip, forces that are typical under ambient conditions. The contact force F can be calculated from the distance between the crossing with the setpoint and the jump-out Δz (430 nm) and the spring constant of the cantilever k (0.05 Nm^{-1}): $F = 21.5$ nN. To eliminate the influence of the capillary forces, experiments were done in water with different contact forces (Fig. 12.18 and 12.19). The "high" contact force (Fig. 12.18B), calculated as the distance between the beginning of the deflection and the setpoint crossing (140 nm), is $F = 7$ nN. It should be noted that no adhesive force is detected now due to the absence of capillary forces. In this case the surface shows no significant changes (Fig. 12.18A). Figure 12.19 shows that as

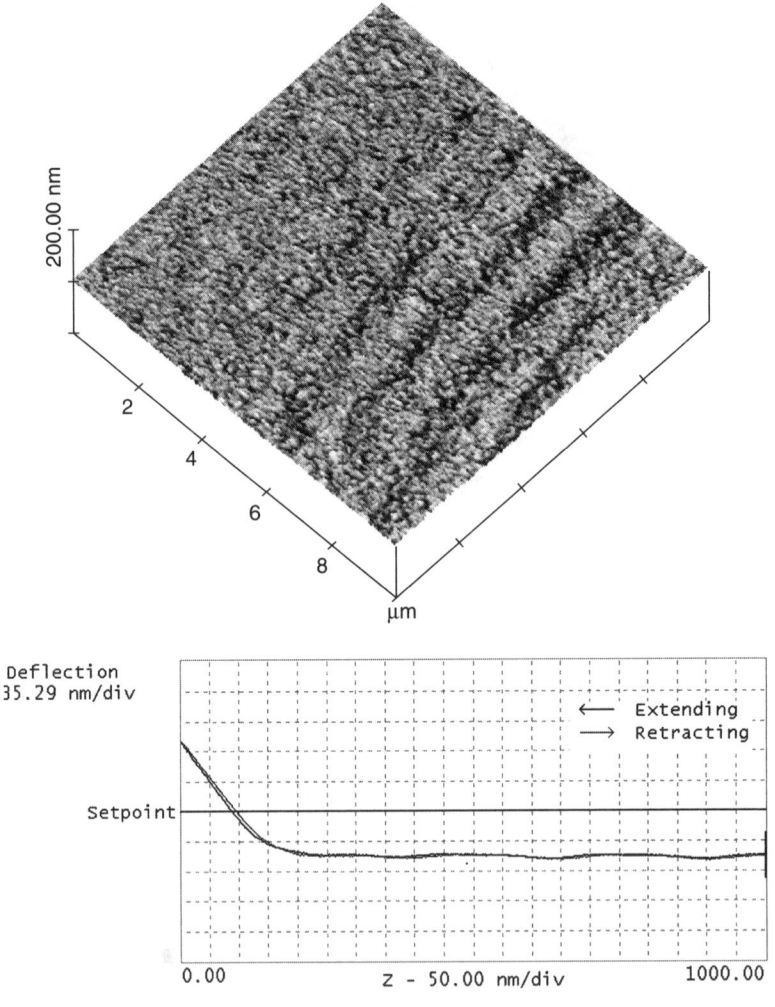

Figure 12.18. (**A**) Surface of a 10/90 PAA/PA6 blend, contact mode in water, "high" contact force. (**B**) Associated force–distance curve.

the force was further lowered, a significant change in roughness and structure is seen (Fig. 12.19A), and the swollen PAA domains are clearly detected. The contact force is $F = 2.5$ nN (Fig. 12.19B). This example shows that small changes in the scan conditions can lead to very different results, and therefore soft surfaces should be scanned at the lowest possible contact force.

The second example shows that even forces at contact mode imaging in liquids can be too high. The sample is a specially prepared hydrogel, poly-N-isopropylacrylamid (PNIPAAm), spin-coated on a Teflon-like film on silicon wafer and stabilized by a treatment in Ar^+-plasma (Nitschke, 2001). The thickness of the dry film was 10 nm; the thickness of the swollen film (water) was 30 nm. AFM was done in water. Figure 12.20 juxtaposes the results in contact mode (left) and in tapping mode (right). Both images of the swollen state show a net-like structure, which in the case of contact mode is a slightly disturbed, such that the scanning tip relocates the highly flexible structure. The tapping mode image reveals a finer network

EXAMPLES AND SPECIAL PROBLEMS   217

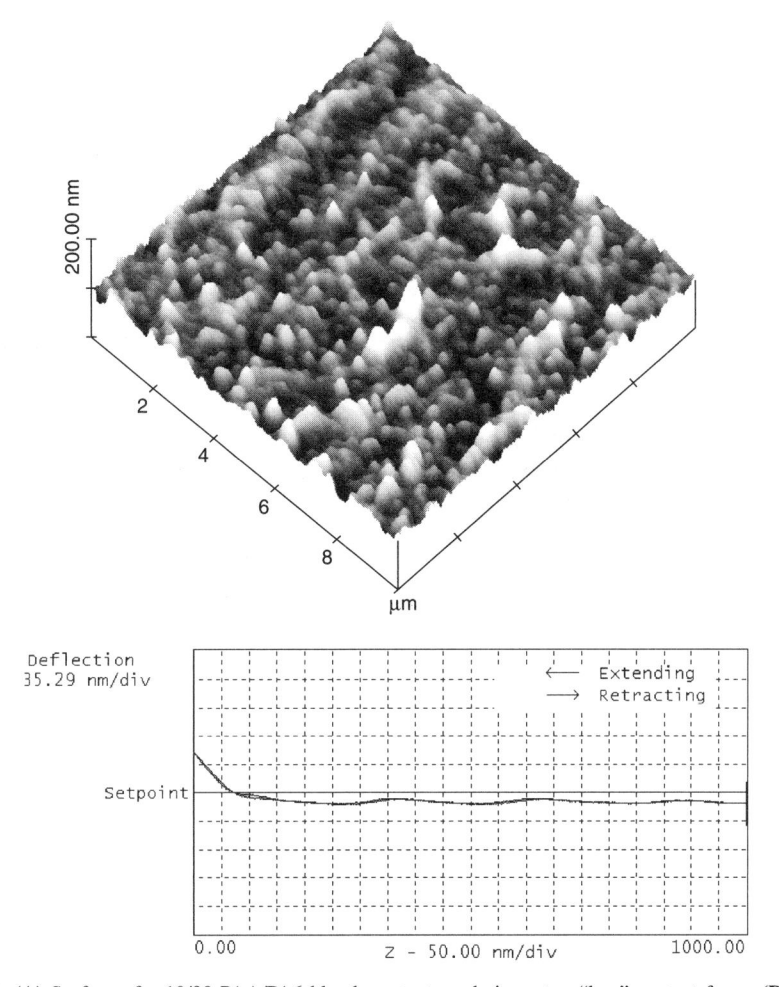

Figure 12.19. (**A**) Surface of a 10/90 PAA/PA6 blend, contact mode in water, "low" contact force. (**B**) Associated force–distance curve.

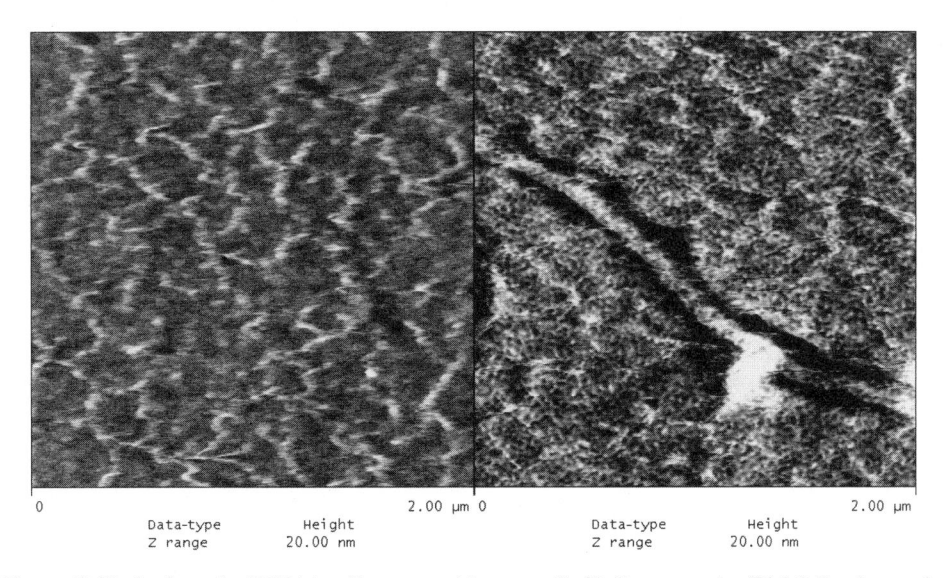

Figure 12.20. Surface of a PNIPAAm film, scanned in water. (**Left**) Contact mode. (**Right**) Tapping mode.

structure of about 30 nm. In contact mode, this structure was smeared out by the tip. In tapping mode, only the contact force was small enough not to disturb these structures.

12.4. CONCLUSIONS

This chapter presented an outline of the basics of AFM and its application for imaging soft surfaces. Some technical details relevant for investigations of sensitive, soft, and flexible surfaces were described. The purpose of the chapter is to show that AFM is a very powerful tool for the examination of soft surfaces with very high resolution and under ambient conditions. The softness of biological structures, however, requires keen awareness of certain precautions and technical details. Due to its adaptability to living culture conditions, AFM is an excellent instrument for the imaging of living tissues and for spectroscopy of biological, chemical, and physical characteristics of cells and biomolecules.

REFERENCES

Anczykowski, B., Cleveland, J. P., Krüger, D., Elings, V. B., and Fuchs, H. (1998). Analysis of the interaction mechanisms in dynamic mode SFM by means of experimental data and computer simulation. *Appl. Phys.* **A66**:885–889.

Binning, G., Quate, C., and Gerber, C. (1986). Atomic force microscope. *Phys. Rev. Lett.* **56**(9):930–933.

Digital Instruments (1993). NanoScope III Multimode Scanning Probe Microscope Instruction Manual, Version 1.5.

Fery, A., Pompe, T., and Herminghaus, S. (1999). Nanometer resolution of liquid surface topography by scanning force microscopy. *J. Adhes. Sci. Technol.* **13**(10):1071–1083.

Herminghaus, S., Fery, A., and Reim, D. (1997). Imaging of droplet of aqueous solutions by TappingMode Scanning Force. *Ultramicroscopy* **69**(3):211–217.

Hills, M. M. (1995). Carbon dioxide jet spray cleaning molecular contaminants, *Vacuum Sci.* **A13**(1):30–34.

Humphris, A. D. L., Tamayo, J., and Miles, M. J. (2000). Active quality factor control in liquids for force spectroscopy. *Langmuir* **16**:7891–7894.

Luginbuhl, R., Szuchmacher, A., Garrison, M. D., Lhoest, J. B., Overney, R. M., and Ratner, B. D. (2000). Comprehensive surface analysis of hydrophobically functionalized SFM tips. *Ultramicroscopy* **82**:171–179.

Magonov, S. N., Elings, V., and Whangbo, M.-H. (1997). Phase imaging and stiffness in tapping-mode atomic force microscopy. *Surf. Sci. Lett.* **375**:L385–L391.

Magonov, S. N., Cleveland, J., Elings, V., Denley, D., and Whangbo, M.-H. (1997). Tapping-mode atomic force microscopy study of the near-surface composition of a styrene-butadiene-styrene triblock copolymer film. *Surf. Sci.* **389**(1–3):201–211.

Nitschke, M. (2001). Institut für Polymerforschung Dresden, Germany, private communication.

Ohnesorge, F., and Binnig, G. (1993). True atomic resolution by atomic force microscopy through repulsive and attractive forces. *Science* **260**:1451–1456.

Pompe, T., and Herminghaus, S. (2000). Three-phase contact line energetics from nanoscale liquid surface topographies, *Phys. Rev. Lett.* **85**(9):1930–1933.

Sheiko, S. S., Eckert, G., Ignateva, G., Muzafarov, A. M., Spickermann, J., Rader, H. J., and Moller, M. (1996). Solid-like states of a dendrimer liquid displayed by scanning force. *Macromol. Rapid Commun.* **17**(5):283–297.

Spatz, J. P., Sheiko, S., Moller, M., Winkler, R. G., Reineker, P., and Marti, O. (1995). *Nanotechnology* **6**:40–44.

Steinert, V. (2000). Institut für Polymerforschung Dresden, Germany, private communication.

Tamayo, J., and R. Garcia. (1996). Deformation, contact time, and phase contrast in tapping mode scanning force microscopy. *Langmuir* **12**(18):4430–4435.

Tsukruk, V. V. (1997). Scanning probe microscopy of polymer surfaces. *Rubber Chem. Technol.* **70**:430–467.

Winkler, R. G., Spatz, J. P., Sheiko, S., Moller, M., Reineker, P., and Marti, O. (1996). *Phys. Rev.* **B54**:8908–8912.

Wong, S. S., Joselevich, E., Woolley, A. T., Cheung, C. L., and Lieber, C. M. (1998). Covalently functionalized nanotube nanometresized probes in chemistry and biology. *Nature* **394**:52–55.

Zhong, Q., Jennis, D., Kjoller, K., and Elings, V. B. (1993). *Surf. Sci. Lett.* **290**:688.

CHAPTER 13

HIGH-SPEED ATOMIC FORCE MICROSCOPY OF BIOMOLECULES IN MOTION

TILMAN E. SCHÄFFER

Center for Nanotechnology and Institute of Physics, Westfälische Wilhelms-Universität Münster, Gievenbecker Weg 11, 48149 Münster, Germany

13.1. INTRODUCTION

"Time is the Life of the Soul," H. W. Longfellow wrote in *Hyperion* (Longfellow, 1839). With regard to atomic force microscopy (AFM) in biology and medicine, "Time is the Soul of Life" may hold a truth of its own. Life embraces motion, development, and growth: the progression of events with time. Such a progression cannot be captured in a single "snapshot" AFM image. The "soul" of life is instead revealed in image sequences. This is why fast and time-resolved AFM imaging is of particular importance in the study of biological systems.

Soon after its invention, AFM (Binnig et al., 1986) was applied to the investigation of biomolecules in solution (Drake et al., 1989). Its capability of imaging with molecular and even submolecular resolution in numerous environmental conditions including biological buffer solutions distinguishes it from other high-resolution techniques that are used to study biomolecules (e.g., x-ray diffraction, nuclear magnetic resonance, cryo-electron microscopy). As compared to these complementary techniques, little sample preparation is required and the sample can be imaged in its native environment over and over again, allowing for the recording of image sequences.

In an AFM, a sharp tip is mounted onto the free end of a flexible cantilever. The tip is brought into contact with a sample surface and is scanned across it. Interaction forces between the tip and the sample cause the tip and thus the cantilever to deflect. This deflection is detected by focusing an optical beam onto the cantilever and directing the beam reflected from it onto a position-sensitive detector. This optical beam deflection ("optical lever") detection method allows the remote and noninvasive detection of tip deflections in biological (transparent) buffer solution with sub-angstrom accuracy.

A key improvement to the imaging of soft samples in liquid was made by the development of "tapping mode" in liquid (Hansma et al., 1994; Putman et al., 1994). In tapping mode, the cantilever is externally driven such that the tip undergoes vertical sinusoidal oscillations. Traditionally, the external drive in liquid is of acoustic nature (Schäffer et al., 1996), but the cantilever can also be driven by magnetic (Han et al., 1997; Revenko and Proksch, 2000) or thermal (Marti et al., 1992) forces. Instead of being dragged over the sample as

---

Force Microscopy: Applications in Biology and Medicine, edited by Bhanu P. Jena and J.K. Heinrich Hörber.
Copyright © 2006 John Wiley & Sons, Inc.

in contact mode, the tip now simply "taps" the surface once during each oscillation cycle. The oscillation amplitude is used as input to a feedback loop keeping the mean tip–sample separation constant. This mode exhibits low shear forces and has enabled the destruction-free imaging of soft biological samples such as single molecules on a routine basis (Fritz et al., 1995; Hansma, 2001).

While the spatial resolution that can be obtained on biomolecules might be higher when imaging in gaseous media, in vacuum, or at cryogenic temperatures (Shao and Zhang, 1996), dynamic processes involving biomolecules generally occur only in liquid environments. Therefore, this chapter focuses on the imaging of dynamic processes in liquids. Although AFM imaging in liquids is still considered somewhat of an "art" due to its technical difficulty, much progress has been made in the investigation of biological processes. Some examples of studies that are concerned with time-resolved AFM imaging of biomolecular interactions are now given.

One of the first studies of dynamic protein–DNA interactions was the observation of enzymatic degradation of DNA by DNase I (Bezanilla et al., 1994), demonstrating that the AFM can image dynamic interactions on a molecular level. Endonucleolytic cleavage of DNA by the restriction enzyme EcoKI was imaged in the presence of ATP (Ellis et al., 1999). Transcription of double-stranded DNA by RNA polymerase (RNAP) was detected after adding nucleoside triphosphates (NTPs) to stalled DNA–RNAP complexes adsorbed onto mica, and transcription rates of 0.5–2 bases/s were observed (Kasas et al., 1997; Guthold et al., 1999). Nonspecific photolyase–DNA complexes were imaged, showing association, dissociation, and movement of photolyase along the DNA, suggesting that photolyase uses a sliding mechanism to scan the DNA for damaged sites (van Noort et al., 1998). Dynamic interactions of the tumor-suppressor protein p53 with DNA were observed by time-lapse AFM imaging, revealing dissociation/reassociation, sliding, and possibly direct binding to the specific binding site, whereby two different modes of target recognition were detected (Jiao et al., 2001) (discussed in more detail in Section 13.2.2).

Protein–protein interactions were visualized by imaging the association and dissociation of the chaperonin GroEL–GroES complex, the lifetime of which was determined as ≈ 5 s (Viani et al., 2000). Protein–lipid interactions were visualized by imaging the degradation (hydrolysis) of membrane phospholipids by phospholipase A2 with a lateral resolution of less than 10 nm (Grandbois et al., 1998). The interactions of proteins with an inorganic crystal were investigated by adsorbing soluble proteins extracted from abalone nacre onto a growing calcite crystal. Dependent on their crystallographic orientation, the atomic step edges changed in shape and growth speed as the proteins attached to the crystal surface (Walters et al., 1997a).

Enzyme activity was directly observed by measuring the height fluctuations of the protein lysozyme as it was exposed to an oligoglycoside substrate (Radmacher et al., 1994a). Dynamic movements of the arms of laminin-1 that might contribute to the diversity of laminin functions were observed (Chen et al., 1998). Calcium-mediated structural changes of native nuclear pore complexes such as the opening and closing of individual nuclear baskets were observed, indicating that cytoplasmic plugging and unplugging of the complex is insensitive to calcium in the absence of ATP (Stoffler et al., 1999). The movements of single sodium-driven rotor proteins from a bacterial ATP synthase, embedded in a lipid membrane, were monitored and the principal modes of the motion were distinguished (Müller et al., 2003).

Conformational changes in supercoiled DNA induced by decreasing the ionic strength of the imaging buffer were imaged (Nagami et al., 2002). Bidirectional growth of polymorphic amylin protein fibrils was monitored *in vitro* yielding an elongation rate of 1.1 ± 0.5 nm/min (Goldsbury et al., 1999).

The disassembly dynamics of microtubules was observed with an average rate of 1–10 tubulin monomers per second. The large fluctuations in this rate might be caused either by multiple pathways in the disassembly kinetics or by stalling due to defects in the microtubule lattice (Thomson et al., 2003). The motion of myosin V on mica in buffer solution was visualized (Ando et al., 2001). The activation process of human platelets could be followed (Fritz et al., 1993), and dynamic changes in the stiffness of the cortex of adherent cultured cells were tracked during cell division (Matzke et al., 2001).

One of the main limitations common to these investigations is the limited time resolution of the imaging. In general, an AFM image requires on the order of minutes to be acquired, and processes occurring faster than that cannot be investigated with conventional AFM imaging. This chapter discusses some of the topics and issues associated with fast AFM imaging of biomolecules in motion.

13.2. TIME-RESOLVED IMAGING

13.2.1. Recording Image Sequences

Real-time biological processes can be investigated by recording AFM image sequences. In principle, this amounts to no more than scanning the same image area over and over again. In practice, however, there are a number of issues that become important when recording image sequences. The most obvious issue is that the time needed to acquire an image sequence is proportional to the number of images in the sequence. For sequences with a lot of images, the total duration of imaging can amount to many minutes or even hours. Drift effects therefore become especially significant.

There are several kinds of drift issues that need to be considered. Mechanical drift can be grouped into two categories: (a) lateral drift and (b) vertical drift. (a) Lateral drift can arise from two different effects. On the one hand, drift in the microscope (due to thermal equilibration processes or due to creep in the piezoelectric scanner) causes the image area to slowly shift in time. On the other hand, the object of interest might move itself, driven for example by thermal diffusion or by conversion of energy. Both effects have the same consequence: moving the object of interest out of the image area. Section 13.2.3 presents a method for (partially) compensating for lateral drift by aligning the images after the recording. (b) Vertical drift can be caused by microscope drift, too, which moves the cantilever support relative to the sample or by drift in the cantilever deflection that moves the tip relative to the cantilever support. Drift in cantilever deflection might be caused by slow thermal bending of the cantilever due to temperature variations in the imaging environment, or by relaxation of the internal stress in the cantilever material. While vertical microscope drift is usually taken care of by the feedback loop that keeps the cantilever deflection or amplitude constant, the drift in cantilever deflection causes the optical beam to move on the detector. This causes a problem in both contact and tapping mode operation: In contact mode, such a deflection is interpreted as a change in imaging force. Such a change causes the feedback loop to decrease the tip–sample distance, resulting in increased imaging forces that can damage the sample, or it causes the feedback loop to increase the tip–sample distance, which can result in a failure of accurate surface tracking. In tapping mode, cantilever drift is less critical since the cantilever oscillation amplitude and not the cantilever deflection is used as error signal in the feedback loop, and the (ac) amplitude is effected less by (dc) drift. But drift in cantilever deflection, which causes the optical beam to move on the detector, results in a situation where the beam is not centered on the detector any more, causing additional detection noise in both contact and tapping mode.

There are yet some other issues to consider when recording image sequences. For example, the quality of the scanning tip can degrade over time, especially when the tip is continually being contaminated with organic debris. And, last but not least, a fragile sample might not "enjoy" being imaged over and over again.

13.2.2. Dynamic Interactions of p53 with DNA

As an example of time-resolved imaging, dynamic interactions of p53 with DNA are now discussed. p53 is a tumor-suppressor protein that is involved in the regulation of cell growth (Levine, 1997). Over 50% of all human cancers (e.g., skin, colon, and breast cancer) are related to nonfunctional p53. The name, p53, stands for "p" as in protein and for "53" as in 53 kD, the molecular weight of the protein. p53 has multiple functions, but mostly it serves as a transcription factor. In solution, it forms tetramers that bind specifically to the consensus binding site consisting of two decameric consensus units that are separated by 0–21 base pairs. It also binds nonspecifically to DNA with a lower affinity. Since most mutations that render the protein dysfunctional occur in its DNA-binding region, it is important to study the interactions of p53 with DNA. The AFM is a valuable tool in this endeavor, since it can image single molecules in solution at molecular resolution. Both p53 and DNA need to be attached to an underlying imaging surface. Mica is usually chosen as the imaging surface for high-resolution imaging, since it is atomically flat and can easily be cleaved. One of the difficulties of using mica as the imaging surface for DNA is that both mica and DNA are negatively charged for typical buffer conditions. A method that promotes binding of DNA to mica is adding divalent cations to the solution that act as molecular crosslinkers (Fig. 13.1). But care must be taken to add just the right concentration of divalent cations: Too little does not attach DNA tightly enough for stable imaging by the AFM tip. Too much binds DNA down too tightly so that it might not be able to move and interact with p53. It was found that

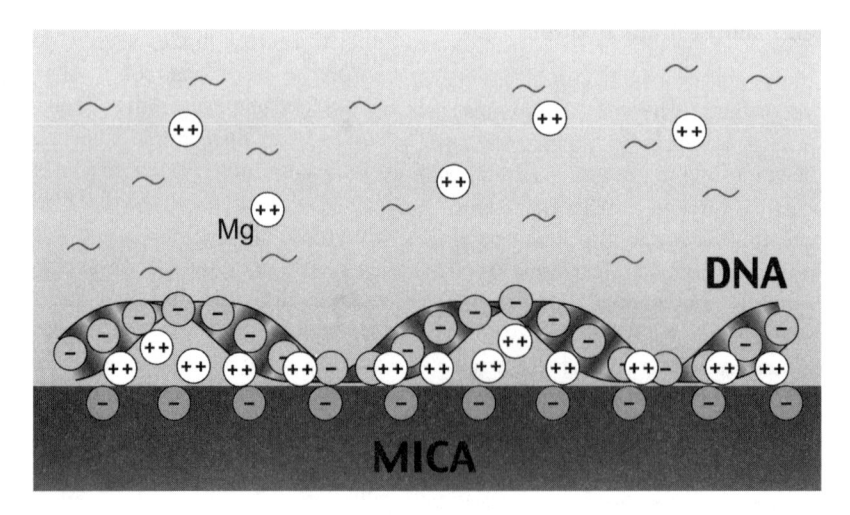

Figure 13.1. Schematic illustrating a difficulty when imaging dynamic interactions involving negatively charged biomolecules such as DNA. The negative charge of both DNA and the underlying mica surface usually prevents adsorption of DNA to the surface. Divalent cations such as magnesium are therefore commonly used as molecular "glue" between DNA and mica. With increasing concentration of divalent cations, the repulsive force between DNA and mica is reverted and becomes attractive, allowing DNA to adsorb to the mica surface. The difficulty is finding the right concentration: Too little attaches DNA to the surface too weakly for AFM imaging, while too much binds the DNA too tightly to the surface so that it might not be able to move and participate in dynamic interactions.

a divalent cation concentration of 20 mM Mg^{2+} in the presence of Na-HEPES allowed both DNA and p53 to loosely attach to mica so that (a) they still interact both specifically and nonspecifically (however with a reduced affinity) and (b) their interactions could be imaged by the AFM (Jiao et al., 2001). Several different types of p53–DNA interaction events could be imaged, including dissociation/reassociation and sliding (Fig. 13.2).

One of the issues that this study addressed is the particular target recognition mode with which p53 binds to DNA. The question is, How does a protein find its target site? In the simplest mode, a protein diffuses three-dimensionally in solution until it binds to the target site directly from solution (Fig. 13.3a). It was, however, shown for a number of proteins other than p53 that such a binding mode can not account for the relatively high binding rates that were experimentally observed for those proteins. A second mode has therefore been proposed that relies on a two-step process: Initially, the protein binds nonspecifically to DNA and then subsequently diffuses one-dimensionally along the DNA until it finds the target site (Fig. 13.3b). Both modes of target recognition were detected in the interactions of p53 with DNA (Jiao et al., 2001).

13.2.3. Image Alignment

After an AFM image sequence of some process is recorded, it needs to be analyzed with the goal of extracting information about the process. A fruitful way to display the image sequence is to animate it, to "make a movie."

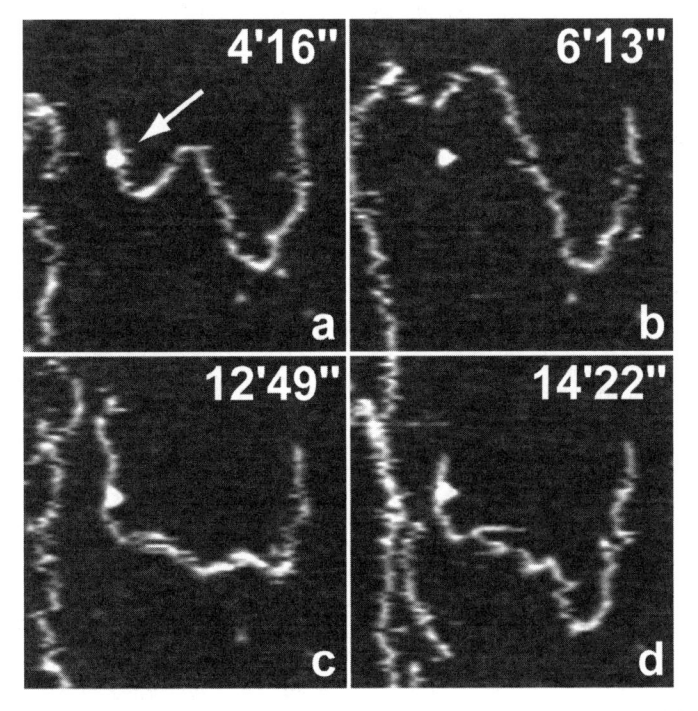

Figure 13.2. Dynamic p53–DNA interactions observed by time-resolved tapping mode AFM imaging in solution. The buffer conditions were adjusted such that both p53 protein and DNA are weakly adsorbed to the mica surface. (a) A p53 protein molecule (arrow) is bound to a DNA fragment. The protein (b) dissociates from and then (c) reassociates with the DNA fragment. (d) A downward movement of the DNA with respect to the protein occurs, constituting a "sliding" event whereby the protein changes its position on the DNA. Image size: 620 nm. Grayscale (height) range: 4 nm. Time units: minutes, seconds. [Reprinted from Jiao et al. (2001), with permission from Elsevier.]

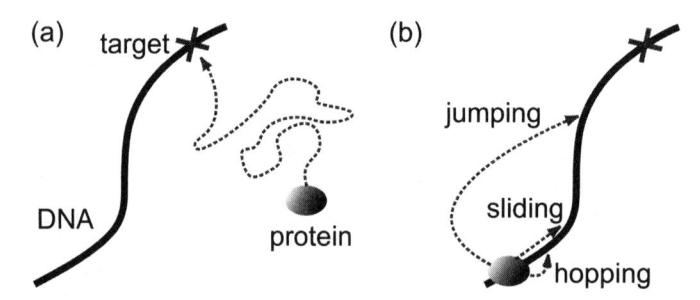

Figure 13.3. Schematic illustrating different protein–DNA interaction modes. (a) The protein diffuses three-dimensionally in solution until it comes close to the specific interaction target site to which it binds. The protein–DNA binding rate in this mode is low because the protein is "unguided" in its search. (b) An alternative binding mode consists of a two-step process: In the first step, the protein binds nonspecifically anywhere to the DNA. In the second step, it undergoes one-dimensional diffusion along the DNA, involving macroscopic dissociation/reassociation ("jumping"), microscopic dissociation/reassociation ("hopping"), intersegment transfer, and sliding, until it finds the target site to which it binds specifically. This interaction mode results in higher binding rates because the protein diffuses in a reduced (one-dimensional) space.

One of the problems of animating an image sequence is that the features in consecutive images are often not aligned with respect to each other. For example, thermal drift causes the imaging area to shift in position, causing the features in the animation to "jump" around. The simplest solution to this problem is to align the images "by hand"—that is, to manually offset consecutive images such that the features of choice overlap as best as possible. This certainly is a possibility, but becomes tedious when the sequence consists of hundreds of images. In the case that the features in the image sequence only change slightly between images, it is possible to align consecutive images automatically by using cross-correlation (van Noort et al., 1999). The cross-correlation of two consecutive images (Image$_{n-1}$ and Image$_n$) is defined as

$$\text{Corr}[\text{Image}_{n-1}, \text{Image}_n](p, q)$$
$$= \sum_{i,j} \text{Image}_{n-1}(i+p, j+q)$$
$$\text{Image}_n(i, j) \quad (13.1)$$

where the parentheses contain the pixel coordinates of the respective image. Equation (13.1) can be expressed in terms of discrete Fourier transforms, $\mathcal{F}$, using the correlation theorem (yielding computational advantages):

$$\text{Corr}[\text{Image}_{n-1}, \text{Image}_n]$$
$$= \widetilde{\mathcal{F}}[\mathcal{F}[\text{Image}_{n-1}]$$
$$(\mathcal{F}[\text{Image}_n])^*] \quad (13.2)$$

The asterisk denotes the complex conjugate and $\widetilde{\mathcal{F}}$ denotes the inverse discrete Fourier transform. To increase the accuracy of the cross-correlation alignment procedure, the images are typically background-subtracted before applying Eq. (13.2). Apart from producing smooth movies, the cross-correlation alignment technique can be used to further analyze the motion of the object of interest, such as the lateral diffusion of DNA fragments in buffer solution that are weakly attached to a mica surface (Jiao et al., 2001). Figure 13.4a shows one of 30 images in the image sequence. A typical cross-correlation between two consecutive images exhibits a well-pronounced peak (Fig. 13.4b). The position of this peak can easily be determined and directly gives the offset by which the two images need to be shifted with respect to each other. Repeating this procedure for all consecutive image pairs yields

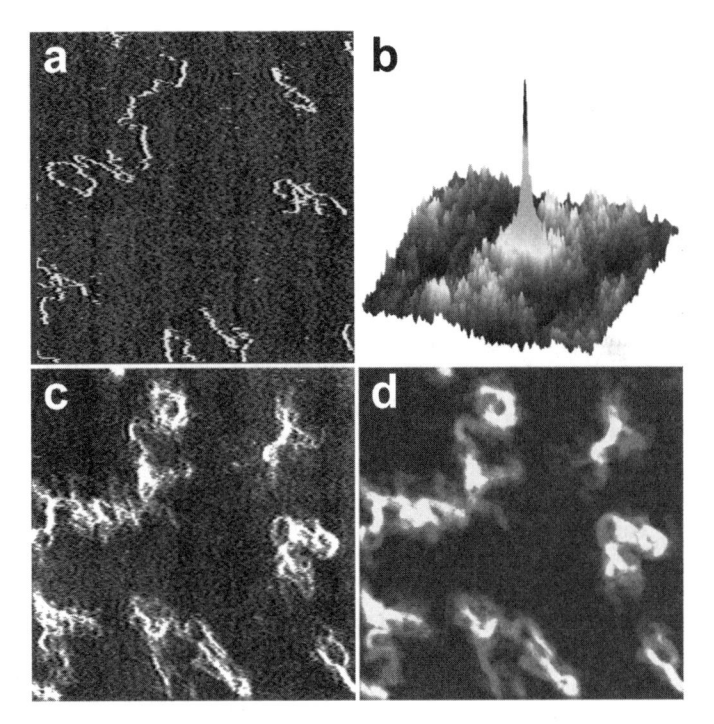

Figure 13.4. Image alignment by cross-correlation of consecutive images. (**a**) One of the original images in the image sequence: DNA fragments that are weakly attached to a mica surface. (**b**) Typical cross-correlation of two consecutive images. The cross-correlation shows a clearly defined peak that is used to determine the lateral offset between the consecutive images and to subsequently align them. (**c**) Average of 30 aligned images. Some parts of the individual DNA fragments exhibit a high mobility and appear as fuzzy areas, whereas other parts are immobile and appear as bright areas. (**d**) Same as (**c**), but the images were binarized and further processed prior to averaging in order to reduce background noise and to more clearly reveal the DNA motion. Image size: 2.3 µm. Grayscale (height) range: 4 nm. [Parts a and c are reprinted from Jiao et al. (2001), with permission from Elsevier.]

a well-aligned image sequence, allowing for further analysis. For example, by averaging all 30 aligned images in the sequence (Fig. 13.4c), it is possible to pinpoint areas of high and low DNA mobility: The mobile sections appear fuzzy while the immobile ones appear bright. It is interesting to note that the DNA fragments seem to be subjected to segmented motion that affect only parts of the molecule, probably caused by local variations on the mica surface (van Noort et al., 1998; Jiao et al., 2001). Using binary thresholding and morphology techniques, the noise in the images can be reduced, and the motion of the DNA fragments can be seen even clearer in the averaged image (Fig. 13.4d).

While the cross-correlation alignment method generally works well for samples that exhibit well-defined features on a flat background, there are some cases where the alignment algorithm is thrown off. This can happen, for example, when the object moves too much between images, or when image distortion or scan errors (e.g., erroneous scan lines) build up. Also, since this alignment algorithm considers all features in the image with equal weight, it might "lock on" to prominent but irrelevant image features.

13.2.4. AFM Images are "Movies" by Themselves

When viewing an AFM image, one is tempted to consider it as a momentary "snapshot" of the sample. Even a single AFM image, however, is a movie by itself in the sense that the scanning tip probes different areas on the sample at different times. In the case of a typical rectangular raster scan, the pixel with pixel coordinates (p, q) is acquired at the time

$$t(p, q) = \frac{1}{f_i n_q}\left(\frac{p}{2n_p} + q\right) \qquad (13.3)$$

where p and q are the (zero-based) pixel coordinates in the fast and slow scan direction, respectively, n_p and n_q are the number of pixels in the direction of p and q, respectively, and f_i is the image (frame) scan rate. This assumes that the tip velocity is constant and that each scan line is scanned once in the forward direction ("trace") and once in the backward direction ("retrace"). The last pixel in an image might be acquired minutes after the first pixel was. An AFM image should therefore rather be considered as a sequence of "pixel snapshots" with a pixel acquisition rate of $f_p = 2n_p n_q f_i$. This mechanism is contrary to that of an image acquired by a CCD camera in optical wide field microscopy, for example. In such a camera, there is a shutter that exposes the whole CCD chip to the light from the sample at the same time. All pixels are therefore acquired simultaneously. (The readout of the individual pixels occurs in a sequential fashion, but this is solely a technical issue.)

There are a number of consequences of this sequential nature of single AFM images. Thermal drift in the microscope that moves the scanning tip relative to the sample might cause distortion in the image, especially when the sample is scanned slowly. Furthermore, in the case where a sample object moves or changes its shape during image acquisition, the object might appear distorted even in the absence of thermal drift. Motion of biomolecules is not uncommon: Certain proteins, for example, change their configuration as they hydrolyze ATP. In general, an object appears distorted when its motion occurs on a timescale that is comparable to or faster than the timescale with which the AFM image is being acquired. The basic properties of such an imaging artifact can be understood with the help of a simple model (Kasas et al., 2000). An object that is modeled as an oscillating half-sphere, for example, appears with horizontal stripes in the image (Fig. 13.5). In the ideal case, the AFM would acquire images fast enough so that the moving object can be assumed stationary in each image resulting in a sequence of undistorted images.

13.3. INCREASING THE IMAGING SPEED

The time resolution in AFM investigations of biological processes is limited by the speed with which image data can be acquired. For biological samples in liquid, a typical image rate corresponds to one image every 1–10 min. Bottlenecks in the path to faster imaging are manifold: the limited detection bandwidth of the sensing cantilever, the limited actuation bandwidth of the scanning piezoelectric elements, and the limited bandwidth of the data acquisition system, to name just a few. There are multiple ways by which the imaging speed can be increased. One possibility is to modify the AFM hardware configuration—for example, by providing compatibility with small cantilevers (Section 13.4). It is possible, however, to scan faster with existing commercial AFMs by employing smart scanning software. Such "indirect" ways are discussed in this section.

13.3.1. Image Tracking

When imaging an object that exhibits lateral mobility, the object might move out of the image area quickly. The object might also be lost when there is a large thermal sample drift. In order to keep the object within the image area during a long period of time, it is therefore necessary to scan an area much larger than the area that is initially occupied by the object. If the motion of the object

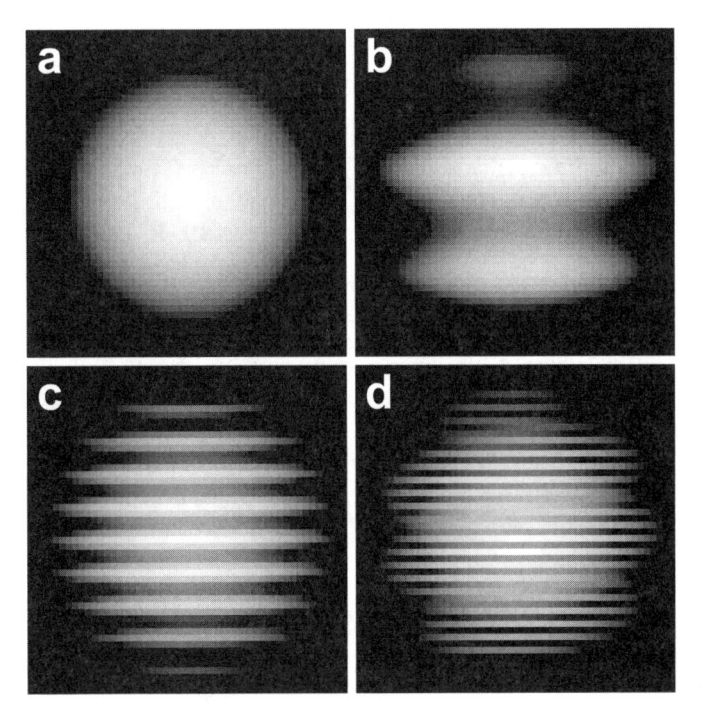

Figure 13.5. Simulation of a moving protein in an AFM. The protein is modeled as a half-sphere that undergoes periodic oscillations in its radius with an amplitude of 20% of the stationary radius and with a frequency of varying multiples, f, of the AFM image frame rate. **(a)** Stationary protein ($f = 0$). There are no imaging artifacts. **(b)** $f = 3$ (i.e., the protein undergoes three cycles while the image is being acquired). The outlines of the protein appear curved. **(c)** $f = 10$. About 10 fairly defined horizontal stripes appear, each one representing a protein cycle. The protein oscillation frequency can therefore be inferred by simply counting the horizontal stripes. **(d)** $f = 27$. This frequency is higher than the "Nyquist frequency" that may be defined as half the number of pixels in vertical direction that cover the protein ($\approx 50/2 = 25$). "Undersampling" therefore occurs, and it no longer is possible to uniquely infer the protein oscillation frequency from the number of horizontal stripes in the image. The protein oscillation frequency is so high that the protein appears as a fuzzy sphere. Moiré patterns appear in the simulation as a consequence of the perfect periodicity of the protein oscillation. These simulations are based on the assumption that the AFM tip does not interfere with the moving protein.

could be accounted for, however, a smaller scan area could be selected which would increase the image rate significantly.

Van Noort et al. (1999) devised a method by which the sample object is tracked and thus kept in the center of the image at all times. They used this method to determine the lateral diffusion coefficient of DNA fragments on a surface. The method is based on a real-time drift correction algorithm using the same cross-correlation image alignment procedure that was discussed in Section 13.2.3: After an image is acquired, it is compared to the preceding image by cross-correlation by which the lateral shift of the object is determined. The voltages that drive the lateral scanner are then offset on-line to shift the scan area such that the object is kept in the center of the image. A small scan area therefore suffices to image the moving object, resulting in higher image rates.

13.3.2. Reducing the Number of Lateral Dimensions

AFM images usually are quadratic in shape. The features of interest in an image, however, might be distributed over just a few scan lines. It might therefore suffice to reduce the number of scan lines and scan a smaller area with a higher image rate. For example, one might be interested in how the height of single molecules that are distributed over a sample surface changes with time. Then it might not be necessary to image a large sample area with a lot of molecules. Instead, it might be adequate to scan even just one line on the sample over and over again, thereby effectively reducing the number of lateral dimensions from two to one. A time series of cross sections is thus created off which the height of the molecules can be read. This method was utilized in investigations of GroEL/GroES chaperonin proteins (Viani et al., 2000). GroEL and GroES form a complex in order to help other proteins fold into their native state. By scanning in one lateral dimension only, it was possible to record protein "tubes," thereby resolving the association and dissociation of individual complexes (Fig. 13.6).

An imaging technique that especially benefits from a reduced number of lateral dimensions is the so-called "force mapping" technique, in which thousands of force spectra are recorded and the image is reconstructed from a quantitative analysis of each of the individual force spectra (Radmacher et al., 1994b; Baselt and Baldeschwieler, 1994; Jiao and Schäffer, 2004). The force mapping technique is capable of measuring the locally resolved adhesion or elasticity of the sample, in addition to the topography. By limiting the scan area to a single repetitive line, it was possible to investigate the process of furrow stiffening during cell division (Matzke et al., 2001).

The reduction in the number of lateral dimensions can be pursued one step further by giving up the second lateral dimension as well and keeping the tip stationary at a fixed lateral position. The tip can then monitor the variations in sample height at that position with an even higher temporal resolution. The motion of a single protein, for example, can be monitored in real time

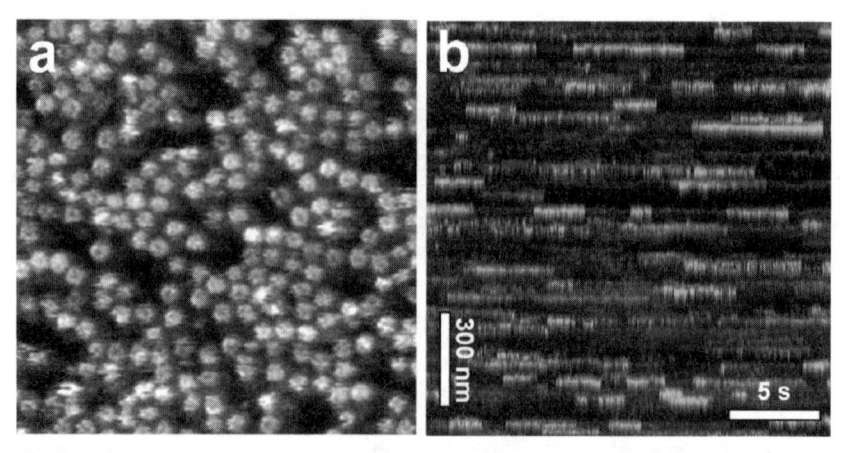

Figure 13.6. Example where the time resolution of the measurement was increased by reducing the number of lateral dimensions in the AFM image. (**a**) Single GroEL protein molecules adsorbed to a mica surface in buffer solution. The proteins appear as (bright) rings surrounding the (dark) central channel. Small cantilevers were used to provide a high imaging speed. Image size: 300 nm; grayscale range: 15 nm. (**b**) Motion along the slow (here: horizontal) scan axis was disabled. The tip thus repeatedly scanned the same (vertical) line, over the same proteins. Those consecutive, one-dimensional vertical line scans were charted next to each other in horizontal direction, effectively generating protein "tubes." Each protein "tube" corresponds to a single protein molecule whose height fluctuations can now be monitored with a $\approx 250 \times$ higher temporal resolution (≈ 100 ms) than what would be possible via regular images having two spatial dimensions (≈ 25 s). By using protein "tubes," it was possible to monitor the association and dissociation of the GroEL–GroES complex in the presence of ATP in real time. [Reprinted from Viani et al. (2000), by permission of Nature Publishing Group.]

by this approach (Radmacher et al., 1994a). One problem with this approach is that the tip eventually drifts away from its initial lateral position. For typical globular proteins, for example, a few nanometers of drift is enough to move the tip off the protein. To increase the duration with which the motion of a single protein can be monitored, a technique dubbed "protein tracking" has been developed (Thomson et al., 1996). In this technique, lateral feedback is applied to the tip: In certain time intervals, the tip is moved in a single horizontal and subsequently in a single vertical scan line over the protein, whereby the new lateral position of the protein is determined. Then the tip is offset to this new position and therefore is located on top of the protein again, where it resumes monitoring the height variations of the protein. This technique is particularly effective since it combines the advantages of reducing the number of lateral dimensions and of image tracking.

13.3.3. Exceeding the Feedback Limit

In regular contact or tapping mode imaging, there is a feedback loop trying to keep the so-called error signal (tip deflection in the case of contact mode and tip amplitude in the case of tapping mode, subtracted by the setpoint) zero, thereby keeping the scanning tip at a constant deflection/amplitude above the sample surface. If the sample surface is scanned at high speeds, the bandwidth of the feedback system might not be large enough to provide perfect surface tracking. Instead, the error signal deviates further from zero, reminiscent of varying imaging forces. If the sample is robust enough, however, such that it is not significantly deformed or damaged by larger imaging forces, the error signal can be combined with the feedback signal (the voltage applied to the z-scanner) in order to reconstruct a better estimation of the true sample topography (Ookubo and Yumotoa, 1999).

13.3.4. Video-Rate Imaging

While the methods that were discussed so far offer some useful advantages, it seems desirable that eventually an AFM could be designed that is capable of imaging with video rate. Video rate corresponds to 25 frames per second and 576 visible horizontal lines per frame for the PAL video standard (29.97 frames per second and 480 visible horizontal lines per frame for the NTSC video standard, respectively). The uncompressed video data bit rate is on the order of 200 Mbps for 24-bit color. Using digital data compression (MPEG), the data rate can be reduced by a factor of 20–50 down to 4–10 Mbps. In one of the six relatively new high-definition TV (HDTV) standards, there are 30 frames per second and a frame consists of 1920 lines with 1080 pixels ("pels") each. The uncompressed bit rate, $\approx$1500 Mbps, is significantly higher than that for the PAL/NTSC standards. With compression, however, the HDTV video signal bit rate can be reduced to about 20 Mbps.

An AFM that could image with video rate would need to scan 25–30 images per second, each having on the order of 500×500 pixels with 24-bit resolution each (depending on the particular video standard). The data rate would be up to 200 Mbps ($\approx$25 Mbytes per second). High-speed data acquisition systems are currently being developed approaching this goal (Fantner et al., 2005; Rost et al., 2005). With such a data rate, a 1-GB hard disk could store only about 40 s of uncompressed image data. It therefore seems clear that image compression will play a significant role in video-rate AFM imaging. But while video-rate compression and data handling technology is already developed and could easily be adapted, there are limitations of current AFM hardware that need to be overcome before video-rate AFM imaging will be possible. The next section introduces small cantilevers as a key toward high-speed imaging of biological samples in liquid.

13.4. SMALL CANTILEVERS

13.4.1. Limits of Conventional Cantilevers

A typical conventional AFM cantilever that is used for imaging in liquid is made of silicon or silicon nitride and is 80–200 μm in length, 20–40 μm in width, and 0.4–1 μm in thickness. Resonant frequencies of those cantilevers in air or vacuum are typically 10–60 kHz, consistent with the prediction from continuum mechanics,

$$f_{\text{res}} \cong \frac{1.02}{2\pi} \sqrt{\frac{E}{\rho}} \frac{T}{L^2} \quad (13.4)$$

where f_{res} is the resonant frequency of the first (fundamental) transverse vibration mode in vacuum and E, ρ, T, and L are the Young's modulus, the density, the thickness, and the length of the cantilever, respectively. Typical quality (Q) factors are 10–100. Upon immersion in aqueous solution, the resonant frequencies drop by a factor of 3–5 to 2–30 kHz. This drop is caused by the additional mass of the liquid in a boundary layer around the cantilever that the cantilever needs to displace during its oscillation. The drop in the Q factor is even more dramatic because typical Q factors in water are on the order of one.

Typical imaging speeds in tapping mode in liquid with conventional cantilevers are 5 min per image consisting of 2×512^2 pixels (trace and retrace). This translates to a pixel acquisition rate of about 2000 pixels per second. In tapping mode in liquids, the cantilever is typically driven on or below its resonant frequency in the liquid. Driving above the resonant frequency—for example, by utilizing higher-order vibrational modes—does not usually allow for stable imaging. Driving below the resonant frequency, on the other hand, is easily possible since the low Q factors cause the amplitude response function of the cantilever to be approximately flat below the resonant frequency. In fact, multiple sharp resonance peaks in the cantilever oscillation spectrum ("forest of peaks") that stem from resonances in the acoustic driving system (Schäffer et al., 1996) usually influence the choice of the particular tapping frequency. For a typical tapping frequency of 10 kHz, there are about 5 "taps" per pixel; that is, the tip probes the sample surface five times before "deciding" which value the pixel should get. Ideally, one tap per pixel should be about sufficient, because the Q factor of the cantilever, a measure of how many oscillations are needed for the cantilever amplitude to adjust to a new value, is on the order of one. Nevertheless, these imaging speeds are close to the limit of what is physically possible with conventional cantilevers.

13.4.2. Increasing the Resonant Frequency of Cantilevers

In order to image faster, cantilevers with higher resonant frequencies are required. This can be achieved by varying the parameters in Eq. (13.4). The material parameters E and ρ have only a weak (square-root dependency) effect on the resonant frequency. The most direct way of increasing the resonant frequency is therefore either increasing the thickness or decreasing the length of the cantilever. An increase in thickness, however, strongly increases the spring constant of the cantilever:

$$k = \frac{E}{4} W \left(\frac{T}{L}\right)^3 \quad (13.5)$$

where W is the cantilever width. Large spring constants ($k > \approx 1$ N/m) are undesirable for imaging soft samples, since they cause higher imaging forces that more easily destroy the sample. The requirement that the spring constant be unchanged upon a change in cantilever dimensions results in the requirement that $T \propto L$ (neglecting the rather weak influence of the width), and consequently $f_{\text{res}} \propto L^{-1}$. Therefore, only by a reduction of the cantilever length (together with a simultaneous reduction of the cantilever thickness) can one obtain high-resonant-frequency cantilevers suitable for the imaging of soft biological samples. It should be noted that if all spatial dimensions are reduced simultaneously by the same ratio, the spring constant scales with length, and the resonant frequency scales with reciprocal length.

Two prominent materials of which small cantilevers have been fabricated is silicon nitride (Reiley et al., 1995; Walters et al., 1996; Viani et al., 1999a) and silicon (Hosaka et al., 2000). Calculated spring constants [Eq. (13.5)] and resonant frequencies [Eq. (13.4)] are displayed for silicon nitride cantilevers of various dimensions (Table 13.1). The resonant frequency increases by a factor of over 150 from 18 kHz for the longest to 2.9 MHz for the shortest cantilever. The spring constant increases by a factor of $\approx$ 6 only. It becomes apparent that the smaller the cantilever, the more difficult it is to keep its spring constant small. This is because there are several factors that set a lower limit to the cantilever thickness. One is of technical nature, because thin free-standing structures generally impose a challenge in microfabrication processes. Another one is connected to the optical detection mechanism of cantilever deflections: The reflectivity of a silicon or silicon nitride cantilever becomes small for small thicknesses, especially when the cantilever is submerged in a liquid. To increase the reflectivity, the cantilever could be coated with a thin (10–30 nm) layer of a metal. But a thin silicon/silicon nitride–metal bilayer is highly susceptible to temperature variations which cause artifact deflections of the cantilever ("bimetallic strip effect") that are undesirable for imaging applications. A solution to this problem is to coat the cantilever on both sides with a metal layer of equal thickness. In either case, the mass of the cantilever increases significantly, thereby decreasing the resonant frequency. The bimetallic strip effect can be minimized by coating the cantilever on an area close to the tip only, whereby creating a "reflector pad" on which the laser can be positioned (Viani et al., 1999a). A coating process can be avoided by fabricating the cantilever entirely out of metal in the first place (Schäffer et al., 1997; Chand et al., 2000).

13.4.3. Properties of Small Cantilevers

Figure 13.7a shows an array of small aluminum cantilevers. The cantilevers are

TABLE 13.1. Calculated Spring Constants and Resonant Frequencies for Silicon Nitride Cantilevers of Various Sizes[a]

Length (μm)	Width (μm)	Thickness (μm)	Spring Constant (N/m)	Resonant Frequency (kHz)
200	20	0.5	0.020	18.2
100	10	0.5	0.078	72.9
20	5	0.1	0.039	364
10	2	0.1	0.125	1458
5	2	0.05	0.125	2916

[a] The calculations were performed using Eqs. (13.4) and (13.5) with $E = 250$ GPa and $\rho = 3100$ kg/m^3.

2.5 μm wide and 8.8–40 μm long. Their resonant frequencies are 0.11–2.5 MHz in air (Fig. 13.7b), and they follow a curve that is well-fitted by the predicted L^{-2} dependency [Eq. (13.4)]. In water, the resonant frequencies drop by a factor of 2.5–4, depending on cantilever shape (Hodges et al., 2001), where this factor is smaller for higher frequencies (Fig. 13.7b). Still, the smallest cantilever has a resonant frequency of about 1 MHz in water. The theory of Sader (1998) describes the vibrational properties of cantilevers in liquids and provides excellent agreement with the measured values.

One of the difficulties in fabricating small cantilevers is designing a microfabrication process by which integrated tips are produced. There have been successful attempts to fabricate metal cantilevers with integrated silicon tips (Chand et al., 2000). Another method is microfabricating tipless cantilevers

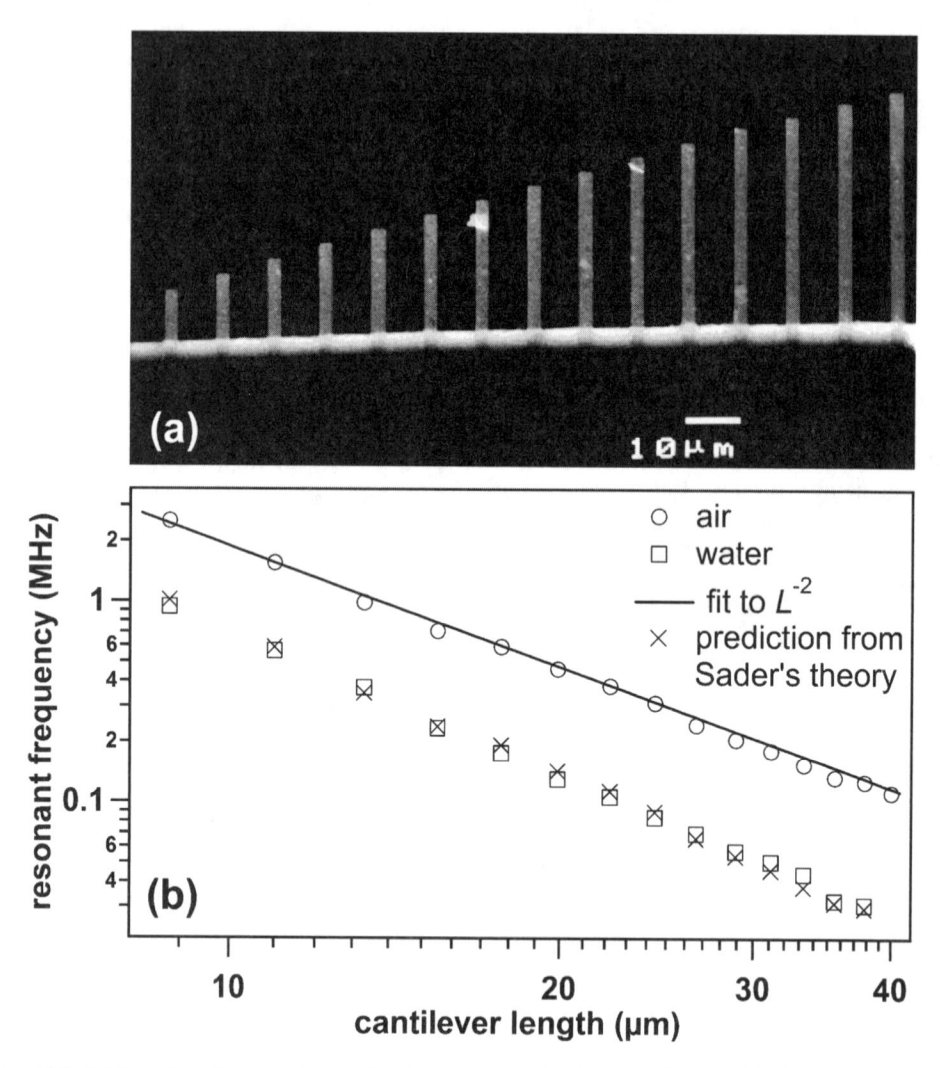

Figure 13.7. (a) Scanning electron micrograph of an array of aluminum cantilevers, 8.8–40 μm in length and 2.5 μm in width. (b) Resonant frequencies of these cantilevers. [Reprinted with modifications from Schäffer et al. (1997), by permission of the International Society of Optical Engineering.].

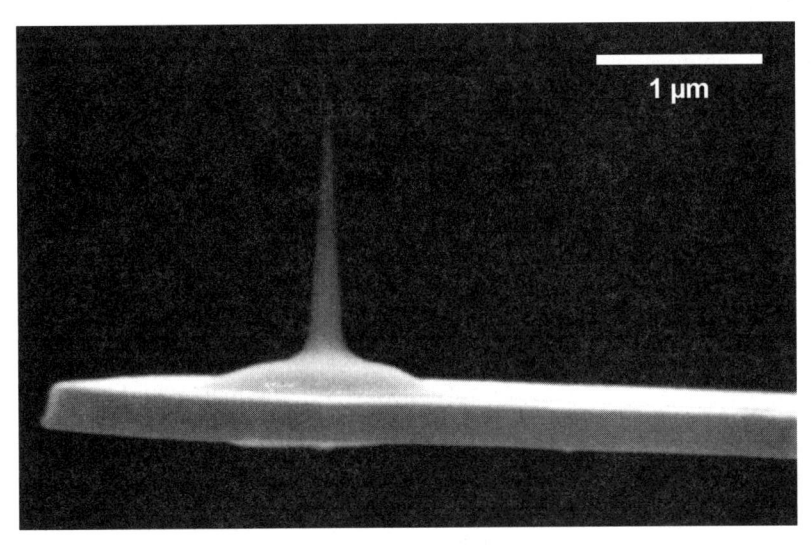

Figure 13.8. Scanning electron micrograph of a tip grown by electron beam deposition (EBD) onto a small cantilever. Such EBD tips have high aspect ratios and small tip radii and are well-suited for the imaging of biomolecules.

first and then manually growing tips on them in an electron microscope by the electron beam deposition (EBD) technique (Fig. 13.8). In this technique, the focused electron beam deposits material at a selected location on the cantilever, thereby growing a tip (Akama et al., 1990), just like a stalagmite in a stalactite cavern is grown. To enable large-scale production of EBD tips, an automated technique based on pattern recognition was developed for the growth of EBD tips on whole wafers of cantilevers (Kindt et al., 2004b).

13.4.4. Thermal Cantilever Noise

Small cantilevers not only have high-speed properties, but also exhibit a small thermal noise level. Thermal noise is one of the limiting factors in many AFM measurements, and so it is important to utilize the low-noise properties of small cantilevers. In order to simplify the discussion of why the thermal noise of a cantilever decreases with cantilever size, the remainder of this section is restricted to applications of contact-mode-type measurements. For the purpose of this section, the cantilever is modeled as a simple harmonic oscillator. Such a model is fairly accurate when the interactions with the sample surface and with the surrounding medium are weak. A simple harmonic oscillator responds to a periodic excitation force, $F(f)$, with an oscillation of amplitude

$$A(f) = G(f)F(f) \quad (13.6)$$

$G(f)$ is the transfer function of the oscillator that gives the (positional) amplitude per unit excitation force as a function of excitation frequency:

$$G(f) = \frac{1/k}{\sqrt{\left(1 - \frac{f^2}{f_{\text{res}}^2}\right)^2 + \left(\frac{f}{f_{\text{res}}Q}\right)^2}} \quad (13.7)$$

Q is the quality factor of the simple harmonic oscillator. It is related to the coefficient of viscous damping, γ, by

$$Q = k/(2\pi f_{\text{res}}\gamma) \quad (13.8)$$

For an excitation force at dc ($f = 0$), $G(f) = 1/k$, and Eq. (13.6) transforms into Hooke's law, $A = F/k$, relating the static position of the oscillator to the applied static force. For an excitation force at the resonant frequency, we have $G(f) = Q/k$, and the amplitude response becomes $A(f_{res}) = QF(f_{res})/k$, expressing the fact that a weakly ($Q \gg 1$) damped oscillator reacts to forces especially well on resonance.

In thermal equilibrium, a simple harmonic oscillator undergoes Brownian motion. This motion can be modeled by assuming a random thermal excitation force with a spectral force density of

$$\overline{F}_{noise} = \sqrt{4k_B T \gamma} \quad (13.9)$$

representing the so-called Nyquist relation, where k_B is the Boltzmann constant and T is the absolute temperature. $\overline{F}_{noise}$ has units of [N/$\sqrt{Hz}$] and describes the amount of thermal excitation force per unit of the square root of the bandwidth. This thermal excitation force density is independent of frequency ("white" spectrum): $\overline{F}_{noise}(f) = \overline{F}_{noise}$. Just like in Eq. (13.6), the transfer function $G(f)$ relates the thermal excitation force density to the thermal response amplitude density: $\overline{A}_{noise}(f) = \sqrt{4k_B T \gamma} G(f)$. At frequencies sufficiently below the resonant frequency, the transfer function is approximately constant and the total RMS thermal noise in a given measurement bandwidth, Δf, is given by $A_{noise} = \sqrt{4k_B T \gamma \Delta f}/k$. The signal-to-noise ratio (SNR) of a measurement of a force signal, F, sufficiently below the resonant frequency therefore becomes

$$\text{SNR} = \frac{A_{signal}}{A_{noise}} = \frac{F}{\sqrt{4k_B T \gamma \Delta f}} \quad (13.10)$$

Since the measurement bandwidth is proportional to the measurement speed, it is instructive to rewrite Eq. (13.10) as

$$\text{SNR}\sqrt{\Delta f} = \frac{F}{\sqrt{4k_B T \gamma}} \quad (13.11)$$

There are two important consequences of this result. First, there is a tradeoff between the signal-to-noise ratio and the measurement bandwidth: Speeding up the measurement by a factor of 2 decreases the signal-to-noise ratio by a factor of $\sqrt{2}$. Second, there are two ways by which the product in Eq. (13.11) can be increased: reducing the temperature or reducing the coefficient of viscous damping. Since the former is not a viable option for most experiments with biomolecules in solution, reducing the coefficient of viscous damping is the only practical way of gaining either signal-to-noise ratio (when the speed is fixed) or speed (when the signal-to-noise ratio is fixed). The most direct way of reducing the coefficient of viscous damping is to reduce the size of the cantilever (Viani et al., 1999a). To demonstrate the advantage of small cantilevers, the amplitude noise density is plotted for two simple harmonic oscillators in thermal equilibrium that model two different cantilevers having the same spring constant but different coefficients of viscous damping (Fig. 13.9). Figure 13.9a shows the amplitude noise density for a simple harmonic oscillator that models a large cantilever, and Fig. 13.9b shows the amplitude noise density for a simple harmonic oscillator that models a small cantilever having a 10× higher resonant frequency and a 10× lower coefficient of viscous damping. In the given measurement bandwidth (0–10 kHz), the thermal RMS amplitude noise is lower by a factor of $\sqrt{10}$ for the small cantilever than it is for the large cantilever.

It should be noted that in Fig. 13.9 we applied the simple harmonic oscillator model despite the fact that the fluid dynamics in the surrounding medium is more complicated, especially when the cantilever experiences squeeze-film damping in proximity to a surface (Serry et al., 1995). Nevertheless, the simple harmonic oscillator model qualitatively describes the vibrational behavior of such cantilevers in a liquid medium and sufficed for the purposes of this section. A better model for cantilever vibrations in a viscous medium, however, does exist (Sader, 1998).

The results of this section give a somewhat complementary reasoning with respect to the previous sections of why small cantilevers perform better than large ones.

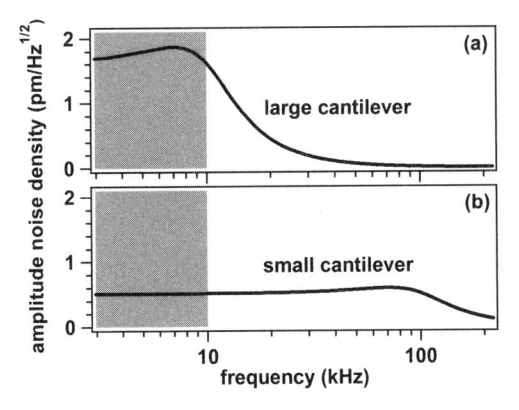

Figure 13.9. Calculated amplitude noise density for a large and for a small cantilever in thermal equilibrium, based on a simple harmonic oscillator model. Compared are the respective fundamental vibration modes, both having the same spring constant ($k = 0.1$ N/m) and temperature ($T = 25°$C), but different coefficients of viscous damping and resonant frequencies. (a) Large cantilever [$\gamma = 1.59 \times 10^{-6}$; $f_{res} = 10$ kHz ($Q = 1$)]. (b) Small cantilever [$\gamma = 1.59 \times 10^{-7}$; $f_{res} = 100$ kHz ($Q = 1$)]. In a given measurement bandwidth (here: 0–10 kHz, highlighted area), the thermal amplitude noise of the small cantilever (≈ 0.51 Å$_{RMS}$) is smaller by a factor of $\sqrt{10}$ than that of the large cantilever (≈ 1.6 Å$_{RMS}$).

Finally, it should be noted that the reduced viscous drag coefficient of smaller cantilevers also reduces artifact deflections due to viscous drag effects during scanning. Such artifact deflections are particularly critical in force spectroscopy experiments (Rief and Grubmüller, 2002; Viani et al., 1999a, b; Janovjak et al., 2005).

13.4.5. Small Focused Spot Size

Current commercial microscopes that use the optical beam deflection ("optical lever") detection method have focused spot diameters on the order of 20–50 μm. With such large diameters, not enough light is reflected off a cantilever onto the detector when using cantilevers as small as 5 μm in length and 2 μm in width. In order for all (or most) of the incident beam power to actually reach the detector, the incident beam needs to be focused to the size of the cantilever or smaller, imposing new design criteria for the construction of AFMs for small cantilevers. A basic relationship between the $1/e^2$ diameter of the collimated incident Gaussian beam, d, and the $1/e^2$ diameter of the focused spot, w_0, is given by

$$w_0 = \frac{4\lambda f}{\pi n d} \quad (13.12)$$

where λ is the wavelength (in vacuum) of the incident beam, f is the effective focal length of the lens system, and n is the index of refraction of the medium (Fig. 13.10). For a focused spot diameter of 1–2 μm, an effective numerical aperture (with respect to the $1/e^2$–beam diameter) of 0.4–0.2 is required when using red light with $\lambda = 670$ nm. Standard commercial microscope objectives easily achieve such numerical apertures and can be used for generating small spots, but their large physical size makes them not optimal for use in an AFM setup that should be mechanically as rigid and compact as possible. Using raytracing software, a compact lens system with small spherical aberrations was designed that provides a nearly diffraction-limited spot diameter of about 1.6 μm (Schäffer et al., 1997).

There are a number of consequences when requiring a small focused spot diameter in an AFM for small cantilevers. A relatively large numerical aperture, for example, is associated with a relatively large half-opening angle,

$$\Theta = \arctan \frac{d}{2f} \quad (13.13)$$

of the incident beam (Fig. 13.10). The incident and the reflected beam therefore occupy a large angular space. One way of dealing with this issue is spatially overlapping the beams and using polarizing optics to separate them again (Schäffer et al., 1996). Another consequence of having a small focused spot diameter is that a small diameter causes a small depth of focus. The depth of focus is the range in propagation direction of the

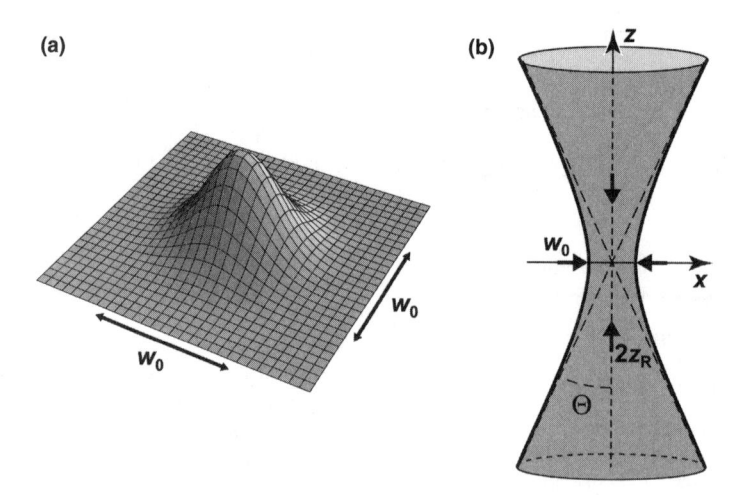

Figure 13.10. Diffraction of a focused circular Gaussian beam. (a) Gaussian intensity distribution in the focal plane ($z = 0$). The focused spot diameter, w_0, is defined as the diameter of the circular line in which the intensity is $1/e^2$ times the maximum intensity. (b) $1/e^2$ intensity profile along the propagation direction of the beam (z axis). The beam does not converge to a point in the focal (xy) plane but remains finite in size (w_0 is marked by the two opposing horizontal arrows). The Rayleigh range, $2z_R$, is the length in the z direction over which the beam diameter remains below $\sqrt{2}w_0$ (marked by the two opposing vertical arrows). The larger the half-opening angle of the incident beam, Θ, the smaller the focused spot diameter and the smaller the Rayleigh range.

beam in its focus over which the beam diameter remains approximately collimated. The Rayleigh range, $2z_R$, serves as a measure for the depth of focus and is defined as the propagation length in which the beam diameter remains below $\sqrt{2}w_0$:

$$2z_R = \frac{\pi n w_0^2}{2\lambda} \quad (13.14)$$

The Rayleigh range defines the "diameter" of the focused spot in axial direction and depends quadratically on w_0. For $w_0 = 2$ μm, $n = 1.33$, and $\lambda = 670$ nm, the Rayleigh range is $2z_R \approx 12$ μm. It therefore becomes important to refocus the incident beam onto a newly mounted cantilever on a one-by-one basis, whereby an optical viewing system with a CCD camera can be of help. It should be mentioned that when using a laser diode as light source, the incident beam is not circular as in Fig. 13.10, but elliptical in shape (unless optical elements are used to artificially circularize the beam). The consequence of this is that the focused spot is elliptical in shape, too. This does, however, not usually cause a problem, as long as the long axis of the focused spot is aligned along the cantilever length. So it usually is the small diameter of the focused spot (along the cantilever width) that is critical and relevant in Eqs. (13.12) and (13.14). A design of a compact AFM for small cantilevers that produces a focused spot of 1.6 μm in diameter is shown in Fig. 13.11.

When the focused spot diameter is a significant fraction of the cantilever length, the deflected cantilever acts as a curved mirror for the incident beam and therefore distorts the reflected beam. In this case, diffraction theory needs to be used for a quantitative analysis of the detection sensitivity (Schäffer and Hansma, 1998; Schäffer, 2002). This becomes especially important when analyzing the motion of higher-order cantilever vibration modes (Stark, 2004; Schäffer and Fuchs, 2005), when assessing the amount of thermal noise in AFM measurements (Schäffer, 2005), or when using thermal noise for a calibration of the cantilever spring constant (Proksch et al., 2004).

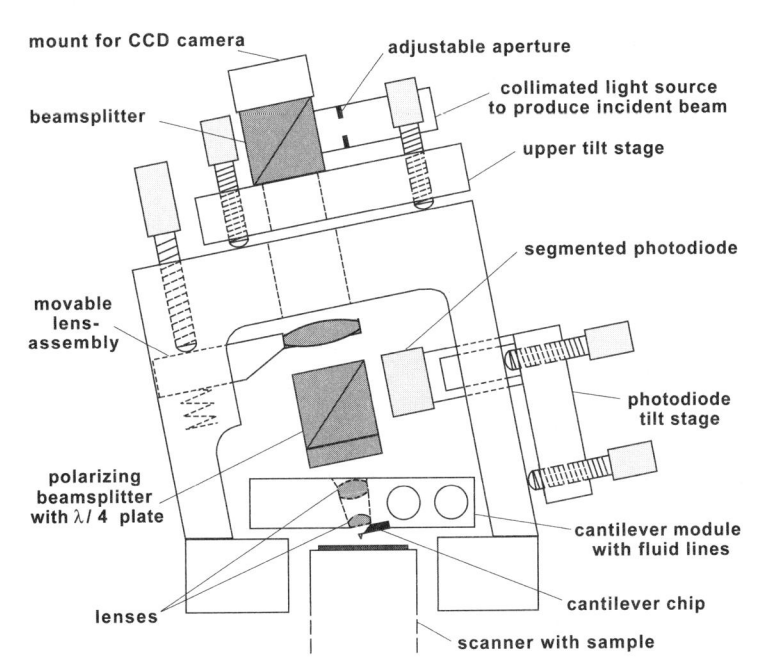

Figure 13.11. Schematic of a head for small cantilevers. A collimated and polarized light source provides the incident beam that is focused onto the cantilever by a movable lens assembly. The reflected beam is separated from the incident beam by the polarizing beamsplitter–$\lambda/4$-plate assembly and is projected onto the photodiode detector. A top view mount provides optical access and aids in the focusing process. [Reprinted from Schäffer et al. (1997), by permission of the International Society of Optical Engineering.]

13.4.6. Fast, Low-Noise Detector

In Section 13.4.4, the result was derived that the thermal cantilever noise depends only on the coefficient of viscous damping. In particular, it is independent of the spring constant. For the detection of small cantilever deflections, however, a smaller spring constant is of advantage, since it increases the magnitude of the transfer function [Eq. (13.7)], which therefore causes larger deflections for a given force acting on the cantilever. But it was shown above that there is a lower limit to the thickness with which small cantilevers can be fabricated, thereby setting a lower limit to the spring constant as well. Also, for some applications it is desirable to have a large spring constant (e.g., for "position-clamp" type of experiments). For small cantilevers, low-noise detection therefore becomes particularly important, because only with low-noise detection can the low-noise properties of small cantilevers be harvested.

For the widespread optical beam deflection ("optical lever") detection method, the significant noise sources are connected with the generation and the detection of the light beam: (a) time-correlated fluctuations/drift in the power of the beam ("intensity noise"), (b) time-correlated fluctuations/drift in the shape/direction of the beam ("pointing noise"), and (c) time-uncorrelated fluctuations in the photodetection process (shot noise). Other noise sources (dark current noise, electronic Johnson noise) are usually not limiting, but electronic $1/f$ noise can be a factor in dc measurements (contact mode, force spectroscopy). Intensity and pointing noise is caused by instabilities in the laser cavity, resulting in mode-hopping and mode competition. Pointing noise can be transformed into intensity noise by the use of a single-mode optical fiber (Cleveland et al.,

1995), and intensity noise can be minimized by the detection circuitry with the help of a fast divider (although analog dividers introduce noise, too). Shot noise arises from the fact that photons are emitted at random points in time (resulting in Poisson statistics). The current produced by the electrons in the photodetection process, where photons cause the separation of electron–hole pairs, therefore exhibits random statistical fluctuations in its magnitude [with a flat ("white") frequency distribution]. These fluctuations increase with the square root of the beam power and are a fundamental limit to the detection sensitivity. They also impose a fundamental limit on the scan speed, since the faster the measurement, the larger the required bandwidth, and thus the higher the noise level in the measurement.

Several methods have been devised to increase the sensitivity of the photodetection in an AFM. Since the beam deflection detection method measures the lateral shift of a light beam, the particular size and shape of the beam is important. The size and shape of the focused spot should be tailored to the size and shape of the particular cantilever for optimum signal-to-noise ratio. One way of achieving this is the use of an adjustable aperture in the incident beam path. By using such an adjustable aperture, the signal-to-noise ratio of a particular measurement of cantilever deflections was increased by a factor of three (Schäffer and Hansma, 1998).

The detection sensitivity can further be increased by the use of an array detector (Schäffer et al., 2000). Such an array detector consists of an array of photodetector segments over which the reflected beam is distributed (Fig. 13.12). The individual signals from each segment are first weighted by amplifying them with individual gain factors. Then all the weighted signals from all the segments are added, producing the cantilever deflection signal that is further used in the feedback or data acquisition system. The interesting

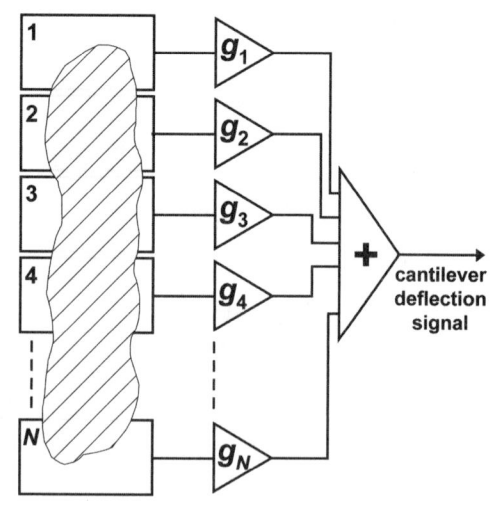

Figure 13.12. Schematic of the array detector. The reflected beam (highlighted area) is spread across an array of photodetector segments (1, 2, ..., N). The signal from each segment is amplified by an individual weighting factor ($g_1, g_2, ..., g_N$) that can be set dynamically. The weighted signals are then added to form the cantilever deflection signal. Such an array detector can be programmed to dynamically adapt to the experimental conditions. For each condition, a set of weighting factors can be found that maximizes the detection sensitivity.

property of such a detector is that a set of gain factors can be found that optimizes the detection sensitivity for any given experimental condition. The array detector can be programmed to dynamically adapt to varying experimental conditions. Depending on the particular shape of the spot on the detector, significant improvements in the signal-to-noise ratio of measurements of cantilever deflection can be obtained. The more distorted and jagged the beam on the detector, the larger the improvement. In a particular case of a 12-μm-long cantilever, an increase in the signal-to-noise ratio by a factor of 5 was achieved (Fig. 13.13). Apart from this increased signal-to-noise ratio, an array detector also provides an increased measurement range, a property that is especially important for the recording of force-distance spectra (Schäffer, 2002).

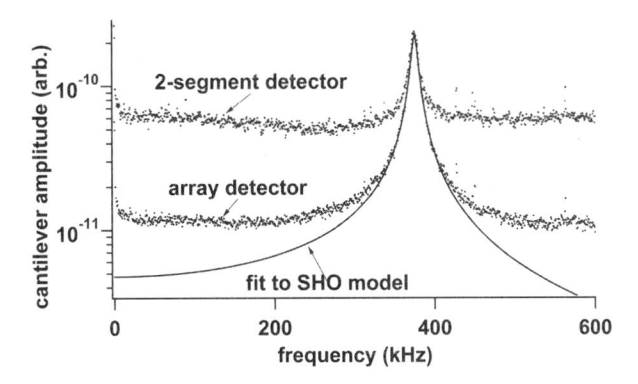

Figure 13.13. Thermal noise amplitude density of a small cantilever ($L = 12$ µm), measured with a two-segment detector and an array detector. Both detectors faithfully identify the thermal resonance peak, but the detection noise level (off-resonance) is lower by a factor of 5 in case of the array detector. [Reprinted from Schäffer et al. (2000), by permission of the American Institute of Physics.]

13.4.7. Fast Scanner

In the AFM, piezoelectric scanners are routinely used for the generation of the scan motion. Those scanners typically consist of ferroelectric lead zirconate titanate (PZT) ceramics that is made piezoelectric by "poling" [applying a large electric field ($> \approx 2000$ V/mm) while the material is heated above the Curie temperature ($T_C \approx 200$–$400°$C)]. A piezoelectric material expands or contracts upon application of a voltage due to the (inverse) longitudinal or transverse piezoelectric effect. Tube scanners are conventionally used since they can be designed such that they independently bend and change their length. By mounting one end of such a tube scanner to a rigid surface and attaching a sample to the other end, the sample can be scanned in three dimensions. A limit to the speed with which the sample can be moved is set by the resonant frequency of the scanner. The fundamental longitudinal resonant frequency of a prismatic (constant cross section) elastic rod of length L that is clamped at one end and free at the other end is

$$f_{\text{res}}^{\text{long}} = \sqrt{\frac{E}{\rho}} \frac{1}{4L} \qquad (13.15)$$

where E and ρ are Young's modulus and the density of the rod, respectively. For typical PZT ceramics, $\rho \approx 7700$ kg/m^3 and $E = E_{11} \approx 65$ GPa (Young's modulus in the direction of the rod elongation, which is usually perpendicular to the polarization axis of the ceramics). A typical piezo tube that may be 5 cm in length therefore has a free resonant frequency of about 14.5 kHz. The extra mass, M, of the sample and mounting devices like magnets that are added to the free end of the tube lowers the resonant frequency of the tube by a factor of about $(1 + 3M/m)^{-1/2}$, where m is the mass of the free tube (this uses the fact that the effective mass of the tube is about one-third of its true mass). In the end, a typical longitudinal resonant frequency of such a scanner may be on the order of 10 kHz, and the resonant frequencies in the bending modes may even be smaller. The most direct way to create fast scanners, just like in the case of cantilevers, is to make them smaller [note that the resonant frequency in Eq. (15) inversely scales with the length]. If the size of the scanner is reduced by a factor of 10, the resonant frequency increases to the order of 100 kHz. Small multilayer PZT scanners (so-called "piezo stacks") with free resonant frequencies of 260 kHz have been used to build a scanner with a mechanical resonant frequency above 60 kHz that was used to scan consecutive 100 × 100 pixel images in a liquid environment at an image rate of 12.5 Hz (Ando et al., 2001). System resonances could be eliminated by an active damping method (Kodera et al., 2005). Using finite element analysis, it was possible to significantly increase the resonant frequency of a mechanical scanning structure to 23.4 kHz (Kindt et al., 2004a). An image rate of 80 Hz (128 × 128 pixels) was demonstrated on highly oriented pyrolytic graphite (HOPG) with a scanning tunneling microscope (STM) at atomic resolution in vacuum (Rost et al., 2005). By using a miniature scanner based on a mechanical microresonator, images of soft molten polymer surfaces in air could be acquired at a rate of 70 Hz (Humphris et al., 2005). Even smaller scanners can be made by integrating a zinc oxide piezo actuator directly on the cantilever (Sulchek et al., 2000a). Such self-actuated cantilevers offer a large imaging bandwidth, but are currently not well-suited for imaging delicate biological samples due to their large spring constants. Recently, however, an insulated piezoelectric cantilever was used to produce tapping mode images of *E. coli* bacteria in liquid with a tip speed of 75.5 μm/s (Rogers et al., 2003).

13.4.8. Fast Feedback Loop

In tapping mode, the cantilever is oscillated at the drive frequency. The signal from the optical detector therefore is also oscillatory. The amplitude of the oscillations is obtained by feeding the signal into an RMS-DC converter. This amplitude accounts for the imaging force and is used as input to the subsequent feedback loop that tries to keep the amplitude constant ("constant amplitude mode"). A conventional RMS-DC converter works by first rectifying the input signal and subsequently low-pass filtering it. The low-pass filter needs to be designed such that it completely separates the modulation signal from the carrier signal (at the drive frequency), typically requiring a low-pass cutoff frequency substantially lower than the drive frequency. (Actually, the RMS-DC converter shifts the carrier signal completely to twice the drive frequency when the deflection signal is ac-filtered.) In other words, a conventional RMS-DC converter typically requires several carrier cycles for the conversion. To circumvent this limitation, a novel RMS-DC converter was designed that is based on peak detection, achieving higher conversion rates (Ando et al., 2002).

In conventional AFMs, a proportional-integral-differential (PID) feedback controller is used to keep the error signal (deflection in contact mode, amplitude in tapping mode) constant. A PID feedback loop, however, is limited to a speed set by the resonant frequency of the (z-) scanner. Additional resonances in the closed-loop system which cause additional phase shifts reduce the maximum scan speed further. Modern high-performance feedback controllers that are based on H_∞ theory account for the presence of system resonances, resulting in an increased scan rate (Schitter et al., 2001, 2004; Salapaka et al., 2005). Also, actively controlling the cantilever dynamics via Q control allows the scan rate to be increased (Mertz et al., 1993; Sulchek et al., 2000b) or allows scanning in the attractive mode to be stabilized (Anczykowski et al., 1998).

13.4.9. Biological Applications

The following is a list of different applications where small cantilevers have been used so far. The first biomolecules imaged with a small cantilever ($L = 26$ μm) were IgG antibodies adsorbed to mica in liquid (Walters et al., 1996). The image rate was 0.5 Hz for 128 × 128 pixels. A cantilever with $L = 10$ μm was used to image the surface of cleaved and bleached nacre, a high-performance biomineralized material from sea shells (Schäffer et al., 1997). The motion of calcite crystal step edges (Walters the chaperonin GroEL–GroES complex was observed, and its lifetime (on a mica surface) was determined as ≈ 5 s (Viani et al., 2000). Myosin V was imaged in the presence of ATP with an image rate of 12.5 Hz at 100×100 pixels, revealing a relative motion between the head/neck region, the long tail, and the globular tail end (Ando et al., 2001). Caged compounds were used to confirm that this motion is likely to be caused by ATP hydrolysis (Ando et al., 2003).

13.5. CONCLUSION

While a photographic snapshot can provide a beautiful image, it is not well-suited for reconstructing the timely succession of events because it preserves a much reduced picture of reality. Image sequences or movies, on the other hand, carry more information about the succession of events and, in a sense, leave less room for subjective interpretation. In just the same way, AFM image sequences allow us to directly address questions in biology and medicine that single AFM images cannot. This chapter presented an overview of some of the issues that are connected with high-speed imaging of biomolecules. It started by giving some examples of time-resolved AFM imaging. Then, a number of indirect ways for increasing the time resolution were presented. Finally, small cantilevers were shown to be a key for increasing the scan speed. It is expected that small cantilevers will contribute extensively in future investigations of dynamic biological processes.

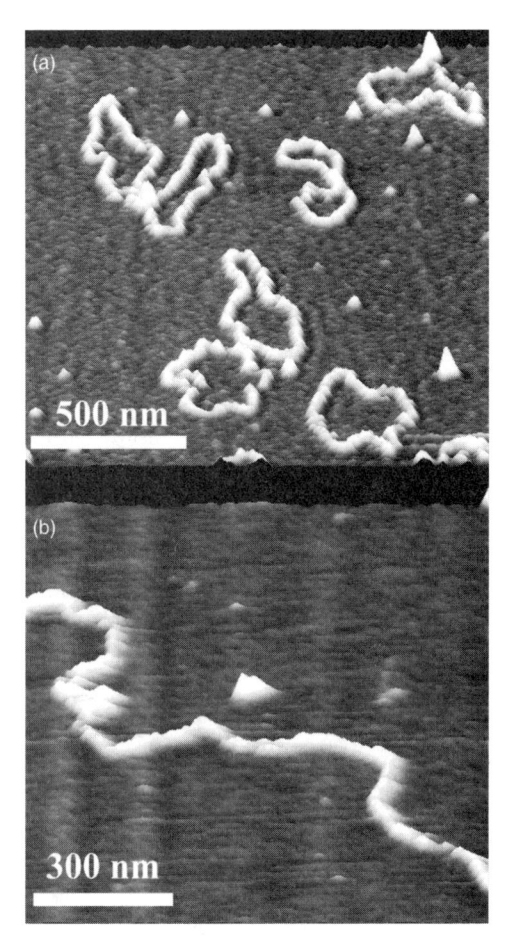

Figure 13.14. High-speed image of DNA adsorbed to mica in buffer solution. (a) Height image of DNA plasmids, acquired in 5.6 s (256×256 pixels) with a tapping frequency of 129 kHz. (b) Height image of linear DNA fragments, acquired in 1.7 s (128×128 pixels) with a tapping frequency of 191 kHz. [Reprinted from Viani et al. (1999), by permission of the American Institute of Physics.]

et al., 1997b) and the step growth close to screw dislocations (Paloczi et al., 1998) were imaged, providing insight into the growth of sea shells that partially consist of calcite. Small cantilevers were used to produce fast images of DNA (Fig. 13.14) and to perform fast force spectroscopy experiments (Viani et al., 1999a, b). The degradation of DNA by DNase I was imaged with an image rate of 0.5 Hz (Hansma, 2001). The association and dissociation of

Acknowledgments

I would like to thank Paul Hansma, Sandor Kasas, Thomas Jovin, Yuekan Jiao, John Sader, and Harald Fuchs for helpful discussions and support. I am also grateful to Jason Cleveland and Mario Viani of Asylum Research for valuable technical and scientific support. This work was supported financially by the Gemeinnützige Hertie–Stiftung in the Stifterverband für die Deutsche Wissenschaft. I am also grateful to Jason Cleveland, Roger Proksch, and Mario Viani of Asylum Research.

REFERENCES

Akama, Y., Nishimura, E., Sakai, A., and Murakami, H. (1990). New scanning tunneling microscopy tip for measuring surface topography. *J. Vac. Sci. Technol. A* **8**(1):429–433.

Anczykowski, B., Cleveland, J. P., Krüger, D., Elings, V., and Fuchs, H. (1998). Analysis of the interaction mechanisms in dynamic mode SFM by means of experimental data and computer simulation. *Appl. Phys. A* **66**:885–889.

Ando, T., Kodera, N., Takai, E., Maruyama, D., Saito, K., and Toda, A. (2001). A high-speed AFM for studying biological macromolecules. *Proc. Natl. Acad. Sci. USA* **98**:12468–12472.

Ando, T., Kodera, N., Maruyama, D., Takai, E., Saito, K., and Toda, A. (2002). A high-speed atomic force microscope for studying biological macromolecules in action. *Jpn. J. Appl. Phys.* **41**:4851–4856.

Ando, T., Kodera, N., Naito, Y., Kinoshita, T., Furuta, K., and Toyoshima, Y. Y. (2003). A high-speed atomic force microscope for studying biological macromolecules in action. *Chem. Phys. Chem.* **4**:1196–1202.

Baselt, D. R., and Baldeschwieler, J. D. (1994). Imaging spectroscopy with the atomic force microscope. *J. Appl. Phys.* **76**:33–38.

Bezanilla, M., Drake, B., Nudler, E., Kashlev, M., Hansma, P. K., Hansma, H. G. (1994). Motion and enzymatic degradation of DNA in the atomic force microscope. *Biophys. J.* **67**:2454–2459.

Binnig, G., Quate, C. F., and Gerber, C. (1986). Atomic force microscope. *Phys. Rev. Lett.* **56**:930–933.

Chand, A., Viani, M. B., Schäffer, T. E., and Hansma, P. K. (2000). Microfabricated small metal cantilevers with silicon tip for atomic force microscopy. *J. Microelectromech. Syst.* **9**:112–116.

Chen, C., Clegg, D. O., and Hansma, H. G. (1998). Structures and dynamic motion of laminin-1 as observed by atomic force microscopy. *Biochemistry* **37**:8262–8267.

Cleveland, J. P., Schäffer, T. E., and Hansma, P. K. (1995). Probing oscillatory hydration potentials using thermal–mechanical noise in an atomic force microscope. *Phys. Rev. B.* **52**:8692–8695.

Drake, B., Prater, C. B., Weisenhorn, A. L., Gould, S. A. C., Albrecht, T. R., Quate, C. F., Cannell, D. S., Hansma, H. G., and Hansma, P. K. (1989). Imaging crystals, polymers, and processes in water with the atomic force microscope. *Science* **243**:1586–1589.

Ellis, D. J., Dryden, D. T. F., Berge, T., Edwardson, J. M., and Henderson, R. M. (1999). Direct observation of DNA translocation and cleavage by the EcoKI endonuclease using atomic force microscopy. *Nature Struct. Biol.* **6**:15–17.

Fantner, G. E., Hegarty, P., Kindt, J. H., Schitter, G., Cidade, G. A. G., and Hansma, P. K. (2005). Data acquisition system for high speed atomic force microscopy. *Rev. Sci. Instrum.* **76**:026118.

Fritz, M., Radmacher, M., and Gaub, H. E. (1993). *In vitro* activation of human platelets triggered and probed by time-lapse atomic force microscopy. *Exp. Cell Res.* **205**:187–190.

Fritz, M., Radmacher, M., Cleveland, J. P., Allersma, M. W., Steward, R. J., Gieselmann, R., Janmey, P., Schmidt, C. F., and Hansma, P. K. (1995). Imaging globular and filamentous proteins in physiological buffer solutions with tapping mode atomic force microscopy. *Langmuir* **11**:3529–3535.

Goldsbury, C., Kistler, J., Aebi, U., Arvinte, T., and Cooper, G. J. S. (1999). Watching amyloid fibrils grow by time-lapse atomic force microscopy. *J. Mol. Biol.* **285**:33–39.

Grandbois, M., Clausen-Schaumann, H., and Gaub, H. E. (1998). Atomic force microscope imaging of phospholipid bilayer degradation by phospholipase A$_2$. *Biophys. J.* **74**:2398–2404.

Guthold, M., Zhu, X., Rivetti, C., Yang, G., Thomson, N. H., Kasas, S., Hansma, H. G., Smith, B., Hansma, P. K., and Bustamante, C. (1999). Direct observation of one-dimensional diffusion and transcription by escherichia coli RNA polymerase. *Biophys. J.* **77**:2284–2294.

Han, W., Lindsay, S. M., Dlakic, M., and Harrington, R. E. (1997). Kinked DNA. *Nature* **386**:563.

Hansma, P. K., Cleveland, J. P., Radmacher, M., Walters, D. A., Hillner, P. E., Bezanilla, M., Fritz, M., Vie, D., Hansma, H. G., Prater, C. B., Massie, J., Fukunaga, L., Gurley, J., and Elings, V. (1994). Tapping mode in liquids. *Appl. Phys. Lett.* **64**:1738–1740.

Hansma, H. G. (2001). Surface biology of DNA by atomic force microscopy. *Annu. Rev. Chem.* **52**:71–92.

Hodges, A. R., Bussmann, K. M., and Hoh, J. H. (2001). Improved atomic force microscope cantilever performance by ion beam modification. *Rev. Sci. Instrum.* **72**:3880–3883.

Hosaka, S., Etoh, K., Kikukawa, A., and Koyanagi, H. (2000). Megahertz silicon atomic force microscopy (AFM) cantilever and high-speed readout in AFM-based recording. *J. Vac. Sci. Technol. B* **18**:94–99.

Humphris, A. D. L., Miles, M. J., and Hobbs, J. K. (2005). A mechanical microscope: High-speed atomic force microscopy. *Appl. Phys. Lett.* **86**:034106.

Janovjak, H., Struckmeier, J., and Müller, D. J. (2005). Hydrodynamic effects in fast AFM single-molecule force measurements. *Eur. Biophys. J.* **34**:91–96.

Jiao, Y., and Schäffer, T. E. (2004). Accurate height and volume measurements on soft samples with the atomic force microscope. *Langmuir* **20**:10038–10045.

Jiao, Y., Cherny, D. I., Heim, G., Jovin, T. M., and Schäffer, T. E. (2001). Dynamic interactions of p53 with DNA in solution by time-lapse atomic force microscopy. *J. Mol. Biol.* **314**:233–243.

Kasas, S., Thomson, N. H., Smith, B. L., Hansma, H. G., Zhu, X., Guthold, M., Bustamante, C., Kool, E. T., Kashlev, M., and Hansma, P. K. (1997). *Escherichia coli* RNA polymerase activity observed using atomic force microscopy. *Biochemistry* **36**:461–468.

Kasas, S., Thomson, N. H., Schäffer, T. E., Dietler, G., Catsicas, S., and Hansma, P. K. (2000). Simulation of an atomic force microscope imaging a moving protein. *Probe Microsc.* **2**:37–44.

Kindt, J. H., Fantner, G. E., Cutroni, J. A., and Hansma, P. K. (2004a). Rigid design of fast scanning probe microscopes using finite element analysis. *Ultramicroscopy* **100**:259–265.

Kindt, J. H., Fantner, G. E., Thompson, J. B., and Hansma, P. K. (2004b). Automated wafer-scale fabrication of electron beam deposited tips for atomic force microscopes using pattern recognition. *Nanotechnology* **15**:1131–1134.

Kodera, N., Yamashita, H., and Ando, T. (2005). Active damping of the scanner for high-speed atomic force microscopy. *Rev. Sci. Instrum.* **76**(5):53708.

Levine, A. J. (1997). p53, the cellular gatekeeper for growth and division. *Cell* **88**:323–331.

Longfellow, H. W. (1839). Hyperion, 1st ed., Book II, Chapter VI, Samuel Colman, Boston.

Marti, O., Ruf, A., Hipp, M., Bielefeldt, H., Colchero, J., and Mlynek, J. (1992). Mechanical and thermal effects of laser irradiation on force microscope cantilevers. *Ultramicroscopy* **42**:345–350.

Matzke, R., Jacobson, K., and Radmacher, M. (2001). Direct, high-resolution measurement of furrow stiffening during division of adherent cells. *Nature Cell Biol.* **3**:607–610.

Mertz, J., Marti, O., and Mlynek, J. (1993). Regulation of a microcantilever response by force feedback. *Appl. Phys. Lett.* **62**:2344–2346.

Müller, D. J., Engel, A., Matthey, U., Meier, T., Dimroth, P., and Suda, K. (2003). Observing membrane protein diffusion at subnanometer resolution. *J. Mol. Biol.* **327**:925–930.

Nagami, F., Zuccheri, G., Samori, B., and Kuroda, R. (2002). Time-lapse imaging of conformational changes in supercoiled DNA by scanning force microscopy. *Anal. Biochem.* **300**:170–176.

Ookubo, N., and Yumotoa, S. (1999). Rapid surface topography using a tapping mode atomic force microscope. *Appl. Phys. Lett.* **74**:2149–2151.

Paloczi, G. T., Smith, B. L., Hansma, P. K., Walters, D. A., and Wendman, M. A. (1998). Rapid imaging of calcite crystal growth using atomic force microscopy with small cantilevers. *Appl. Phys. Lett.* **73**:1658–1660.

Proksch, R., Schäffer, T. E., Cleveland, J. P., Callahan, R. C., and Viani, M. B. (2004). Finite optical spot size and position corrections in thermal spring constant calibration. *Nanotechnology* **15**:1344–1350.

Putman, C. A. J., van der Werf, K. O., de Grooth, B. G., and Van Hulst, N. F. (1994). Tapping mode atomic force microscopy in liquid. *Appl. Phys. Lett.* **64**:2454–2456.

Radmacher, M., Fritz, M., Hansma, H. G., and Hansma, P. K. (1994a). Direct observation of enzyme activity with the atomic force microscope. *Science* **265**:1577–1579.

Radmacher, M., Cleveland, J. P., Fritz, M., Hansma, H. G., and Hansma, P. K. (1994b). Mapping interaction forces with the atomic force microscope. *Biophys. J.* **66**:2159–2165.

Reiley, T. S., Fan, L. S., and Mamin, H. J. (1995). Micromechanical structures for data storage. *Microelectron. Eng.* **27**:495–498.

Revenko, I., and Proksch, R. (2000). Magnetic and acoustic tapping mode microscopy of liquid phase phospholipid bilayers and DNA molecules. *J. Appl. Phys.* **87**:526–533.

Rief, M., and Grubmüller, H. (2002). Force spectroscopy of single biomolecules. *Chem. Phys. Chem.* **3**:255–261.

Rogers, B., Sulchek, T., Murray, K., York, D., Jones, M., Manning, L., Malekos, S., Beneschott, B., Adams, J. D., Cavazos, H., and Minne, S. C. (2003). High speed tapping mode atomic force microscopy in liquid using an insulated piezoelectric cantilever. *Rev. Sci. Instrum.* **74**:4683–4686.

Rost, M. J., Crama, L., Schakel, P., van Tol, E., van Velzen-Williams, G. B. E. M., Overgauw, C. F., ter Horst, H., Dekker, H., Okhuijsen, B., Seynen, M., Vijftigschild, A., Han, P., Katan, A. J., Schoots, K., Schumm, R., van Loo, W., Oosterkamp, T. H., and Frenken, J. W. M. (2005). Scanning probe microscopes go video rate and beyond. *Rev. Sci. Instrum.* **76**:053710.

Sader, J. E. (1998). Frequency response of cantilever beams immersed in viscous fluids with applications to the atomic force microscope. *J. Appl. Phys.* **84**:64–76.

Salapaka, S., De, T., and Sebastian, A. (2005). Sample-profile estimate for fast atomic force microscopy. *Appl. Phys. Lett.* **87**:53112.

Schäffer, T. E., Richter, M., and Viani, M. B. (2000). Array detector for the atomic force microscope. *Appl. Phys. Lett.* **76**:3644–3646.

Schäffer, T. E. (2002). Force spectroscopy with a large dynamic range using small cantilevers and an array detector. *J. Appl. Phys.* **91**:4739–4746.

Schäffer, T. E. (2005). Calculation of thermal noise in an atomic force microscope with a finite optical spot size. *Nanotechnology* **16**:664–670.

Schäffer, T. E., and Fuchs, H. (2005). Optimized detection of normal vibration modes of atomic force microscope cantilevers with the optical beam deflection method. *J. Appl. Phys.* **97**:083524.

Schäffer, T. E., and Hansma, P. K. (1998). Characterization and optimization of the detection sensitivity of an atomic force microscope for small cantilevers. *J. Appl. Phys.* **84**:4661–4666.

Schäffer, T. E., Cleveland, J. P., Ohnesorge, F., Walters, D. A., and Hansma, P. K. (1996). Studies of vibrating atomic force microscope cantilevers in liquid. *J. Appl. Phys.* **80**:3622–3627.

Schäffer, T. E., Viani, M., Walters, D. A., Drake, B., Runge, E. K., Cleveland, J. P., Wendman, M. A., and Hansma, P. K. (1997). An atomic force microscope for small cantilevers. *Proc. SPIE* **3009** (Micromachining and Imaging):48–52.

Schitter, G., Menold, P., Knapp, H. F., Allgöwer, F, and Stemmer, A. (2001). High performance feedback for fast scanning atomic force microscopes. *Rev. Sci. Instrum.* **72**:3320–3327.

Schitter, G., Allgöwer, F., and Stemmer, A. (2004). A new control strategy for high-speed atomic force microscopy. *Nanotechnology* **15**:108–114.

Serry, F. M., Neuzil, P., Vilasuso, R., and Maclay, G. J. (1995). Air-damping of resonant AFM microcantilevers in the presence of a nearby surface. *Proceedings of the Second International Symposium on Microstructures and Microfabricated Systems*, pp. 83–89, Electrochemical Society, Chicago.

Shao, Z., and Zhang, Y. (1996). Biological cryo atomic force microscopy: A brief review. *Ultramicroscopy* **66**:141–152.

Stark, R. W. (2004). Optical lever detection in higher eigenmode dynamic atomic force microscopy. *Rev. Sci. Instrum.* **75**(11):5053–5055.

Stoffler, D., Goldie, K. N., Feja, B., and Aebi, U. (1999). Calcium-mediated structural changes of native nuclear pore complexes monitored by time-lapse atomic force microscopy. *J. Mol. Biol.* **287**:741–752.

Sulchek, T., Hsieh, R., Adams, J. D., Minne, S. C., Quate, C. F., and Adderton, D. M.

(2000a). High-speed atomic force microscopy in liquid. *Rev. Sci. Instrum.* **71**:2097–2099.

Sulchek, T., Hsieh, R., Adams, J. D., Yaralioglu, G. G., Minne, S. C., Quate, C. F., Cleveland, J. P., Atalar, A., and Adderton, D. M. (2000b). High-speed tapping mode imaging with active *Q*-control for atomic force microscopy. *Appl. Phys. Lett.* **76**:1473–1475.

Thomson, N. H., Fritz, M., Radmacher, M., Cleveland, J. P., Schmidt, C. F., and Hansma, P. K. (1996). Protein tracking and detection of protein motion using atomic force microscopy. *Biophys. J.* **70**:2421–2431.

Thomson, N. H., Kasas, S., Riederer, B. M., Catsicas, S., Dietler, G., Kulik, A. J., and Forró, L. (2003). Large fluctuations in the disassembly rate of microtubules revealed by atomic force microscopy. *Ultramicroscopy* **97**:239–247.

van Noort, S. J. T., van der Werf, K. O., Eker, A. P. M., Wyman, C., de Grooth, B. G., van Hulst, N. F., and Greve, J. (1998). Direct visualization of dynamic protein–DNA interactions with a dedicated atomic force microscope. *Biophys. J.* **74**:2840–2849.

van Noort, S. J. T., van der Werf, K. O., de Grooth, B. G., and Greve, J. (1999). High speed atomic force microscopy of biomolecules by image tracking. *Biophys. J.* **77**:2295–2303.

Viani, M. B., Schäffer, T. E., Chand, A., Rief, M., Gaub, H. E., and Hansma, P. K. (1999a). Small cantilevers for force spectroscopy of single molecules. *J. Appl. Phys.* **86**:2258–2262.

Viani, M. B., Schäffer, T. E., Paloczi, G. T., Pietrasanta, L. I., Smith, B. L., Thompson, J. B., Richter, M., Rief, M., Gaub, H. E., Plaxco, K. W., Cleland, A. N., Hansma, H. G., and Hansma, P. K. (1999b). Fast imaging and fast force spectroscopy of single biopolymers with a new atomic force microscope designed for small cantilevers. *Rev. Sci. Instrum.* **70**:4300–4303.

Viani, M. B., Pietrasanta, L. I., Thompson, J. B., Chand, A., Gebeshuber, I. C., Kindt, J. H., Richter, M., Hansma, H. G., and Hansma, P. K. (2000). Probing protein–protein interactions in real time. *Nature Struct. Biol.* **7**:644–647.

Walters, D. A., Cleveland, J. P., Hansma, P. K., Wendman, M. A., Gurley, G., and Elings, V. (1996). Short cantilevers for atomic force microscopy. *Rev. Sci. Instrum.* **67**:3583–3589.

Walters, D. A., Smith, B. L., Belcher, A. M., Paloczi, G. T., Stucky, G. D., Morse, D. E., Hansma, P. K. (1997a). Modification of calcite crystal growth by abalone shell proteins: An atomic force microscope study. *Biophys. J.* **72**:1425–1433.

Walters, D. A., Viani, M., Paloczi, G. T., Schäffer, T. E., Cleveland, J. P., Wendman, M. A., Elings, V., and Hansma, P. K. (1997b). Atomic force microscopy using short cantilevers. *Proc. SPIE* **3009** (Micromachining and Imaging): 43–47.

CHAPTER 14

ATOMIC FORCE MICROSCOPY IN CYTOGENETICS

S. THALHAMMER

Department für Geo-und Umweltwissenschaften, University of München, Theresienstr.41, Munich Germany; GSF — National Center for Environment and Health, Institute of Radiation Protection, Ingolstädter Landstrasse 1, 85764 Neuherberg, Germany

W. M. HECKL

Department für Geo-und Umweltwissenschaften, University of München, Theresienstr.41, Munich Germany

I must beg leave to propose a separate technical name "chromosome" for those things have been called by Boveri "chromatic elements", in which there occurs one of the most important acts in karyokinesis, viz., the longitudinal splitting. They are so important that a special and shorter name appears useful. If the term I propose is practically applicable; it will become familiar, otherwise it will soon sink into oblivion.—Waldeyer (1890)

14.1. INTRODUCTION

The beginning of modern human cytogenetics dates back to 1956. In this year, scientists from Indonesia, Sweden, and England showed the number of human chromosomes to be 46 and not 48 as previously estimated (Tjio and Levan, 1956; Ford and Hamerton, 1956). Chromosomes were a late discovery, and were first described in 1842 by Nägeli (Nägeli, 1842).

Advances in human cytogenetics have been largely dependent on technical factors, such as the availability of suitable tissue containing an adequate number of cells with the potential for cell division. In the late 1950s, metaphase chromosomes were obtained from skin fibroblasts. The use of phytohemagglutinin (PHA) to stimulate cell division in peripheral blood lymphocytes (Moorhead et al., 1960), along with hypotonic treatment for obtaining better metaphase spreads (Hsu, 1952), was a major breakthrough that increased the number of metaphases available for analysis after 69–72 h. These metaphases had chromosomes of optimal quality that could clearly be classified into seven groups, A to G. The X chromosome was identified within the C group, whereas the Y chromosome was placed in the G group. The identification of individual chromosomes was an arduous task, because chromosomes within a group resembled one another morphologically and specific staining techniques were not yet available.

In the 1960s, investigators started to identify individual single chromosomes and

---

Force Microscopy: Applications in Biology and Medicine, edited by Bhanu P. Jena and J.K. Heinrich Hörber.
Copyright © 2006 John Wiley & Sons, Inc.

structural abnormal chromosomes using autoradiographic techniques (German, 1964). However, neither chromosome morphology nor autoradiography provided unequivocal identification of all chromosomes in the human genome. In this field the first advances came from the Caspersson group. They described chromosome staining by quinacrine mustard (QM), a technique they termed Q-banding, since chromosomes were differentiated into bright and dark regions, termed "bands" (Caspersson et al., 1964). After that, more than a dozen new staining techniques for identifying individual chromosomes and abnormalities were developed. An overview is listed by Verma and Babu (1989).

Fluorescence *in situ* hybridization (FISH) has become a popular technique over the past years, often applied to gene mapping and molecular cytogenetics (Adionolfi and Davies, 1994). Biotin or digoxigenin-labeled probes have been used most commonly in conjunction with secondary reagents for probe detection. Nucleoside triphosphate analogs directly coupled to fluorophores that can be incorporated into DNA or RNA specimens by conventional enzymatic reactions are now available from a variety of commercial available sources, as are chemically synthesized oligonucleotide labels with one or more fluorochrome molecules.

In 1982 the scanning tunneling microscope (STM; Binnig et al., 1982) was invented. This instrument, which can image solid surfaces with atomic resolution, was revolutionary for microscopy and surface analysis. In the STM a sharp conducting tip is brought into close proximity of the sample surface and a current of electrons flows from the tip to sample, or vice versa depending on the sign of the bias voltage. Scanning the tip in a grid pattern over the surface yields a topographic image of the sample surface under investigation. The main limitation of the STM is the fact that only conducting or semiconducting samples can be imaged. To circumvent this problem of sample conductivity, Binnig, Quate and Gerber came up with a new microscope, the atomic force microscope (AFM; Binnig et al., 1986).

In the atomic force microscope a sharp tip is mounted on the free end of a cantilever. While scanning the tip over the sample surface, forces acting between tip and sample deflect the cantilever. A two-dimensional map of the local interactions between tip and sample surface can be obtained, by measuring these deflections from point to point with a beam-bounce detection method. With this technique, as well as utilizing various different scanning modes, a wide range of data (e.g., the surface relevant structure or local measurements of the physical and chemical properties of the sample) can be recorded (Colton et al., 1998). A lateral resolution of 1 nm and vertical resolution of 0.1 nm can be achieved. Furthermore, the AFM can operate under fluid. It is an essential step for biological applications to work and to image under physiological environment (Marti et al., 1987). Structural analysis at high resolution not only provides more information about molecular complexes but is also used for *in vivo* experiments using biological systems. Data can be recorded in real time. Besides structural information of biological systems, three-dimensional data, micromechanical behaviors, dynamic processes, and molecular interactions can be recorded. Table 14.1 shows a short comparison of AFM to other microscopic techniques together with the necessary sample preparation procedures.

14.2. IMAGING OF GENETIC MATERIAL

Cytogenetics is basically a visual science. Established microscopic techniques, such as light and electron microscopy, have been widely used for the study of chromosomes. After the invention of the atomic force microscope, it was applied in different fields of genetic applications. Double-stranded DNA on freshly cleaved mica in air was imaged by several groups (Thundat et al., 1992; Yang and

TABLE 14.1. Comparison of Different Microscopic Techniques to Image the Sample Topography—For Example, Metaphase Chromosome.

	Conventional Optical Microscopy	Scanning Electron Microscopy (SEM)	Field Emission in Lens Scanning Electron Microscope (FEISEM)	Atomic Force Microscopy (AFM)
Microscopic environment	Ambient liquid vacuum	Vacuum	Vacuum	Ambient liquid vacuum
Focus depth	Medium	Small		Small
Resolution x,y,z	100 nm n/a	5 nm n/a		0.1–1.0 nm 0.01 nm
Magnification	$1 \times -2 \times 10^3 \times$	$10 \times -10^6 \times$		$5 \times 10^2 \times -10^8 \times$
Necessary sample preparation	Low	Freeze-drying, coating	Low	Low
Necessary sample properties	Samples do not have to be completely transparent for visible light	Samples should not charge and have to be vacuum compatible		Samples should do not have excessive changes in height compared to tip geometry

Shao, 1993). Hansma et al. (1992 and 1993) successfully imaged plasmid-DNA fixed on mica in propanol. By adding new spreading chemicals (e.g., quaternary ammonium salts), it is possible to reduce surface impurities, and the surface density of the molecules could be reproducibly measured (Schaper et al., 1994). After the introduction of the tapping mode, it became possible to image DNA with lower shear forces during scanning, which resulted in being able to obtain much detailed images of the macromolecule (Delain et al., 1992). In Fig. 14.1 we have imaged a double-stranded plasmid, pUC19, in AFM tapping mode in air. DNA in a highly condensed state in sperm cells was imaged by Allen et al. (1993) in air and in liquids. Furthermore, Fritzsche and co-workers performed structural experiments on chromatin fibers (Fritzsche et al., 1997), as well as determined the volume of metaphase chromosomes (Fritzsche and Henderson, 1996). Structural examinations on metaphase chromosomes were also performed by de Grooth and Putman (de Grooth and Putman, 1992; Rasch et al., 1993). Figure 14.2 shows untreated human metaphase chromosomes. Using chemically and enzymatically untreated metaphase chromosomes, a GTG-like banding pattern—G-bands by trypsin using Giemsa—could be observed in the topographic images (Musio et al., 1994). Metaphase chromosomes imaged by AFM have revealed structures similar to those reported in light and electron microscopy. Depending on the preparation technique, substructural details can be recorded in metaphase chromosomes (Tamayo et al., 1999; Gobbi et al., 2000). After pepsin digestion of the metaphase chromosomes, a granular substructure was detected using AFM in contact mode. In such cases, not only the covering plasma layer but also the scaffold stabilizing proteins were digested. The recorded details represent a nucleosomal structure, which has been discussed by several investigators (Winfield et al., 1995; de Grooth and Putman, 1992; Fritzsche et al., 1994). The recorded data are comparable to those generated by scanning electron microscopy (Wanner et al., 1991). Metaphase chromosomes consist of 30-nm fibers folded in a tandem array of radial loops, which are packaged into a fiber with an overall diameter between 200 and 250 nm. High-resolution AFM images of metaphase chromosomes revealed structural

252 ATOMIC FORCE MICROSCOPY IN CYTOGENETICS

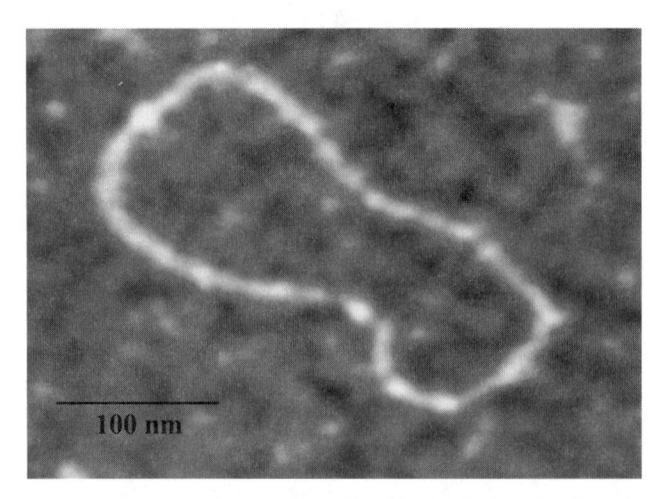

Figure 14.1. AFM image of plasmid pUC19 (~2.7 kbp) diluted in dH$_2$O; deposited by spin stretching and imaged in tapping mode in air.

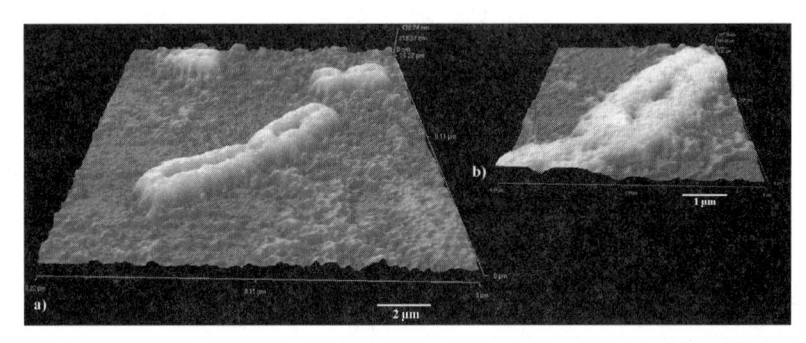

Figure 14.2. (a) Three-dimensional AFM image, recorded in contact mode, of untreated human chromosomes. Bar represents 2 μm; (b) AFM enhancement of the p-arm of the recorded chromosome. Bar represents 1 μm.

features in the size range of 30–100 nm, which correspond to the loops of the 30-nm fiber (de Grooth and Putman, 1992). Other, (Winfield et al., 1995; McMaster et al., 1996) have reported features as small as 10–20 nm, which could correspond to individual nucleosomes. While scanning in the contact mode, the tip of the AFM can be contaminated with chromosomal material. This material, adhering to the tip, can limit the use of such tips for manipulation and microdissection experiments.

In noncontact mode, the tip scans at a distance of a few hundred angstroms over the chromosomal surface. The tip is not in contact with the sample surface and therefore does not get contaminated while scanning.

The advantages and disadvantages of these two operation modes are combined in the tapping mode (Hansma, 1993), where the tip comes in contact with the chromosomal surface for a very short period. In summary, these three operating modes can be used for imaging chromosomal material (see **Fig. 14.3**), and to record substructures depending on sample preparation. Care is taken so that the topography of the metaphase chromosomes are preserved and not deformed (de Grooth and Putman, 1992; Musio et al., 1994). Table 14.2 summarizes the methodical properties of contact, noncontact, and tapping mode for high-resolution imaging and manipulation of metaphase chromosomes.

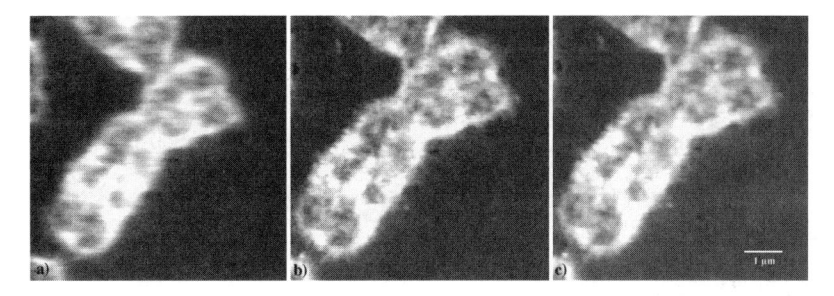

Figure 14.3. Comparison of different operating modes. (**a**) AFM image in contact mode. (**b**) Noncontact mode. (**c**) Tapping mode of a GTG-banded human metaphase chromosome 11. Bar represents 1 μm.

TABLE 14.2. Methodical Properties of the Different Operation Modes in AFM for High-Resolution Imaging, Manipulation, and Microdissection of Metaphase Chromosomes.

Operation Mode	Contact Mode	Noncontact Mode	Tapping Mode
Tip loading force	Low → high	Low	Low
Contact with sample surface	Yes	No	Periodical
Manipulation of sample	Yes	No	Yes
Contamination of AFM tip	Yes	No	Yes
Microdissection	Yes	No	No

Using specific *in situ* hybridization techniques, we can detect distinct areas of hybridization in metaphase chromosomes. Biotinylated DNA probes were used for mapping, and the specific sites were visualized by detecting changes in topography induced by a peroxidase–diaminobenzidine reaction (Putman et al., 1993; Rasch et al., 1993). Using the same detection technique, Kalle et al. (1995) were able to identify specific signals after RNA *in situ* hybridization. In studies on cereal chromosomes, a genome specific probe has been used. Using AFM, we can detect changes in height of biotin–avidin–fluorescein isothiocyanate complexes formed as a consequence of fluorescence *in situ* hybridization procedures (McMaster et al., 1996).

The AFM has also been used to image genetic material in liquids. In this case, fixed metaphase chromosomes swell depending on the buffer used (Fig. 14.4). The viscoelastic properties of rehydrated chromosomes and their volume, have been determined in liquids using AFM (Frizsche, 1999). In comparison to light or electron microscopy, the AFM is able to operate in liquids and to perform local measurements on any point of the sample surface (Stark et al., 1998 and Jiao Y. et al., 2004; Hoshio et al., 2004). Due to this attribute, data the biophysical properties of the metaphase chromosome can be obtained using the AFM.

14.3. KARYOTYPING OF METAPHASE CHROMOSOMES

Chromosome banding techniques have facilitated the precise identification of individual chromosomes. The GTG banding obtained by digesting chromosomes with proteolytic trypsin, followed by Giemsa staining, is most widely used in routine chromosome analysis. However, the interpretation of the GTG

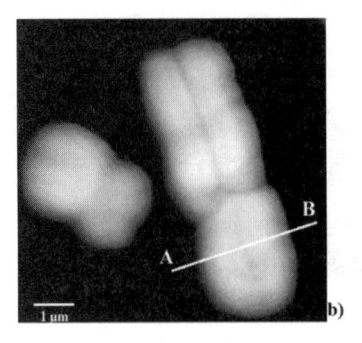

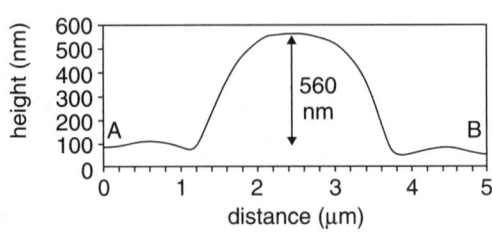

Figure 14.4. (a) AFM image in contact mode of a rehydrated human metaphase chromosomes in PBS buffer. (b) A cross-sectional analysis from point A to B. The swelling in PBS buffer was 560 nm in comparison to TE buffer with 900 nm (data not shown). Bar represents 1 μm.

bands is still unclear. A direct role of Giemsa stain in producing the GTG bands was suggested by McKay (1973). Several authors inferred that chromosomes contain a preexisting structure, which is enhanced by GTG-banding. It is, however, still unclear how this enhancement occurs (Comings, 1978; Ambros and Sumner, 1987). It is hypothesized that the differences between positive and negative GTG bands may be induced by the spatial organization of chromosomal protein and DNA.

In atomic force microscopy, as opposed to light microscopy, changes in color is undetectable. Differentiation can only occur by topographical information of the metaphase chromosome. The resulting image is not only the result of topographical changes in the chromosome surface, but also a result of the interaction of the tip with the viscoelastic properties of the chromosome. In Fig. 14.5A, a topographic AFM image of a GTG-banded chromosome 7 homologue is shown. The morphology of the chromosome is preserved, and both the banding pattern and the fibrous nature are detectable. Structural protrusions along the chromosome, corresponding to the dark bands in Fig. 14.5A, are also detectable. A linescan of the q-arm (point A to B in Fig. 14.5B) shows differences in height between dark and light bands of about 90 nm. The length of the ridges is about 540 nm. The corresponding bands are marked in the idiogram (Fig. 14.5C). It is known from chromosomes imaged by scanning electron microscopy that the Giemsa light and dark bands differ in height (Harrison et al., 1981). One must be aware that the AFM image represents not only the topology of the sample surface but also the compressibility of the sample; therefore height is partially expressed as topography. Figure 14.6 shows a human $2n = 46$, XX female metaphase spread. The light and dark bands are clearly detectable and all chromosomes are identifiable. The corresponding karyotype is shown in Fig. 14.7.

Furthermore, it is possible to identify features equivalent to G-banding pattern in untreated chromosomes and to use these for classification (Musio et al., 1994; Tamayo J. 2003). As in light microscopy, dark and light bands can be correlated in GTG-banded chromosomes and can be classified accordingly (Thalhammer et al., 2001). In earlier AFM studies, G-positive bands were determined to be areas with a higher surface relief (Rasch et al., 1993). After more than 20 years, the discussion about the banding mechanism is still unsolved. Also, not all related biochemical and physical reactions are understood. The comparison of unstained and Giemsa-stained chromosomes by phase contrast microscopy (Yunis and

KARYOTYPING OF METAPHASE CHROMOSOMES 255

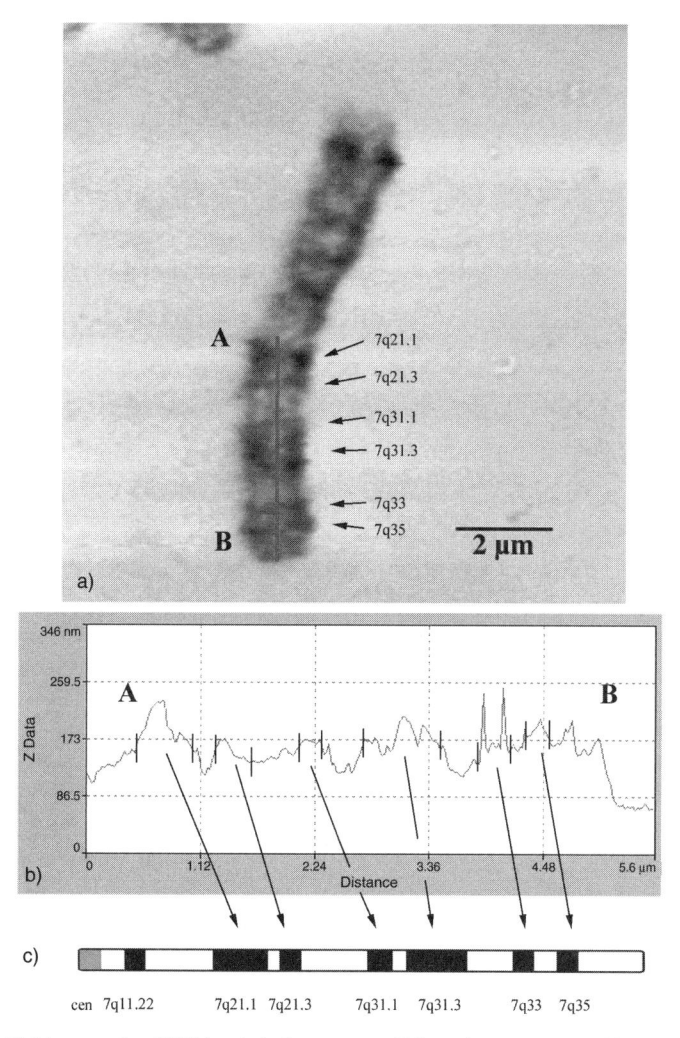

Figure 14.5. (a) AFM image of a GTG-banded chromosome 7 homologue, topographic AFM image, grayscale inverted. The bright and dark banding pattern is detectable (bar represents 2 μm). (b) Line measurement through point A to B of the q-arm of chromosome 7. (c) Idiogram of the q-arm of chromosome 7.

Sanchez, 1993; McKay, 1973) set up the hypothesis that staining techniques amplify a preexsisting structure of incomplete structural organization in chromatin. Electron microscopy studies support this hypothesis (Heneen and Caspersson, 1973; Bahr and Larsen, 1974). McMaster et al. (1994) suggested an influence of the stains to structure and morphology, based on their work on untreated metaphase chromosomes. In scanning electron microscopy, light and dark bands in the R and G banding pattern can be differentiated by changes in height (Harrison et al., 1981). This suggests that the high resolution of the AFM has allowed an intrinsic banding pattern to be visualized which otherwise would not be possible without the enhanced accumulation of stains for viewing using light microscopy. The accuracy of chromosomal banding is strongly related to DNA organization with the associated proteins.

256 ATOMIC FORCE MICROSCOPY IN CYTOGENETICS

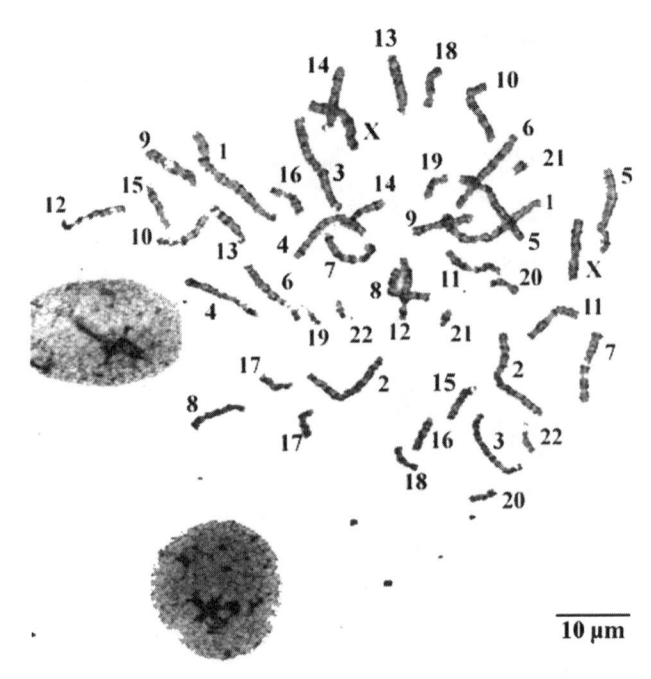

Figure 14.6. AFM image of a $2n = 46$, XX female GTG-banded metaphase spread. Bar represents 10 μm.

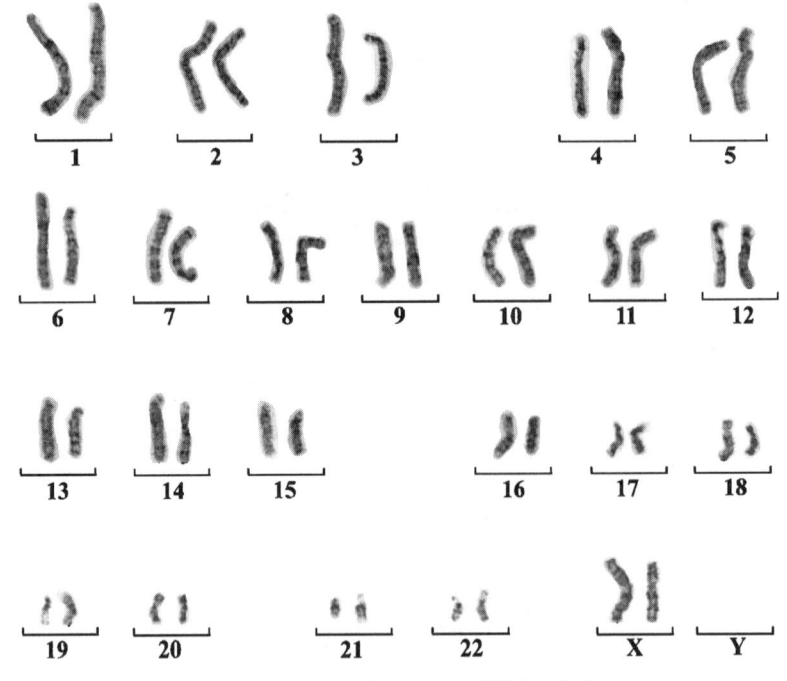

Figure 14.7. AFM image of a $2n = 46$, XX female karyotype.

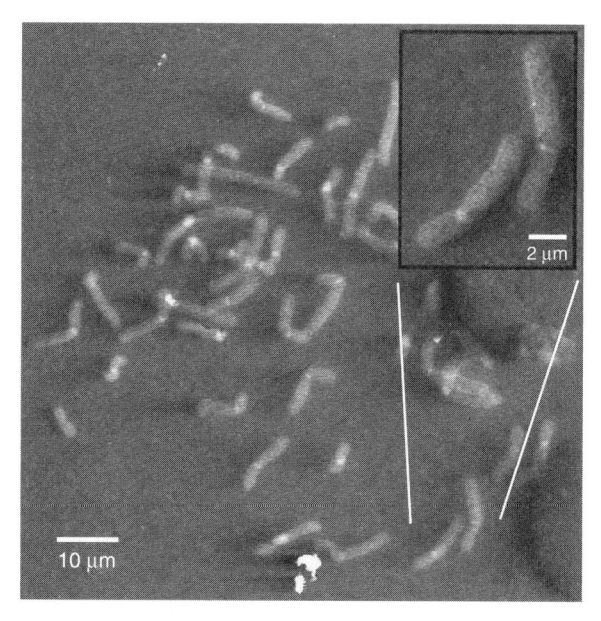

Figure 14.8. AFM image of a $2n = 46$, XY male CBG-banded metaphase spread. Bar represents 10 μm. Enlargement shows two chromosomes with highlighted centromeric region. Bar 2 represents μm.

14.3.1. CBG-Banding

The C-banding technique produces selective staining of constitutive heterochromatin. These bands are mostly located at the centromeric regions of chromosomes, and hence they are known as C-bands. The original method described by Arrighi and Hsu (1971) primarily involved treatment with an alkali (sodium hydroxide) to denature the chromosomal DNA, with subsequent incubation in a salt solution. An alternate method, described by Sumner (1972), utilizes a milder alkali, (barium hydroxide). However, both methods produce similar characteristic C-banding patterns (Fig. 14.8) observed using AFM in the contact mode. In contrast to light microscopic images, the stained centromeric regions are clearly detectable in the AFM micrographs. Similarly, C-banding for studying chromosome rearrangements near centromeres and for investigating polymorphisms have also been performed using AFM by Tan et al. (2001) and Fukushi et al. (2005).

14.4. CORRELATIVE HIGH-RESOLUTION MORPHOLOGICAL ANALYSIS OF THE THREE-DIMENSIONAL ORGANIZATION OF HUMAN METAPHASE CHROMOSOMES

An understanding of the cell's nuclear functions requires an accurate knowledge of the spatial organization of nuclear structures. For many years, the study of human metaphase chromosomes has been carried out using light microscopy following staining protocols. Scanning electron microscopy (SEM) provides higher-resolution imaging compared to light microscopy, and it permits surface analysis of the chromosomal structure, which cannot be adequately obtained from transmission electron microscopy (TEM). Nevertheless, in order to obtain high-resolution SEM images, the use of a high electron accelerating voltage (up to 30 kV) has been used (Sumner and Ross, 1989;

Sanchez-Sweatman, 1993; Sumner, 1996). Under these experimental conditions, sputter-coating or conductive staining of the samples is generally required (Sumner, 1994; Wanner and Formanek, 1995). However, since both procedures allow electron-charging dispersion from the sample, sputtering or staining obscures fine details and may produce alterations to the sample (Hermann and Müller, 1992).

Today, only a few techniques are available for high-resolution imaging of chromosomal material with reduced artifacts. These include the field emission in lens scanning electron microscope (FEISEM) and the atomic force microscope (AFM). The FEISEM represents a special kind of SEM, fitted with a cold cathode field emission electron gun (Nagatani et al., 1987; Pawley, 1997) that can operate at low accelerating voltage with reduced electron charging of the sample. In fact, the low-voltage and low-current electron beam of the FEISEM, together with a liquid nitrogen anticontamination device in correspondence to the specimen area and an "in lens" assembly of the electron-optic column, allows high-resolution imaging of the biological sample without any conductive staining or metal coating. Hence, contamination of the specimen is greatly reduced, compared to conventional SEM imaging (Nagatani et al., 1987). The sample location between the objective pole pieces limits the dispersion of the secondary electrons collected by the magnetic field of the lens. These characteristics allow the observation of uncoated biological samples with a higher resolution than with conventional SEM (Rizzoli et al., 1994; Rizzi et al., 1995; Lattanzi et al., 1998; Gobbi et al., 1999).

FEISEM and AFM microscopy can also be combined to observe the same metaphase chromosome prepared from human HL 60 cells, by standard cytogenetic method after cleaning the metaphase spreads (Rizzoli et al., 1994). The analysis of the same samples can be facilitated by the use of conductive glass (ITO glass) for the chromosome map preparation. These two different technical approaches show a high correlation of the respective morphological information, in both normal and treated samples. The high resolution potential of the FEISEM, together with the possibility of observing hydrated samples and/or nanomanipulating the specimen with the AFM, confirms morphological data and offers an enhanced information (Figs. 14.9 and 14.10) (Gobbi et al., 2000).

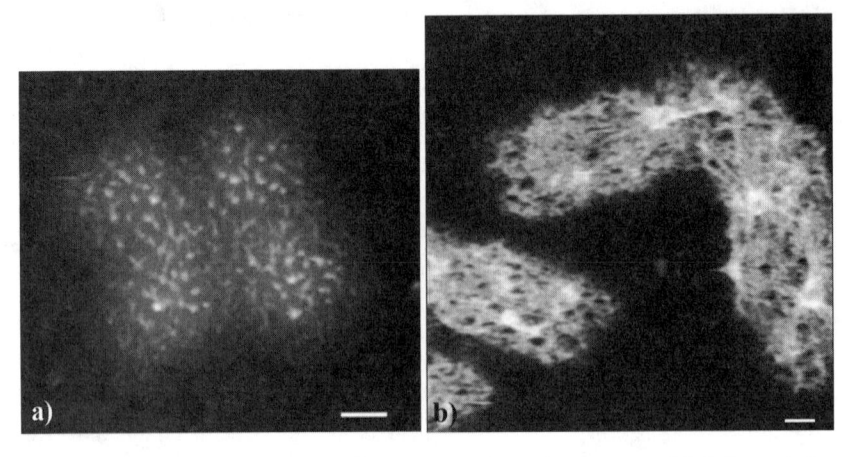

Figure 14.9. Atomic force microscopy images of human metaphase chromosomes. (**a**) A flat and short untreated chromosome is well-identifiable on the ITO glass. (**b**) Present a more defined network structure. Bar represents 1 μm.

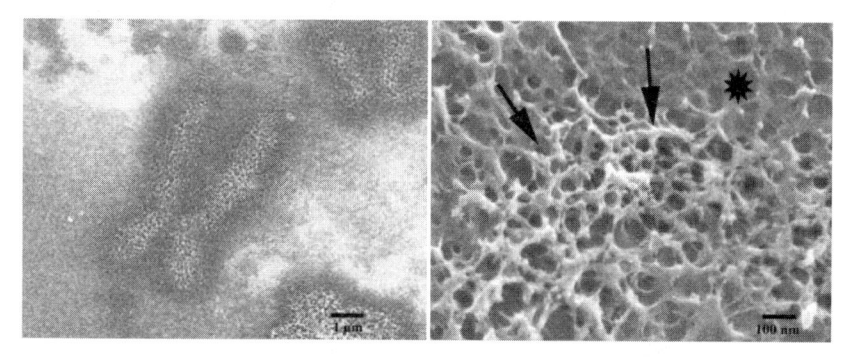

Figure 14.10. Field emission in-lens scanning electron microscopy analysis of protease K-treated samples. (**a**) the centromeric region (arrows) and the chromatids are well-recognizable. A dark halo surrounds the entire chromosome (bar represents 1 μm). (**b**) Increasing the magnification, the chromosomal surface appears to be constituted of a network. Some fibrillar structures parallel to the axis of the chromatid are well-detectable (arrows). The halo around the chromatid appears to be formed by a mix of fibers and homogeneous phase (asterisk) (Bar represents 100 nm).

14.5. AFM AS A NANOMANIPULATOR AND DISSECTING TOOL

By combining high structural resolution with the ability to change the image parameters at any place of the scan area, it is possible to use the AFM as a manipulation tool. In 1991, Hoh et al. (1991) demonstrated the use of AFM as a microdissection device. They performed microdissection on cellular tight junctions. Controlled nanomanipulation of biomolecules have also been performed on genetic material (Hansma et al., 1992) where 100- to 150-nm fragments were cut-out circular plasmid rings. Isolated DNA adsorbed on a mica surface have been dissected both in air (Henderson, 1992; Vesenka et al., 1992; Geissler et al., 2000) and in liquids—for example, propanol (Hansma et al., 1992). These experiments demonstrate the feasibility of microdissection in the nanometer range using the AFM. Combining AFM imaging and microdissection capability, the organization of bovine sperm nuclei has been investigated, showing small protein and DNA containing subunits of 50–100 nm in diameter (Allen et al., 1993). Similarly, tobacco mosaic viruses have been dissected and displaced on a graphite surface to record mechanical properties of the virus (Falvo et al., 1997).

Microdissection of genetic material in a different condensation states (polytene chromosomes of *Drosophila melanogaster*) have been performed by Henderson and coworkers (Mosher et al., 1994; Jondle et al., 1995). Metaphase chromosomes have been microdissected, and the extracted material was used for biochemical evaluation (Thalhammer et al., 1997; Xu and Ikai, 1998). As illustrated in Fig. 14.11, chromosomes can be microdissected at selected regions, and the process can be documented (Thalhammer et al., 1997; Stark et al., 1998; Iwabuchii, S. et al., 2002). In Figure 14.11c, an electron microscopic image of an AFM tip following microdissection is shown. The extracted genetic material adhering to the tip can be amplified by polymerase chain reaction and used as a probe for fluorescence *in situ* hybridization (Fig. 14.12) (Thalhammer et al., 1997). Importantly, the AFM can also operate in liquids. While performing microdissection in liquids on rehydrated chromosomes, only uncontrolled dissections can be produced (Fig. 14.13) (Stark et al., 1998).

14.6. AFM NANOEXTRACTION

The procedure for AFM-based microdissection is described in this section. All steps should be performed under sterile conditions

260 ATOMIC FORCE MICROSCOPY IN CYTOGENETICS

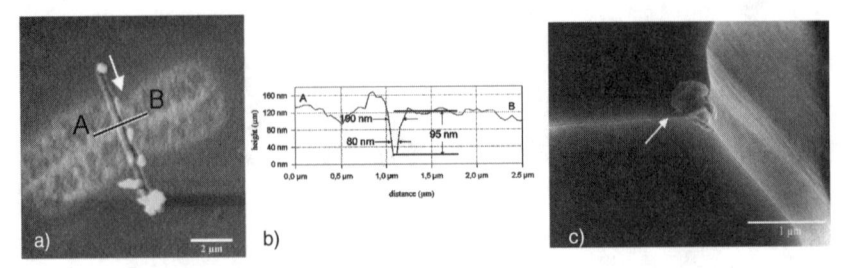

Figure 14.11. (a) AFM image after microdissection of band 2q12 of human chromosome 2. The arrow indicates the scan orientation. (b) Cross-sectional analysis along the cut site A to B. The cut width is 95 nm. (c) Electron microscopic image of the AFM tip after microdissection. The arrow indicates the extracted DNA, bar represents 1 μm.

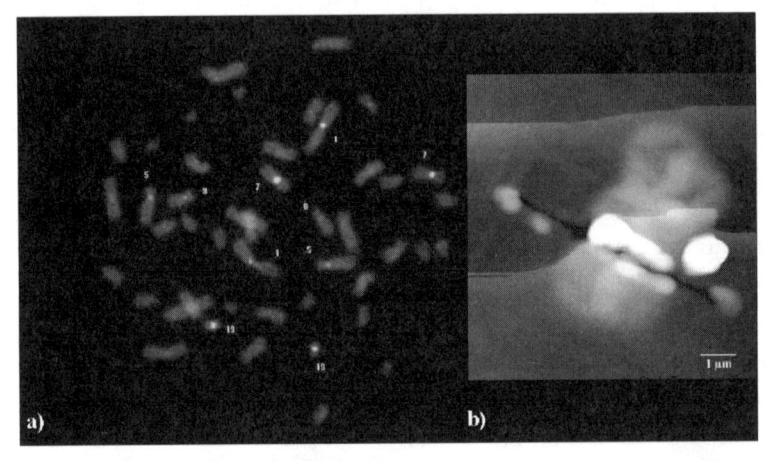

Figure 14.12. Fluorescence *in situ* hybridization (FISH) after microdissection of the centromeric region of a human C group chromosome: (a) The hybridization signals are visible in the centromeric regions of chromosome 1, 5, 7, 8, and 19. (b) AFM control image of the microdissected chromosome. Bar represents 1 μm.

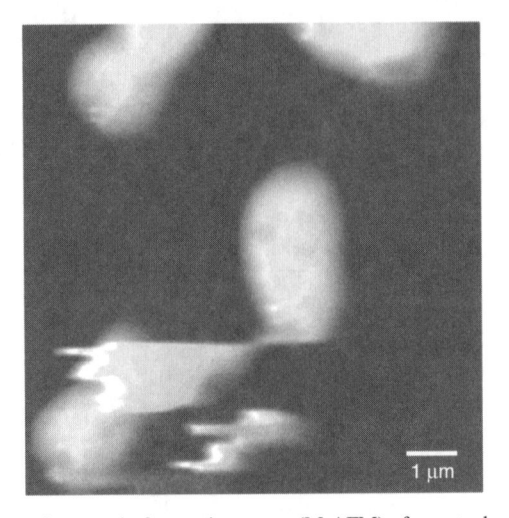

Figure 14.13. Microdissection using atomic force microscopy (M-AFM) of a metaphase chromosome rehydrated in TE-Puffer. For dissection the loading force was increased to 1 μN, and a line scan at 1 μm/s was performed without *z*-modulation. A large part of the chromosome was pushed to the side in an uncontrolled manner.

to avoid contamination. To identify the chromosomal region of interest and to minimize contamination of the AFM tip during scanning the area of interest, GTG-banded metaphase chromosomes should be imaged in noncontact mode in ambient air. The chromosome can also be identified using a "preset" fluorescence *in situ* hybridization of chromosome-specific painting probes (Thalhammer et al., 1997) and fluorescence microscopy. This "preset" hybridization also increases the amount of extracted genetic material. For microdissection, the chromosome is placed at a 90° angle to the scan direction and the chromosomal area is zoomed into. For distance control, amplitude detection is used and the damping level is set to 50% of the amplitude of free oscillation for imaging before extraction. After identification of the extraction site, the scan is stopped and the feedback is turned off. The loading force of the tip onto the sample is increased. Figures 14.14 and 14.15 show the results of AFM microdissection by applying

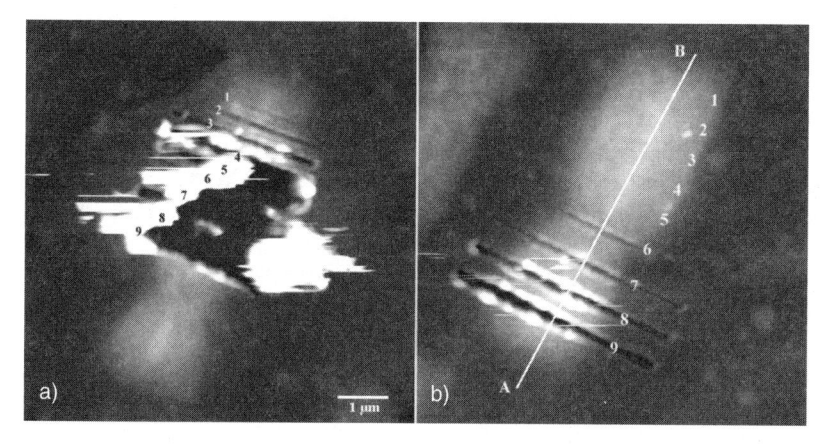

Figure 14.14. Comparison of two different AFM dissection modes in air (z-scale 0–180 nm). **(a)** A series of cuts has been made without z-modulation. Each cut was performed by scanning one line scan at 1 μm/s at a certain loading force: #1, 2.8 μN; #2, 5.6 μN; #3, 8.5 μN; #4, 11.2 μN; #5, 14.0 μN; #6, 16.8 μN; #7, 19.6 μN; #8, 22.4 μN; #9, 25.6 μN. **(b)** The human metaphase chromosome was imaged by AFM in ambient conditions after a series of dissections made by AFM. For dissection z-modulation (~5 nm) has been used. The oscillation amplitude of the cantilever was smaller than 1% of the amplitude of free oscillation for all cuts.

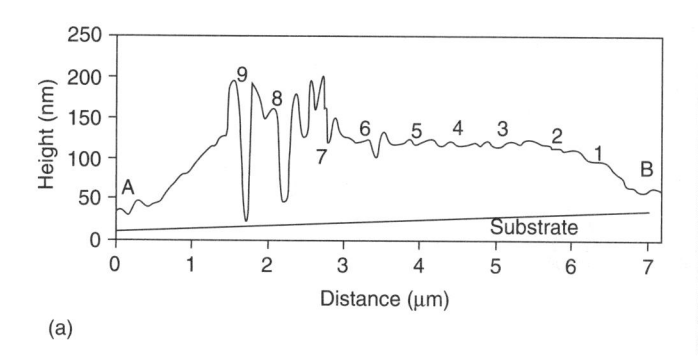

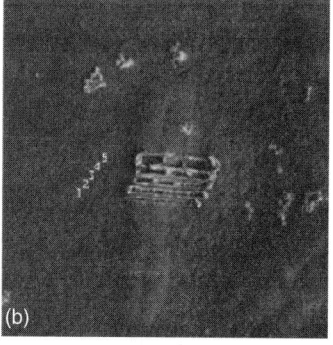

Figure 14.15. **(a)** Cross-sectional analysis as indicated in Fig. 14.13B. The positions of the different cuts are marked by the numbers. **(b)** Electron microscopic image of human metaphase chromosome after microdissection with vertical modulation of the z-piezo. Loading forces are: # 1, 16.8 μN; #2, 19.6 μN; #3, 22.4 μN; #4, 25.6 μN; #5, > 27 μN.

different loading forces to the specimen and by controlling the modulation of the z-piezo.

To extract DNA, a one-line scan with 1 μm/s is performed at the site. During dissection of the chromosome, lateral forces play an important role. The tip performs a stick–slip movement, and the forces between the cantilever tip and the chromosome are reduced while operating with z modulation. The shear forces of the tip are reduced during dissection and reproducible cuts of 100 nm can be made, depending on the geometry of the tip. During microdissection, not only the apex but also the flank of the tip is in contact with the chromosome, and the loading area to the chromosome is increased. The chromosomal material is not dissected in a first step but pushed like a snowplough. Parts of the chromosomal material adhere by van der Walls interaction and unspecific adsorption to the tip. To increase the dissection efficiency, electron-beam-deposited tips (EBD) can be used. EBD tips with a rough surface can be used to increase the extraction efficiency (Fig. 14.16A), while EBD tips with predetermined breaking points minimize the risk of contamination (Fig. 14.16B) (Thalhammer et al., 2002). Thus, the modified AFM tips can be used like a mechanical "nanoscalpel" or a "nanoshovel". After the tip is retracted from the sample surface, the cantilever is transferred into a reaction tube. A new cantilever is used to check the cut at the nanoextraction site on the chromosome. The reaction tube contains a collection buffer to stabilize the extracted genetic material. Enzymatic digestion of the chromosome stabilizing and covering proteins is performed to increase primer binding and therefore the efficiency of the polymerase chain reaction (PCR). Unspecific amplification can also be performed with PCR techniques using degenerated primers (Thalhammer et al., 1997) or adaptor-linked PCR (Thalhammer et al., 2005). The generated genetic samples can further be used for cytogenetic studies—for example, FISH (Thalhammer et al., 1997).

14.7. CONCLUSION

It is clear that based on the working principles of the AFM, it may be used not only for high-resolution imaging of genetic material, but also as a nanosurgical tool. In addition to high structural three-dimensional analysis of genetic material, the AFM microdissection can be applied to isolate smaller cytogenetic samples for biochemical analysis. These can be used in combination with highly sensitive polymerase chain reactions and fluorescence *in situ* hybridization, for either physical mapping of the genome, evolutionary studies, or diagnostic research. Furthermore, it will be interesting to implement a near-field optical microscope in order to identify a

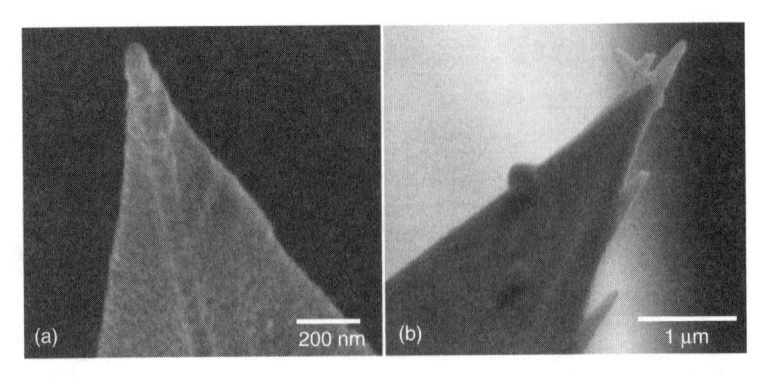

Figure 14.16. Scanning tunneling microscopic image of electron beam deposited AFM tips to increase the extraction efficiency after M-AFM. **(a)** Rough AFM tip, bar represents 200 nm; tip with syringe tip and predetermined breaking point, bar represents 1 μm.

particular genomic region labeled with only few dye molecules for subsequent nanodissection using the AFM.

REFERENCES

Adinolfi, M., and Davies, A. F. (1994). Non-isotopic *in situ* hybridization. In: Applications to Clinical Diagnosis and Molecular Genetics, Medical Intelligence Unit series, R.G. Landes, Austin, Texas.

Allen, M. J., Lee, C., Lee, J. D., IV, Pogany, G. C., Balooch, M., Siekhaus, W. J., and Balhorn, R. (1993). Atomic force microscopy of mammalian sperm chromatin. *Chromosoma* **102**:623–630.

Ambros, P. F., and Sumner, A. T. (1987). Correlation of pachytene chromomeres and metaphase bands of human chromosomes and distinctive properties of telomeric regions. *Cytogenet. Cell Genet* **44**:223–228.

Arrighi, F. E., and Hsu, T. E. (1971). Localization of heterochromatin in human chromosomes. *Cytogenetics* **10**:81–86.

Bahr, G. F., and Larsen, P. M. (1974). Structural bands in human chromosomes. *Adv. Cell. Mol. Biol.* **3**:191–212.

Binnig, G., Rohrer, H., Gerber, C., and Weibl, E. (1982). Surface studies by scanning tunneling microcopy. *Phys. Rev. Lett.* **49**:57–61.

Binnig, G., Quate, C. F., and Gerber, C. (1986). Atomic force microscope. *Phys. Rev. Lett.* **56**:930–933.

Caspersson, T., Farber, S., Foley, G. E., Kudynowski, J., Modest, E. J., Simonsson, E., Wagh,
U., and Zech, L. (1964). Chemical differentiation along metaphase chromosomes. *Exp. Cell Res.* **49**:219–222.

Colton, R. J., Engel, A., Frommer, J. E., Gaub, H. E., Gewirth, A. A., Guckenberger, R., Rabe, J., Heckl, W. M., and Parkinson, B. (1998). Procedures in Scanning Probe Microscopies. John Wiley & Sons, Chichester.

Comings, D. E. (1978). Mechanisms of chromosomes banding and implications for chromosome structure. *Annu. Rev Genet.* **12**:25–46.

de Grooth, B. G., and Putman, C. A. J. (1992). High-resolution imaging of chromosome related structures by atomic force microscopy. *J. Microsc.* **168**:239–247.

Delain, E., Fourcade, A., Poulin, J. C., Barbin, A., Coulaud, D., Lecam, E., and Paris, E. (1992). Comparative observations of biological specimens, especially DNA and filamentous actin molecules in atomic force, tunneling and electron microscopes. *Microsc. Microanal. Microstruct.* **3**:457–470.

Falvo, M. R., Washburn, S., Superfine, R., Finch, M., Brooks, F. P., Chi, V., Taylor, R. M. (1997). Manipulation of individual viruses: Friction and mechanical properties. *Biophys. J.* **72**:1396–1403.

Ford, C. E., and Hamerton, J. L. (1956). The chromosomes of man. *Nature* **178**:1020–1023.

Fritzsche, W. (1999). Salt-dependent chromosome viscoelasticity characterized by scanning force microscopy-based volume measurements. *Microsc. Res. Tech.* **44**:357–362.

Fritzsche, W., and Henderson, E. (1996). Volume determination of human metaphase chromosomes by scanning force microscopy. *Scanning Microsc.* **10**:1–7.

Fritzsche, W., Schaper, A., and Jovin, T. M. (1994). Probing chromatin with the scanning force microscope. *Chromosoma* **103**:231–236.

Fritzsche, W., Takac, L., and Henderson, E. (1997). Application of atomic force microscopy to visualization of DNA, chromatin and chromosomes. *Crit. Rev. Eukaryot. Gene Expr.* **7**:231–240.

Fukushi, D., and Ushiki, T. (2005). The structure of C-banded human metaphase chromosomes as observed by atomic force microscopy. *Arch Histol Cytol.* **68**(1):81–7.

Geissler, B., Noll, F., and Hampp, N. (2000). Nanodissection and noncontact imaging of plasmid DNA with an atomic force microscope. *Scanning* **22**:7–11.

German, J. L. (1964). The pattern of DNA synthesis in the chromosomes of human blood cells. *J. Cell Biol.* **20**:37–55.

Gobbi, P., Falconi, M., Vitale, M., Galanzi, A., Artico, M., Martelli, A. M., and Mazzotti, G. (1999). Scanning electron microscopic detection of nuclear structures involved in DNA replication. *Arch. Histol. Cytol.* **62**:317–326.

Gobbi, P., Thalhammer, S., Falconi, M., Stark, R., Heckl, W. M., and Mazzotti, G. (2000). Correlative high resolution morphological analysis of the three-dimensional organization of human metaphasechromosomes. *Scanning* **22**:273–281.

Hansma, H. G., Vesenka, J., Siegerist, C., Kelderman, G., Morrett, H., Sinsheimer, R. L., Elings, V., Bustamante, C., and Hansma, P. K. (1992). Reproducible imaging and dissection of plasmid DNA under liquid with the atomic force microscope. *Science* **256**:1180–1183.

Hansma, H. G., Sinsheimer, R. L., Groppe, J., Bruice, T. C., Elings, V., Gurley, G., Bezanilla, M., Mastrangelo, I. A., Hough, P. V., and Hansma, P. K. (1993). Recent advances in atomic force microscopy of DNA. *Scanning* **15**(5):296–299.

Harrison, C. J., Britch, M., Allen, T. D., and Harris, R. (1981). Scanning electron microscopy of the G-banded human karyotype. *Exp. Cell Res.* **134**:141–153.

Henderson, E. (1992). Imaging and nanodissection of individual supercoiled plasmids by atomic force microscopy. *Nucleic Acids Res.* **20**:445–447.

Heneen, W. K., and Caspersson, T. (1973). Identification of chromosomes of rye by distribution patterns of DNA. *Hereditas* **74**:259–272.

Hermann, R., and Müller, M. (1992). Towards high resolution SEM of biological objects. *Arch. Histol. Cytol.* **55**:17–25.

Hoh, J. H., Lal, R., John, S. A., Revel, J. P., and Arnsdorf, M. F. (1991). Atomic force microscopy and dissection of gap junctions. *Science* **253**:1405–1408.

Hoshi, O., Owen, R., Miles, M., and Ushiki, T. (2004). Imaging of human metaphase chromosomes by atomic force microscopy in liquid. *Cytogenet Genome Res.* **107**(1–2):28–31.

Hsu, T. C. (1952). Mammalian chromosomes *in vitro*. 1. The karyotype of man. *J. Hered.* **43**:167–172.

Iwabuchii, S., Mori, T., Ogawa, K., Sato, K., Saito, M., Morita, Y., Ushiki, T., and Tamiya, E. (2002). Atomic force microscope-based dissection of human metaphase chromosomes and high resolutional imaging by carbon nanotube tip. *Arch Histol Cytol.* **Dec;65**(5):473–9.

Jiao, Y., and Schaffer, T. E. (2004). Accurate height and volume measurements on soft samples with the atomic force microscope. *Langmuir.* **9;20**(23):10038–45.

Jondle, D. M., Ambrosio, L., Vesenka, J., and Henderson, E. (1995). Imaging and manipulating chromosomes with the atomic force microscope. *Chromosome Res.* **3**:239–244.

Kalle, W. H. J., Macville, M. V. E., van de Corput, M. P. C., de Grooth B. G., Tanke, H. J., and Raap, A. K. (1995). Imaging of RNA *in situ* hybridization by atomic force microscopy. *J. Microsc.* **182**:192–199.

Lattanzi, G., Galanzi, A., Gobbi, P., Falconi, M., Matteucci, A., Breschi, L., Vitale, M., and Mazzotti, G. (1998). Ultrastructural aspects of the DNA polymerase a distribution during the cell cycle. *J. Histochem. Cytochem.* **46**:1435–1442.

Marti, O., Drake, B., and Hansma, P. K. (1987). Atomic force microscopy of liquid-covered surfaces: Atomic resolution images. *Appl Phys Lett.* **51**:484–486.

McKay, R. D. (1973). The mechanism of G and C banding in mammalian metaphase chromosome. *Chromosoma* **44**:1–14.

McMaster, T. J., Hickish, T., Min, T., Cunningham, D., and Miles, M. J. (1994). Application of scanning force microscopy to chromosome analysis. *Cancer Genet. Cellgenet.* **76**:93–95.

McMaster, T. J., Winfield, M. O., Karp, A., and Miles, M. J. (1996). Analysis of cereal chromosomes by atomic force microscopy. *Genome* **39**:439–444.

Moorhead, P. S., Nowell, P. C., Mellman, W. J., Battips, D. M., and Hungerford, D. A. (1960). Chromosome preparations of leucocytes cultured from human peripheral blood. *Exp. Cell Res.* **20**:613–616.

Mosher, C., Jondle, D., Ambrosio, L., Vesenka, J., and Henderson, E. (1994). Microdissection and measurement of polytene chromosomes using the atomic force microscope. *Scanning Microsc.* **8**(3):491–497.

Musio, A., Mariani, T., Frediani, C., Sbrana, I., and Ascoli, C. (1994). Longitudinal patterns similar to G-banding in intreated human chromosomes: Evidence from atomic force microscopy. *Chromosoma* **103**:225–229.

Nägeli, K. (1842). *Zur Entwicklungsgeschichte der Pollen*. (Ilse Jahn, ed.) Spektrum Akademischer Verlag Gustav Fischer, Berlin, Zürich.

Nagatani, T., Saito, S., Sato, M., and Yamada, M. (1987). Development of an ultra high resolution scanning electron microscope by means of a field emission source and in-lens system. *Scanning Microsc.* **1**:901–909.

Pawley, J. (1997). The development of field-emission scanning electron microscopy for imaging biological surfaces. *Scanning* **19**: 324–336.

Putman, C. A. J., de Grooth, B. G., Wiegant, J., Raap, A. K., van der Werf, K. O., van Hulst, N. F., and Greve, J. (1993). Detection of *in situ* hybridization to human chromosomes with the atomic force microscope. *Cytometry* **14**:356–361.

Rasch, P., Wiedemann, U., Wienberg, J., and Heckl, W. M. (1993). Analysis of banded human-chromosomes and *in situ* hybridization patterns by scanning force microscopy. *Proc. Natl. Acad. Sci. USA* **90**:2509–2511.

Rizzi, E., Falconi, M., Baratta, B., Manzoli, L., Galanzi, A., Lattanzi, G., and Mazzotti, G. (1995). High-resolution FEISEM detection of DNA centromeric probes in HeLa metaphase chromosomes. *J. Histochem. Cytochem.* **43**: 413–419.

Rizzoli, R., Rizzi, E., Falconi, M., Galanzi, A., Baratta, B., Lattanzi, G., Vitale, M., Manzoli, L., and Mazzotti, G. (1994). High resolution detection of uncoated metaphase chromosome by means of field emission scanning electron microscopy. *Chromosoma* **103**:393–400.

Schaper, A., Pascual, S. J. P., and Jovin, T. M. (1994). The scanning force microscopy of DNA in air and in *n*-propanol using new spreading agents. *FEBS Lett.* **355**:91–95.

Stark, R., Thalhammer, S., Wienberg, J., and Heckl, W. M. (1998). The AFM as a tool for chromosomal dissection—the influence of physical parameters. *Appl. Phys.* **A66**:579–584.

Sumner, A. T. (1972). A simple technique for demonstrating centromeric heterochromatin. *Exp. Cell Res.* **75**:304–306.

Sumner, A. T. (1996). Problems in preparation of chromosomes for scanning electron microscopy to reveal morphology and to permit immunocytochemistry of sensitive antigens.. *Scan Microsc.* **10**:165–176.

Sumner, A. T., and Ross, A. (1989). Factors affecting preparation of chromosomes for scanning electron microscopy using osmium impregnation. *Scanning Microsc. (Suppl.)* **3**:87–97.

Sumner, A. T., Ross, A. R., and Graham, E. (1994). Preparation of chromosomes for scanning electron microscopy. *Methods Mol. Biol.* **29**:41–50.

Tamayo, J., Miles, M., Thein, A., and Soothill, P. (1999). Selective cleaning of the cell debris in human chromosome preparations studied by scanning force microscopy. *J. Struct. Biol.* **128**:200–210.

Tamayo, J. (2003). Structure of human chromosomes studied by atomic force microscopy. Part II. Relationship between structure and cytogenetic bands. *J Struct Biol.* **141**(3):189–97.

Tan, E., Iffet, F., Ergün, M. A., Ercan, I., and Menevse, A. (2001). C-banding visualized by atomic force microscopy. *Scanning* **23**:32–35.

Thalhammer, S., Stark, R., Müller, S., Wienberg, J., and Heckl, W. M. (1997). The atomic force microscope as a new microdissecting tool for the generation of genetic probes. *J. Struct. Biol.* **119**(2):232–237.

Thalhammer, S., Köhler, U., Stark, R., and Heckl, W. M. (2001). GTG banding pattern on human metaphase chromosomes revealed by high resolution atomic-force microscopy. *J. Microsc.* **202**(3):464–467.

Thalhammer, S., and Heckl, W. M. (2005). Manipulation of genetic material. *Nano Today* **5**:40–49.

Thundat, T., Allison, D. P., Warmack, R. J., and Ferrell, T. L. (1992). Imaging isolated strands of DNA molecules by atomic force microscopy. *Ultramicroscopy* **42**44:1101–1106.

Tjio, J. H., and Levan, A. (1956). The chromosome number of man. *Hereditas* **42**:1–6.

Verma, R. S., and Babu, A. (1989). Human Chromosomes: Manual of Basic Techniques, Pergamon Press, New York.

Vesenka, J., Guthold, M., Tang, C. L., Keller, D., Delaine, E., and Bustammante, C. (1992). Substrate preparation for reliable imaging of DNA molecules with the scanning force microscope. *Ultramicroscopy* **42**44:1243–1249.

Waldeyer, W. (1890). Karyokinesis and its relation to the process of fertilization. *Q. J. Microsc. Sci.* **30**:159–281.

Wanner, G., and Formanek, H. (1995). Imaging of DNA in human and plant chromosomes by high-resolution scanning electron microscopy. *Chromosome Res.* **3**:368–374.

Wanner, G., Formanek, H., Martin, R., and Herrmann, R. G. (1991). High resolution scanning electron microscopy of plant chromosomes. *Chromosoma* **100**:103–109.

Winfield, M., McMaster, T. J., Karp, A., and Miles, M. J. (1995). Atomic force microscopy of plant chromosomes. *Chromosome Res.* **3**: 128–131.

Xu, X. M., and Ikai, A. (1998). Retrieval and amplification of single-copy genomic DNA from a nanometer region of chromosomes: A new and potential application of atomic force microscopy in genomic research. *Biochem. Biophys. Res. Commun.* **248**:744–748.

Yang, J., and Shao, Z. (1993). Effect of probe force on the resolution of atomic force microscopy of DNA. *Ultramicroscopy* **50**:157–170.

Yunis, J. J., and Sanchez, O. (1993). G-banding and chromosome structure. *Chromosoma* **44**: 15–23.

CHAPTER 15

ATOMIC FORCE MICROSCOPY IN THE STUDY OF MACROMOLECULAR INTERACTIONS IN HEMOSTASIS AND THROMBOSIS: UTILITY FOR INVESTIGATION OF THE ANTIPHOSPHOLIPID SYNDROME

WILLIAM J. MONTIGNY

Microscopy Imaging Center, College of Medicine, University of Vermont, Bulington, Vermont

ANTHONY S. QUINN

Department of Pathology and Microscopy Imaging Center, College of Medicine, University of Vermont, Burlington, Vermont

XIAO-XUAN WU

Department of Pathology, Montefiore Medical Center, Albert Einstein College of Medicine, Bronx, New York

EDWIN G. BOVILL

Department of Pathology, College of Medicine, University of Vermont, Burlington, Vermont

JACOB H. RAND

Department of Pathology, Montefiore Medical Center, Albert Einstein College of Medicine, Bronx, New York

DOUGLAS J. TAATJES

Department of Pathology and Microscopy Imaging Center, College of Medicine, University of Vermont, Burlington, Vermont

15.1. INTRODUCTION

This chapter demonstrates the unmatched efficacy of atomic force microscopy (AFM) for the investigation of interactions occurring between biological cell-surface macromolecules on artificial phospholipid membranes. Specifically, the delineation of a series of complex interactions that take place among factors believed to have seminal roles in mediating increased thrombosis in a clinically defined syndrome, the antiphospholipid syndrome (APS), clearly illustrates the extraordinary versatility and flexibility of this technique (Marchant et al., 2002; Rand et al., 2003). The unique ability of the AFM to image objects of an unprecedented range of size and complexity (from single

---

Force Microscopy: Applications in Biology and Medicine, edited by Bhanu P. Jena and J.K. Heinrich Hörber.
Copyright © 2006 John Wiley & Sons, Inc.

small molecules, to complex multicomponent enzymes, to whole cells) *in situ*, in a temporal fashion and within ambient physiological buffers that preserve biological form and function, renders this technique invaluable in studies such as this, that seek to explore delicate and ephemeral biochemical relationships (Bustamante and Rivetti, 1996; Dunlap et al., 1997; Bustamante et al., 1999; Hansma et al., 1999). In concert with other techniques such as ellipsometry, the AFM provides a valuable method to link nanoscale cell-surface molecular changes with the increased thrombosis and concomitant etiology of the antiphospholipid syndrome (Rand et al., 2003).

15.2. ATOMIC FORCE MICROSCOPY

The development of the atomic force microscope has engendered a paradigm shift in the way that the structure and function of biological macromolecules and the complex relationships that are inherent among them are examined. The AFM (Binnig et al., 1986) was originally conceived and developed as an outgrowth of scanning probe microscopy, which includes near-field scanning optical microscopy (NSOM) (Betzig et al., 1986; Durig et al., 1986), scanning tunneling microscopy (STM) (Binnig et al., 1982a, b), cryo AFM (Mou et al., 1993; Han et al., 1995; Shao and Zhang, 1996), and magnetic force AFM (MAC mode AFM) (Han et al., 1996). While other atomic (angstrom) or molecular (nanometer) scale imaging techniques such as x-ray crystallography, nuclear magnetic resonance, and transmission electron microscopy are dependent on the detection of transmission, diffraction, or deflection of various wavelengths of light, electron beams, and x-ray beams, the AFM relies on a technology that is based entirely on tactile sensing. In addition to its ability to convert extremely subtle nuances of surface depth and texture into high-resolution three-dimensional images, many other parameters, such as biomechanical characteristics, force measurements, and enzymatic processes, can be explored by AFM. The tactile quality of AFM, in which a nanoscale probe makes physical contact with the sample, means that this technique can accurately quantify inter- and intramolecular force interactions as small as the breaking of hydrogen bonds and as subtle as differences in antibody–antigen binding energies (Stroh et al., 2004). The level of sensitivity that can be achieved allows detection and accurate measurement of pico-newton (pN) levels of force under ambient aqueous conditions that closely replicate *in vivo* environments. In fact, molecular interactions such as DNA–protein binding, intramolecular stretching and unwinding, and receptor–ligand coupling can be measured to tolerances of several pico-newtons (Bustamante et al., 2000; Marchant et al., 2002; Allemand et al., 2003). In imaging mode, because experiments can be performed under physiologically relevant conditions with minimal sample processing, AFM can develop dramatic images of dynamic or enzymatic processes such as transcription, replication, DNA bending, DNA–protein, and DNA–lipid interactions in a temporal fashion, at submolecular resolution (Dunlap et al., 1997; Hansma et al., 1999; Montigny et al., 2001, 2003; Abdelhady et al., 2003).

The working principles for AFM, while complex in execution, are simple in conception and are as follows (see Fig. 15.1): A nanoscale-dimension, oxide-sharpened silicon nitride probe is mounted at the end of a silicon nitride cantilever that, depending on various imaging modalities, can be of several different configurations, lengths, and spring constants. The nominal radius of curvature at the apex of the probe has been estimated at about 30 nm, although the ability to achieve resolution that may exceed 1 nm argues for a much sharper, higher aspect shape. This may in part be due to the presence of very fine asperities formed during probe fabrication that act as the primary scanning point (Bustamante and Rivetti, 1996; Taatjes et al., 1999). The back of the flexible cantilever is coated with gold and is highly reflective. The

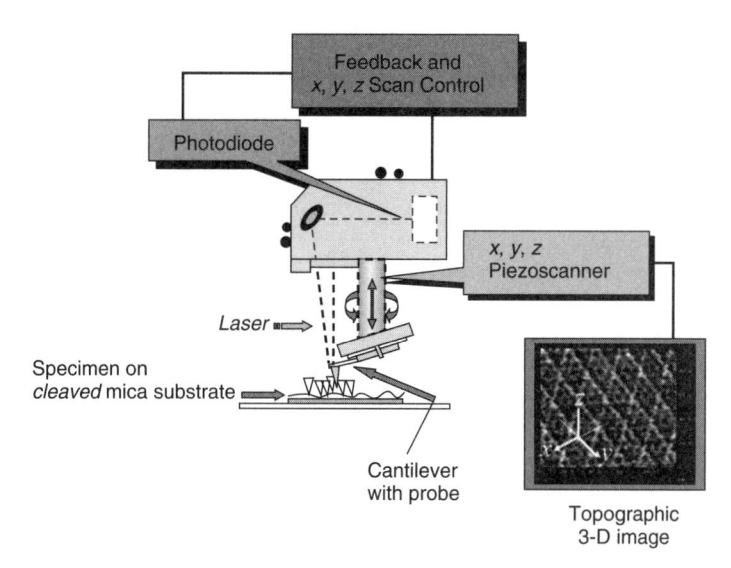

Figure 15.1. Schematic representation of the principles of atomic force microscopy.

cantilever-probe unit is attached to a block affixed to a supporting structure that is in turn mounted on the piezo tube. The design and configuration of this support is dependent on the mode of imaging. In air, macromolecules are deposited (with or without fixation) on simple metal or mica discs, and the cantilever mount can be quite simple. Imaging in ambient room atmosphere (i.e., "contact mode in air") is the simplest mode of AFM and was the first method developed. However, a plethora of uncontrollable variables, which are treated in more detail below, render this method problematic for imaging most biological samples. The existence of strong adhesive forces (such as capillary attraction), engendered by the nucleation of fluid layers between tip and sample, requires the use of extremely high tip–sample force levels that often produce severe sample damage (Bustamante and Rivetti, 1996; Morris et al., 1999). The use of buffered fluid environments for AFM imaging, while more difficult, is far more amenable for preserving delicate biological molecules or for capturing dynamic biochemical processes. It has largely replaced air imaging for most biological applications (Bustamante et al., 1997; Kindt et al., 2002).

For imaging in fluid environments, a Plexiglas fluid cell with a watertight gasket is used. The sample chamber used under these conditions can be as simple as a water meniscus created by a drop of buffer applied directly to a freshly cleaved mica disc epoxied to a slide; more complex arrangements have been devised that are sealed with O-rings to prevent evaporation and consequent changes in the ionic strength of imaging buffers. The most sophisticated units permit the introduction of buffer through external ports while simultaneously scanning, and they may utilize an automated pump delivery system for continuous replenishment and/or instantaneous modification of imaging buffer solutions *in situ*. Also, finely calibrated temperature control can be implemented, allowing a level of approximation and modulation of the physiological environment that is incomparable at such high (submolecular) levels of resolution (Kindt et al., 2002).

The probe support assembly (i.e., the fluid cell in fluid mode) with its attached cantilever/probe unit is mounted at the end of a tube constructed of piezoelectric ceramic (lead titanium zirconate/PTZ) material. Electrodes are attached to the column in such a fashion as to create a bias voltage across the piezoelectric tube at sets of horizontally opposed points located 90° apart and at the top and bottom of the tube. The piezoelectric effect refers to the ability of these materials to generate voltage upon compression and deformation; the reverse piezoelectric effect causes the piezoelectric tube to deform in atomically accurate increments upon administration of a bias voltage, allowing precise movement in the x, y, and z directions of the scanning plane (Taylor, 1993; Bustamante and Rivetti, 1996; Morris et al., 1999). During operation, samples are affixed to freshly cleaved muscovite mica, which has an atomically flat surface with an RMS surface roughness of 0.06 nm. This surface provides a low-relief nonreactive substrate that easily exceeds the surface tolerances needed to image the smallest biological molecules, and it can even be utilized for atomic-scale imaging (Hansma and Laney, 1996; Kindt et al., 2002). The probe is scanned across the surface in raster fashion in the x and y axis in response to the electrical voltage applied to the electrodes. As the probe encounters anomalies such as molecular shapes that perturb the uniform mica surface during the scan, the cantilever is deflected and the degree of deflection is detected by use of an optical lever effect, created by focusing a laser on the back of the highly reflective cantilever surface. The laser is reflected up to a four-quadrant photodiode, which converts laser movement on its surface into a measurable electrical voltage change, which is in turn translated via computer software into sensitive height and amplitude information. Changes in the laser spot position on the vertical axis of the photodiode indicate height differences, while horizontal changes indicate twisting or torque-like movements of the cantilever which furnish information about surface characteristics such as charge, malleability, and adhesion. In addition, a feedback loop can be used to maintain preset force and cantilever deflection levels; z-axis movements to raise or lower the piezo in response to surface height changes are monitored, and these data provide an alternative to direct recording of changes in photodiode voltage to generate data about the scanned surface. The optical lever effect vastly amplifies the sensitivity of detection, with very small low-angle laser deflections at the probe–surface interface translated into comparatively large movements on the photodiode surface (Bustamante and Rivetti, 1996; Morris et al., 1999). The *theoretical* detection limits of cantilever displacement are infinitesimally small, on the order of 4×10^{-4} Å, with a signal-to-noise ratio of 1. However, this has been achieved only when extremely short, stiff cantilevers are utilized in rigidly controlled settings that are different from those found in biochemical AFM experiments (Meyer and Amer, 1988). Under actual experimental conditions, where random thermal excitation of the cantilever and instrumental noise are unavoidable, the cantilever displacement detection sensitivity still falls well within atomic scale. That is, a 0.01-nm displacement can generate a detectable voltage change (Rugar and Hansma, 1990; Bustamante and Rivetti, 1996; Morris et al., 1999). The lower limit of resolution for biological specimens is dependent on a myriad of factors, and ranges from 10^{-10} to 10^{-9} m (angstrom–nanometer range). All cantilever deflection and piezo movement information from a scan is converted into a digital point map of the surface, and these height and amplitude data are used to generate a three-dimensional topographical surface rendering with precise images of sample shape at submolecular resolution.

The AFM can be operated in several different modes (tapping, contact, fluid imaging, etc.) that have varying degrees of utility depending on experimental conditions and sample types. The deciding factors in choosing a particular imaging mode are concerned with attaining balance between often competing parameters that include (a) achieving reproducible, high-resolution images, (b) preserving delicate and complex structures, (c) simulating physiological conditions, and (d) achieving binding conditions that permit surface equilibration and movement. Contact mode in air is relatively straightforward and simple, with molecules bound tightly to the mica surface, providing consistent reproducible images. However, forces generated by the probe in this mode are large enough to do severe damage to biological samples, especially due to lateral shearing, and molecules may be swept completely off the mica substrate. Under normal ambient environmental conditions in air, capillary forces and hard-core repulsions dominate the interaction landscape between the probe tip and the sample (Bustamante and Keller, 1994; Bustamante and Rivetti, 1996).

A shell of hydration several water molecules thick forms on all objects, particularly on sharp tips with small radii of curvature, which act as nucleation points for the deposition of water molecules. Capillary interactions (Laplace's force) are relatively enormous (on the order of hundreds of nano-newtons), dwarfing other forces, and necessitating the use of extremely high piezo forces to overcome the attractive interaction (Israelachvili, 1985; Morris et al., 1999). In some cases, the probe may actually become pinned to the surface due to this attractive force (Rugar and Hansma, 1990). Forces in contact mode can be minimized by stringently controlling atmospheric humidity levels to less than 35%. At this level, forces in contact mode can be limited to 1–10 nN, but this still puts a high lower limit on the tip–sample force, often causing structural damage to delicate samples or multisubunit assemblies. In some cases this force is still high enough to sweep samples off of the mica surface (Bustamante and Rivetti, 1996).

The AFM tapping mode seeks to ameliorate some of these concerns, and it is successful in most respects. In this mode the cantilever is excited (by various electrical, magnetic, piezo, or acoustic stimuli) at close to its resonance frequency. This induces high-frequency oscillation of the cantilever, causing the tip to make a series of extremely brief contacts on the sample surface, which is far less damaging and invasive, and generates lower surface force (Hansma et al., 1993; Putman et al., 1994). However, due to the attractive capillary forces that still exist between tip and sample, a relatively high lower limit is still set on the force needed to keep the tip in oscillation and not bound to the sample surface (Bustamante et al., 1994). Consequently, tapping force levels, while lower than those found in contact mode, are still high enough to cause damage to "soft" biological macromolecules or assemblies, especially due to shearing forces. Also, at the force levels needed to prevent probe capture, probe oscillation and amplitude are much more difficult to control, leading to difficulties in achieving consistent high-resolution images. Finally, both contact and tapping modes (in air) may induce artifacts, severely alter structural relationships and molecular interactions, and completely abrogate dynamic enzymatic processes, due to sample dehydration and desiccation (Engel et al., 1999).

Use of the tapping mode, in concert with a buffered liquid environment, eliminates nearly all of the above problems while introducing several other advantages that make this mode of operation far more amenable for the analysis of most biological samples. Most importantly, imaging in liquid abrogates the shell of hydration at the interface of sample surface and atmosphere. This nearly eliminates capillary interactions, which dwarf all other types of atomic interactions, causing the problems noted above. Now the force interaction landscape is dominated by van der Waals forces, electrostatic attractive forces, and hard core repulsions, all of which are relatively small (0.1–1 nN), allowing the force that is applied during imaging to be much smaller (Bustamante and Rivetti, 1996). Even these lower levels of force are not ideal for use with biological macromolecules, and so tapping in conjunction with a liquid environment, while difficult, was developed to provide the optimum combination of lowest force levels, high-quality resolution, and physiological environment (Hansma et al., 1993; Putman et al., 1994; Kindt et al.,

2002). Since the lower limits of the force which must be applied are greatly reduced (because there is no possibility for capillary entrapment of the probe by the surface) to 0.01–0.1 nN, the amplitude of oscillation is concomitantly reduced by a significant degree, allowing far greater control and far less damage in tip–sample interactions. The use of fluids in the imaging environment also eliminates the artifacts and alterations introduced by sample dehydration, and it allows for the design and maintenance of specific highly controlled conditions that closely approximate the biochemistry and physiology of the cellular milieu. Imaging under such conditions allows for extremely sensitive modulation of surface binding energies by varying the ionic concentration (primarily through divalent cations) of both deposition and imaging solutions (Hansma and Hoh, 1994; Hansma and Laney, 1996; Thompson et al., 1996; Morris et al., 1999). More sophisticated and subtle details can be controlled, such as the ability to allow for movement and equilibration of molecules as they transition from a three-dimensional solution structure to a two-dimensional surface shape (Bustamante and Rivetti, 1996; Guthold et al., 1999). This ensures that lowest-energy conformations approximating the solution structure of the molecule or molecular assembly are accurately preserved, without artificial conformational constraints that might have major consequences in studying inter- and intramolecular relationships (Rivetti et al., 1996, 1998). Such an ambient, physiologically buffered environment allows for enzymatic or dynamic processes including transcription, DNA looping and bending, and protein-induced conformational changes to be captured *in situ*, in dynamic temporal scale, essentially "live" (Guthold et al., 1999; Houchens et al., 2000). There is no other imaging technology that can do all of these things within the specific constraints that apply to the analysis of biological systems, while achieving consistently high resolution over an enormous range of scale, complexity, and conditions.

Several concerns about potential limitations during the initial forays into this technique have proved to be of little significance, or even to be advantageous. The first concern was the perception that, due to the unique manner in which samples are prepared prior to AFM imaging (by binding to atomically flat mica surfaces), conditions would be less physiologic than those found in conventional "solution/test tube" environments. However, as noted elsewhere, cellular biochemistry often takes place when molecules are localized on surfaces of one type or another (inner and outer surfaces of membranes, ribosomes, extracellular matrix, cytoskeleton, interacting macromolecular surfaces, etc); therefore, AFM in fluid tapping mode generates conditions that are certainly as valid as the proverbial test tube, as a simulacrum of the biological environment (Hansma, 2001; Kindt et al., 2002). The second caveat was the initial disappointment among scientists when it appeared that achieving atomic (i.e., angstrom-level) resolution (a "holy grail" of structural biochemistry) might not be technically feasible for the AFM when applied to biological macromolecules under aqueous physiological conditions. While it has since been demonstrated that such resolution is indeed possible, albeit only under several types of highly constrained conditions (Engel et al., 1999; Muller et al., 1999; Engel and Muller, 2000; Stahlberg et al., 2001; Horber and Miles, 2003), it is far more important to note the incomparable levels of resolution that are possible for a variety of substrates and conditions that cannot be imaged under any circumstances by such "atomic level" techniques as x-ray crystallography, nuclear magnetic resonance, and cryo-EM. These include the ability to achieve nanoscale molecular-level resolution of objects too large, complex, and/or delicate to be analyzed by the aforementioned atomic-scale techniques. This size/complexity range, described as the "mesoscopic" realm (Bustamante and Rivetti, 1996), includes macromolecules between 10 nm and 200 nm in size, such as multicomponent polypeptide assemblages, multisubunit enzymes, and dynamic protein–protein or protein–DNA complexes, whose analysis has proven highly intractable to other techniques. The ability to achieve very high levels of resolution with very low signal-to-noise ratios, without resorting to damaging fixation or staining techniques, in ambient physiological buffers and in real time, allows (a) the preservation and analysis of essential structural, functional, and enzymatic characteristics and (b) delicate inter- and intramolecular relationships that were previously unavailable to other forms of analysis.

15.3. AFM AND LIPIDS

AFM has been used with much success to image and analyze the structure and conformation of lipid films, surfaces, and bilayers having a broad range of different chemical compositions. Because such work can be done in fluid tapping mode utilizing ambient physiological buffers, the experimental conditions can replicate the cellular environment to a far greater degree than any other technique having comparable resolution. Any analysis of lipid bilayers that is not conducted in a buffered liquid environment will have serious issues of relevance from a biological standpoint. The interactions of the hydrophobic and hydrophilic components of the phospholipid bilayers that form biological membranes, and aspects of their two- and three-dimensional structure, are created and constrained by their relationships with intra- and extracellular fluid interfaces. Conventional techniques such as transmission electron microscopy must resort to dehydration, staining, or freeze-fracture techniques that can profoundly affect the most basic characteristics of lipid bilayers or membranes. These treatments thus remove the aqueous partner in the powerful and delicate interplay of the hydrophobic interactions that maintain the integrity of lipid layers. In addition, dehydration may also introduce imaging artifacts that are difficult to interpret. In contrast, the AFM is particularly well-suited for the examination of lipid surfaces, which can be imaged at very high resolution. Using fluid tapping mode, the resolution that can be achieved under certain experimental conditions is unmatched, often approximating atomic-scale (angstrom) levels. In these cases, protein components within lipid layers may manifest as "two-dimensional crystals" (Fig. 15.2) (Reviakine et al., 1998; Muller et al., 1999, 2000; Stahlberg et al., 2001; Rand et al., 2003), which form densely packed or regularly repeated arrays that are amenable to high-resolution imaging, followed by Fourier transform image processing. Defects that involve the absence of several atoms can sometimes be visualized (Engel and Muller, 2000). By varying the composition of the lipids through the introduction of different ratios and types of phospholipids, and by subsequent addition of other membrane components such as proteins, it is possible to construct an infinite variety of imaging substrates with a high degree of physiological fidelity to biological membranes. These lipid surfaces provide for dynamic imaging and force experiments with the AFM that cannot be achieved with other experimental methods.

Mica surfaces prepared with planar phospholipid bilayers must be continually submerged in buffers in fluid chambers during incubations and imaging to minimize oxidation, which can result in structural defects. Fortuitously, lipid bilayer preparations with void defects on atomically flat mica provide a means of measuring layer thicknesses (vertical displacement) via sectional analysis. Such baseline measurements are valuable for future structural analyses and volumetric measurements of membrane proteins and subsequent complexes (i.e., two-dimensional crystals and antibody–cofactor toroids) that form upon their interaction with the lipid planar surface and other components in, or subsequently added to, a fluid environment. Visualizing dynamic protein–lipid molecular interactions in an artificial environment under physiological conditions has

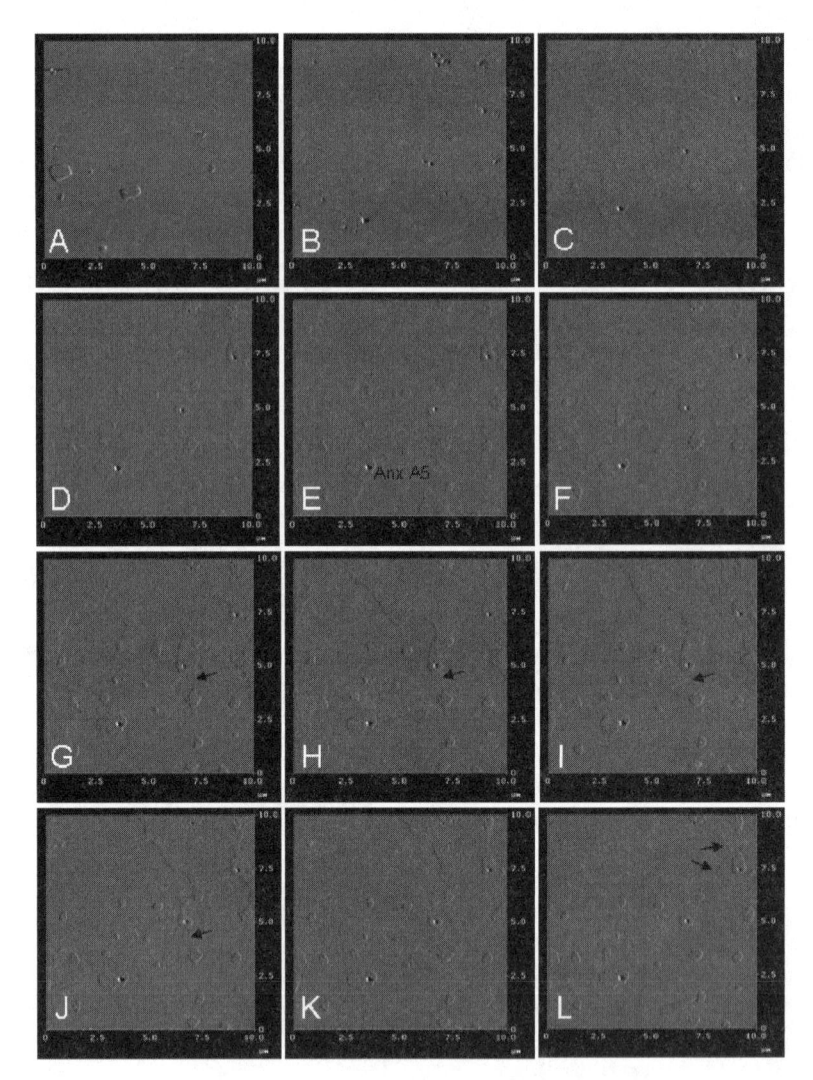

Figure 15.2. An example of the temporal formation of a two-dimensional crystalline lattice of the anticoagulant protein AnxA5 on an artificial planar lipid surface composed of 30% phosphatidylserine and 70% phosphatidylcholine (PSPC). (**A**) The planar lipid membrane (only) showing several defects. (**B**) Bovine serum albumin is present in the defects. Between images **B** and **C**, AnxA5 had been added. Images **C** and **D** display a pebbled and raised surface appearance indicative of the anionic phospholipid-containing planar bilayer conformationally changing in response to AnxA5 and physiological 1.25 mM $CaCl_2$ being present. Within an hour the AnxA5 forms nucleation sites and coalesces into a two-dimensional crystalline lattice over the planar lipid membrane as shown in the following temporal sequence of images **E–L**. Two-dimensional AnxA5 crystal growth displays junctions termed furrows (black arrows), which are suspected to represent weak points in the AnxA5 lattice. The images depicted in **A–L** are original 10.0-μm × 10.0-μm topographical amplitude scans with no off-line processing. Vertical scan lines in the left side of the images result from the "ringing" effect, which is not uncommon at a scan rate of 3.05 kH or higher.

greatly improved the present understanding of morphological changes in lipid bilayers during peptide–membrane interactions (Janshoff and Steinem, 2001).

15.4. AFM AND HAEMOSTATIC PROTEINS

Fibrinogen is a large multifunctional glycoprotein with a molecular weight of 340 kDa.

In the blood plasma, fibrinogen participates in cardiovascular events through its conversion by thrombin to the insoluble fibrin complex, as well as via its binding to activated platelets leading to cross-linking and platelet aggregation. We have previously imaged purified human fibrinogen bound to poly-L-lysine coated mica by fluid tapping-mode AFM (Taatjes et al., 1997). Using both conventional silicon nitride pyramidal tips and electron-beam-deposited carbon tips, we observed that human fibrinogen in fluid appeared as a bi- or trinodular, slightly curved linear molecule. These fibrinogen molecules averaged 65.8 nm in length and presented as a "dumbbell" shape. This appearance was in good agreement with prior images generated of fixed, dehydrated fibrinogen by transmission electron microscopy (Hall and Slayter, 1959). Similar AFM imaging of fibrinogen has been reported by Marchant and co-workers (Marchant et al., 1997; Sit and Marchant, 1999).

15.5. THE ANTIPHOSPHOLIPID SYNDROME (APS)

APS is an enigmatic autoimmune disorder that is defined by the presence of circulating antiphospholipid (aPL) antibodies in patients who also have evidence of thrombosis, embolism, or recurrent spontaneous pregnancy losses (for reviews see Rand et al., 1997a; Macik et al., 2001; Warkentin et al., 2003; Rand, 2003). The epitopes recognized by aPL antibodies are believed to be phospholipid-binding cofactor proteins, of which the major one is ß$_2$-glycoprotein I (ß$_2$GPI). The mechanism for the syndrome has been particularly elusive since the antibodies generated by patients with the disorder frequently have anticoagulant properties *in vitro*. The anticoagulant effects of the antibodies—known as the "lupus anticoagulant" (LA) phenomenon—are due to their avid binding to cofactor–phospholipid complexes, thereby reducing the availability of phospholipid for coagulation reactions.

Remarkably, the positive LA tests are associated with thrombosis, rather than bleeding, *in vivo*.

15.6. MECHANISM FOR THE ANTICOAGULANT EFFECTS OF aPL ANTIBODIES

Phospholipids play a critical role in the following four enzymatic reactions that lead to the generation of fibrin, the end product of blood coagulation, from fibrinogen: (**1 and 2**) the activations of factors IX and X by the tissue factor-VIIa complex, (**3**) the activation of factor X by the factor IXa–VIII complex, and (**4**) the activation of factor II by the factor Xa–V complex. aPL antibody-binding to domains of ß$_2$GPI other than its carboxyterminal phospholipid binding site increases the avidity of the ß$_2$GPI dimers and pentamers for the phospholipid, thereby reducing their availability for coagulation reactions.

15.7. ß$_2$GPI COFACTOR

Human ß$_2$GPI, also known as apolipoprotein H, is a 42- to 70-kDa membrane-adhesion plasma glycoprotein that was discovered in 1990 to be a cofactor for anticardiolipin antibodies from patients with APS (McNeil et al., 1989; Galli et al., 1990; Matsuura et al., 1992). Under native conditions and with no detergent and reducing agents present, ß$_2$GPI shows a molecular mass of approximately 320 kDa and is dissociable by boiling in 6 M urea (Gushiken et al., 2003). ß$_2$GPI in a dissociated, purified form varies slightly in molecular weight (kDa) as indicated by SDS-PAGE, depending on source and preparation(s) (Arvieux et al., 1996; Mori et al., 1996; Galazka et al., 1998; Harper et al., 1998; Horbach et al., 1998; Ma et al., 2000; Willems et al., 2000; Gushiken et al., 2003). Therefore, purified ß$_2$GPI will be presented as a monomer, dimer or an oligomer of a few to several subunits. *In vivo*, it has been suggested that

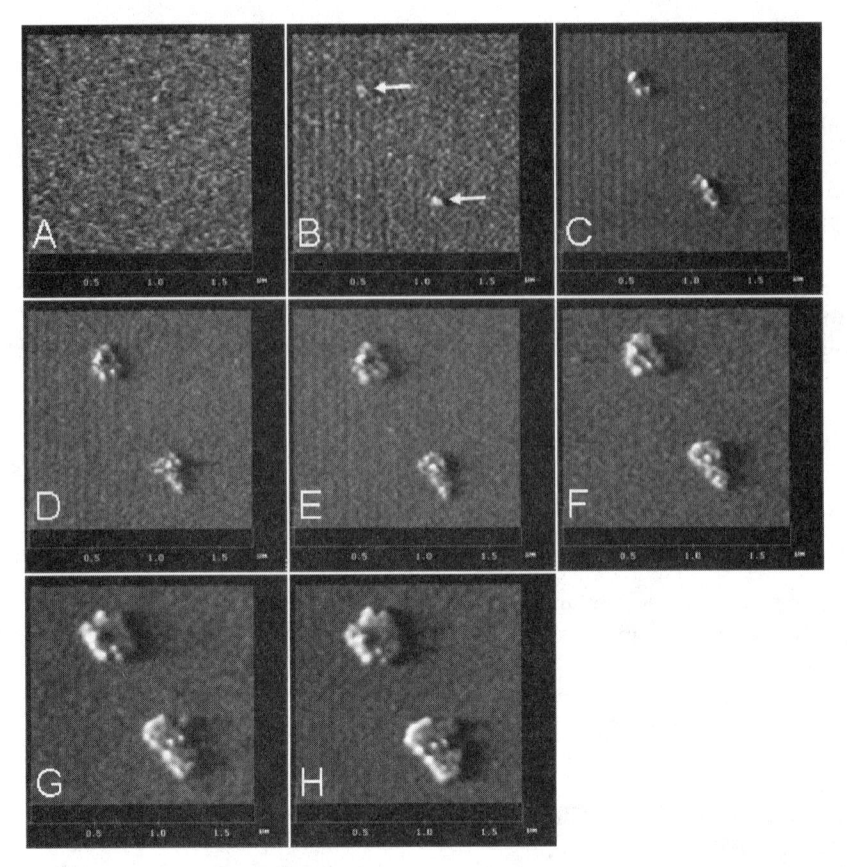

Figure 15.3. Sequential time-captured three-dimensional amplitude images of dynamic visualization of complex formation between the cofactor β$_2$GPI and human aPLmAb IS2 on lipid (PSPC). **(A)** PSPC alone on freshly cleaved mica. **(B)** Following addition of β$_2$GPI, globular structures representing β$_2$GPI multimers form (white arrows). **(C–H)** Temporal sequence showing the effect of the addition of the monoclonal antibody IS2 to the cofactor multimers present on the planar lipid layer. A conformational rearrangement to form toroid-shaped structures occurs. The 1.75-μm × 1.75-μm images were electronically zoomed from original 10-μm × 10-μm images. The three-dimensional amplitude images are displayed at 70° pitch and 0° rotation and have been subjected to fast Fourier transformation with the 2d spectrum off-line module of the NanoScope IIIa software.

β$_2$GPI may circulate as a multimer and may bind in an aggregate form, which may present a much needed higher density form for high-affinity antibody binding (Harper et al., 1998). Harper et al. (1998) presented data indicating that β$_2$GPI binding to membranes with physiological anionic content is relatively weak compared to that of plasma coagulation proteins. Increasing the lipid vesicle content of phosphatidylserine (PS) from 5% to 20% resulted in a twofold increase in binding affinity. Comparatively, prothrombin and factor X presented an approximate 10-fold increase over the same PS concentration range. Anti β$_2$GPI and anti-prothrombin antibodies have been suggested as probable risk factors for arterial and venous thrombosis [for a recent review see Galli et al(2003)]. During the last decade, work from various laboratories convincingly showed that aPL antibodies do not recognize anionic phospholipids without a plasma protein cofactor (Fig. 15.3). In other words, aPL antibodies require another plasma protein such as β$_2$GPI, prothrombin, protein C, factor Va, and so on, to bind to suitable anionic (not necessarily phospholipid) surfaces, and the degree of such binding is influenced by the binding affinities of such phospholipid-binding proteins (Walker, 1981; Krishnaswamy and Mann, 1988; Cutsforth et al., 1989; Tait et al., 1989; Harper et al., 1998). These antibodies may result in a multiplicity of effects, and the potential roles of these various aPL antibody specificities in the pathophysiology of APS has been previously reviewed (Rand, 2003). These have included mechanisms affecting platelet function, prostaglandin synthesis, cell signaling, apoptosis, and many more. Since phospholipids play so many key biologic roles, it is not surprising that antibodies that bind to them, either directly or via cofactor proteins, can, in experimental conditions, perturb virtually any of those activities. The problem is determining which of these effects are relevant to the mechanisms of thrombosis in the disease state. This difficulty is highlighted by the lupus anticoagulant phenomenon, which has no *in vivo* consequence with respect to a bleeding tendency.

Among the questions that remain to be elucidated are: Do antibodies recognizing epitopes on various plasma proteins affect or influence the binding capabilities of aPL antibodies? What differences may exist between artificial phospholipid bilayers and cellular membranes? How might various electrolytes and physiological constituents influence binding capabilities? What components interact with the membrane first? Does binding of components happen in the circulating environment initially and subsequently bind to anionic surfaces, and what are the biologic effects of these interactions?

15.8. ANNEXIN A5

Annexin A5 (AnxA5) (previously known as annexin-V) is a particularly attractive candidate for studying aPL antibody–cofactor interactions because it is itself a phospholipid-binding protein with potent anticoagulant properties. The anticoagulant properties of the protein are a consequence of its high affinity for anionic phospholipid (Funakoshi et al., 1987; Tait et al., 1988). AnxA5 forms two-dimensional crystals on phospholipid surfaces (Fig. 15.2) (Mosser et al., 1991; Voges et al., 1994; Reviakine et al., 1998; Rand et al., 2003) that shield the phospholipids from availability for phospholipid-dependent coagulation reactions (Andree et al., 1992). AnxA5 is highly expressed by human endothelial cells (Flaherty et al., 1990) and trophoblasts (Kirkun et al., 1994) and is present on the surfaces of these cells (Kirkun et al., 1994; Rand et al., 1997b). AnxA5 is required for the maintenance of placental integrity in mice (Wang et al., 1999). Infusion of anti-AnxA5 IgG antibodies into pregnant animals decreased the availability of AnxA5 to bind to the trophoblast surfaces and caused placental thrombosis, necrosis, and fetal loss.

15.9. DISRUPTION OF THE AnxA5 ANTICOAGULANT SHIELD

The first evidence to suggest that aPL may affect AnxA5 was the demonstration, by immunohistochemistry and ELISA, that the protein is markedly reduced on apical membranes of placental villi exposed to aPL antibodies (Rand et al., 1994, 1997b). The antibodies also reduced the quantity of AnxA5 on cultured trophoblasts and endothelial cells and accelerated the coagulation of plasma exposed to these cells (Rand et al., 1997b). It was then established via ellipsometry, binding assays, and immunoassays that aPL reduced the binding of AnxA5 on reconstituted membranes—that is, artificial phospholipid bilayers, coated microtiter plates, and phospholipid suspensions (Rand et al., 1998; Hanly and Smith, 2000).

Based upon the above work, we proposed the "Disruption of the AnxA5 Anticoagulant Shield" hypothesis for thrombosis in APS. This hypothesis also offers a plausible explanation for the 50-year-old LA paradox (Rand et al., 1998); a model for the concept is depicted in Fig. 15.4. Within this construct, aPL antibodies (mediated by cofactors) can increase the availability of phospholipid by causing defects in AnxA5 crystallization. Remarkably, the combination of two competing phospholipid-binding moieties—aPL antibodies and AnxA5—does not eliminate the availability of phospholipids for coagulation reactions. Rather, the presence of the antibodies causes defects in the AnxA5 crystalline array, thereby increasing the exposure of unshielded phospholipids, which, in turn, promote coagulation reactions. Firm evidence for the concept that aPL antibodies can actually disrupt the AnxA5 anticoagulant was provided by AFM imaging (Rand et al., 2003) (see below).

The above studies were followed by an initial translational study and showed that disruption of AnxA5 binding and reduction

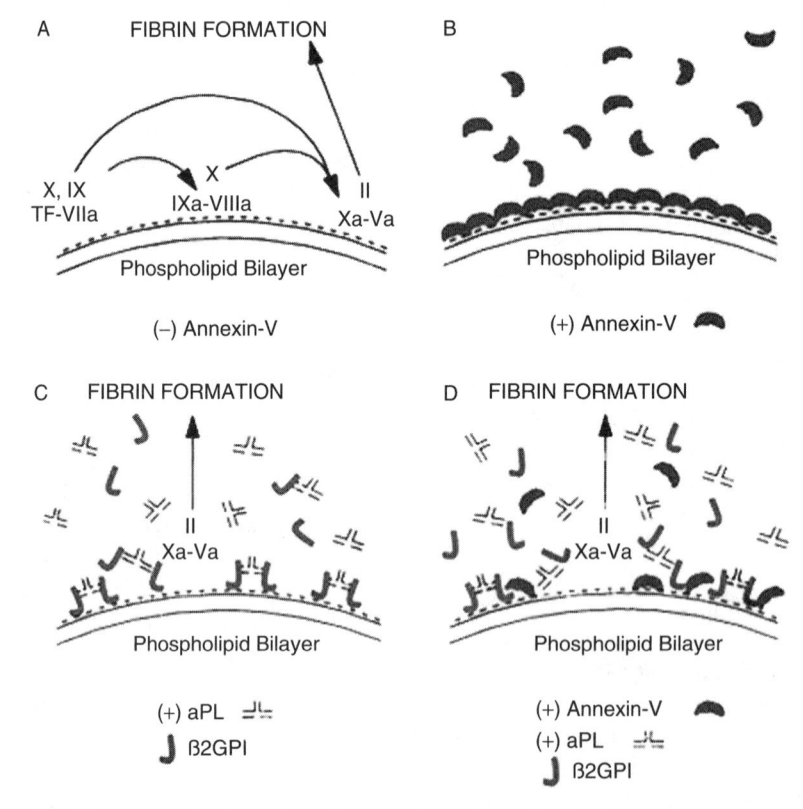

Figure 15.4. Model for the mechanisms of the "lupus anticoagulant effect" and for a "lupus procoagulant effect." (A) Anionic phospholipids serve as potent cofactors for the assembly of three different coagulation complexes—the tissue factor (TF)–VIIa complex, the IXa–VIIIa complex, and the Xa–Va complex—and thereby accelerate blood coagulation. (B) AnxA5, in the absence of aPL antibodies, is a potent anticoagulant that forms a crystal lattice over the anionic phospholipid surface, shielding it from availability for assembly of the phospholipid-dependent coagulation complexes. (C) In the absence of AnxA5, aPL antibody–β_2GPI complexes can prolong the coagulation times, compared to control antibodies. This occurs via antibody recognition of domains I or II on the β_2GPI, which results in dimers and pentamers of antibody–β_2GPI complexes having high affinity for phospholipid via domain V. These high-affinity complexes serve to reduce the access of coagulation factors to anionic phospholipids. This could result in a "lupus anticoagulant" effect in conditions where there are limiting quantities of anionic phospholipids. (D) In the presence of AnxA5, antiphospholipid antibodies, either directly or via interaction with protein–phospholipid cofactors, disrupt the ability of AnxA5 to form two-dimensional crystals on the phospholipid surface. This results in a net increase in the amount of anionic phospholipid available for promoting coagulation reactions. (From *Biochimica et Biophysica Acta* **1498**:169–173, 2000.)

of its anticoagulant effect could also be observed in plasmas from patients with aPL antibodies (Rand, 2003). The aPL-mediated effects on AnxA5 were confirmed by other investigators using phospholipid-coated microtiter plates (Hanly and Smith, 2000) and with platelets (Tomer, 2002).

15.10. CONTRIBUTIONS OF AFM TO THE STUDY OF THE ANTIPHOSPHOLIPID SYNDROME

To investigate the seemingly contradictory effects of aPL antibodies and annexin A5, we reconstituted an *in vitro* model of the AnxA5 two-dimensional crystal on a phospholipid bilayer, to be imaged by tapping mode AFM in fluid. Mica disks were incubated with a 30% phosphatidylserine:70% phosphatidylcholine phospholipid mixture. A planar lipid layer formed after 60 min at room temperature. To this planar lipid layer, purified human AnxA5 was added, and two-dimensional crystals formed on the lipid layer over the course of an hour (Fig. 15.2). This two-dimensional AnxA5 crystal structure was observed to grow from nucleation sites until the separately forming crystals met at furrow sites, as previously described

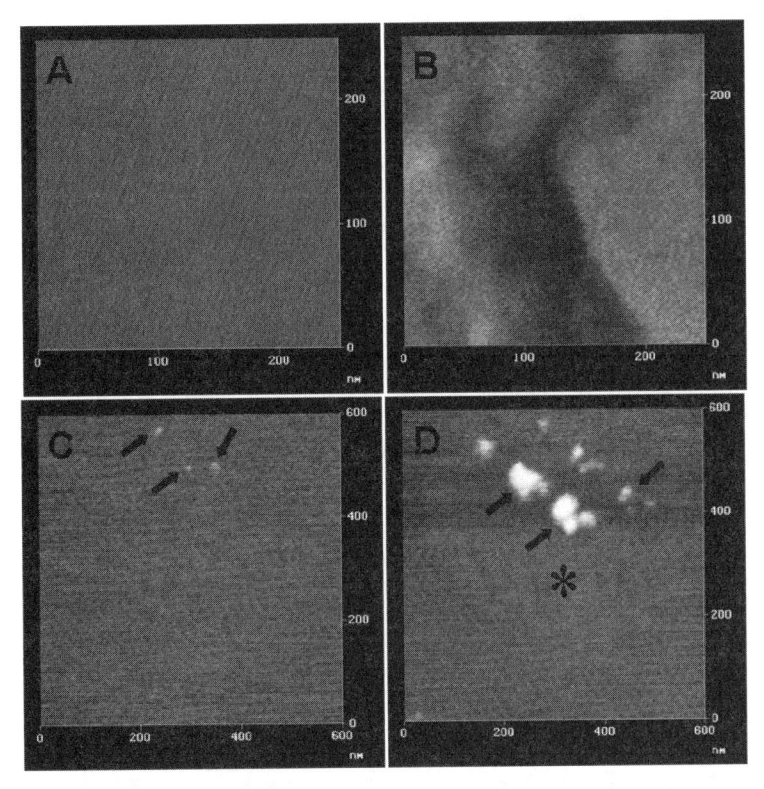

Figure 15.5. Effect of human monoclonal antibodies on previously formed AnxA5 crystal structure. (**A, C**) AnxA5 surface in the presence of monoclonal antibodies CL15 and CL1, respectively, prior to the addition of β_2GPI cofactor. Note the two-dimensional lattice structure of the AnxA5. After the addition of the required β_2GPI cofactor, CL15 grossly disrupts the AnxA5 crystal lattice (**B**), forming large furrows where the AnxA5 has been displaced by the antibody. Note that in **C**, the addition of monoclonal antibody CL1 to the AnxA5 crystal lattice resulted in the deposition of some globules (arrows), but the crystal structure remains intact. When the β_2GPI cofactor is added, the globules enlarge (arrows in **D**) and the AnxA5 crystal structure is minimally disrupted (asterisk). Images **B–D** are height images with off-line zero-flatten and low-pass filtering applied; image in **A** is an amplitude image with no processing. (From *American Journal of Pathology* **163**:1193–1200, 2003.)

(Reviakine et al., 1998). The thickness of the formed two-dimensional AnxA5 crystal was measured to be 2.44 ± 0.13 nm on top of the planar lipid layer (Rand et al., 2003). After confirming that AnxA5 formed a crystal lattice on top of the planar lipid layer attached to mica, we sought to investigate the effect of several human antiphospholipid antibodies on the integrity of this crystal layer. The addition of ß$_2$GPI cofactor alone had no effect on the crystalline lattice structure (Fig. 15.5). However, the addition of a monoclonal antiphospholipid antibody in conjunction with the cofactor led to disruption of the crystal lattice. The severity and overall appearance of the disruption was determined by the antibody applied. For instance, incubation with the antibody designated CL15 led to a gross disruption of the crystal lattice (Fig. 15.5), while incubation with the antibody designated IS3 resulted in the appearance of circular pits and some vacancy defects (Fig. 15.6). This mechanism can also be well-visualized by analyzing a series of height images during the course of an experiment. This is depicted in Fig. 15.7, where a series of height images, together with height measurements, are shown. As the AnxA5 crystal forms over the planar lipid layer, the height increases from 4.503 nm to 6.949 nm. Following addition of an aPL antibody, the height measurement decreases to 4.372 nm, representing the removal of the AnxA5 crystal layer from the underlying

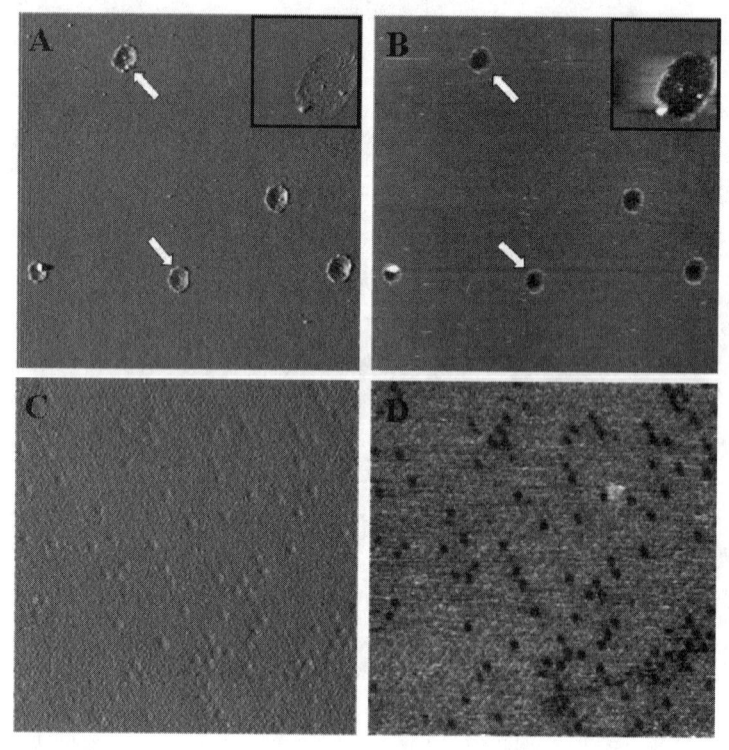

Figure 15.6. Effect of monoclonal antibody IS3 on preformed AnxA5 crystal two-dimensional lattice. When IS3 and β$_2$GPI were added to the AnxA5 crystal lattice formed on a planar lipid layer, circular pits (arrows in **A** and **B**) appeared, indicating disruptions in the crystal lattice. A representative pit is shown at higher magnification in the inset to **A** and **B**. At higher resolution (**C,D**), multiple vacancy defects (small, round dark holes) in the crystalline lattice are apparent. The amplitude images (**A,C**) were processed with a 1 × convolution filter, and the height images (**B,D**) were processed by a zero-order flatten filter. Original magnifications: 10-μm × 10-μm scan (**A,B**); 500 nm × 500 nm (**C,D**). (From *American Journal of Pathology* **163**:1193–1200, 2003.)

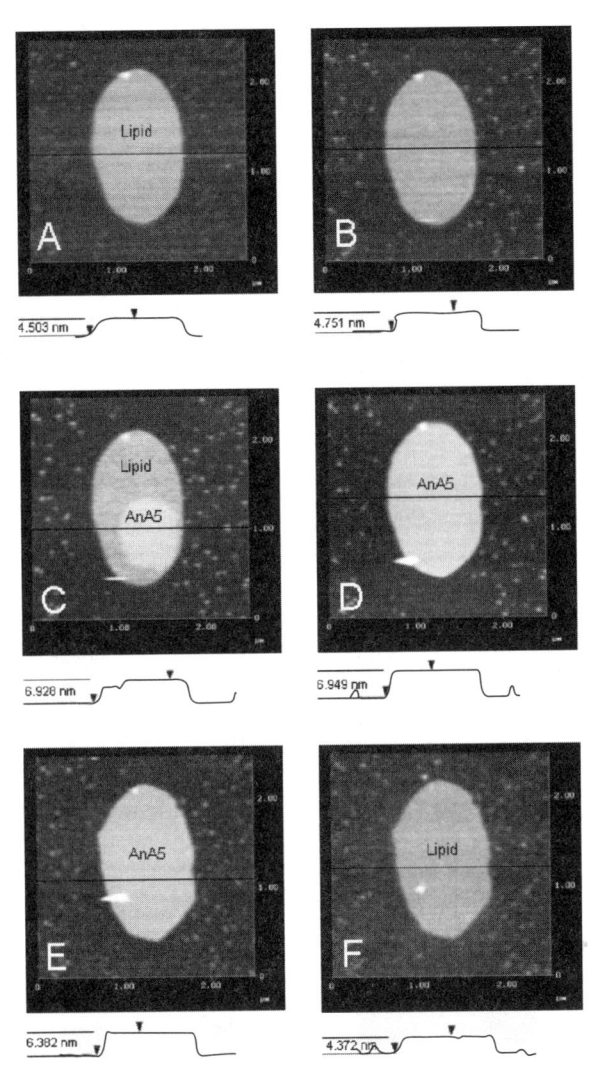

CONTRIBUTIONS OF AFM TO THE STUDY OF THE ANTIPHOSPHOLIPID SYNDROME 281

Figure 15.7. Temporal height mode images of the dynamic visualization of the structural alteration of AnxA5 crystalline lattice on a planar lipid layer by cofactor β_2GPI and human aPLmAb IS4. (**A**) An isolated lipid (PSPC) patch on freshly cleaved mica. (**B**) The first of three steps of AnxA5 formation; the "pebbled" granulated appearance represents the binding of monomeric AnxA5 molecules to the planar lipid surface. Membrane trimers begin forming during the second step in the process (**C**). The third step (**D**) of two-dimensional crystalline arrays of trimers represents complete coalescence with no visible furrows. Note the increase in mass thickness of the indicated lipid patch in **A–D**, as the AnxA5 crystal forms on top of the lipid layer. (**E**) Following the addition of β_2GPI cofactor and incubation for 1 h, no significant change in vertical measurements is apparent. (**F**) Finally, following subsequent addition of aPLmAb IS4, a dramatic decrease in lipid/AnxA5 patch thickness is revealed, representing removal of the AnxA5 crystal by the aPL/cofactor complex. While comparing the images in **A–F**, note the differences in surface image contrast; surface is a light gray (**A**) and changes to a "pebbled" white (**B**), to all white (**D**), and returns to a grayish tone (**F**). The contrast is a representation of the vertical height: The more toward the white the image tends, the greater the height (as demonstrated by the linear height measurements represented below each image). The 2.5-μm × 2.5-μm images were electronically zoomed from 10-μm × 10-μm scanned originals. No other processing of images was performed. The corresponding cross-sectional profile plots (below each image) were acquired utilizing the off-line Sectional Analysis module of the NanoScope IIIa software.

planar lipid layer. This change in height can also be visualized as a change in gray level on the images themselves (Fig. 15.7). Additionally, the incubation of a planar lipid layer with aPL antibodies and cofactor prevented the formation of a crystal lattice upon the subsequent addition of AnxA5.

These studies represent the first morphological evidence that aPL antibodies disrupt the AnxA5 anticoagulant shield. This may therefore represent a pathophysiological mechanism for thrombosis and spontaneous pregnancy loss in the aPL syndrome, as depicted in the model in Fig. 15.4.

15.11. SUMMARY

AFM offers a novel approach for investigating macromolecular interactions in a temporal fashion at high resolution. In the case of the enigmatic disorder known as the APS, AFM has provided novel insights into the mechanism by which aPL antibodies may provoke a thrombotic condition. Although not intended as a diagnostic, clinical tool, the AFM should continue to provide the means by which valuable mechanistic information underlying pathological conditions may be garnered.

ACKNOWLEDGMENT

This work was supported in part by a grant from the National Heart, Lung, and Blood Institute to JHR (R01 HL061331).

REFERENCES

Abdelhady, H., Allen, S., Davies, M., Roberts, C., Tendler, S. and Williams, P. (2003). Direct real-time molecular scale visualization of the degradation of condensed DNA complexes exposed to DNase I. *Nucleic Acids Res.* **31**:4001–4005.

Allemand J.-F., Bensimon, D. and Croquette, V. (2003). Stretching DNA and RNA to probe their interactions with proteins. *Curr. Opin. Struct. Biol.* **13**:266–274.

Andree, H. A. M., Stuart, M. C., Hermens, W. T., Reutelingsperger, C. P., Hemker, H. C., Frederik, P. M., and Willems, G. M. (1992). Clustering of lipid-bound annexin V may explain its anticoagulant effect. *J. Biol. Chem.* **267**: 17907–17912.

Arvieux, J., Darnige, L., Hachulla, E., Roussel, B., Bensa, J. C., and Colomb, M. G. (1996). Species specificity of anti-beta 2 glycoprotein I autoantibodies and its relevance to anticardiolipin antibody quantitation. *Thromb. Haemost.* **75**: 725–730.

Betzig, E., Lewis, A., Harootuniam, A., Isaacson, M., and Kratschmer E. (1986). Near-field scanning optical microscopy (NSOM): Development and biophysical applications. *Biophys. J.* **49**:269–279.

Binnig, G., Rohrer, H., Gerber, C. and Weibel, E. (1982a). Tunneling through a controllable vacuum gap. *Phys. Rev. Lett.* **40**:178–180.

Binnig, G., Rohrer, H., Gerber, C., and Weibel, E. (1982b). Surface studies by scanning tunneling microscopy. *Phys. Rev. Lett.* **49**:57–61.

Binnig, G., Quate, C. F., and Gerber, C. (1986). Atomic force microscope. *Phys. Rev. Lett.* **56**:930–933.

Bustamante, C., and Keller, D. (1994). Scanning force microscopy in biology. *Phys. Today* **48**:32–38.

Bustamante, C. and Rivetti, C. (1996). Visualizing protein–nucleic acid interactions on a large scale with the scanning force microscope. *Annu. Rev. Biophys. Biomol. Struct.* **25**:395–429.

Bustamante, C., Erie, D. A., and Keller, D. (1994). Biochemical and structural applications of scanning force microscopy. *Curr. Opin. Struct. Biol.* **4**:750–760.

Bustamante, C., Rivetti, C., and Keller, D. (1997). Scanning force microscopy under aqueous solutions. *Curr. Opin. Struct. Biol.* **7**:709–716.

Bustamante, C., Guthold, M., Zhu, X., and Yang, G. (1999). Facilitated target location on DNA by individual *Escherichia coli* RNA polymerase molecules observed with the scanning force microscope operating in liquid. *J. Biol. Chem.* **274**:16665–16668.

Bustamante, C., Smith, S., Liphardt, J., and Smith D. (2000). Single molecule studies of

DNA mechanics. *Curr. Opin. Struct. Biol.* **10**:279–285.

Cutsforth, G. A., Whitaker, R. N., Hermans, J., and Lentz, B. R. (1989). A new model to describe extrinsic protein binding to phospholipid membranes of varying composition: Application to human coagulation proteins. *Biochemistry* **28**:7453–7461.

Dunlap, D., Maggi, A., Soria, M., and Monaco, L. (1997). Nanoscopic structure of DNA condensed for gene delivery. *Nucleic Acids Res.* **25**:3095–3101.

Durig, U., Pohl, D. W., and Rohner, F. (1986). Near-field optical-scanning microscopy. *J Appl Phys* **59**:3318–3327.

Engel, A., and Muller, D. J. (2000). Observing single biomolecules at work with the atomic force microscope. *Nature Struct. Biol.* **7**:715–718.

Engel, A., Lyubchenko, Y., and Muller, D. (1999). Atomic force microscopy: A powerful tool to observe biomolecules at work. *Trends Cell Biol.* **9**:77–80.

Flaherty, M. J., West, S., Heimark, R. L., Fujikawa, K., and Tait, J. F. (1990). Placental anticoagulant protein-I: Measurement in extracellular fluids and cells of the hemostatic system. *J. Lab. Clin. Med.* **115**:174–181.

Funakoshi, T., Heimark, R. L., Hendrickson, L. E., McMullen, B. A., and Fujikawa, K. (1987). Human placental anticoagulant protein: Isolation and characterization. *Biochemistry* **26**:5572–5578.

Galazka, M., Keil, L. B., Kohles, J. D., Li, J., Kelty, S. P., Petersheim, M., and DeBari, V. A. (1998). A stable, multi-subunit complex of beta2glycoprotein I. *Thromb. Res.* **90**:131–137.

Galli, M., Comfurius, P., Maassen, C., Hemker, H. C., de Baets, M. H., van Breda-Vriesman, P. J., Barbui, T., Zwaal, R. F., and Bevers, E. M. (1990). Anticardiolipin antibodies (ACA) directed not to cardiolipin but to a plasma protein cofactor. *Lancet* **335**:1544–1547.

Galli, M., Luciani, D., Bertolini, G., and Barbui, T. (2003). Anti-beta 2-glycoprotein I, antiprothrombin antibodies, and the risk of thrombosis in the antiphospholipid syndrome. *Blood* **102**:2717–2723.

Gushiken, F. C., Le, A., Arnett, F. C., and Thiagarajan, P. (2003). Polymorphisms of beta2-glycoprotein I: Phospholipid binding and multimeric structure. *Thromb Res* **108**:175–180.

Guthold, M., Xingshu, Z., Rivetti, C., Yang, G., Thompson, N., Kasas, S., Hansma, H., Smith, B., Hansma, P., and Bustamante, C. (1999). Direct observation of one-dimensional diffusion and transcription by *E. coli* RNA polymerase. *Biophys. J.* **77**:2284–2294.

Hall, C. E., and Slayter, H. S. (1959). The fibrinogen molecule: Its size, shape and mode of polymerization. *J. Biophys. Biochem. Cytol.* **5**:11–16.

Han, W., Mou, J., Sheng, J., and Shao, Z. (1995). Cryo-atomic force microscopy: A new approach for biological imaging at high resolution. *Biochemistry* **34**:8215–8220.

Han, W., Lindsay, S. M., and Jing, T. (1996). A magnetically driven oscillating probe microscope for operation in liquids. *Appl. Phys. Lett.* **69**:1–3.

Hanly, J. G., and Smith, S. A. (2000). Anti-beta2-glycoprotein I (GPI) autoantibodies, annexin V binding and the anti-phospholipid syndrome. *Clin. Exp. Immunol.* **1120**:537–543.

Hansma, H. (2001). Surface biology of DNA by atomic force microscopy. *Annu. Rev. Phys. Chem.* **52**:71–92.

Hansma, H., and Ho, J. (1994). Biomolecular imaging with the atomic force microscope. *Annu. Rev. Biophys. Biomol. Struct.* **23**:115–139.

Hansma, H. G., and Laney, D. E. (1996). DNA binding to mica correlates with cationic radius: Assay by atomic force microscopy. *Biophys. J.* **70**:1933–1939.

Hansma, H., Bezanilla, M., Zenhausen F., Adrian, M., and Sinsheimer, R. (1993). Atomic force microscopy of DNA in aqueous solutions. *Nucleic. Acids. Res.* **21**:505–512.

Hansma, H., Golan, R., Hsieh, W., Daubendiek, S., and Kool, E. (1999). Polymerase activities and RNA structures in the atomic force microscope. *J. Struct. Biol.* **127**:240–247.

Harper, M. F., Hayes, P. M., Lentz, B. R., and Roubey, R. A. (1998). Characterization of beta-2-glycoprotein I binding to phospholipid membranes. *Thromb. Haemost.* **80**:610–614.

Horbach, D. A., van Oort, E., Tempelman, M. J., Derksen, R. H., and de Groot, P. G. (1998). The prevalence of a non-phospholipid-binding form of beta2-glycoprotein I in human plasma—consequences for the development of anti-beta2-glycoprotein I antibodies. *Thromb. Haemost.* **80**:791–797.

Horber, J., and Miles, M. (2003). Scanning probe evolution in biology. *Science* **302**:1002–1005.

Houchens, C. R., Montigny, W., Zeltser, L., Dailey, L., Gilbert, J. M., and Heintz, N. H. (2000). The dhfr oribeta-binding protein RIP60 contains 15 zinc fingers: DNA binding and looping by the central three fingers and an associated proline-rich region. *Nucleic Acids Res.* **28**:570–581.

Israelachvili, J. (1985). Intermolecular and Surface Forces, Academic Press, New York.

Janshoff, A., and Steinem, C. (2001). Scanning force microscopy of artificial membranes. *ChemBioChem* **2**:798–808.

Kindt, J., Sitko, J., Pietrasanta, L., Oroudjev, E., Becker, N., Viani, M. and Hansma, H. (2002). Methods for biological probe microscopy in aqueous fluids. In: Atomic Force Microscopy in Cell Biology: Methods in Cell Biology, Vol. 68 (B. P. Jena and J. K. Horber, eds.), pp. 214–229, Academic Press, New York.

Kirkun, G., Lockwood, C. J., Wu,, X. X., Zhou, X. D., Guller, S., Calandri, C., Guha, A., Nemerson, Y., and Rand, J. H. (1994). The expression of the placental anticoagulant protein, annexin V, by villous trophoblasts: Immunolocalization and *in vitro* regulation. *Placenta* **15**:601–612.

Krishnaswamy, S., and Mann, K. G. (1988). The binding of factor Va to phospholipid vesicles. *J. Biol. Chem.* **263**:5714–5723.

Ma, K., Simantov, R., Zhang, J. C., Silverstein, R., Hajjar, K. A., and McCrae, K. R. (2000). High affinity binding of beta 2-glycoprotein I to human endothelial cells is mediated by annexin II. *J. Biol. Chem.* **275**:15541–15548.

Macik, B. G., Rand, J. H., and Konkle, B. A. (2001). Thrombophilia: What's a practitioner to do? *Am. Soc. Hematol. Educ. Program* 322–338.

Marchant, R. E., Barb, M. D., Shainoff, J. R., Eppell, S. J., Wilson, D. L., and Siedlecki, C. A. (1997). Three dimensional structure of human fibrinogen under aqueous conditions visualized by atomic force microscopy. *Thromb. Haemost.* **77**:1048–1051.

Marchant, R. E., Kang, I., Sit, P. S., Zhyou, Y., Todd, B. A., Eppell, S. J., and Lee, I. (2002). Molecular views and measurements of hemostatic processes using atomic force microscopy. *Curr. Protein Pept. Sci.* **3**:249–274.

Matsuura, E., Igarashi, Y., Fujimoto, M.,Ichikawa, K., Suzuki, T., Sumida T, Yasuda, T., and Koike, T. (1992) Heterogeneity of anticardiolipin antibodies defined by the anticardiolipin cofactor. *J. Immunol.* **148**:3885–3891.

McNeil, H. P., Chesterman, C. N., and Krilis, S. A. (1989). Anticardiolipin antibodies and lupus anticoagulants comprise separate antibody subgroups with different phospholipid binding characteristics. *Br. J. Haematol.* **73**:506–513.

Meyer, G., and Amer, N. (1988). Novel approach to atomic force microscopy. *Appl. Phys. Lett.* **53**:1045–1047.

Montigny, W. J., Houchens, C. R., Illenye, S., Gilbert, J., Coonrod, E., Chang, Y. C., and Heintz, N. H. (2001). Condensation by DNA looping facilitates transfer of large DNA molecules into mammalian cells. *Nucleic Acids Res.* **29**:1982–1988.

Montigny, W. J., Phelps, S. F., Illenye, S., and Heintz, N. H. (2003). Parameters influencing high-efficiency transfection of bacterial artificial chromosomes into cultured mammalian cells. *Biotechniques* **35**:796–807.

Mori, T., Takeya, H., Nishioka, J., Gabazza, E. C., and Suzuki, K. (1996). Beta 2-glycoprotein I modulates the anticoagulant activity of activated protein C on the phospholipid surface. *Thromb. Haemost.* **75**:49–55.

Morris V., Kirby, A., and Gunning, A. (1999). Atomic Force Microscopy for Biologists, Imperial College Press, London.

Mosser, G., Ravavat, C., Freyssinet, J. M., and Brisson, A. (1991). Sub-domain structure of lipid-bound annexin-V resolved by electron image analysis. *J. Mol. Biol.* **217**:241–245.

Mou, J., Yang, J. and Shao, Z. (1993) An optical detection low temperature atomic force microscope at ambient pressure for biological research. *Rev. Sci. Instrum.* **64**:1483–1488.

Muller, D. J., Fotiadis, D., Scheuring, S., Muller, S. A., and Engel, A. (1999). Electrostatically balanced subnanometer imaging of biological specimens by atomic force microscope. *Biophys. J.* **76**:1101–1111.

Muller, D. J., Heymann, J. B., Oesterhelt, F., Moller, C., Gaub, H., Buldt, G., and Engel, A.

(2000). Atomic force microscopy of native purple membrane. *Biochim. Biophys. Acta* **1460**:27–38.

Putman, C., Van der Werf, K., De Grooth, B., Van Hulst, N. and Greve, J. (1994). Tapping mode atomic force microscopy in liquid. *Appl. Phys. Lett.* **64**:2454–2456.

Rand, J. H. (2003). The antiphospholipid syndrome. *Annu. Rev. Med.* **54**:409–424.

Rand, J. H., Wu, X. X., Guller, S., Gil, J., Guha, A, Scher, J., and Lockwood, C. J. (1994). Reduction of annexin-V (placental anticoagulant protein-I) on placental villi of women with antiphospholipid antibodies and recurrent spontaneous abortion. *Am. J. Obstet. Gynecol.* **171**:1566–1572.

Rand, J. H., Wu, X. X., Andree, H. A., Lockwood, C. J., Guller, S., Scher, J., and Harpel, P. C. (1997a). Pregnancy loss in the antiphospholipid-antibody syndrome—a possible thrombogenic mechanism. *N. Engl. J. Med.* **337**:154–160.

Rand, J. H., Wu, X. X., Guller, S., Scher, J., Andree, H. A. M., and Lockwood, C. J. (1997b). Antiphospholipid immunoglobulin G antibodies reduce annexin-V levels on syncytiotrophoblast apical membranes and in culture media of placental villi. *Am. J. Obstet. Gynecol.* **177**:918–923.

Rand, J. H., Wu, X. X., Andree, H. A. M., Alexander Ross, J. B., Rosinova, E., Gascon-Lema, M. G., Calandri, C., and Harpel, P. C. (1998). Antiphospholipid antibodies accelerate plasma coagulation by inhibiting annexin-V binding to phospholipids: A "lupus procoagulant" phenomenon. *Blood* **92**:1652–1660.

Rand, J. H., Wu, X.-X., Quinn, A. S., Chen, P. P., McCrae, K. R., Bovill, E. G., and Taatjes, D. J. (2003). Human monoclonal antiphospholipid antibodies disrupt the annexin A5 anticoagulant crystal shield on phospholipid bilayers. Evidence from atomic force microscopy and functional assay. *Am. J. Pathol.* **163**:1193–1200.

Reviakine, I., Bergsma-Schutter, W., and Brisson, A. (1998). Growth of protein 2-D crystals on supported planar lipid bilayers imaged *in situ* by AFM. *J. Struct. Biol.* **121**:356–361.

Rivetti, C., Guthold, M., and Bustamante, C. (1996). Scanning force microscopy of DNA deposited onto mica: Equilibration versus kinetic trapping studied by statistical polymer chain analysis. *J. Mol. Biol.* **264**:919–932.

Rivetti, C., Walker, C., and Bustamante, C. (1998). Polymer chain statistics and conformational analysis of DNA molecules with bends or sections of different flexibility. *J. Mol. Biol.* **280**:41–59.

Rugar, D., and Hansma, P. K. (1990). Atomic force microscopy. *Phys. Today* **43**:23–30.

Shao, Z., and Zhang, Z. (1996). Biological cryo atomic force microscopy: A brief review. *Ultramicroscopy* **66**:141–152.

Sit, P. S., and Marchant, R. E. (1999). Surface-dependent conformations of human fibrinogen observed by atomic force microscopy under aqueous conditions. *Thromb. Haemost.* **82**:1053–1060.

Stahlberg, H., Fotiadis, D., Scheuring, S., Remigy, H., Braun, T., Mitsuoka, K., Fujiyoshi, Y., and Engel, A. (2001). Two-dimensional crystals: A powerful approach to assess structure, function and dynamics of membrane proteins. *FEBS Lett.* **504**:166–172.

Stroh, C., Wang, H., Bash, R., Ashcroft, B., Nelson, J., Gruber, H., Lohr, D., Lindsay, S. and Hinterdorfer, P. (2004). Single-molecule recognition imaging microscopy. *Proc. Natl. Acad. Sci.* **34**:12503–12507.

Taatjes, D. J., Quinn, A. S., Jenny, R. J., Hale, P., Bovill, E. G., and McDonagh, J. (1997). Tertiary structure of the hepatic cell protein fibrinogen in fluid revealed by atomic force microscopy. *Cell Biol. Int.* **21**:715–726.

Taatjes, D. J., Quinn, A. S., Lewis, M. R., and Bovill, E. G. (1999). Quality assessment of atomic force microscopy probes by scanning electron microscopy: Correlation of tip structure with rendered images. *Microsc. Res. Tech.* **44**:312–326.

Tait, J. F., Gibson, D., and Fujikawa, K. (1989). Phospholipid binding properties of human placental anticoagulant protein-I, a member of the lipocortin family. *J. Biol. Chem.* **264**:7944–7949.

Tait, J. F., Sakata, M., McMullen, B. A., Maio, C. H., Funakoshi, T., Hendickson, L. E., and Fujikawa, K. (1988). Placental anticoagulant proteins: Isolation and comparative characterization of four members of the lipocortin family. *Biochemistry* **27**:6268–6276.

Taylor, M. (1993). Dynamics of piezoelectric tube scanners for scanning probe microscopy. *Rev. Sci. Instrum.* **64**:154–158.

Thompson, N., Kasas, S., Smith, B., Hansma, H. and Hansma, P. (1996). Reversible binding of DNA to mica for imaging. *Langmuir* **12**:5905–5908.

Tomer, A. (2002). Antiphospholipid antibody syndrome: Rapid, sensitive, and specific flow cytometric assay for determination of antiplatelet phospholipid autoantibodies. *J. Lab. Clin. Med.* **139**:147–154.

Voges, D., Berendes, R., Burger, A., Demange, P., Baumeister, W., and Huber, R. (1994). Three-dimensional structure of membrane-bound annexin V. A correlative electron microscopy–X-ray crystallography study. *J. Mol. Biol.* **238**:199–213.

Walker, F. J. (1981). Regulation of activated protein C by protein S. The role of phospholipid in factor Va inactivation. *J. Biol. Chem.* **256**:11128–11131.

Wang, X., Campos, B., Kaetzel, M. A., and Dedman, J. R. (1999). Annexin V is critical in the maintenance of murine placental integrity. *Am. J. Obstet. Gynecol.* **180**:1008–1016.

Warkentin, T. E., Aird, W. C., and Rand, J. H. (2003). Platelet–endothelial interactions: Sepsis, HIT, and antiphospholipid syndrome. *Hematology (Am. Soc. Hematol. Educ. Program)* 497–519.

Willems, G. M., Janssen, M. P., Comfurius, P., Galli, M., Zwaal, R. F., and Bevers, E. M. (2000). Competition of annexin V and anticardiolipin antibodies for binding to phosphatidylserine containing membranes. *Biochemistry* **39**:1982–1989.

INDEX

Actin regulation:
  atomic force microscopic imaging, inverted optical microscopic assembly, 140–141
  cell secretion and, 8–17
Adenylate cyclase, signal transduction, 55–56
Adhesion mapping:
  microbial cell surface imaging, atomic force microscopy analysis, 81–82
  sonicated unilamellar vesicle bilayers, 192–197
Adsorption strategies:
  microbial cell surface imaging, atomic force microscopy analysis, 71
  polymer-based nanodrug delivery, 120–121
Alkanethiol self-assembled monolayers, microbial cell surface imaging:
  atomic force microscopy analysis, 71
  probe surface chemistry, 80–81
Ambisome, nanodrug delivery with, 131–132
Annexin A5 (AnxA5), antiphospholipid syndrome, 277–279
  atomic force microscopy, 279–282
Antibiotics, microbial cell surface effects, atomic force microscopy analysis, 78–79
Anticoagulant effects, antiphospholipid syndrome, 275
  annexin A5, 277–279
Antigen presenting cell (APC), avidity modulation, basic principles, 169–171
Antiphospholipid syndrome (APS), atomic force microscopic imaging:
  annexin A5 shield, 277–279
  aPL antibodies, anticoagulation effects, 275
  clinical applications, 279–282
  β_2GPI cofactor, 275–277
  hemostatic proteins, 274–275
  lipid analysis, 273–274
  theoretical background, 267–272
Apolipoprotein (APL) antibodies:
  anticoagulant effects, 275
  antiphospholipid syndrome, annexin A5, 278–279

Aquaporins:
  microbial cell surface imaging, atomic force microscopy analysis, 74–75
  immobilization strategies, 71
  secretory vesicle content expulsion, basic principles, 37–38, 40
Array detectors, atomic force microscopy cantilevers, 240–241
Atomic-based membrane fusion processes, calcium participation in, 27–34
Atomic force microscopy (AFM):
  antiphospholipid syndrome:
    annexin A5 shield, 277–279
    aPL antibodies, anticoagulation effects, 275
    clinical applications, 279–282
    β_2GPI cofactor, 275–277
    hemostatic proteins, 274–275
    lipid analysis, 273–274
    theoretical background, 267–272
  biomolecular motion:
    basic principles, 221–223
    biological applications, 242–243
    cantilever properties:
      limits, 232
      resonant frequency, 232–233
      small cantilevers, 233–235
      thermal noise, 235–237
    fast, low-noise detector, 239–241
    fast feedback loop, 242
    fast scanner, 241–242
    focused spot size, 237–239
    imaging speed, 228–231
      feedback limit, 231
      lateral dimension reduction, 229–230
      tracking mechanisms, 228–229
      video-rate imaging, 231
    time-resolved imaging, 223–228
      alignment, 225–227
      moving images, 227–228

Atomic force microscopy (AFM) (*continued*)
  p53-DNA dynamic interactions, 224–225
  recording sequences, 223–224
 cytogenetics applications:
  C-banding technique, 257
  genetic material imaging techniques, 250–253
  metaphase chromosome karyotyping, 253–257
  metaphase chromosome three-dimensional analysis, 257–259
  nanoextraction, 259–262
  nanomanipulation and dissection, 259
  theoretical background, 249–250
 electrophysiological techniques, 137–150
  inner ear tissue sections, 144–150
  inverted optical microscope, 137–141
 growth hormone fusion pore structure and dynamics, 59–60
 growth hormone secretory vesicles, 61–63
 historical evolution of, 2–3
 leukocyte adhesion molecules, 160
  avidity modulation:
   adhesive force measurements, 172–173
   basic principles, 169–171
   cells and reagents, 171
   clinical applications, 176–178
   elasticity measurements, 173
   integrin lateral redistribution, 173–175
   ionomycin/thapsigargin stimulation, 175–176
   protein immobilization, 172
   3A9 cell crosslinking, 171–172
  integrin activation, 162–166
  single-molecule unbinding, 160–166
 lipid bilayers, thickness and micromechanical properties, 181–197
  supported thickness, 182–183
 lipid membrane preparation, porosome reconstitution, 4
 microbial cell surface property analysis, 69–90
  bacillus S-layers, 73
  basic principles, 69–70
  cell surface layer imaging, 73–75
  elasticity, 85–89
   layers stretching, 87
   single cell stretching, 87–89
   single macromolecules, 86–87
  external agent effects, 78–79
  future applications, 89–90
  hexagonally packed intermediate layer, 73
  imaging techniques, 70–79
   immobilization strategies, 71
   selection criteria, 71–72
   substrate requirements, 70–71
  microbial biofilms, 75–76
  nanostructures, 76–78

  physical properties measurement, 79–89
   force-distance curves, 80
   force measurements, 80–81
   functionalized probes, 80–81
  physicochemical properties and molecular interactions, 81–85
   adhesion mapping, 81–82
   cell probe technique, 85
   surface charges and electrostatic interactions, 83–85
   surface energy and solvation interactions, 82–83
  physiological changes, 78
  porin crystals, 74–75
  purple membranes, 73–74
 pancreatic acinar cell isolation, 3–4
 pancreatic plasma membrane isolation, 4
 plant cell wall imaging, 96–97
  cellulose, 99–100
  lignins, 103–108
  pectins, 97–99
 porosome imaging, 4–5
  structural analysis, 6–17
 secretory vesicle content expulsion, vesicle swelling in pancreatic acinar cells, 38–44
 SNARE-induced membrane fusion, bilayer interaction and conducting channel formation, 26–27
 soft tissue imaging, swollen polymer surfaces, 214–218
 sonicated unilamellar vesicle:
  bilayer thickness and morphology, 187–191
  micromechanical properties, 191–197
 synaptosome/synaptosomal/synaptic vesicle isolation, 4
Avidity modulation, leukocyte adhesion molecules:
 adhesive force measurements, 172–173
 basic principles, 169–171
 cells and reagents, 171
 clinical applications, 176–178
 elasticity measurements, 173
 integrin lateral redistribution, 173–175
 ionomycin/thapsigargin stimulation, 175–176
 protein immobilization, 172
 3A9 cell crosslinking, 171–172

Bacillus S-layers, microbial cell surface imaging, atomic force microscopy analysis, 73
 immobilization strategies, 71
Bell single-molecule unbinding model, leukocyte adhesion molecules, 160–166
 integrin activation, 165–166
Bending modulus, sonicated unilamellar vesicle bilayer micromechanics, 192–197
"Bimetallic strip effect," atomic force microscopy cantilevers, 233

INDEX

Biocompatibility, polymer-based nanodrug delivery, 131–132
Biocorrosion studies, microbial cell surface imaging, atomic force microscopy analysis, 76
Biofilms, microbial cell surface imaging, atomic force microscopy analysis, 75–76
Biomembrane force probe (BFP), leukocyte adhesion molecules, 160
Biomolecular motion, atomic force microscopy (AFM):
  basic principles, 221–223
  biological applications, 242–243
  cantilever properties:
    limits, 232
    resonant frequency, 232–233
    small cantilevers, 233–235
    thermal noise, 235–237
  fast, low-noise detector, 239–241
  fast feedback loop, 242
  fast scanner, 241–242
  focused spot size, 237–239
  imaging speed, 228–231
    feedback limit, 231
    lateral dimension reduction, 229–230
    tracking mechanisms, 228–229
    video-rate imaging, 231
  time-resolved imaging, 223–228
    alignment, 225–227
    moving images, 227–228
    p53-DNA dynamic interactions, 224–225
    recording sequences, 223–224
Bis(sulfosuccinimidyl) suberate (BS^3), avidity modulation:
  integrin lateral redistribution, 174–175
  LFA-1 receptor cross-linking, 176–178
Boltzmann constant:
  atomic force microscopy, thermal cantilever noise, 236–237
  leukocyte adhesion molecules, single-molecule unbinding theory, 160–166
  photonic force microscopy, 152–157
Brownian motion, atomic force microscopy, thermal cantilever noise, 236–237
Buffered fluid environment, atomic force microscopy, antiphospholipid syndrome, 269–272

Calcium ions:
  leukocyte adhesion:
    avidity modulation, 170–172
    ionomycin/thapsigargin stimulation, 175–176
    signal transduction mechanisms, 55–56
  SNARE-induced membrane fusion:
    atomic levels for, 27–35
    cell-based mechanisms, 27

Cancer pathophysiology, nanodrug delivery systems, 115–116
  liposome structure, 117–118
Cantilever structures:
  atomic force microscopy:
    antiphospholipid syndrome, 268–272
    biological applications, 242–243
    conventional sizes, 232
    fast, low-noise detector, 239–241
    fast feedback loop, 242
    fast scanner, 241–242
    focused spot size, 237–239
    inverted optical microscope schematic, 137–141
    resonant frequency, 232–233
    small cantilever properties, 233–235
    thermal noise, 235–237
  leukocyte adhesion molecules, avidity modulation, 172–173
  microbial cell surfaces, atomic force microscopy analysis, imaging modes, 71–72
  photonic force microscopy, 151–157
  scanning probe microscopy, soft surface imaging, 203, 210–211
  supported lipid bilayer thickness measurements, 184–186
Capacitance measurements, SNARE-induced membrane fusion, atomic-level calcium participation in, 30–34
C-banding technique, metaphase chromosomes, 257
Cell-based fusion machinery, SNARE and calcium for, 27
Cell clustering, leukocyte adhesion, 177–178
"Cell probe" technique, microbial cell surface imaging, atomic force microscopy analysis, 85
Cell secretion:
  growth hormone secretory vesicles, 61–63
  molecular mechanisms for, 17, 19–20
  porosome functions, 1–3
    AFM imaging of, 6–17
  secretory vesicle content expulsion, molecular mechanisms:
    basic principles, 37–38
    swelling process in cellular secretion, 38–44
    swelling process in neuron secretion, 44–46
Cell spreading, leukocyte adhesion, 177–178
Cellulolytic microorganisms, cellulose digestion, atomic force microscopic imaging, 99–100
Cellulose:
  atomic force microscopic imaging, 99–100
  plant cell walls, basic properties, 96

Charge-coupled device (CCD) camera:
  atomic force microscopy cantilevers, focused spot dimensions, 238–239
  time-resolved imaging, 227–228
Chloride channels, porosome structural analysis, 11–17
Cholesterol incorporation, sonicated unilamellar vesicle bilayer micromechanics, 195–197
Conductance measurements, SNARE-induced membrane fusion, atomic-level calcium participation in, 30–34
Constant-deflection imaging, microbial cell surfaces, atomic force microscopy analysis, 71–72
Constant force mode, scanning probe microscopy, soft surface imaging, 203
Constant-height imaging:
  antiphospholipid syndrome, 281–282
  microbial cell surfaces, atomic force microscopy analysis, 71–72
  scanning probe microscopy, soft surface imaging, 203–205
  sonicated unilamellar vesicle bilayers, thickness and morphology, 187–191
Contact mode imaging:
  cytogenetics research, 251–253
  microbial cell surfaces, atomic force microscopy analysis, 71–72
  scanning probe microscopy:
    liquid droplets, 212–214
    soft surface imaging, 203–205
    swollen polymer surfaces, 214–218
  sonicated unilamellar vesicle bilayers, thickness and morphology, 188–191
Corticotropin-releasing hormone (CRH), growth-hormone release and synthesis, 55
Covalent immobilization, microbial cell surface imaging, atomic force microscopy analysis, 71
Critical micelle concentration (CMC), supported lipid bilayer thickness measurements, 184–186
Cross-correlation alignment:
  real-time correction algorithm, 229
  time-resolved imaging, 225–227
Cross-linking mechanisms, leukocyte adhesion:
  avidity modulation, 171–172
  bis(sulfosuccinimidyl) suberate (BS^3), 176–178
Cyclic AMP (cAMP), signal transduction, 55–56
Cyclooxygenase-2 (COX-2), nanodrug delivery systems, dendrimer-ibuprofen nanodevices, 127
Cytochalasin B, cell secretion and, 8–17
Cytogenetics, atomic force microscopy:
  C-banding technique, 257
  genetic material imaging techniques, 250–253
  metaphase chromosome karyotyping, 253–257
  metaphase chromosome three-dimensional analysis, 257–259
  nanoextraction, 259–262
  nanomanipulation and dissection, 259
  theoretical background, 249–250
Cytotoxic T-lymphocyte response (CTL), nanodrug delivery systems, 129–130

Daunoxome, nanodrug delivery with, 131–132
De-adhesion, leukocyte adhesion, ionomycin/thapsigargin stimulation, 175–176
Degradation mechanisms, polymer-based nanodrug delivery, 120–121
Dendrimers, nanodrug delivery systems, 122–130
  chemistry, 124
  drug conjugation, 126–129
  free dendrimer interactions, 124–126
  ibuprofen nanodevices, 126–127
  methyl prednisolone nanodevices, 127–129
"Depression technique," microbial cell surface elasticity, atomic force microscopy imaging, 85–89
Differential interference contrast (DIC) light microscopy, inner ear tissue studies, 144–150
Diffusion mechanisms:
  membrane viscosity, thermal fluctuation measurements, 155–157
  polymer-based nanodrug delivery, 120–121
Dioleoyl phosphatidyl ethanolamine (DOPE), nanodrug delivery systems, 118–120
Discrete Fourier transform (DFT), time-resolved imaging, alignment techniques, 226–227
DNA:
  atomic force microscopic imaging, 243
  cytogenetics research, 250–253
  metaphase chromosome karyotyping, 255–257
  enzymatic degradation, atomic force microscopic imaging, 222–223
  p53 protein-DNA interaction, time-resolved imaging, 224–226
Drift phenomena:
  atomic force microscopy, real-time correction algorithm, 229
  biomolecular motion, time-resolved imaging, 223–228
  time-resolved imaging:
    alignment techniques, 226–227
    thermal drift, 227–228
Dynamic force spectroscopy, leukocyte adhesion molecules:
  integrin activation, 162–166
  selectin/sLeX complexes, 162
  single-molecule unbinding theory, 161–166

INDEX

Dynamic mode imaging. *See* Tapping mode atomic force microscopy (TMAFM)

Egg yolk phosphatidylcholine (EggPC):
  sonicated unilamellar vesicle bilayers:
    micromechanical measurements, 192–197
    thickness and morphology, 187–191
  supported lipid bilayer thickness, 182–186
Elasticity:
  atomic force microscopy imaging, inner ear tissue studies, 147–150
  leukocyte adhesion, avidity modulation, 173
  microbial cell surfaces, atomic force microscopy imaging, 85–89
    layers stretching, 87
    single cell stretching, 87–89
    single macromolecules, 86–87
  photon force microscopy, single molecules, 154–157
  sonicated unilamellar vesicle bilayer micromechanics, 192–197
Electron beam deposition (EBD), atomic force microscopy cantilevers, 233
  nanoextraction, 262
Electron microscopy:
  metaphase chromosome karyotyping, 255–257
  microbial cell surface imaging, nanostructures, 77–79
  plant cell wall imaging, 97
Electrophysiology:
  atomic force microscopy and, 137–150
    inner ear tissue studies, 144–150
    inverted optical microscope schematic, 137–141
    patch-clamp integration into, 141–144
  microbial cell surface imaging, atomic force microscopy analysis, 83–85
  porosome structure and function, 17
  scanning probe microscopy, soft surface imaging, 211
  secretory vesicle content expulsion:
    basic principles, 37–38
    vesicle swelling in pancreatic acinar cells, 41–44
  SNARE-induced membrane fusion, atomic-level calcium participation in, 30–34
Electrostatic interactions, microbial cell surface imaging, atomic force microscopy analysis, 83–85
Encapsulation efficiency, polymer-based nanodrug delivery, 120–121
Endocrine pancreas, porosome structure and function and, 8–17
Endocytosis, nanodrug delivery systems, 114–116

Enhanced permeability and retention (EPR), nanodrug delivery systems, liposome structure, 117–118
Enzyme activity, atomic force microscopic imaging, 222–223
ESEM imaging, lignin formation, 103–108
Exocrine pancreas, porosome structure and function and, 8–17
Exocytosis, atomic force microscopic imaging, inverted optical microscopic assembly, 140–141
External agents, microbial cell surface effects, atomic force microscopy analysis, 78–79
Extracellular polymeric substances (EPS), microbial cell surface imaging, atomic force microscopy, 76
Extravasation, leukocyte adhesion molecules, 159

Feedback limits, atomic force microscopy, 231
  fast feedback loop, 242
Fibrinogen, atomic force microscopic imaging:
  anticoagulant effects, 275
  antiphospholipid syndrome, 274–275
Field emission in lens scanning electron microscopy (FEISEM), metaphase chromosomes, 258–259
Fluorescein isothiocyanate (FITC), nanodrug delivery systems:
  cellular interactions, 125–126
  dendrimer chemistry, 124
  methyl prednisolone-dendrimer nanodevices, 128–129
Fluorescence *in situ* hybridization (FISH):
  cytogenetics research, 250
  nanoextraction, 261–262
Fluorescence lifetime imaging (FLIM), plant cell walls, 108–109
Fluorimetric fusion assays, SNARE-induced membrane fusion, atomic-level calcium participation in, 30–34
Focused spot dimensions, atomic force microscopy cantilevers, 237–239
Folate receptor ligand, nanodrug delivery systems, 119–120
Force-distance curves:
  atomic force microscopy, low-noise detection, 240–241
  microbial cell surface imaging:
    atomic force microscopy analysis, 80–81
    elasticity measurements, 88
    layered surface applications, 87
    future applications, 89–90
  soft tissue imaging, swollen polymer surfaces, 214–218

Force-distance curves (*continued*)
  sonicated unilamellar vesicle bilayers, 192–197
  supported lipid bilayer thickness, 183–186
Force measurements:
  atomic force microscopic imaging:
    inner ear tissue studies, 148–150
    lateral dimension reduction, 230–231
    lipid bilayers and vesicles, 181–182
  leukocyte adhesion molecules:
    avidity modulation, 172–173
    integrin lateral redistribution, 174–175
    single-molecule unbinding theory, 161–166
  microbial cell surface imaging, atomic force microscopy analysis, 80–81
  sonicated unilamellar vesicle bilayers, 191–197
Fourier Transform infrared (FTIR) imaging, lignin structures, 108
Freely jointed chain (FJC), microbial cell surface elasticity, 87–88
Friction force microscopy (FFM), soft surface imaging, 205
Fusion pores. *See* Porosomes
"Fusogenic liposomes," nanodrug delivery systems, 118–120
  cycotoxic T-lymphocyte response, 129–130

Ghrelin:
  growth-hormone release and synthesis, 53–54
  signal transduction mechanisms, 55–56
Glass substrates, microbial cell surface imaging, atomic force microscopy analysis, 70–71
Glucosides, lignin synthesis, 100–108
Gonadotropin-releasing hormone (GnRH), growth-hormone release and synthesis, 54
β_2GPI cofactor, antiphospholipid syndrome, atomic force microscopy, 275–277
GroEL/GroES chaperonin proteins, atomic force microscopy, 230–231
Growth-hormone (GH), cell fusion pores, pituitary gland:
  atomic force microscopy analysis, 59–60
  hypothalamic hormone control, 50–55
    ghrelin, 53–54
    growth-hormone-releasing factor, 50–52
    peptide regulation, 54–55
    somatostatin, 52–53
  immunocytochemical somatotroph distribution, 57–58
  overview, 49
  post-secretion secretory vesicle amounts, 60–63
  signal transduction mechanisms, 55–56
  in vivo neuroendocrine regulation, 49–50
Growth-hormone-releasing hormone (GHRH):
  growth hormone release and synthesis, 50–52
  growth hormone secretory vesicles, secretion effects, 62–63
  signal transduction mechanisms, 55–56
GTG banding, atomic force microscopic imaging:
  genetics research, 250–253
  metaphase chromosome karyotyping, 253–257
  nanoextraction, 261–263
Guanosine triphosphate (GTP), secretory vesicle content expulsion:
  basic principles, 37–38
  vesicle swelling in neurons, 44–46
  vesicle swelling in pancreatic acinar cells, 41–44

Harmonic oscillation, atomic force microscopy, thermal cantilever noise, 235–237
Hemicelluloses, plant cell walls, basic properties, 96
Hemostatic proteins, atomic force microscopic imaging, antiphospholipid syndrome, 274–275
Herzian model, sonicated unilamellar vesicle bilayers, force mechanics, 191–197
Hexagonally packed intermediate layers, microbial cell surface imaging, atomic force microscopy analysis, 73
  immobilization strategies, 71
  layered surface applications, 87
High-definition television, atomic force microscopy, 231
Highly-oriented pyrolytic graphite, scanning tunneling microscopy, 242
High-pass frequency filter, plant cell walls, atomic force microscopic imaging, 96–97
Hooke's law:
  force-distance curves, microbial cell surface imaging, 80–81
  scanning probe microscopy, soft surface imaging, 211
Hydration mechanisms, SNARE-induced membrane fusion, atomic-level calcium participation in, 30–34
Hydrogels, scanning probe microscopy, swollen polymer surfaces, 216–218
Hydrophilic interactions, plant cell wall imaging, lignin structures, 106–108
Hydrophobic interactions, plant cell wall imaging, lignin structures, 106–108
Hydrophobic substrates, microbial cell surface imaging, atomic force microscopy analysis, 70–71
Hypothalamic deafferentiation, *in vivo* growth-hormone secretion, 49–51

INDEX

Hypothalamic hormone control, growth-hormone release and synthesis, 50–55
  ghrelin, 53–54
  growth-hormone-releasing factor, 50–52
  peptide regulation, 54–55
  somatostatin, 52–53

Ibuprofen, nanodrug delivery systems:
  chemistry, 124
  dendrimer-ibuprofen nanodevices, 126–127
Image alignment techniques, time-resolved imaging, 225–227
Image tracking techniques, atomic force microscopy, biomolecular motion, 228–229
Imaging speed, atomic force microscopy:
  biomolecular motion, 228–231
    feedback limit, 231
    lateral dimension reduction, 229–230
    tracking mechanisms, 228–229
    video-rate imaging, 231
  cantilever resonant frequency, 232–233
Immobilization strategies:
  leukocyte adhesion, avidity modulation, 172
  microbial cell surface imaging, atomic force microscopy analysis, 71
Immuno-atomic force microscopy (ImmunoAFM):
  fixed cells, 5
  live cells, 5
  secretory product imaging and, 10–17
Immunocytochemistry, growth hormone cells, 57
Immunogold localization, live pancreatic acinar cells, 5
Immunoprecipitation:
  pancreatic plasma membrane, 5–6
  porosome structural analysis, 11–17
Inner ear tissue, atomic force microscopy studies on, 144–150
Integrins, leukocyte adhesion molecules:
  activation analysis, 162–166
  basic properties, 159–160
  lateral redistribution, 173–175
Intercellular adhesion molecule-1 (ICAM-1):
  avidity modulation:
    basic principles, 169–171
    cells and reagents, 171
    force measurements, 172–173
    integrin lateral redistribution, 173–175
    3A9 cell line adhesion, 176–178
    ionomycin/thapsigargin stimulation, 175–176
Intermolecular forces:
  leukocyte adhesion molecules:
    basic principles, 159–160
    integrin activation, 162–166
    selectin/sLeX complexes, 162
    single-molecule unbinding, 160–166

  plant cell wall imaging, lignin structures, 107–108
Inverted optical microscope:
  atomic force microscopy on, 137–141
  photonic force microscopy and, 152–157
In vivo studies:
  growth-hormone secretion, neuroendocrine regulation, 49–50
  lignin synthesis, scanning probe microscopic imaging, 100–110
  photon force microscopy, 154–157
Ion channel formation:
  atomic force microscopy, patch-clamp integration, 141–144
  secretory vesicle content expulsion, basic principles, 37–38
  SNARE-induced membrane fusion, 25–27
Ionomycin, leukocyte adhesion and, 175–176

Jump-in point, supported lipid bilayer thickness, 183–186

Karyotyping, atomic force microscopic imaging, metaphase chromosomes, 253–257
Kinesin structures, photonic force microscopy, single molecules, 153–157

Laemmli sample preparation, pancreatic plasma membrane, 5–6
Langmuir-Blodgett films, plant cell wall imaging, lignin structures, 108
Laser instrumentation, photonic force microscopy, 152–155
Lateral dimensions, atomic force microscopy, 229–230
Lateral drift, biomolecular motion, time-resolved imaging, 223–228
Lateral force microscopy (LFM), soft surface imaging, 205–206
LCAO-MO calculations, lignin formation, 101–108
Leptin, growth-hormone release and synthesis, 54
Leukocyte adhesion molecules:
  avidity modulation studies:
    adhesive force measurements, 172–173
    basic principles, 169–171
    cells and reagents, 171
    clinical applications, 176–178
    elasticity measurements, 173
    integrin lateral redistribution, 173–175
    ionomycin/thapsigargin stimulation, 175–176

Leukocyte adhesion molecules (*continued*)
  protein immobilization, 172
  3A9 cell crosslinking, 171–172
  intermolecular forces:
    basic principles, 159–160
    integrin activation, 162–166
    selectin/sLeX complexes, 162
    single-molecule unbinding, 160–166
Leukocyte function-associated antigen-1 (LFA-1):
  avidity modulation:
    basic principles, 169–171
    ionomycin/thapsigargin stimulation, 175–176
    results and discussion, 176–178
  leukocyte adhesion molecules:
    basic properties, 159–160
    integrin activation, 162–166
    lateral redistribution, 173–175
Life mode, scanning probe microscopy, soft surface imaging, 210
Ligand-decorated liposomes, nanodrug delivery systems, 119–120
Light scattering analysis, SNARE-induced membrane fusion, atomic-level calcium participation in, 30–34
Lignin:
  plant cell walls, basic properties, 96
  scanning probe microscopic imaging, 100–108
Lipid bilayers:
  atomic force microscopic imaging, antiphospholipid syndrome, 273–274
  secretory vesicle content expulsion, vesicle swelling in pancreatic acinar cells, 41–44
  SNARE-induced membrane fusion, 25–27
    atomic-level calcium participation in, 27–34
  thickness and micromechanical properties:
    atomic force microscopic imaging, 181–197
    supported thickness, 182–183
Lipid membrane:
  porosome reconstitution on mica, 4
  porosome structure and function and, 12–17
  "raft" structures, viscosity measurements, 156–157
  secretory vesicle content expulsion, vesicle swelling in pancreatic acinar cells, 42–44
  SNARE-induced membrane fusion, bilayer interaction and conducting channel formation, 26–27
Liposomes, nanodrug delivery systems:
  conventional structures, 116–118
  modified liposomes, 118–120
Liquid droplets, scanning probe microscopy, soft surface imaging, 212–214
Lithium chloride, microbial cell surface effects, atomic force microscopy analysis, 79

Loading force measurements, sonicated unilamellar vesicle bilayers, micromechanical measurements, 192–197
Low-noise detection, atomic force microscopy cantilevers, 239–241
L-plastin, leukocyte adhesion, avidity modulation, 170–171
Lupus anticoagulant (LA) phenomenon, antiphospholipid syndrome, 275
  annexin A5, 278–279

MacMARCKS substrate, leukocyte adhesion, avidity modulation, 170–171
Macromolecules:
  atomic force microscopic imaging, hemostasis and thrombosis, 267–283
  microbial cell surface elasticity, atomic force microscopy imaging, 86–87
  pectin "brush structure," atomic force microscopic imaging, 97–98
Macropinocytosis, nanodrug delivery systems, 114–116
Madin-Darby canine kidney (MDCK) cells, nanodrug delivery systems, dendrimer structures, 124
Major histocompatibility complex (MHC) molecules, nanodrug delivery systems, 129–130
Mechanical drift, biomolecular motion, time-resolved imaging, 223–228
Membrane fusion:
  atomic force microscopic imaging, inverted optical microscopic assembly, 140–141
  porosome functions, 1–3
  SNARE-induced, molecular mechanisms, 25–34
    bilayer interaction circular array, conducting channel formation, 25–27
    calcium ion participation, 27–34
    cell fusion machinery with, 27
Membrane viscosity, photonic force microscopy, thermal fluctuation measurements, 155–157
Metaphase chromosomes, atomic force microscopic imaging:
  correlative three-dimensional morphological analysis, 257–259
  genetic material, 250–253
  GTG-banding patterns, 251–253
  historical background, 249–250
  karyotyping of, 253–257
Methyl prednisolone nanodevices, nanodrug delivery systems, dendrimer conjugates, 127–129
Mica substrates:
  atomic force microscopic imaging, antiphospholipid syndrome, 273–274

microbial cell surface imaging, atomic force microscopy analysis, 70–71
Micelles, polymer-based nanodrug delivery, 121–122
Microbial cell surfaces, atomic force microscopy analysis, 69–90
  bacillus S-layers, 73
  basic principles, 69–70
  cell surface layer imaging, 73–75
  elasticity, 85–89
    layers stretching, 87
    single cell stretching, 87–89
    single macromolecules, 86–87
  external agent effects, 78–79
  future applications, 89–90
  hexagonally packed intermediate layer, 73
  imaging techniques, 70–79
    immobilization strategies, 71
    selection criteria, 71–72
    substrate requirements, 70–71
  microbial biofilms, 75–76
  nanostructures, 76–78
  physical properties measurement, 79–89
    force-distance curves, 80
    force measurements, 80–81
    functionalized probes, 80–81
  physicochemical properties and molecular interactions, 81–85
    adhesion mapping, 81–82
    cell probe technique, 85
    surface charges and electrostatic interactions, 83–85
    surface energy and solvation interactions, 82–83
  physiological changes, 78
  porin crystals, 74–75
  purple membranes, 73–74
Microdissection, atomic force microscopy, 259–260
Microfibrils, plant cell walls, basic properties, 95–96
Micromechanical measurements:
  atomic force microscopy imaging, inner ear tissue studies, 146–150
  sonicated unilamellar vesicle bilayers, 191–197
Molecular mechanisms:
  cell secretion, porosome structure and function and, 17, 19–20
  microbial cell surface imaging, atomic force microscopy analysis, 81–85
  photonic force microscopy, thermal fluctuation measurements, 153–157
  plant cell walls, atomic force microscopic imaging, 96–97
  secretory vesicle content expulsion:
    basic principles, 37–38
    cellular secretion swelling, 38–44

    neuron secretion swelling, 44–46
  SNARE-induced membrane fusion, 25–34
    bilayer interaction circular array, conducting channel formation, 25–27
    calcium ion participation, 27–34
    cell fusion machinery with, 27
Molecular weight calculations, lignin formation, 103–108
Monoclonal antibodies, antiphospholipid syndrome, annexin A5, 278–279
Motilin, growth-hormone release and synthesis, ghrelin mediation, 54
Multilamellar vesicle (MLV) solution, sonicated unilamellar vesicle bilayers, thickness and morphology, 188–191

Nanobioscience:
  antiphospholipid syndrome, 268–272
  atomic force microscopy, cytogenetic applications, 259–262
  historical evolution of, 2–3
Nanodrug delivery systems:
  cellular interactions:
    basic principles, 113–116
    dendrimers, 122–130
      chemistry, 124
      drug conjugation, 126–129
      free dendrimer interactions, 124–126
      ibuprofen nanodevices, 126–127
      methyl prednisolone nanodevices, 127–129
    lipid-based systems, 116–120
    polymer-based systems, 120–130
      micelles, 121–122
      nanoparticles, 120–121
  clinical applications and future research, 130–132
Nanoextraction, atomic force microscopy, 259–262
Nanoparticles:
  AFM manipulation and dissection, 259
  polymer-based nanodrug delivery, 120–121
  scanning probe microscopy, liquid droplets, 212–214
Nanostructures:
  microbial cell surface imaging, atomic force microscopy analysis, 76–79
    physiological changes, 78
  photonic force microscopy imaging, 151–157
Near-field scanning optical microscopy (NSOM):
  antiphospholipid syndrome, 268–272
  lignin imaging, 104–108
  plant cell wall imaging, 108–109
Neuroendocrine regulation, *in vivo* growth-hormone secretion, 49–50

Neurons:
  porosome structure and function in, 8–17
  secretory vesicle content expulsion, 44–46
Neuropeptide Y (NPY), growth-hormone release and synthesis, ghrelin mediation, 54
Noncontact mode:
  cytogenetics research, 252–253
  scanning probe microscopy, soft surface imaging, 210
NSF-ATP assays, SNARE-induced membrane fusion, atomic-level calcium participation in, 31–34
Nyquist relation, atomic force microscopy, thermal cantilever noise, 236–237

OmpF porin, microbial cell surface imaging, atomic force microscopy analysis, immobilization strategies, 71
Opposing bilayers, SNARE-induced membrane fusion, 25–27
Optical beam deflection, atomic force microscopy cantilevers, low-noise detection, 239–241
Optical trapping techniques, photonic force microscopy, 151–157

Pancreatic acinar cells:
  immunogold localization, 5
  isolation, 3–4
  porosome structure in, 6–17
  secretory vesicle content expulsion:
    basic principles, 37–38
    vesicle swelling during secretion, 38–44
  transmission electron microscopy imaging, 5
Pancreatic plasma membrane:
  immunoprecipitation and Western blot analysis, 5–6
  isolation, 4
  porosome structural analysis, 10–17
  secretory vesicle content expulsion, basic principles, 37–38
Patch-clamp integration, atomic force microscopy, 141–144
  inner ear tissue studies, 144–150
Pectins:
  atomic force microscopic imaging, 97–98
  basic properties, 96
PEO-b-poly(L-amino acids) (PLAA) block copolymers, polymer-based nanodrug delivery, 122
  multidrug resistance and, 131
Phagocytosis, nanodrug delivery systems, 114–116
  liposome structure, 117–118
Phase contrast microscopy, metaphase chromosome karyotyping, 254–257

Phase shift detection, scanning probe microscopy, soft surface imaging, 209–210
Phenolic alcohols, lignin synthesis, 100–108
Phorbol myristate acetate (PMA), leukocyte adhesion:
  avidity modulation, 170–171
  cell spreading, 177–178
  integrin lateral redistribution, 174–175
Phosphatidylserine (PS), antiphospholipid syndrome, 276–277
Phospholipase C-IP3-protein kinase C, signal transduction, 55–56
Photodetection sensitivity, atomic force microscopy cantilevers, 240–241
Photonic force microscopy (PFM):
  development of, 151–157
  position-sensing systems, 151–153
  thermal fluctuation measurements:
    membrane viscosity, 155–157
    single molecules, 153–155
Physicochemical properties, microbial cell surface imaging, atomic force microscopy, 81–85
Physiological changes, microbial cell surface imaging, atomic force microscopy analysis, 78
Phytohemagglutinin (PHA), cytogenetics applications, atomic force microscopy, 249
Piezoelectric crystals:
  atomic force microscopy:
    antiphospholipid syndrome, 269–272
    fast scanners, 241–242
  scanning probe microscopy, soft surface imaging, 202–203
Pipette structures, atomic force microscopy, 138–141
Pituitary adenylate cyclase-activating peptide (PACAP):
  growth-hormone release and synthesis, 54
  signal transduction mechanisms, 55–56
Pituitary gland, growth-hormone cell fusion pores:
  atomic force microscopy analysis, 59–60
  hypothalamic hormone control, 50–55
    ghrelin, 53–54
    growth-hormone-releasing factor, 50–52
    peptide regulation, 54–55
    somatostatin, 52–53
  immunocytochemical somatotroph distribution, 57–58
  overview, 49
  post-secretion secretory vesicle amounts, 60–63
  signal transduction mechanisms, 55–56
  in vivo neuroendocrine regulation, 49–50
Pixel acquisition, time-resolved imaging, 227–228

Planck's constant, leukocyte adhesion molecules, single-molecule unbinding theory, 160–166
Plant cell wall:
  atomic force microscopy (AFM), 96–97
    cellulose, 99–100
    pectins, 97–99
  scanning probe microscopic imaging:
    basic properties, 95–96
    future research techniques, 108–109
    lignin, 100–108
Plasma membrane:
  growth hormone secretory vesicles, secretion effects, 61–63
  porosome structure and function and, 8–17
Poisson's ratio, sonicated unilamellar vesicle bilayer micromechanics
Poly(acrylic acid) (PAA), soft tissue imaging, swollen surfaces, 214–218
Polyamidoamine (PAMAM) dendrimers, nanodrug delivery systems:
  basic properties, 123–124
  cellular interactions, 125–126
  dendrimer-drug conjugates, 126–129
Poly(DL-lactide-co-glycolide) (PLGA), polymer-based nanodrug delivery, 120–121
Poly(ethylene oxide) (PEO):
  polymer-based nanodrug delivery, 122
  supported lipid bilayer micromechanics, 195–197
  supported lipid bilayer thickness measurements, 184–186
Polylactides, polymer-based nanodrug delivery, 120–121
Polymer-based systems:
  nanodrug delivery, cellular interactions, 120–130
    micelles, 121–122
    nanoparticles, 120–121
  soft tissue imaging, swollen surfaces, 214–218
Polysaccharides:
  pectin macromolecules, atomic force microscopic imaging, 97–98
  plant cell walls, basic properties, 95–96
Porin crystals, microbial cell surface imaging, atomic force microscopy, 74–75
Porosomes:
  atomic force microscopy, 4–5
  basic properties, 1–3
  growth-hormone secretion in pituitary gland:
    atomic force microscopy analysis, 59–60
    hypothalamic hormone control, 50–55
      ghrelin, 53–54
      growth-hormone-releasing factor, 50–52
      peptide regulation, 54–55
      somatostatin, 52–53
    immunocytochemical somatotroph distribution, 57–58
    overview, 49

  post-secretion secretory vesicle amounts, 60–63
  signal transduction mechanisms, 55–56
  in vivo neuroendocrine regulation, 49–50
immuno-atomic force microscopy:
  fixed cells, 5
  live cells, 5
immunoprecipitation and Western blot analysis, 5–6
lipid membrane reconstitution, mica-based, 4
molecular-based cell secretion, 17–22
pancreatic acinar cell isolation, 3–4
pancreatic plasma membrane preparation, 4
secretory vesicle content expulsion, vesicle swelling in pancreatic acinar cells, 41–44
structural properties, 6–17
synaptosome/synaptosomal/synaptic vesicle isolation, 4
transmission electron microscopy, 5
zymogen granule isolation, 5
PPO complexes, supported lipid bilayer micromechanics, 195–197
p53 protein-DNA interaction, time-resolved imaging, 224–226
Probe surface chemistry:
  atomic force microscopy, antiphospholipid syndrome, 269–272
  microbial cell surface imaging, atomic force microscopy, 81–82
Proportional-integral-differential (PID) feedback loop, atomic force microscopy, 242
Prostaglandin (PGE$_2$), nanodrug delivery systems:
  dendrimer-ibuprofen nanodevices, 127
  methyl prednisolone-dendrimer nanodevices, 128–129
Protein immobilization, leukocyte adhesion, avidity modulation, 172
Protein kinase A, signal transduction, 55–56
Protein kinase C:
  leukocyte adhesion, avidity modulation, 170–171
  signal transduction, 55–56
Protein-protein interactions, atomic force microscopic imaging, 222–223
Proton nuclear magnetic resonance, plant cell wall imaging, lignin structures, 106–108
Pulronic copolymers, supported lipid bilayer thickness measurements, 184–186
Purple membranes, microbial cell surface imaging, atomic force microscopy analysis, 73–74
  immobilization strategies, 71

Q-banding, cytogenetics research, 249–250

Q control:
  atomic force microscopy:
    cantilever resonant frequency, 232–233
    thermal cantilever noise, 235–237
  scanning probe microscopy:
    liquid droplets, 214
    soft surface imaging, 207–209
Quantitative analysis, growth hormone cells, 57–58

Rack1 phosphorylation, leukocyte adhesion, avidity modulation, 170–171
"Raft" lipids, viscosity measurements, 156–157
Raman spectroscopy, plant cell wall imaging, lignin structures, 106–108
Rayleigh range, atomic force microscopy cantilevers, focused spot dimensions, 238–239
Receptor affinity modulation, leukocyte adhesion, 169–171
Receptor-mediated endocytosis, nanodrug delivery systems, 114–116
Recrystallization techniques, microbial cell surface imaging, atomic force microscopy analysis, 71
Repulsive force, sonicated unilamellar vesicle bilayer micromechanics, 195–197
Resonant frequencies, atomic force microscopy cantilevers, 232–233
Reverse hemolytic plaque assay, growth hormone fusion pore structure and dynamics, 59–60
RMS-DC converter, atomic force microscopy:
  antiphospholipid syndrome, 270–272
  fast feedback loop, 242
RPMI 1640 medium, leukocyte adhesion, avidity modulation, 171

Scanner design, atomic force microscopy, 241–242
Scanning electron microscopy (SEM):
  metaphase chromosomes, 257–258
  simultaneous AFM/patch-clamp recordings, 145–147
Scanning probe microscopy (SPM):
  plant cell wall:
    basic properties, 95–96
    future research techniques, 108–109
    lignin, 100–108
  soft surface imaging:
    basic principles, 201–202
    cantilevers and probes, 210–211
    contact mode, 203–205
    dynamic mode, 205–209
    experimental setup, 202–203
    liquid droplets, 212–214

phase shift technique, 209–210
physics and methods, 202–211
swollen polymer surfaces, 214–218
Scanning tunneling microscopy (STM):
  antiphospholipid syndrome, 268–272
  cytogenetics research, 250
  lignin structure, 101–108
  scanner speed, 242
Secretagogue exposure, cell secretion and, 6–17
Secretory vesicles:
  content expulsion, molecular mechanisms:
    basic principles, 37–38
    swelling process in cellular secretion, 38–44
    swelling process in neuron secretion, 44–46
  growth hormone cells, secretion effects, 60–63
Selectin/sLeX complexes, atomic force microscopic imaging, 162
Signal-to-noise ratio, atomic force microscopy:
  low-noise detection, 240–241
  thermal cantilever noise, 236–237
Signal transduction:
  adenylate cyclase/cAMP/protein kinase A and phospholipase C-IP3-protein kinase C, 55–56
  growth-hormone release and synthesis:
    growth-hormone-releasing hormone, 51–52
    somatostatin, 52–53
Silicon nitride cantilevers:
  atomic force microscopy:
    antiphospholipid syndrome, 268–272
    resonant frequency, 233
  scanning probe microscopy, soft surface imaging, 210–211
Silicon oxide substrates, microbial cell surface imaging, atomic force microscopy analysis, 70–71
Single molecule force spectroscopy, microbial cell surface elasticity, 86–88
Single-molecule unbinding theory, leukocyte adhesion molecules, 160–166
SNARE complex:
  membrane fusion, molecular mechanisms, 25–34
    bilayer interaction circular array, conducting channel formation, 25–27
    calcium ion participation, 27–34
    cell fusion machinery with, 27
  porosome structure and function and, 13–17
Soft surface imaging, scanning probe microscopy:
  basic principles, 201–202
  cantilevers and probes, 210–211
  contact mode, 203–205
  dynamic mode, 205–209
  experimental setup, 202–203
  liquid droplets, 212–214

phase shift technique, 209–210
physics and methods, 202–211
swollen polymer surfaces, 214–218
Solvation interactions, microbial cell surface imaging, atomic force microscopy analysis, 82–83
Somatostatin (SRIF), growth-hormone release and synthesis, 52–53
Somatotrophs:
  growth-hormone release and synthesis, 55
  immunocytochemical distribution, 57–58
Sonicated unilamellar vesicle (SUV) bilayers:
  micromechanical properties, 191–197
  thickness and morphology, 187–191
Stalk sectioning techniques, *in vivo* growth-hormone secretion, 49–50
"Stealth" liposomes, nanodrug delivery systems, 118–120
Stereocilia properties, atomic force microscopy imaging, inner ear tissue studies, 144–150
"Sterically stabilized" liposomes, nanodrug delivery systems, 118–120
Stiffness measurement, atomic force microscopy imaging, inner ear tissue studies, 148–150
Stokes drag, membrane viscosity, thermal fluctuation measurements, 156–157
Substrate requirements:
  leukocyte adhesion, avidity modulation, 170–171
  microbial cell surface imaging, atomic force microscopy analysis, 70–71
Supercritical fluids, polymer-based nanodrug delivery, 120–121
Supermodule structural unit, lignin formation, 101–108
Supported lipid bilayer (SLB), thickness measurements, 182–186
Supramolecular activation clusters, leukocyte adhesion, avidity modulation, 170–171
Surface charges, microbial cell surface imaging, atomic force microscopy analysis, 83–85
Surface energy, microbial cell surface imaging, atomic force microscopy analysis, 82–83
Surface-enhanced Raman spectroscopy, plant cell wall imaging, 108–109
Synaptic vesicles:
  isolation, 4
  porosome structure and function in, 12–17
Synaptosomal membrane:
  isolation, 4
  porosome structural analysis, 11–17
Synaptosome:
  isolation, 4
  porosome structure and function in, 12–17

Tapping mode atomic force microscopy (TMAFM):

antiphospholipid syndrome, 271–272
biomolecular motion, basic principles, 221–223
feedback limits, 231
genetic material imaging, 251–253
microbial cell surface imaging:
  biofilms, 76
  external agent effects, 79
  purple membranes, 74
pancreatic acinar cells, 5
porosome imaging, 5
soft surface imaging, 205–209
  swollen polymer surfaces, 216–218
sonicated unilamellar vesicle bilayers, thickness and morphology, 188–191
Thapsigargin, leukocyte adhesion stimulation, 175–176
Thermal cantilever noise, atomic force microscopy, 235–237
  low-noise detection, 240–241
Thermal fluctuation measurements, photonic force microscopy, single molecules, 153–157
3A9 cell line, leukocyte adhesion:
  avidity modulation, 171–172
  force measurements, 172–173
  integrin clustering, 176–178
Three-dimensional morphological analysis, atomic force microscopic imaging, metaphase chromosomes, 257–259
Three-dimensional topographic imaging, supported lipid bilayer thickness, 183–186
Thyrotropin-releasing hormone (TRH), growth-hormone release and synthesis, 54
Time-resolved imaging, biomolecular motion, 223–228
  alignment, 225–227
  moving images, 227–228
  p53-DNA dynamic interactions, 224–225
  recording sequences, 223–224
Tip effect, sonicated unilamellar vesicle bilayers, thickness and morphology, 190–191
Topographic imaging, scanning probe microscopy, soft surface imaging, 205
Transmission electron microscopy (TEM):
  growth hormone fusion pore structure and dynamics, 59–60
  growth hormone secretory vesicles, 61–63
  metaphase chromosomes, 257–258
  pancreatic acinar cell isolation, 5
  plant cell walls, 97
  porosome structure and function and, 10–17

Ultraviolet (UV) radiation, plant cell wall imaging, lignin structures, 106–108

Vertical drift, biomolecular motion, time-resolved imaging, 223–228
Very late antigen-4 (VLA-4), leukocyte adhesion molecules:
    basic properties, 159–160
    integrin activation, 165–166
Vesicles:
    secretory vesicle content expulsion, swelling mechanisms:
        basic principles, 37–38
        cellular secretion, 38–44
        neuron secretion, 44–46
    sonicated unilamellar vesicle:
        bilayer thickness and morphology, 187–191
        micromechanical properties, 191–197
    synaptic vesicles:
        isolation, 4
        porosome structure and function in, 12–17
Video-rate imaging, atomic force microscopy, 231
"Virosome," nanodrug delivery systems, 118–120
Viscosity of membranes:
    photonic force microscopy, thermal fluctuation measurements, 155–157
    scanning probe microscopy, soft surface imaging, 212–214
Viscous damping, atomic force microscopy, thermal cantilever noise, 235–237

Water transport, plant cell wall imaging:
    basic properties, 96–97
    lignin mechanisms for, 105–108
Western blot analysis, pancreatic plasma membrane, 5–6
Worm-like chain (WLC) model, microbial cell surface elasticity, 87–88

X-ray diffraction, SNARE-induced membrane fusion, atomic-level calcium participation in, 27–34

Young's modulus:
    leukocyte adhesion molecules:
        elasticity, 173
        integrin lateral redistribution, 174–175
    sonicated unilamellar vesicle bilayer micromechanics, 191–197

Zymogen granules (ZG):
    isolation, 5
    porosome structure and function and, 10–17
    secretory vesicle content expulsion:
        basic principles, 37–39
        neuron vesicle swelling, 44–46
        pancreatic acinar cell vesicle swelling, 38–44